TAURINE 4

Taurine and Excitable Tissues

ADVANCES IN EXPERIMENTAL MEDICINE AND BIOLOGY

Editorial Board:

NATHAN BACK, *State University of New York at Buffalo*

IRUN R. COHEN, *The Weizmann Institute of Science*

DAVID KRITCHEVSKY, *Wistar Institute*

ABEL LAJTHA, *N. S. Kline Institute for Psychiatric Research*

RODOLFO PAOLETTI, *University of Milan*

A Continuation Order Plan is available for this series. A continuation order will bring delivery of each new volume immediately upon publication. Volumes are billed only upon actual shipment. For further information please contact the publisher.

TAURINE 4

Taurine and Excitable Tissues

Edited by

Laura Della Corte
University of Florence
Florence, Italy

Ryan J. Huxtable
University of Arizona College of Medicine
Tucson, Arizona

Giampietro Sgaragli
University of Siena
Siena, Italy

and

Keith F. Tipton
Trinity College
Dublin, Ireland

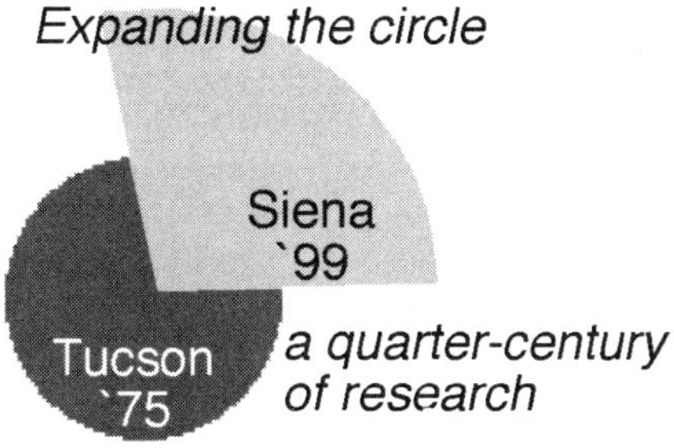

Expanding the circle

Siena
`99

Tucson
`75

a quarter-century
of research

Kluwer Academic / Plenum Publishers
New York, Boston, Dordrecht, London, Moscow

QP801
T3
T37
2000

Library of Congress Cataloging-in-Publication Data

Taurine 4 : taurine and excitable tissues / edited by Laura Della Corte ... [et al.].
    p. ; cm. -- (Advances in experimental medicine and biology ; v. 483)
    Includes bibliographical references and index.
    ISBN 0-306-46447-0
    1. Taurine--Physiological effect--Congresses. I. Title: Taurine four. II. Corte, Laure
Della. III. International Taurine Symposium '99: Expanding the Circle (1999 : Siena,
Italy) IV. Series.
    [DNLM: 1. Taurine--physiology--Congresses. QU 60 T2289 2000]
    QP801.T3 T374 2000
    612'.0157--dc21
                                                                        00-059277

Proceedings of the International Taurine Symposium '99, an official Satellite Symposium of the 17th Biennial
Meeting of the International Society for Neurochemistry, held August 3–8, 1999, in Siena, Italy

ISBN 0-306-46447-0

©2000 Kluwer Academic / Plenum Publishers, New York
233 Spring Street, New York, N.Y. 10013

http://www.wkap.nl

10   9   8   7   6   5   4   3   2   1

A C.I.P. record for this book is available from the Library of Congress

All rights reserved

No part of this book may be reproduced, stored in a retrieval system, or transmitted in any form or by any
means, electronic, mechanical, photocopying, microfilming, recording, or otherwise, without written permission
from the Publisher

Printed in the United States of America

PREFACE
Expanding the circle: A quarter century of research

This volume is the selected, edited proceedings of the International Taurine Symposium held at the Certosa di Pontignano, near Siena, Italy in August 1999. The meeting was a satellite to the 17th Biennial Meeting of the International Society of Neurochemistry which was held in Berlin, Germany, immediately afterwards. The subtitle of this meeting was taken from the fact that the first International Symposium was held in Tucson, Arizona, USA in 1975. However, it might equally well apply to expanding ones circle of friends, which was certainly an outcome of this biggest-ever Taurine Symposium.

The Certosa di Pontignano, founded in 1341 as a Carthusian Monastery and now a conference centre of the Universitá degli Studi Siena, proved to be an ideal venue for discussing good science in relaxing and beautiful surroundings.

Sections of the Symposium were dedicated to Emeritus Professors Doriano Cavallini and Alberto Giotti who were among the 'founding-fathers' of taurine research. Appreciations of their contributions are included in this volume. Unfortunately Professor Cavallini was unable to attend, but it was a great pleasure that Alberto Giotti, who was one of the organisers of an international taurine meeting held at San Miniato, Italy, in 1986, was able to be present.

We are grateful to the International Society for Neurochemistry for its continued support for the International Taurine meetings and to the European Union COST D8 and D13 Programmes for sponsoring the lectures on *Taurine and metals* and on *Taurine and analogues-New molecules towards health care*. Gratitude is also due to the following of sponsoring organisations, Taisho Pharmaceutical Co Ltd, Red Bull GmbH, Nestlé SA, Istituto Gentili S.p.A, Shimadzu Italia S.r.l., Noris S.r.l., Bracco S.p.A. and Bruschettini S.r.l. The generous assistance of the Universitá degli Studi di

Siena and the Universitá degli Studi di Firenze is also acknowledged with thanks.

The International Advisory Committee, comprising Ryan Huxtable (USA), Kinya Kuriyama (Japan), Raphael Martin Del Rio (Spain), Simo Oja (Finland) and Keith Tipton (Ireland) made useful contributions to the planning and organisation of the meeting but it was the Organising Committee, Silvestro Duprè (Roma), Mitri Palmi (Siena), Giancarlo Pepeu (Firenze) and Massimo Valoti (Siena) who bore a greater burden. They were ably assisted by the meeting secretariat, composed of Loria Bianchi, Maria Frosini and Maria Luisa Valacchi, and many staff and students from the Istituto di Scienze Farmacologiche, Siena who were ever present and cheerful to help with local and travel arrangements. The staff of the Certosa di Pontignano, lead by Sergio Gambassi and Andrea Machetti, are also to be thanked for splendid hospitality. As the meeting organisers, Laura Della Corte and Giampietro Sgaragli, are also co-editors of this volume, it is inappropriate for them to thank themselves in this Preface, but they deserve special thanks and congratulations from the rest of us for immense amount of work they put in to ensuring the success of a memorable meeting.

We have not commented on the exciting developments in our understanding of the functions of taurine that arose during the meeting as we hope the contents of this volume will amply illustrate them.

Laura Della Corte
Ryan Huxtable
Giampietro Sgaragli
Keith Tipton

PARTICIPANTS

Dr. Fernanda Amicarelli
Dip. di Biologia di Base
Università dell' Aquila
Via Vetoio
Coppito - L'Aquila I-67010
Italy

Dr. Junichi Azuma
Dept. Clin. Evalut. Med. &
    Therapeutic
Osaka University
1-6 Yamadaoka, Suita-shi
Osaka 565-0871
Japan

Dr. Steven I. Baskin
USA Medical Research
Inst. of Chemistry
3100 Ricketts Point Road
Aberdeen Proving Grd-EA., MD
    21010-5400
USA

Dr. Jennifer J. Bedford
Dept. of Physiology
University of Otago
P.O. Box 913
Dunedin
New Zealand

Ms. Lidia Bellik
Ist. Scienze Farmacologiche
Università di Siena
Via ES Piccolomini 170
Siena I-53100
Italy

Ms. Francesca Bellucci
Ist. Scienze Farmacologiche
Università di Siena
Via ES Piccolomini 170
Siena I-53100
Italy

Mr. Alberto Benocci
Ist. Scienze Farmacologiche
Università di Siena
Via ES Piccolomini 170
Siena 1-53100
Italy

Dr. Loria Bianchi
Dip. di Farmacologia Preclinica e
    Clinica
Università di Firenze
V.le Pieraccini, 6
Firenze I-50139
Italy

Dr. Andrea Budreau Patters
Dept. of Pediatrics
University of Tennessee
Le Bonheur Children's Med. Center
50 N. Dunlap - Rm. 306
Memphis, TN 38103-4909
USA

Dr. Julian Bustamante
Depto. de Fisiologia
Facultad de Medicina
Universidad Computense
Madrid E-28040
Spain

Dr. Selene Capodarca
Dip. Farmacologia Preclinica e Clinica
Università di Firenze
V.le Pieraccini. 6
Firenze I-50139
Italy

Dr. Kyung Ja Chang
Department of Food and Nutrition
Inha University
253 Yonghyen-Dong, Nam-Ku
Inchon 402-753
Korea

Dr. Russel W. Chesney
Dept. of Pediatrics
University of Tennessee
Le Bonheur Children's Medical
    Center
50 N. Dunlap - Room 306
Memphis, Tennessee 38103-4909
USA

Ms. Sun-Young Cho
Dept. of Food and Nutrition
Seoul National University
Shillim Doug
Seoul - 151742
Korea

Ms. Yewon Chun
Department of Food and Nutrition
Seoul National University
Shillim Dong
Seoul 151-742
Korea

Dr. M. Alessandra Colivicchi
Dip. di Farmacologia Preclinica e
    Clinica
Università di Firenze
V. le Pieraccini, 6
Firenze I-50139
Italy

Prof. Diana Conte Camerino
Dip. Farmacobiologico, Facoltà di
    Farmacia
Università di Bari
Via Orabona 4 (Campus)
Bari I-70125
Italy

Dr. Claire Cuisinier
Physical Education Institute
Catholic University of Louvain
1 Place P. de Coubertin
Louvain-la-Neuve B-1348
Belgium

Dr. Colm M. Cunningham
CNS Inflammation Group, School of
    Biol. Sci.
Biomedical Science Building
University of Southampton
S016 7 PX - Southampton
UK

Dr. Ewart Davies
Pharmacology Dept., Medical School
University of Birmingham
Edgbaston
Birmingham B15 2TT
UK

Prof. Ralph Dawson, Jr.
Department of Pharmacodynamics
University of Florida
JHMHC Box 100407
Gainesville, Florida 32610
USA

Dr. Enrico De Micheli
Divisione di Neurochirurgia
Ospedale di Lecco
Via Ghislanzoni 29
Lecco I-39100
Italy

Dr. Philippe De Witte
Biologie du Comportement
Unite de Biologie
Univ. de Louvain-la-Neuve
1, Place Croix du Sud
Louvain-la-Neuve B-1348
Belgium

Dr. Nuria Del Olmo
Depto. Investigacion
Hospital Ramon y Cajal
Ctra. Colmenar Km 9
Madrid E-28034
Spain

Prof Laura Della Corte
Dip. Farmacologia Preclinica e Clinica
Università di Firenze
V.le Pieraccini, 6
Firenze I-50139
Italy

Ms. Stefania Dragoni
Ist. Scienze Farmacologiche
Università di Siena
Via ES Piccolomini 170
Siena I-53100
Italy

Prof. Silvestro Dupré
Dip. di Scienze Biochimiche
Università "La Sapienza"
Piazzale Aldo Moro 5
Roma I-00185
Italy

Ms. Lucia Esposito
Ist. di Scienze Farmacologiche
Università di Siena
Via E.S. Piccolomini 170
Siena I-53100
Italy

Dr. Paola Failli
Dip. di Farmacologia Preclinica e
    Clinica
Università di Firenze
V.le Pieraccini, 6
Firenze I-50139
Italy

Dr. Maria Frosini
Ist. Scienze Farmacologiche
Università di Siena
Via ES Piccolomini 170
Siena I-53100
Italy

Dr. Renato Frosini
Direzione Scientifica
Bruschettini Srl
Via Isonzo 6
Genova I-16147
Italy

Dr. Kjell Fugelli
Dept. of Biology
Div. of Gen. Physiology
University of Oslo
PO Box 1051, Blindern
Oslo N-0316
Norway

Dr. Fabio Fusi
Ist. Scienze Farmacologiche
Università di Siena
Via ES Piccolomini 170
Siena I-53100
Italy

Dr. Carla Ghelardini
Dip. di Farmacologia Preclinica e
    Clinica
Università di Firenze
Viale G. Pieraccini 6
Firenze I-50139
Italy

Emeritus Prof. Alberto Giotti
Dip. di Farmacologia Preclinica e
    Clinica
Università di Firenze
V. le Pieraccini, 6
Firenze I-50139
Italy

Prof. Shri N. Giri
Dept. of Molecular Bioscience
University of California
Schields Avenue One
Davis, CA 95616
USA

Dr. José Maria Gonzalés-Vigueras
Neurobiologia-Investigacion
Hospital Ramon y Cajal
Ctra. de Colmar Km 9
Madrid E-28034
Spain

Ms. Beatrice Gorelli
Ist. Scienze Farmacologiche
Università di Siena
Via ES Piccolomini 170
Siena I-53100
Italy

Dr. Ramsen C. Gupta
Dept. of Chemistry
SASRD
Nagaland University
Medziphema 797106
India

Dr. Xiaobin Han
Dept. of Pediatrics
Univ. of Tennessee
50 N. Dunlap St. Room 336
Memphis, Tennessee 38103
USA

Dr. Yasmine E. Hazoutouniau
Inst. of Biochemistry
National Academy of Armenia
Baghramain St., Passage 1, Ft. 29
Yerevan, 44-375019
Armenia

Dr. Wojciech Hilgier
Dept. of Neurotoxicology
Medical Research Center
Polish Academy of Sciences
Pawinskiego 5
Warsaw 02-106
Poland

Dr. Nicolas Hussy
Biologie des Neurones Endocrine
CNRS-UPR 9055, CCIPE
141 Rue de la Cardonille
Montpellier, cedex 5 F-34094
France

Prof. Ryan J. Huxtable
Dept. of Pharmacology
Univ. of Arizona College of Medicine
Tucson, Arizona 85724
USA

Dr. Keisuke Imada
Pharmacological Evaluation Lab.
Taisho Pharmaceutical Co., Ltd.
1-403, Yoshino-Cho
Ohmiya 330-8530
Japan

Prof. Sadanobu Kagamimori
Dept. of Welfare Promotion &
    Epidemiology
Toyama Medical and
Pharm. University
Toyama 930-0194
Japan

Prof. Young-Sook Kang
College of Pharmacy
Sookmyung Women's University
Chungpa-dong, Yongsan-ku
Seoul 151-742
Korea

Prof. Etsuko Kibayashi
Dept. of Life & Culture
Sonoda Woman's College
Minamitukaguchi-cho 7-29-1
Amagasaki, Hyogo 661 8520
Japan

Dr. An-Keun Kim
Dept. of Food Science & Nutrition
Dankook University
College of Science
Hannam-Dong 8, Yongsan-Gu
Seoul 140-714
Korea

Dr. Byong-Kak Kim
College of Pharmacy
Seoul National University
Seoul 151-742
Korea

Dr. Eul-Sang Kim
Dept. of Food Science & Nutrition
Dankook University, College of
  Science
Hannam-Dong 8, Yongsan-Gu
Seoul 140-714
Korea

Dr. Ha Won Kim
Dept. of Life Science
Seoul City University
90 Jeonnong-Dong, Dongdaemu-Gu
Seoul 130-743
Korea

Prof. Harriet Kim
Department of Food and Nutrition
Seoul National University
Shillim Dong
Seoul 151-742
Korea

Dr. Yukiko Kondo
Medical Res. Laboratories
Taisho Pharmaceutical Co., Ltd.
1-403 Yoshino-Cho
Ohmiya, Saitama 330-8530
Japan

Dr. Sanna Kotisari
Dept. of Pharmacology and
  Toxicology
University of Kuopio
Hazjuleutiç 1A, 1kzs
Kuopio FIN-70219
Finland

Prof. Arnold R. Kriegstein
Dept. of Pathology
Center for Neurobiology and Behavior
630 W. 168th St., Box 31
New York, NY 10032
USA

Prof. Robert Kroes
Res. Inst. for Toxicology
Fac. Veterinary Med.
RITOX - Utrecht University
Yalelaan 2, P.O. Box 80176
Utrecht NL-3508 TD
The Nederlands

Dr. Frédéric Lallemand
Bâtiment De Serres B. -1
UCL - Biologie du Comportement
1, Place Croix du Sud
Louvain-la-Neuve B-1348
Belgium

Dr. Ian H. Lambert
Institute of Biological Chemistry
August Krogh Institute
Universitetsparken 13
Copenhagen DK-2100
Denmark

Dr. Cesar A. Lau-Cam
Pharmaceutical Sciences
College of Pharmacy
St. John's University
8000 Utopia Parkway
Jamaica, New York 11439
USA

Dr. Robert O. Law
Dept. Cell Physiology &
  Pharmacology
University of Leicester
P.O. Box 138
Leicester LE1 9HN
UK

Dr. John Leader
Dept. of Physiology
University of Otago
P.O. Box 913
Dunedin
New Zealand

Prof. Eun Bang Lee
National Products Research Institute
Seoul National University
#28 Yunkun-Dong, Jongno-Ku
Seoul 110-460
Korea

Dr. Lucimey Lima Perez
Lab. de Neuroquimica, Centro Biofis
    y Bioquim
Inst. Venezolano de Invest. Cientifica
Apdo. 21827
Caracas DF 1020-A
Venezuela

Prof. John B. Lombardini
Dept. of Pharmacology
Texas Tech. University
Health Sciences Center
Lubbock, Texas 79430
USA

Dr. Fabrizio Machetti
Dip. Chimica Organica "U. Schiff"
Università di Firenze
V.G. Capponi 9
Firenze 50121
Italy

Dr. Marina Marangolo
Dept. Biochemistry
Trinity College
University of Dublin
Dublin 2
Ireland

Prof. Rafael Martin Del Rio
Dpto. de Investigacion
Hospital Ramon Y Cajal
Ctra. de Colmenar Km 9
Madrid E-28034
Spain

Ms. Antonella Meini
Ist. di Scienze Farmacologiche
Università di Siena
Via E.S. Piccolomini 170
Siena I-53100
Italy

Prof. Dietrich V. Mitchalk
Universtitat Kinder Klinik
Universtitat Kohl
Joseph-Stelzmann. Str. 9
Koln D-50931
Germany

Dr. Sandra Morales Mulia
Inst. of Cellular Physiology
National University of Mexico
Apartado Postal 70-253
Mexico D.F. 04510
Mexico

Dr. Shigeru Murakami
Medical Res. Lab.
Taisho Pharmaceutical Co., Ltd.
1-403, Yoshino-Cho
Ohmiya, Saitama 330-8530
Japan

Prof. Kevin B. Nolan
Dept. of Chemistry
Royal College of Surgeons in Ireland
Dublin 2
Ireland

Dr. Nina Novoselova
Sechenov Inst. of Evolutionary
Physiology & Biochemistry, RAS
M. Thorez 44
St. Petersburgh 194223
Russia

Dr. Simo S. Oja
Tampere Brain Research Center
University Tampere
Box 607
Tampere FIN-33101
Finland

Dr. James E. Olson
Dept. of Emergency Medicine
Cox Institute
3525 Southern Boulevard
Kettering, Ohio 45429
USA

Dr. Toshiya Ona
Pharmacological Evaluation Lab.
Taisho Pharmaceutical Co., Ltd.
24-1 Takata, 3-Chome Toshimaku
Tokyo 170-8633
Japan

Dr. Susumu Otomo
Pharmacological Evaluation Lab.
Taisho Pharmaceutical Co., Ltd.
1-403, Yoshino-Cho
Ohmiya, Saitama 330-8530
Japan

Dr. Mitri Palmi
Ist. Scienze Farmacologiche
Università di Siena
Via ES Piccolomini 170
Siena I-53100
Italy

Dr. Enkyue Park
Dept. of Immunology
New York State Inst. for Basic
    Research
1050 Forest Hill Road
Staten Island, New York 10314-6399
USA

Prof. Taesun Park
Dept. of Food & Nutrition
Coll. Human Ecology
Yonsei University
134 Shinchon-dong, Sudaemun-ku
Seoul 120749
Korea

Dr. Federica Pessina
Ist. Scienze Farmacologiche
Università di Siena
Via ES Piccolomini 170
Siena I-53100
Italy

Dr. Andranik M. Petrosian
Inst. of Biochemistry
National Academy of Sciences of
    Armenia
Proezd 1, House 1, Ft. 29
Baghramain St., Passage 1
Yerevan 19 -375019
Armenia

Dr. Giusi Pitari
Dip. di Biologia di Base ed Applicata
Università di L'Aquila
Via Vetoio
Coppito - L'Aquila 67010
Italy

Dr. Yudhachai Rajatasereekul
Osotspa Co., Ltd.
White Group Bldg. 2
2100 Ramkhamhaeng Rd., Huamak
Bangkapi, Bangkok BKK 10420
Thailand

Dr. Gerard Rebel
Centre De Neurochimie
IRCAD-Hopital Civil
BP 426
Strasbourg F-67090
France

Mr. Gianmarco Rocco
Ist. Scienze Farmacologiche
Università di Siena
Via ES Piccolomini 170
Siena I-53100
Italy

Dr. Stephen J. Rose
Dept. of Pediatrics
Hartlands Hospital
Bordesley Green East
Birmingham B9 5SS
UK

Dr. Antonio Sanchez Herranz
Neurobiologia-Investigacion
Hospital Ramon y Cajal
Ctra. de Colmar Km 9
Madrid E-28034
Spain

Prof Chaican Sangdee
Dept. Pharmacology
Medical Faculty
Chiang Mai University
Chiang Mai 50200
Thailand

Dr. Francesco Santangelo
Project Manager
Zambon Group Spa
Via Lillo del Duca 10, Bresso
Milano I-20091
Italy

Ms. Simona Saponara
Ist. Scienze Farmacologiche
Università di Siena
Via ES Piccolomini 170
Siena I-53100
Italy

Prof. Pirjo Saransaari
Tampere Brain Res. Center
Dept. of Biomedical Sci.
University of Tampere
Box 607, Tampere SF-33101
Finland

Dr. Stephen Wallace Schaffer
Dept. of Pharmacology
University of South Alabama
College of Medicine
Mobile, Alabama 36688
USA

Dr. Dieter Scheller
Drug Discovery
JANSSEN-CILAG
Raiffeisenstr. 8
Neuss D-41470
Germany

Dr. Georgia Schuller-Levis
Depts. of Neurovirology & Develop.
   Biochem.
Inst. for Basic Res. in Develop.
   Disabilities
1050 Forest Hill Road
Staten Island, NY 10314
USA

Dr. Casilde Sesti
Ist. di Scienze Farmacologiche
Università di Siena
Via ES Piccolomini 170
Siena I-53100
Italy

Prof. Giampietro Sgaragli
Ist. di Scienze Farmacologiche
Università di Siena
Via ES Piccolomini 170
Siena I-53100
Italy

Dr. Jose Maria Solis
Dept. Investigacion
Hospital Ramon y Cajal
Ctra. de Colmenar Km 9
Madrid 28034
Spain

Ms. Hee-Young Son
Department of Food and Nutrition
Seoul National University
Shillim Dong
Seoul 151-742
Korea

Prof. Martha H. Stipanuk
Division of Nutritional Science
Cornell University
225 Savage Hall
Ithaca, NY 14853
USA

Dr. M.-Saadeh Suleiman
Bristol Heart Institute
University of Bristol
Bristol Royal Infirmary
Bristol, Avon BS2 8HW
UK

Dr. Yasuhide Tachi
Clinical Research Division
Taisho Pharmaceutical Co., Ltd.
24-1 Takata 3-Chome Toshimaku
Tokyo 170-8633
Japan

Dr. Kyoko Takahashi
Dept. Clin. Evaluation Med. &
    Therapeutics
Osaka University
1-6 Yamada-oka - Suita-shi
Osaka 565-0871
Japan

Dr. Taka-aki Takenaga
Pharmacological Evaluation Lab.
Taisho Pharmaceutical Co., Ltd.
1-403, Yoshino-Cho
Ohmiya 330-8530
Japan

Dr. Etsuko Tamura
Manager Taurine Project Team
Taisho Pharmaceutical Co., Ltd.
24-1, Takata 3-ChomeToshima-Ku
Tokyo 170-8633
Japan

Dr. Marcel Tappaz
Directeur de Recerche CNRS
INSERM - Unitè 433, Fac. de Mèd.
    Laennec
Rue Gullaume Paradin
Lyon, Cedex 08 F-69373
France

Dr. Gisele Tchuisseu-Youmbi
Ist. Scienze Farmacologiche
Università di Siena
Via ES Piccolomini 170
Siena I-53100
Italy

Prof. John Timbrell
Pharmacy Dept.
King's College London
Cornwall House, Stamford Str.
London SE1 8WA
UK

Prof. Keith F. Tipton
Dept. Biochemistry
Trinity College
University of Dublin
Dublin 2
Ireland

Dr. Pavel Torkounov
Dept. of Military Medicine
Scientific Research Institute
Lesoparkovaya str. 4
Saint Petersburg 195043
Russia

Dr. Katsuharu Tsuchida
Research Center
Taisho Pharmaceutical Co., Ltd.
1-403, Yoshino-cho
Ohmiya 330-8530
Japan

Ms. Maria Luisa Valacchi
Ist. Scienze Farmacologiche
Università di Siena
Via ES Piccolomini 170
Siena I-53100
Italy

Dr. Massimo Valoti
Ist. Scienze Farmacologiche
Università di Siena
Via ES Piccolomini 170
Siena I-53100
Italy

Dr. Volker Viechtbauer
Director
Red Bull GmBH
Brunn 115
Fuschl am See A-5330
Austria

Dr. Roberta J. Ward
Biologie du Comportement
Unite de Biologie
Univ. de Louvain-la-Neuve
1, Place Croix du Sud
Louvain-la-Neuve 1348
Belgium

Dr. Catherine J. Waterfield
School of Pharmacy, Toxicology
Dept.
University of London
Brunswick Square
London WC1N 1AX
United Kingdom

Dr. Marsha C. Wertzberger Gardner
Arent Fox Kintner
Plotkin & Kahn, PLLC
1050 Connecticut Avenue NW
Washington DC 20036-5339
USA

Dr. Jang-Yen Wu
Dept. of Physiology & Cell Biology
University of Kansas
Lawrence
Kansas 66045-2106
USA

Prof. Hidchiko Yokogoshi
Lab. of Nutritional Biochem.
Shizuoka University
School of Food & Nutrit. Science
52-1 Yada
Shizuoka 422-8526
Japan

CONTENTS

Keynote Presentation

Appreciations

Plenary Presentations

Part 1. Basic Aspects and Peripheral Actions of Taurine

Part 2. Metabolic and Nutritional Aspects

Part 3. Taurine in the CNS

Part 5. Taurine and Vision

Part 6. Taurine and Cell Homeostatic Mechanisms

Part 7. Clinical Aspects

Appendix

EXPANDING THE CIRCLE 1975-1999: SULFUR BIOCHEMISTRY AND INSIGHTS ON THE BIO-LOGICAL FUNCTIONS OF TAURINE

Ryan J. Huxtable
Department of Pharmacology, University of Arizona College of Medicine, Tucson, Arizona 85724

INTRODUCTION

Laura Della Corte suggested the title of this lecture. It refers to the first symposium devoted to taurine[38], held in Tucson, Arizona in 1975. As an indication of how the area has changed, of the participants in this meeting, apart from myself, only Steve Baskin and Barry Lombardini were at that first meeting, and only Barry and I have participated in all the taurine symposia. Compared to 25 years ago, there is much more general interest in taurine among the public and the biomedical community. The taurine derivative, acamprosate, has been approved in some countries as an adjunct in the treatment of alcoholism. Taurine-containing drinks, such as Red Bull™ in Europe or Lipovitan™ in Japan, are increasingly popular. There are numerous approved or not-so-approved uses of taurine ranging from supplementation of baby foods, adjunctive treatment of heart failure, and treatments for epilepsy, hyperactivity and chemical dependencies.

Figure 1 shows the Aztec gods of life and death, taken from the Codex Borgia, which is now in Rome. Quetzalcoatl, the God of Life, is on the right, and Mictlanecuhtli, the God of death, on the left. This image metaphorically indicates the inevitable relationship between past and future, the one arising from the other only to fade back into it. I suppose that my scripted function in this keynote lecture that opens the meeting is to behave like these Siamese-twinned gods, and to review in an hour all the work done on taurine in the last 25 years, triumphantly to foresee the future, and then to fade quietly into oblivion. However, I cannot even predict what the weather will be like tomorrow morning, much less foresee how a complex area of research in a

Taurine 4, edited by Della Corte *et al.*
Kluwer Academic / Plenum Publishers, New York, 2000.

changing social setting will evolve. And as for the past, that is largely the accumulated achievements of those of you sitting here.

Figure 1. A metaphor of past and future: The Aztec Quetzalcoatl, the God of Life (right), and Mictlanecuhtli, the God of death (left). From the Codex Borgia.

I would, therefore, prefer to talk in more general terms about taurine. First, however, permit me to say that having been associated with the taurine field for 30 years, I have a strong sense of the continuity of research in this area, and the contributions of many fine scientists and humans. In particular, I would like to pay tribute to Italian contributions. My introduction to taurine dates from 1970, with studies on the relationship between taurine and isethionic acid, and the role of taurine in muscle function[34]. A stimulus for this work was the various publications from Italian investigators, particularly of the group from Florence. It is a pleasure to start this talk by acknowledging the pharmacological contributions of Dr. Giotti and his collaborators[20,21,29,30,66,67].

It is also a pleasure to acknowledge the enormous contributions of Dr. Cavallini and his group to the biochemistry of sulfur metabolites related to taurine[12-17]. Dr. Cavallini, incidentally, participated in the first taurine meeting in Tucson, as well as the first meeting devoted to sulfur biochemistry, which was held in Roscoff, France, in 1956[27]. Dr. Cavallini has spent his career researching unstable sulfur compounds with complex biochemistries and chemistries.

Despite the progress in taurine research in the last 25 years, basic questions as to biological mechanisms still remain. It may be useful to start the meeting by stepping back and putting taurine into a larger and more coherent picture, to see what patterns emerge and what deductions may be posited about the biological significance of taurine. With that in mind, I will begin by putting taurine into a biochemical context, and reviewing the comparative biochemistry of sulfur with relationship to taurine.

THE BIOLOGICAL SULFUR CYCLE

Unlike its group VI congener, oxygen, sulfur has d orbitals available for use. As a result, sulfur exists in oxidation states ranging from -2 to $+6$. The biological sulfur cycle involves the cyclic oxidation and reduction between the most reduced and most oxidized states (Figure 2). Oxidation involves dropping down a free energy ladder, and hence is energy-producing. Conversely, reduction is energy-requiring.

Sulfur from the geosphere enters the biological sulfur cycle at three major points: as sulfide, elemental sulfur and sulfate. Bacteria are the major source of organic sulfur for the biological cycle. However, plants can transport and use inorganic sulfate (assimilatory reduction, where the sulfur is incorporated into organic molecules).

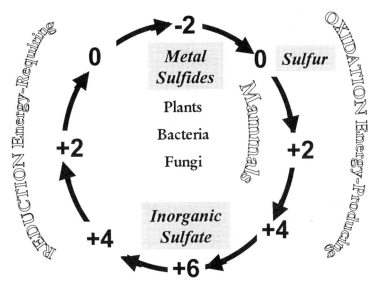

Figure 2. The biological sulfur cycle, showing the various oxidation states for reduction and oxidation (odd-numbered oxidation states require dimeric intermediates). Sample oxidation states are -2: sulfide; 0: sulfur; +2: sulfinate, thiosulfide; +4: sulfonate, taurine; +6 sulfate. Major entry points of inorganic sulfur into the cycle are indicated in gray boxes. Mammals are capable only of sulfur oxidation. Plants, bacteria and fungi can perform the complete cycle.

Unlike the rest of the biosphere, animals, or more specifically mammals, are able only to carry out the oxidative side of the cycle. Mammals must rely on bacteria and plants to reduce sulfur back again. The energies involved in sulfur oxidation are large, around 840 kJ/mol for complete oxidation. This may be compared to the energy of hydrolysis of ATP (31.2 kJ/mol). The oxidation of 1 mol of sulfide releases energy equivalent to the energy of 25 ATP molecules. Reduction of nitrate to ammonia requires 142 kJ/mol and of CO_2 to carbohydrate 460 kJ/mol. However, in mammals the energy released by the oxidation of sulfur is not coupled to ATP production. It goes to waste, appearing as heat.

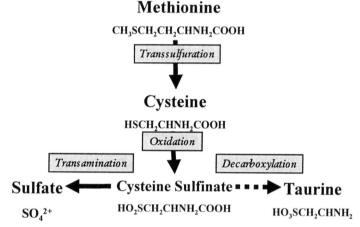

Figure 3. Major flow of sulfur metabolism in mammals. In most mammals, the branch to taurine is quantitatively minor or quantitatively insignificant.

SULFUR METABOLISM AND FUNCTION IN MAMMALS

Methionine and the Methionine-Homocysteine Cycle

As sulfur reduction does not occur in mammals, sulfide, at an oxidation state of −2, is a dietary requirement. This requirement is met with cysteine and methionine. A high availability of methionine "spares" cysteine, as it can be converted to cysteine by transsulfuration. Transsulfuration is irreversible in mammals, although not in fungi which can interconvert methionine and cysteine in either direction. Enteric bacteria synthesize cysteine *de novo* and convert it to methionine. The main path of sulfur metabolism in mammals involves the four processes of transsulfuration, oxidation, transamination and decarboxylation (Figure 3).

Although dietary cysteine can lower the dietary requirement for methionine, there is an absolute requirement for methionine as it has functions in addition to transsulfuration and it is also needed for protein synthesis. These functions include methylation and polyamine synthesis. However, methionine is one of the most toxic amino acids[7]. It causes growth retardation, tissue damage and iron deposition in the spleen. Tissue levels must therefore be kept low, and a dietary load must be processed rapidly. A further problem with methionine is its similarity to a football mom, that overworked denizen of American suburbs. Methionine has too much to do and some of the responsibilities are mutually conflicting.

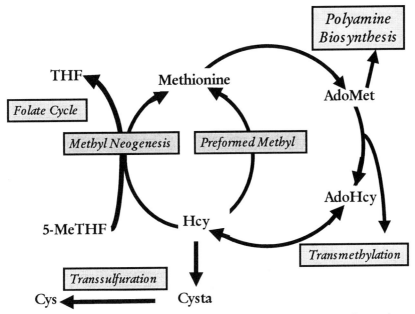

Figure 4. Methionine metabolism and the methionine-homocysteine cycle. AdoHcy: S-Adenosylhomocysteine; AdoMet: S-Adenosylmethionine; Cys: Cysteine; Cysta: Cystathionine; Hcy: Homocysteine; 5-MeTHF: 5-Methyltetrahydrofolate; THF: Tetrahydrofolate.

The major metabolic activities of methionine involve the methionine-homocysteine cycle (Figure 4): Adenosylation of methionine activates both the methyl group for transmethylation and the aminopropyl moiety for transfer in polyamine synthesis. Transmethylation yields S-adenosylhomocysteine, which is further converted to homocysteine. This is a substrate for three processes: transsulfuration, methyl neogenesis (from the tetrahydrofolate path), and methyl recycling (from choline via betaine, dimethylglycine, sarcosine to glycine). The conflicting demands of these processes create a traffic control problem in the cycle. Polyamine synthesis and transsulfuration drain intermediates away from the cycle, although polyamine synthesis is a quantitatively minor pathway for methionine. A given molecule of methionine survives only two turns of the cycle. Transsulfuration and remethylation compete for homocysteine. Transmethylation must occur for transsulfuration to occur, as that is the only source of homocysteine.

Homocysteine is toxic, producing cardiovascular disease and neural tube defects. Despite the high flux through the homocysteine pool, therefore, the pool size must be kept small (i.e. there must be rapid remethylation or rapid movement into transsulfuration). Serious toxicity has not been reported for S-adenosylmethionine[1]. However, the current interest in S-adenosylhomocysteine as a dietary supplement is disturbing[18] in light of the toxicity of the closely related compounds methionine and homocysteine. S-adenosylhomocysteine is being widely used as an antidepressant and in treatment of arthritis and liver disease in recommended doses of up to 1.6 g/day.

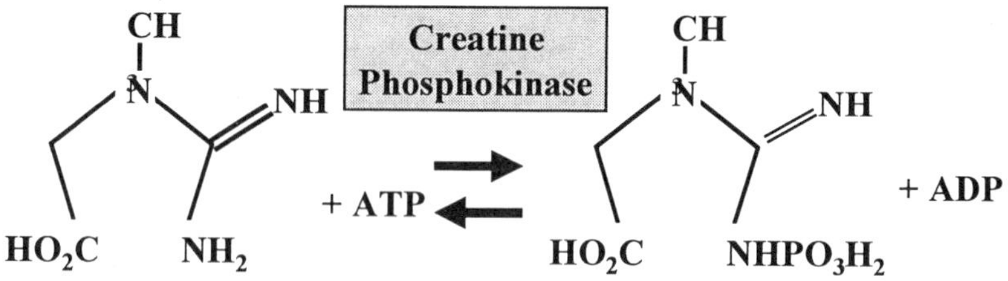

$$H_2NCH(=NH)NHCH_2CH_2SO_3H \qquad H_2NCH(=NH)NHCH_2CH_2CO_2H$$

Guanidinoethane sulfonate   **Guanidinopropionate**

Figure 5. Methylation and muscle contraction. The methyl group of creatine is not necessary for its function as a phosphokinase substrate.

The requirement that transmethylation precede transsulfuration may explain a mystery of muscle biochemistry. Quantitatively, the most significant methylation process is the conversion of guanidinoacetate to creatine (Figure 5). The methylated compound creatine is a phosphagen that maintains myofibrillar ATP levels. Muscle levels of creatine are around 3x higher than

ATP levels. Biosynthesis of this one compound accounts for 80% of mammalian methylation[57]. The other 20% is consumed in epinephrine synthesis, carboxymethylation of proteins and phospholipid methylation.

But why is the methyl group of creatine needed for muscle function? It probably is not. In invertebrates such as polychaetes (marine worms), various sponges and sea anemones, the unmethylated compound, guanidino-ethane sulfonate, serves as the phosphagen[2,8,55,76]. Guanidinopropionate and guanidinoethane sulfonate are substrates for creatine phosphokinase. In mammals, creatine can be replaced by guanidinopropionate without major problems for muscle function[25,68]. In human males, there is a shortfall of around 5.1 mmol of methyl per day between methyl groups available from diet and methyl groups consumed in transmethylation or methyl oxidation. This shortfall is met by methyl neogenesis via 5,10-methylenetetrahydrofolate and 5-methyltetrahydrofolate (Figure 4). The use of guanidinoacetate as a phosphagen would alleviate the need for this expensive reductive neogenesis. However, it is only by the passage of methyl groups through the methionine-homocysteine cycle that transsulfuration can proceed. This process is required for the production of cysteine under conditions of low dietary intake of cysteine, and for the removal of toxic methionine under conditions of high dietary intake. The use of creatine, therefore, may be an inefficient biochemical way of producing homocysteine needed for transsulfuration: creatine simply provides a sink for the methyl groups that must be consumed in order for the methionine-homocysteine cycle to operate.

Transsulfuration

Transsulfuration is so called because the sulfur of homocysteine is transferred to the carbon chain of serine, yielding cysteine (Fig. 6). Transsulfuration serves the functions of removing toxic methionine and producing cysteine required for GSH and protein synthesis. In the brain, for example, methionine levels are below 1 μM. Cerebral cortical synaptosomes from adult rats contain 0.2 μmol methionine/g protein compared with 25 μmol taurine/g protein[53]. In biopsied human brain, methionine and phenylalanine were present in the lowest concentration of all amino acids examined[50]. Both enzymes involved in transsulfuration are pyridoxal phosphate-dependent. The significance of transsulfuration is indicated by the consequences of genetic deficiency in cystathionine ß-synthase deficiency. Deficiency is associated with homocystinuria, hypermethioninemia, and decreased conversion of methionine to sulfate. Clinically, there are vascular, mental, skeletal and visual abnormalities.

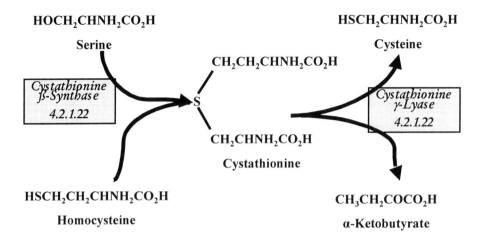

Figure 6. Mammalian transsulfuration.

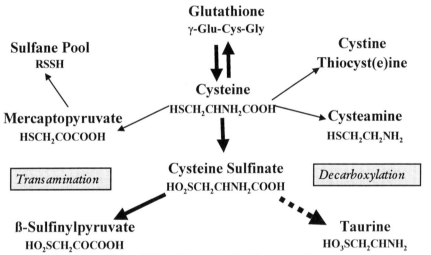

Figure 7. Cysteine metabolism in mammals.

The Functions and Metabolism of Cysteine

The functions of cysteine include protein synthesis and GSH synthesis for redox and conjugation functions. Cysteine is both neurotoxic and readily oxidized to cystine, a poorly soluble compound. GSH serves as a transportable form of extracellular cysteine as it is less susceptible to oxidation. Cell cysteine levels are lower than GSH levels. Cysteine acts at the NMDA receptor[58], being ten times more toxic to neuronal cell lines than its oxidized analog, cysteine sulfinic acid[60]. Local increases in cysteine in the brain (perhaps due to low levels of cysteine dioxygenase) may be involved in neurodegenerative disorders[58]. Plasma cysteine levels are increased in Parkinson's disease, Alzheimer's disease and motor neuron disease.

Cysteine is catabolized by oxidation to cysteine sulfinic acid via the action of cysteine dioxygenase. This enzyme raises the oxidation state of sulfur from -2 to +2. Hepatic cysteine dioxygenase activity increases after dietary loading with cysteine or methionine, but is decreased by cyclic AMP[3].

Both cysteine and cysteine sulfinate can undergo either transamination or decarboxylation (Figure 7). Transamination of cysteine yields mercaptopyruvate, which is metabolized further to compounds such as mercaptolactate, thiocyanate, thiosulfite and other compounds. These are quantitatively insignificant pathways for cysteine, however. The bulk of cysteine metabolism is oxidative. Urinary excretion of sulfur compounds in humans reflects transamination of CSA almost completely (taurine in the urine being primarily dietary in origin). Less than 0.7% of sulfur metabolism reflects cysteine transamination[70].

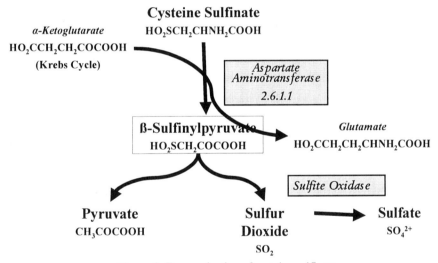

Figure 8. Transamination of cysteine sulfinate.

Cysteine sulfinate is an excitatory amino acid, increasing neuronal cyclic AMP levels. Taurine blocks the cysteine sulfinate-induced rise in cyclic AMP, but not the rise induced by glutamate or aspartate[9]. Cysteine sulfinate appears to act at a unique receptor[10]. Cysteine sulfinate blocks glutamate uptake into astrocytes with an IC_{50} of 114 μM[6].

Even in species capable of decarboxylating cysteine sulfinate to hypotaurine and taurine, transamination is quantitatively the major route of sulfur catabolism (Figure 8). Aspartate aminotransferase appears to be the most significant decarboxylase involved, yielding an enzyme-bound unstable intermediate, ß-sulfinylpyruvate, which has never been isolated. This spontaneously decomposes to sulfur dioxide which, following oxidation with sulfite oxidase, is excreted in the urine as sulfate or used as sulfonation substrate by sulfotransferases (Figure 9).

Advantages of the transamination pathway include the salvaging of carbon as pyruvate, and the reuse of sulfate in sulfonation pathways, following its activation to phosphoadenosine phosphosulfate (Figure 9). However, despite this reutilization pathway, the most sulfur excretion is in the form of sulfate, indicating that the flux of sulfur through transamination exceeds that needed for sulfonation. Despite this, increased sulfate in the diet decreases the need for methionine: sulfate spares methionine. In the taurine pathway (Figure 10), both sulfur and carbon are excreted without further metabolic employment apart from bile salt synthesis and a limited use in conjugation.

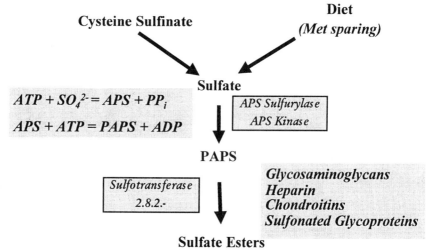

Figure 9. Sulfate activation and sulfonation. Inorganic sulfate derives from the diet or from sulfur amino acid catabolism. It is converted to phosphoadenosine phosphosulfate (PAPS; "active sulfur") by a two-step process. PAPS is a sulfonate donor for numerous processes.

Mammals either synthesize taurine from cysteine sulfinate, get it in the diet, or both. Species with a high dietary intake seem to have lost the ability for ready biosynthesis. Even species with a high synthetic capacity process most sulfur via transamination. The enzyme involved in the first step - cysteine sulfinate decarboxylase - has now been well characterized and studied, thanks to the contributions of some of the people at this meeting. This is in major contrast to our confusion and lack of knowledge 24 years ago. The second step, the oxidation of hypotaurine to taurine, is still not well understood.

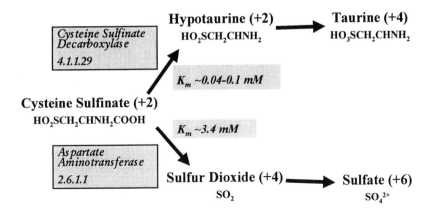

Figure 10. Metabolic choices for cysteine sulfinate.

Increased protein in the diet leads to increased cysteine oxidase and transaminase activities but decreased cysteine sulfinate decarboxylase activity. The decarboxylase saturates at lower levels of cysteine sulfinate than the transaminase (Figure 10)[4]. This means that the greater the flux of sulfur, the more passes through transamination[23]. Increasing the level of sulfur amino acids in the diet (from 0.36; 0.54; to 1.06%) leads to decreased activity of liver cysteine sulfinate decarboxylase (from 30 to 20 to 12 nmol/min/mg, respectively[46]). However, taurine production increases markedly. For example, on the same three diets urinary taurine changes from 0.27 μmol/day on the 0.36% sulfur amino acid diet, to 6.0 μmol/day on the 0.54% diet to 80 μmol/day on the 1.06% diet. It must be concluded that falling enzyme activity is not incompatible with increased metabolism, perhaps because the higher substrate level leads to better saturation of the enzyme. Stipanuk also finds that taurine production increases with increased casein in the diet[3].

In rat hindquarter muscle, most cysteine sulfinate metabolism occurs via transamination. Perfusion for 2 h with 0.2 mM cysteine sulfinate leads to taurine production of 18 nmol/g muscle and sulfate production of 101 nmol/g/muscle[5]. There is an even more marked difference in the enzyme activities controlling the two pathways. Cysteine sulfinate decarboxylase activity is 0.69 μmol/h/g wet weight tissue, compared to 3495 μmol/h/g wet weight for aspartate aminotransferase. Liver cysteine sulfinate decarboxylase has a low K_m of 0.045 mM, indicating that saturation occurs at relatively low concentrations of substrate[75]. In the brain, cysteine sulfinate decarboxylase is activated by phosphorylation[74].

The product of cysteine sulfinate decarboxylase is hypotaurine, a substance readily oxidized to taurine. The enzymology of oxidation is still not understood. This is partially due to the chemical instability of hypotaurine.

In the presence of oxygen and traces of various metals, it readily oxidizes to taurine. Fellman and coworkers[24] proposed an elegant one electron oxidation of hypotaurine at the last taurine meeting in Italy, in 1986. This involved the intermediacy of a disulfone that dismutated to one molecule of taurine and one of hypotaurine. Fellman advanced experimental evidence in support of this scheme. However, it awaits confirmation. At the meeting in Tucson in 1997, Duprè and coworkers[22] reported their inability to reproduce Fellman's findings.

Tissue hypotaurine levels are typically higher than cysteine sulfinate levels[77], the ratio ranging from 10:1 in the brain, 25:1 in kidney, and over 300:1 in liver. Hypotaurine has been proposed as an antioxidant. It is present in high levels in seminal fluid of various species[48] and in regenerating rat liver[72]. The spotty and variable distribution of taurine biosynthetic pathways, however, militates against a generalized antioxidant function of hypotaurine.

The fact that the animal kingdom, although not the bacterial or plant, retain large amounts of taurine from the oxidation of sulfur amino acids - whether an individual species makes it for itself or gets it from a dietary source - implies that taurine confers an evolutionary advantage. There are other pathways of sulfur metabolism that provide clear advantages, yet the animal kingdom has "chosen" to retain a plurality of metabolic routes.

THE FUNCTIONS OF TAURINE

Bile Salt Conjugation

One clear mammalian function of taurine is a metabolic one: that of bile salt synthesis. In addition, taurine can conjugate with various xenobiotics. In mammals, xenobiotic conjugation seems to be a relict pathway, significant for certain compounds in certain species but in general quantitatively minor compared with GSH, sulfate, and glucuronide conjugation.

Invertebrates contain surface tension-lowering substances in the alimentary canal. These are typically sulfate or taurine conjugates with various lipophilic molecules. Vertebrates conjugate only with taurine, except for mammals. In mammals, taurine conjugation is the norm (Figure 11). Glycine conjugation evolved for placental mammals, and seems more common in herbivores. However, taurine is a more efficient conjugator for bile salts, as it remains ionized under the episodic high acidity of the upper intestine.

As well as varying in the use of taurine as a bile salt conjugator, mammalian species vary in their reliance on biosynthetic versus dietary taurine. As much of the clinical and nutritional interest in taurine stems from the significance of a dietary source, studies on species other than humans have to be carefully interpreted.

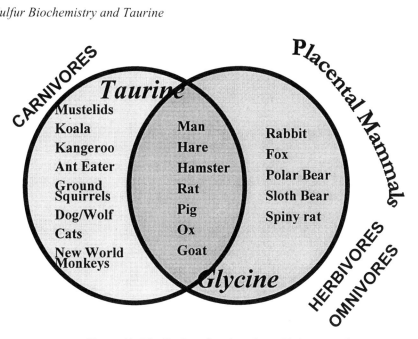

Figure 11. Distribution of conjugating acids in mammals.

Table 1. Diet, conjugation and biosynthesis

Species	Diet	Taurine Synthesis	Taurine Conjugation	Dietary Dependence on Taurine
Guinea Pig	Herbivore	High	Glycine	None
Rat	Herbivore	High	Taurine	Low
Old World Monkey		Poor	Glycine/Taurine	?Moderate*
New World Monkey		Poor	Taurine	?High
Human	Omnivore	Very poor	Glycine/Taurine	High*
Cat	Carnivore	Very poor	Taurine	Absolute

*These species can increase glycine/taurine ratio in response to a dietary deficiency in taurine

Even in a given species, taurine balance and source are sensitive to dietary and other conditions, as shown by the examples of the rat and mouse. Mice are taurine conjugators. Mice were given a diet containing 0.06% [^{3}H]taurine and 0.74% [^{35}S]methionine as sole sources of sulfur amino acids[40]. Under such conditions, diet and biosynthesis contribute about equally to the taurine pool, which shows a whole body half-life of 18.6 days. The protein content of the diet had little effect on the sources of taurine.

In a similar experiment, male rats were given diets containing 0.58% [^{35}S]methionine with or without the addition of 0.4% [^{3}H]taurine[41]. In the absence of dietary taurine, "depot" taurine (i.e. taurine present in the animal at the beginning of the experiment) is held tenaciously, being 46% of the total body pool of taurine even after 87 days. The contribution of biosynthesis to the total body pool also increases in animals on a taurine-free diet. This does not seem to be due to increased biosynthesis, however. Animals on a taurine-free diet have twice as much biosynthetic taurine after 63 days as do animals receiving a 0.4% taurine diet. However, the taurine-deprived animals also excreted less biosynthetically derived taurine in the urine. The sum of biosynthesized taurine excreted in urine and retained in the animal is the same in both groups of animals, suggesting that rats respond to deficiency by controlling excretion rather than by increasing synthesis. In keeping with this, as dietary taurine becomes inadequate, renal reuptake increases[63]. Regulation of reuptake appears to be more significant than regulation of biosynthesis.

Table 2. Effect of taurine and its analogs on the reverse-phase chromatography of neutral phospholipids

Substance	Concentration (mM)	K_{PE}	K_{PC}	ΔK_{PE}	ΔK_{PC}
Control (methanol/water 95:5)		23.3	27.4		
Taurine	1.6	27.3	32.0	+4.0	+4.7
Aminopropane Sulfonate	1.6	25.4	27.9	+2.1	+0.5
Ethanolamine Sulfonate	2.0	18.7	22.0	-4.6	-5.4
Ethanolamine	2.0	5.5	28.2	-17.7	+0.9
Choline chloride	2.0	19.5	22.0	-3.7	-5.4
Phosphorylcholine	0.45		26.1		-1.3
Aminoethane Phosphonate	0.1	24.4	25.6	+1.1	-0.8

Data from Alfredo Cantafora (personal communication). Lipids used were palmitoyloleylphosphatidylethanolamine (PE) and palmitoyloleylphosphatidylcholine (PC). $K_{PE} = (RT_{PE} - RT_0)/RT_0$; $K_{PC} = (RT_{PC} - RT_0)/RT_0$; $\Delta K_{PE} = \Delta K_{PE}$ in presence of substance less ΔK_{PE} in the absence of substance (control).

In rats, even at birth one third of the total body taurine has been synthesized *in utero*[41]. By weaning, total body taurine has increased from 42 μmol at birth to 315 μmol[42]. Most of this derives from biosynthesis. The large synthesis of taurine occurs despite - even in rats - the bulk of sulfur amino acid metabolism occurring via transamination. This is very different from the situation in human babies, where there seems to be insignificant synthetic capacity. Human milk is somewhat higher in taurine (337 μM) that rat milk (152 μM)[61]. However, humans have extremely low cysteine sulfinate decarboxylase activity in liver compared to rat: 0.3 nmol CO_2/mg protein in humans versus 460 nmol CO_2/mg protein in rats[33],[73].

The combination of biosynthetic and conjugating abilities leads to species differences in dietary dependency on taurine (Table 1). Again, the species variation makes for extreme caution in extrapolating from experimental animals to humans.

Phospholipid Interactions

For functions of taurine other than conjugation, there are a number of major phenomena involving membrane effects. These include osmoregulation, calcium availability, channel regulation and phospholipid methylation. In general, taurine is high in species lacking cell walls (animals) and low in species with cell walls (plants; bacteria).

In 1986, we proposed and advanced experimental support for the hypothesis that taurine: phospholipid interactions at the membrane were responsible for many of the biological actions of taurine[44]. We proposed that ion-pair formation between the head groups produced changes in membrane conformation. These changes were suggested to be responsible for the observed modifications in cation binding produced by taurine.

Direct evidence for ion-pair formation between taurine and neutral phospholipids has been found by Cantafora, in Rome (personal communication). Ion-pairing agents interact with the charged head groups of phospholipids to increase the retention time of the latter on reverse phase HPLC (ion-pairing masks the ionic groups allowing better partitioning of the lipophilic chains with the lipophilic resin). Taurine, unlike the other substances listed (Table 2), increases retention time of phosphatidylethanolamine and phosphatidylcholine, thus demonstrating that it is ion pairing with the head groups.

P-32 NMR also provides direct evidence of taurine-phospholipid interactions in cation binding[69]. Phospholipid vesicles generate ^{32}P NMR signals from phospholipids both on the inside and outside of the vesicle. The inside signal can be erased by forming the vesicles in the presence of cobalt chloride. Suspending the vesicles in the presence of the shift reagent, europium chloride, shifts the signal downfield from phosphatidylserine on the outside (Eu^{3+} being a calcium analog and binding to the acidic site on the phospholipid). Phosphatidylcholine is shifted to a lesser extent. In the presence of taurine (20 mM) the europium-induced shifts are antagonized, the signals for phosphatidylcholine and phosphatidylserine shifting upfield and fusing. Again, this is direct evidence for an influence of taurine on the cation-binding characteristics of phospholipid membranes.

Table 3. Binding characteristics of taurine to biological membranes

Taurine binding affinity (mM)	14-33
Hill coefficient	1.9-3.3
Binding antagonists	$Ca^{2+} > Na^+ > K^+$
Effect of taurine on Ca^{2+} binding	Increased affinity (K_D); Decreased binding (B_{max})
Effect of Na^+ on taurine binding	Increased affinity (K_D); Decreased binding (B_{max})
Effect of hypotaurine on taurine binding	Inhibits
Effect of ß-alanine on taurine binding	No effect
Effect of guanidinoethane sulfonate on taurine binding	No effect

Table 4. Summary of binding characteristics of taurine and calcium to artificial phospholipid vesicles

	Phosphatidyl			
	Inositol	Serine	Ethanol-amine	Choline
Calcium binding	Yes	Yes	No	No
Taurine binding	No	Yes	Yes	Yes
Stimulation of calcium binding by taurine	No	Yes*	No	No

*Stimulation by taurine is greatest when the liposome contains a neutral phospholipid in addition to phosphatidylserine

The effects of taurine on cation binding are mediated through a low affinity binding site for taurine. This site can be demonstrated on biological proteolipid membranes and on synthetic phospholipid vesicles[45,64,65]. Binding characteristics to biological membranes are summarized on Table 3. Taurine binding shows positive cooperativity. Cations and taurine mutually inhibit each other's binding.

Using phospholipid vesicles of known composition, it can be demonstrated that taurine binds to neutral phospholipids, calcium binds to acidic phospholipids, and that taurine stimulates the binding of Ca to phosphatidyl-serine when a neutral phospholipid is also present (Table 4)[44]. Other effects of taurine on ion handling by membranes include the taurine-induced increases G_{Cl} in muscle. This is mediated by a low affinity interaction with membrane phospholipids[19].

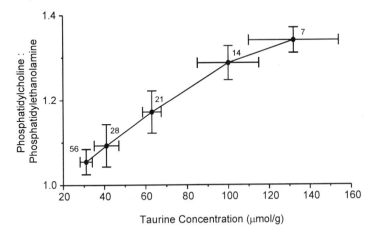

Figure 12. Phosphatidylcholine and phosphatidylethanolamine ratio as a function of taurine concentration in synaptosomes (P$_2$B fraction) from developing rat brain. The day of life is indicated by the number above each datum point[52].

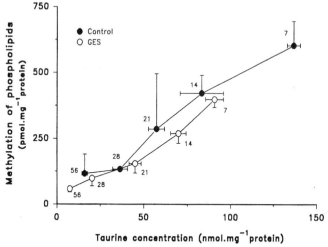

Figure 13. Phospholipid methylation as a function of taurine concentration in synaptosomes (P$_2$B fraction) from developing rat brain. ● Control rats; ○ Taurine-depleted rats (pregnant rats were maintained on drinking water containing 1% guanidinoethane sulfonate for the final two weeks of pregnancy and throughout lactation; pups were weaned onto the same drinking water). The day of life is indicated by the number above each datum point. Phospholipid methylation was determined from the incorporation of [^{3}H-methyl]methionine into phosphatidyl-choline[43].

We subsequently found a remarkable correlation between neutral phospholipid ratio and taurine concentration during brain development in a membrane preparation, the P2B fraction (Figure 12)[52]. Both taurine content and neutral phospholipid ratio change markedly in a correlated manner during the developmental period in P_2B membrane. Such a correlation could, of course, be simple coincidence, of no biological significance.

Of more significance, in the same preparation, when synaptosomal taurine levels are artificially lowered *in vivo* by treating rats with guanidinoethane sulfonate, developmental changes in phospholipid methylation rate are shifted (Figure 13)[43]. In synaptosomes, phospholipid methylation rate falls sharply during development, in keeping with the fall in taurine concentration. Guanidinoethane sulfonate treatment results in lowered synaptosomal content of taurine at all time points. Phospholipid methylation rate shifts in concert. This implies a cause-and-effect relationship between taurine content and methylation rate. As methylation rate affects the neutral phospholipid ratio in the cell membrane, this further suggests that the taurine content of a cell and the membrane phospholipid composition are related.

Cantafora et al.[11] found that 0.4% taurine in drinking water lowered liver phosphatidylcholine but left phosphatidylethanolamine content unchanged (i.e. there was decreased methylation). Guertin et al.[28] found that taurine added to TPN-fed guinea pigs lowered the cholesterol:phospholipid ratio and corrected abnormalities in lipid:protein ratio.

These reports suggest a close relationship between taurine and phospholipid methylation. The neutral phospholipids are metabolically linked by the phospholipid methylation pathway, which adds three methyls sequentially to phosphatidylethanolamine to yield phosphatidylcholine. The change in ratio in these phospholipids during development imply a decrease in rate of phospholipid methylation

Although most phospholipid is synthesized via the CDP-choline pathway, methylation is a crucial regulator of membrane function. In the liver, methylation of phosphatidylethanolamine accounts for 15-20% of phosphatidylcholine synthesis[62]. Ethanol decreases phospholipid methylation at the final step, causing increases in mono and dimethylphosphatidylethanolamine[71].

Schaffer and co-workers found in the heart that exogenous taurine inhibited rather than stimulated phospholipid methylation[31]. These differences from our observations imply there are other variables involved in the interaction of taurine and phospholipid methylation. However, regulation by taurine of phospholipid methylation also provides a potential regulation of homocysteine supply for transsulfuration: the end-product of sulfur amino acid metabolism regulating the initial step that diverts sulfur away from the methionine:homocysteine cycle.

Properties of Taurine Relevant to Biological Function

Taurine is a chemically simple compound. Even so, it has some unique physicochemical and biochemical aspects of relevance to its biological function:

Poor chelation. Association constants for taurine with cations are low[47]. Chelation of metal ions, therefore, is unlikely to be involved in the biological actions of taurine. This implies that taurine does not directly disturb or displace cations.

Low net charge. Taurine has a pK_1 of 1.5 (ionization of sulfonate group) and a pK_2 of 8.82 (ionization of amino group)[36]. At a pH of 7.4, the sulfonate is essentially 100% ionized and the amino group is 96.3% ionized. The net residual charge on taurine is therefore 0.037 negative; i.e. taurine is almost completely zwitterionic. Transport across membranes does not alter charge distribution.

High ionization. As both the acidic SO_3H and basic NH_2 functions of taurine are highly ionized at physiological pH, the compound has low liposolubility. It thus has a low rate of diffusion across membranes compared to more lipophilic analogs such as ß-alanine[35,37,39]. Consequently, high concentration gradients can be maintained without a high energy expenditure to combat diffusion loses.

Metabolic inactivity. Taurine is not involved in primary metabolism. Other amino acids are involved in protein synthesis, the Krebs cycle, gluconeogenesis, and other biochemical pathways. Unlike these amino acids, with taurine there is no competition between substrate requirements and non-metabolic functions. Metabolic demands do not interfere with maintenance of cell levels. This is not the case for other sulfur amino acids, such as cysteine, methionine, cysteine sulfinate or homocysteine. These compounds are both metabolically active and toxic. In general, partially oxidized sulfur intermediates (-2 to +2) are important but, apart from GSH, are toxic. The cell must maintain levels sufficient for metabolic requirements without allowing levels to reach the threshold for toxicity. In general, with these intermediates, the turnover of pools is high. The cell has much more freedom to vary taurine levels before serious consequences ensue.

The highly acidic sulfonate gives taurine a pI lower than that of other neutral amino acids (but higher than the acidic amino acids). Taurine is also unique in the intracellular:extracellular ratio in the brain[51]. The lower the lipophilicity and the higher the charge on a molecule, the less energetically expensive it is to maintain high ratios. Neurotransmitter amino acids maintain high ratios, a decrease in the ratio (increased extracellular concentration) triggering neurotransmission. Neutral or basic proteinogenic amino acids maintain low ratios. Taurine, again, is intermediate.

Enantiostasis

Enantiostatic regulation has been defined as occurring when the effect of a change in a physicochemical property of the internal milieu is opposed by a change in another physicochemical property[56]. Such a regulation protects against the consequences of environmental change. Enantiostasis differs from homeostasis in that function is maintained although physicochemical state is altered. Thus, when taurine antagonizes calcium-induced alterations in cardiac contractility[26], it is regulating enantiostatically, inasmuch as physiologically normal contractility is re-established in the face of physiologically abnormal calcium concentrations. Taurine fits all the requirements for an enantiostatic agent. It is highest in cells with a requirement to accommodate to environmental changes and lowest in cells that are protected from the environment by cell walls. Taurine is present in high concentrations in electrically active cells and cells subject to a fluctuating environment, such as retinal cells and secretory cells. Taurine is present only in trace amounts in bacteria and plants.

Biological systems are regulated in general by the balance of opposing influences. A balance of inhibitory and excitatory neurotransmitters, for example, determines the net excitability of the brain. Taurine and calcium function as opposing principles, both functionally and biochemically, and can be considered as a yin-yang metaphor. Taurine is high inside the cell (mM) and low outside (μM). Calcium is low inside the cell (nM) and high outside (mM). In concentrations higher than a few μM, calcium, a structure-forming element, is inimical to protein tertiary structure, phosphate solubilization (nucleotides, nucleic acids, and inorganic phosphate), and oxyanion solubilization. It displaces Mg^{2+} from Mg^{2+} binding sites with a change in coordination stereochemistry from planar to hexahedral, and with a resulting change in macromolecular conformation. Life as we understand it could only evolve after in the presence of cell membranes for excluding calcium ions from cellular space. Entry of Ca^{2+} is tightly regulated. Even so, it produces profound changes in macromolecular conformations that are used as signalling devices in the cell to catalyze events ranging from excitation-contraction coupling, excitation-secretion coupling, and a host of other phenomena.

Taurine is a passive agent, resisting calcium-induced changes by modifying calcium binding and antagonizing a range of calcium-dependent events, such as contractility. It helps the cell continue functioning in the face of homeostatic changes; that is to say, it functions as an enantiostatic agent.

One site of interaction of taurine and calcium is the membrane, particularly in its phospholipid components. The mechanism appears to involve ion-pair formation that alters the ion-binding characteristics of the membrane and the way the membrane interfaces with an aqueous phase. The consequence is a taurine-modulated modification of membrane properties involving, among other phenomena, an increased affinity and decreased capacity of calcium binding[64],[65], stimulation of high affinity, bicarbonate-dependent Ca^{2+} transport[49],[59], an altered transition temperature for Ca^{2+} transport[54], and increased Cl^- conductance[19]. In rabbit myocytes, taurine depresses K_{ATP} activity[32] with a K_d of 13.5 mM. It decreases lifetime of bursts, increases interburst interval and decreases the number of functional channels.

The allosteric interactions with calcium have a physicochemical basis, but as with other calcium-dependent phenomena these interactions can be incorporated into numerous biological phenomena.

Overall, taurine is a low affinity regulator of membrane function:

(i)   It modifies membrane structure and composition, and hence function, via processes such as phospholipid methylation

(ii)   It modifies membrane function directly, via an action on cation binding, channel properties and membrane enzyme activities

(iii)   It acts as a rapid response force to enable cells to accommodate enantiostatically to changes in the environment (cell concentrations of taurine change more rapidly than membrane composition)

In turn, the phospholipid composition of the membrane modifies the interaction with taurine.

For good historical reasons involving centuries of dogmatism from the medieval Catholic church to 20th century Lysenkoism, scientists are suspicious of over-much theorizing. However, I have tried to present an organizing overview that accommodates many of the known actions of taurine, and I have tried to place the overview in a logical physicochemical and evolutionary biochemical framework.

ACKNOWLEDGEMENTS

I thank Dr. Cantafora for his kindness in allowing me to show the data of Table 2 on reverse-phase chromatography.

REFERENCES

1. Anonymous, 1987, Proceedings of a symposium in SAMe (Osteoarthritis: the clinical picture, pathogenesis, and management with studies on a new therapeutic agent, S-adenosylmethionine. *Am.J.Med.* 83: 1-110.
2. Allen, J.A., and Garrett, M.R., 1971, Taurine in marine invertebrates. *Adv.Mar.Biol.* 9: 205-253.
3. Bagley, P.J., Hirschberger, L.L., and Stipanuk, M.H., 1995, Evaluation and modification of an assay procedure for cysteine dioxygenase activity: High-performance liquid chromatography method for measurement of cysteine sulfinate and demonstration of physiologi-

cal relevance of cysteine dioxygenase activity in cysteine catabolism. *Anal.Biochem.* 227: 40-48.

4. Bagley, P.J., and Stipanuk, M.H., 1994, The activities of rat hepatic cysteine dioxygenase and cysteinesulfinate decarboxylase are regulated in a reciprocal manner in response to dietary casein level. *J.Nutr.* 124: 2410-2421.

5. Bella, D.L., Hirschberger, L.L., Hosokawa, Y., and Stipanuk, M.H., 1999, Mechanisms involved in the regulation of key enzymes of cysteine metabolism in rat liver in vivo. *Am.J.Physiol.Endocrinol.Metab.* 276: E326-E335

6. Bender, A.S., Woodbury, D.M., and White, H.S., 1989, □□-DL-methylene-aspartate, an inhibitor of aspartate aminotransferase, potently inhibits L-glutamate uptake into astrocytes. *Neurochem.Res.* 14: 641-646.

7. Benevenga, N.J., Yeh, M.-H., and Lalich, J.J., 1976, Growth depression and tissue reaction to the consumption of excess dietary methionine and S-methyl-L-cysteine. *J.Nutr.* 106: 1714-1720.

8. Bergquist, P.R., and Hartman, W.D., 1969, Free amino acid patterns and the classification of the Demospongiae. *Mar.Biol.* 3: 247-268.

9. Boss, V., and Boaten, A.S., 1995, An L-cysteine sulfinic acid-sensitive metabotropic receptor mediates increased cAMP accumulation in hippocampal slices. *Neurosci.Lett.* 184: 1-4.

10. Boss, V., Nutt, K.M., and Conn, P.J., 1994, L-cysteine sulfinic acid as an endogenous agonist of a novel metabotropic receptor coupled to stimulation of phospholipase D activity. *Mol.Pharmacol.* 45: 1177-1182.

11. Cantafora, A., Mantovani, A., Masella, R., Mechelli, L., and Alvaro, D., 1986, Effect of taurine administration on liver lipids in guinea pigs. *Experientia* 42: 407-408.

12. Cavallini, D., De Marco, C., and Mondovi, B., 1961, Detection and distribution of enzymes for oxidizing thiocysteamine. *Nature (London)* 192: 557

13. Cavallini, D., De Marco, C., Mondovi, B., and Stirpe, F., 1954, The biological oxidation of hypotaurine. *Biochim.Biophys.Acta* 15: 301-303.

14. Cavallini, D., De Marco, C., Scandurra, R., Duprè, S., and Graziani, M.T., 1966, The enzymatic oxidation of cysteamine to hypotaurine. *J.Biol.Chem.* 241: 3189-3196.

15. Cavallini, D., Federici, G., Ricci, G., Duprè, S., Antonucci, A., and De Marco, C., 1975, The specificity of cysteamine oxygenase. *FEBS Letters* 56: 348-351.

16. Cavallini, D., Mondovi, B., and De Marco, C., 1955, the isolation of pure hypotaurine from the urine of rats fed cystine . *J.Biol.Chem.* 216: 577-582.

17. Cavallini, D., Scandurra, R., and De Marco, C., 1963, The enzymatic oxidation of cystamine to hypotaurine in the presence of sulfide. *J.Biol.Chem.* 238: 2999-3005.

18. Cowley, G., and Underwood, A., 1999, What is SAMe. *Newsweek* 46-50.

19. De Luca, A., Pierno, S., and Camerino, D.C., 1996, Effect of taurine depletion on excitation-contraction coupling and Cl⁻ conductance of rat skeletal muscle. *Eur.J.Pharmacol.* 296: 215-222.

20. Dolara, P., Agresti, A., Giotti, A., and Pasquini, G., 1973, Effect of taurine on calcium kinetics of guinea pig heart. *Eur.J.Pharmacol.* 24: 352-358.

21. Dolara, P., Marino, P., and Buffoni, F., 1973, Effect of 2-aminoethanesulphonic acid (taurine) and 2-hydroxyethane sulphonic acid (isethionic acid) on calcium transport by rat liver mitochondria. *Biochem.Pharmacol.* 22: 2085-2094.

22. Duprè, S., Spirito, A., Sugahara, K., and Kodama, H., 1998, Hypotaurine oxidation: An HPLC-mass approach. In *Taurine 3: Cellular and Regulatory Mechanisms*, (S. Schaffer, J.B. Lombardini, and R.J. Huxtable, eds.), Plenum Press, New York, pp. 3-8.

23. Ensunsa, J.L., Hirschberger, L.L., and Stipanuk, M.H., 1993, Catabolism of cysteine, cystine, cysteinesulfinate, and OTC by isolated perfused rat hindquarter. *Am.J.Physiol.Endocrinol.Metab.* 264: E782-E789

24. Fellman, J.H., Green, T.R., and Eicher, A.L., 1987, The oxidation of hypotaurine to taurine: *Bis*-aminoethyl-□-disulfone, a metabolic intermediate in mammalian tissue. In

The Biology of Taurine: Methods and Mechanisms, (R.J. Huxtable, F. Franconi, and A. Giotti, eds.), Plenum Press, New York, pp. 39-48.

25. Fitch, C.D., Jellinek, M., and Mueller, E.J., 1974, Experimental depletion of creatine and phosphocreatine from skeletal muscle. *J.Biol.Chem.* 249: 1060-1063.

26. Franconi, F., Martini, F., Stendardi, I., Matucci, R., Zilletti, L., and Giotti, A., 1982, Effect of taurine on calcium levels and contractility in guinea pig ventricular strips. *Biochem.Pharmacol.* 31: 3181-3186.

27. Fromageot, C. 1956,. *La Biochimie du Soufre,* CNRS, Paris.

28. Guertin, F., Roy, C.C., Lepage, G., Yousef, I., and Tuchweber, B., 1993, Liver membrane composition after short-term parenteral nutrition with and without taurine in guinea pigs: The effect to taurine. *Proc.Soc.Exp.Biol.Med.* 203: 418-423.

29. Guidotti, A., Badiani, G., and Giotti, A., 1971, Potentiation by taurine of inotropic effect of strophanthin K on guinea pigs isolated auricles. *Pharmacol.Res.Commun.* 3: 29-38.

30. Guidotti, A., Badiani, G., and Pepeu, G., 1972, Taurine distribution in cat brain. *J.Neurochem.* 19: 431-435.

31. Hamaguchi, T., Azuma, J., and Schaffer, S., 1991, Interaction of taurine with methionine: Inhibition of myocardial phospholipid methyltransferase. *J.Cardiovasc.Pharmacol.* 18: 224-230.

32. Han, J., Kim, E., Ho, W.K., and Earm, Y.E., 1996, Blockade of the ATP sensitive potassium channel by taurine in rabbit ventricular myocytes. *J.Mol.Cell.Cardiol.* 28: 2043-2050.

33. Hayes, K.C., Stephan, Z.F., and Sturman, J.A., 1980, Growth depression in taurine-depleted infant monkeys. *J.Nutr.* 110: 119-125.

34. Huxtable, R., and Bressler, R., 1972, Taurine and isethionic acid: distribution and inter-conversion in the rat. *J.Nutr.* 102: 805-814.

35. Huxtable, R., and Chubb, J., 1977, Adrenergic stimulation of taurine transport by the heart. *Science* 198: 409-411.

36. Huxtable, R.J., 1982, Physicochemical properties of taurine: Introduction. In *Taurine in Nutrition and Neurology*, (R.J. Huxtable, and H. Pasantes-Morales, eds.), Plenum Press, New York, pp. 1-4.

37. Huxtable, R.J., 1987, From heart to hypothesis: a mechanism for the calcium modulatory actions of taurine. In *The Biology of Taurine: Methods and Mechanisms*, (R.J. Huxtable, F. Franconi, and A. Giotti, eds.), Plenum Press, New York, pp. 371-388.

38. Huxtable, R.J. and Barbeau, A. 1976, *Taurine,* New York: Raven Press.

39. Huxtable, R.J., Laird, H.E., and Lippincott, S.E., 1979, The transport of taurine in the heart and the rapid depletion of tissue taurine content by guanidinoethyl sulfonate. *J.Pharmacol.Exptl.Therap.* 211: 465-471.

40. Huxtable, R.J., and Lippincott, S.E., 1982a, Diet and biosynthesis as sources of taurine in the mouse. *J.Nutr.* 112: 1003-1010.

41. Huxtable, R.J., and Lippincott, S.E., 1982b, Relative contribution of diet and biosynthesis to the taurine content of the adult rat. *Drug-Nutrient Inter.* 2: 153-168.

42. Huxtable, R.J., and Lippincott, S.E., 1983, Relative contribution of the mother, the nurse and endogenous synthesis to the taurine content of the newborn and suckling rat. *Nutr.Metab.* 27: 107-116.

43. Huxtable, R.J., Murphy, J., and Lleu, P.-L., 1994, Developmental effects of taurine depletion on synaptosomal phospholipids in the rat. In *Taurine in Health and Disease*, (R.J. Huxtable, and D. Michalk, eds.), Plenum Press, New York, pp. 343-354.

44. Huxtable, R.J., and Sebring, L.A., 1986, Towards a unifying theory for the action of taurine. *Trends Pharmacol.Sci.* 7: 481-485.

45. Huxtable, R.J., and Sebring, L.A., 1989, Taurine and the heart: the phospholipid connection. In *Taurine and the Heart* , (H. Iwata, J.B. Lombardini, and T. Segawa, eds.), Kluwer Academic Press, Boston, pp. 31-42.

46. Ide, T. 1997, Simple high-performance liquid chromatographic method for assaying cys-teinesulfinic acid decarboxylase activity in rat tissue. *J.Chromatogr.B Biomed.Sci.Appl.* 694: 325-332.
47. Irving, C.S., Hammer, B.E., Danyluk, S.S., and Klein, P.D., 1982, Coordination and bind-ing of taurine as determined by nuclear magnetic resonance measurements on ^{13}C-labeled taurine. In *Taurine in Nutrition and Neurology*, (R.J. Huxtable, and H. Pasantes-Morales, eds.), Plenum Press, New York, pp. 5-17.
48. Jones, R. 1978, Comparative biochemistry of mammalian epididymal plasma. *Comp.Biochem.Physiol.* 61B: 365-370.
49. Kuo, C.H., and Miki, N., 1980, Stimulatory effect of taurine on Ca uptake by disc mem-branes from photoreceptor cell outer segments. *Biochem.Biophys Res.Commun.* 94: 646-651.
50. Labiner, D.M., Yan, C.C., Weinand, M.E., and Huxtable, R.J., 1999, Disturbances of amino acids from temporal lobe synaptosomes in human complex partial epilepsy. *Neu-rochem.Res.* 24: 1379-1383.
51. Lerma, J., Herranz, A.S., Herreras, O., Abraira, V., and Del Rio, M., 1986, In vivo deter-mination of extracellular concentration of amino acids in the rat hippocampus. A method based on brain dialysis and computerized analysis. *Brain Res.* 384: 145-155.
52. Lleu, P.-L., Croswell, S., and Huxtable, R.J., 1992, Phospholipids, phospholipid methyla-tion and taurine content in synaptosomes of developing rat brain. In *Taurine: Nutritional Value and Mechanisms of Action*, (J.B. Lombardini , S.W. Schaffer, and J. Azuma, eds.), Plenum Publishing, New York, pp. 221-228.
53. Lleu, P.-L., and Huxtable, R.J., 1992, Phospholipid methylation and taurine content of synaptosomes from cerebral cortex of developing rat. *Neurochem.International* 21: 109-118.
54. Lombardini, J.B. 1985, Taurine effects on the transition temperature in Arrhenius plots of ATP-dependent calcium ion uptake in rat retinal membrane preparation. *Bio-chem.Pharmacol.* 34: 3741-3745.
55. Makisumi, S. 1961, Guanidino compounds from a sea-anemone, *Anthopleura japonica* Verrill. *J.Biochem.(Tokyo)* 49: 284-291.
56. Mangum, C., and Towle, D., 1977, Physiological adaptation to unstable environments. *Am.Scient.* 65: 67-75.
57. Mudd, S.H., and Poole, J.R., 1975, Labile methyl balances for normal humans on various dietary regimens. *Metabolism* 24: 721-735.
58. Parsons, R.B., Waring, R.H., Ramsden, D.B., and Williams, A.C., 1997, Toxicity of cys-teine and cysteine sulphinic acid to human neuronal cell-lines. *J.Neurol.Sci.* 152: S62-S66
59. Pasantes-Morales, H., and Ordonez, A., 1982, Taurine activation of a bicarbonate-dependent, ATP-supported calcium uptake in frog rod outer segments. *Neurochem.Res.* 7: 317-328.
60. Pean, A.R., Parsons, R.B., Waring, R.H., Williams, A.C., and Ramsden, D.B., 1995, Tox-icity of sulphur-containing compounds to neuronal cell lines. *J.Neurol.Sci.* 129 Suppl.: 107-108.
61. Rassin, D.K., Sturman, J.A., and Gaull, G.E., 1978, Taurine and other free amino acids in milk of man and other mammals. *Early Human Devel.* 2: 1-13.
62. Ridgway, N.D., and Vance, D.E., 1992, Phosphatidylethanolamine N-methyltransferase from rat liver. *Meth.Enzymol.* 209: 366-374.
63. Rozen, R., and Scriver, C.R., 1982, Renal transport of taurine adapts to perturbed taurine homeostasis. *Proc.Natl.Acad.Sci.USA* 79: 2101-2105.
64. Sebring, L., and Huxtable, R.J., 1985, Taurine modulation of calcium binding to cardiac sarcolemma. *J.Pharmacol.Exptl.Therap.* 232: 445-451.
65. Sebring, L.A., and Huxtable, R.J., 1986, Low affinity binding of taurine to phospholipo-somes and cardiac sarcolemma. *Biochim.Biophys.Acta* 884: 559-566.

66. Sgaragli, G., and Pavan, F., 1972, Effects of amino acid compounds injected into cerebro-spinal fluid on colonic temperature, arterial blood pressure and behavior of the rat. *Neuropharmacology* 11: 45-56.

67. Sgaragli, G.P., Pavan, F., and Galli, A., 1975, Is taurine-induced hypothermia in the rat mediated by 5-HT? *N.-S.Arch.Pharmacol.* 288: 179-184.

68. Shields, R.P., and Whitehair, C.K., 1973, Muscle creatine: *in vivo* depletion by feeding β-guanidinopropionic acid. *Can.J.Biochem.* 51: 1046-1049.

69. Shindo, S., and Huxtable, R.J., 1987, Interaction of taurine with phospholipid vesicles. *Sulfur Amino Acids* 10: 127-132.

70. Sörbo, B., Hannestad, U., Lundquist, P., Mårtensson, J., and Öhmsn, S., 1980, Clinical chemistry of mercaptopyruvate and its metabolites. In *Natural Sulfur Compounds: Novel Biochemical and Structural Aspects*, (D. Cavallini, G.E. Gaull, and V. Zappia, eds.), Plenum Press, New York, pp. 463-470.

71. Srivastava, R., Srivastava, N., and Misra, U.K., 1989, Interaction of ethanol, cholesterol and vitamin A on phospholipid methylation in hepatic microsomes of rat. *Internat.J.Vit.Nutr.Res.* 60: 236-239.

72. Sturman, J.A., and Fellman, J.H., 1982, Taurine metabolism in the rat: effect of partial hepatectomy. *Int.J.Biochem.* 14: 1055-1060.

73. Sturman, J.A., and Hayes, K.C., 1980, The biology of taurine in nutrition and development. In *Advances in Nutritional Research, Volume III*, (H.H. Draper, ed.), Plenum Press, New York, pp. 231-299.

74. Tang, X.W., Hsu, C.C., Schloss, J.V., Faiman, M.D., Wu, E., Yang, C.Y., and Wu, J.Y., 1997, Protein phosphorylation and taurine biosynthesis *in vivo* and *in vitro*. *J.Neurosci.* 17: 6947-6951.

75. Tappaz, M., Almarghini, K., Legay, F., and Remy, A., 1992, Taurine biosynthesis enzyme cysteine sulfinate decarboxylase (CSD) from brain: The long and tricky trail to identification. *Neurochem.Res.* 17: 849-859.

76. Thoai, N.V., and Robin, Y., 1954, Métabolisme des dérivés guanidylés. II. Isolement de la guanidotaurine (taurocyamine) et de l'acide guanidoacétique (glycocyamine) des vers marins. *Biochim.Biophys.Acta* 13: 533-536.

77. Togawa, T., Ohsawa, A., Kawanabe, K., and Tanabe, S., 1997, Simultaneous determination of cysteine sulfinic acid and hypotaurine in rat tissues by column-switching high-performance liquid chromatography with electrochemical detection. *J.Chromatogr.B Biomed.Sci.Appl.* 704: 83-88.

PROFESSOR DORIANO CAVALLINI: A LIFE FOR SULFUR BIOCHEMISTRY

Silvestro Duprè
Dipartimento di Scienze Biochimiche, Università di Roma "La Sapienza", Roma; Centro di Biologia Molecolare, CNR, Roma, Italy

Prof. Doriano Cavallini was present in 1975 at the first Taurine meeting; his high age hampered the participation to the today meeting, but he is still very interested in research work and continues to contribute "expanding the circle", with one of the today posters.

I have the honour, as one of his scholars and as successor on his chair at the Medical Faculty of the First University of Roma, to introduce this session dedicated to him, with a short presentation of his scientific career.

The session's title is "Metabolic and nutritional aspects". Cavallini's work over 60 years has been certainly one of the propellants in this field, giving very important contributions to the knowledge of the metabolic interconnections between sulfur-containing compounds, of the enzymes involved, of metabolic diseases of sulfur metabolism, ranging from cysteine and its derivatives to methionine, pantetheine, glutathione, lipoic acid and to many aspects of enzyme catalysis and active site reactions, as the role of persulfide groups and labile sulfur in rhodanese.

Some highlights of his more than 300 papers may be illustrated following the scientific career, which began in 1938. First are studies on the metabolism of cysteine *in vivo* in the rat. A one year stage in 1947 at the Cornell University in Du Vigneau laboratories, were the long-standing friendship with Lester Reed and Alton Meister began, gave the start to the pioneering work on hypotaurine, taurine, thiotaurine, cysteine disulfoxide

Taurine 4, edited by Della Corte *et al.*
Kluwer Academic / Plenum Publishers, New York, 2000.

and sulfone, which led to the discovery of cysteamine dioxygenase and to the definition of the metabolic map of sulfur fluxes. The outstanding ability in chemical synthesis, in a time where very few compounds were commercially available, allowed the identification of many intermediates. Probably not many of yours know that the Sigma hypotaurine is synthesized following Cavallini's method published in the 40th years. He has been always very attentive to new techniques: from New York, just after the end of 2nd World War, he imported in Italy ion-exchange chromatography, amino acids analysis, the use of labelled compounds for *in vivo* experiments; he began to use techniques like EPR or mass spectrometry, at a time where applications to biochemical problems were rather rare.

Another interesting field, which comes out frequently, is the study of functional outcomes of structural modifications, as substitution of methylene groups with S or S with Se, and the ambiguities of many enzymes in respect to oxygen- or sulfur-containing functional groups. In the same context I would mention how cystamine, for us a molecule with a disulfide bond, was seen by him also as a modified diamine substrate of diamine oxidase.

Very early begun also the study of oxidative model systems (copper two and hydrogen peroxide appears in a paper of 1946) and the role of sulfur with its many oxidation states. The oxidation of hypotaurine to taurine is one of the many mysterious steps which has not been completely unravelled until now and which is the topic of his last paper in this year: the reaction of hypotaurine with singlet oxygen.

A research subject which, once begun, was inherited to other, was pantetheine and the metabolism of CoA. Again it is connected with the discovery of a new enzyme entity, pantetheine hydrolase, a rather strange enzyme, which has been recently involved in control of dislipidemias and was supposed to be one of the pacemaker enzymes in the so-called 'thioester world", which (at least in the view of theoretical biochemists) should have preceded the RNA world, forerunner of the DNA world of today.

A paper of 1981 reports the enzymatic *in vitro* synthesis of cystine ketimine (a seven membered sulfur nitrogen containing cyclic compound), produced by oxidation of L-cystine by L-amino acid oxidase. It is the first paper concerned with ketimines, as this class has been baptized by Cavallini, the first of more than 70 papers and 18 years research of his group. Nowadays this class of natural compounds has been well characterized and studied, although many aspects remain to be disclosed. First a role has been assigned to lanthionine, cystathionine (this latter present in high amount in the brain) and S-aminoethyl cysteine (recently discovered in human urine); ketimines and their reduced stable derivatives have been found and quantitatively determined in tissues and living organisms; strange connections have been established with taurine derivatives (as tauropines); physiological effects have been found, as the priming effect of cystathionine ketimine in human neutrophils, the specific binding of lanthionine ketimine to brain membranes and relaxing effects *in vivo* in the rat, the intense protective effect of the dimer of S-aminoethyl cysteine ketimine on mitochondria in the presence of drugs and antioxidant effect on copper-mediated oxidation of isolated LDL. The intuition of Cavallini has revealed the existence, together with many others, of these six new sulfur-containing metabolites with unknown connections with health and diseases: a good task for a research life.

I would like to mention furthermore that many research themes of Cavallini's younger years, started by him were then continued by his coworkers and became important veins of the department research. The hemoglobin group by Antonini, the transaminase group by Fasella, the amine oxidase research, the rich field of copper proteins, the many possibilities of glutathione research, all them started first with work and intuition of Cavallini. And also now, in the library crowded of young students, you may see his white head when he reads a JBC: I have always the impression he is hunting, he is scenting a trail. His knowledge of the literature is not only profound and really giant, but also his memory is enormous and at the service of intuition and fantasy. I have learned a lot by him: probably not so much in academic policy, but a burden in science and research. And from him I also learned "never to fall in love with your hypothesis".

This brief synopsis can not fully bear witness to the scientific level and to the long academic career of Prof. Cavallini. I tried to emphasize some aspects, and I hope that those who know him will recognize the friend, and those who never met him will, through my words, appreciate this life dedicated to sulfur.

Thank you.

PROFESSOR ALBERTO GIOTTI: A FATHER OF TAURINE RESEARCH

Giampietro Sgaragli
Istituto di Scienze Farmacologiche, Università degli Studi di Siena, Via Piccolomini 170, Siena

It is most appropriate that the first session in this Symposium, "Basic aspects and peripheral actions of taurine", is dedicated to Professor Alberto Giotti, who has contributed so much to the development of "taurinology".

The long and distinguished list of his co-workers and students provides some insight into his achievements in attracting and fostering so many brilliant, young scientists. A central feature of his approach was the integration of pharmacology with cytochemistry, biochemistry, physiology and clinical sciences. When he co-chaired Taurine Meeting in San Miniato, Firenze, Italy, in 1986 he selected the starfish as the logo of the Meeting. In addition to being rich in taurine, the starfish, with its five arms, could represent the emphasis he placed on this concept of the importance of these inter-disciplinary relationships to the development of the pharmacological sciences

Fig. 1. The starfish logo of the International Taurine Meeting, 1986.

Taurine 4, edited by Della Corte *et al.*
Kluwer Academic / Plenum Publishers, New York, 2000.

His interest in Taurine started during the 1960s. His intuition, that further knowledge of its physiological role in mammalian heart could provide valuable therapeutic strategies against cardiovascular diseases, arose from his extensive research in the area of cardio-vascular pharmacology. Much of this was presented in a rigorous and thoughtful overview, *Le basi fisiologiche della farmacologia sperimentale della diastole miocardica*, published 1n 1956[3], which still represents a valuable model for young scientists, as it contains a systematic discussion of the action mechanisms of several cardiotropic drugs used either experimentally or in the clinic at that time. He was before his time in advocating, since the late 1960s, the clinical use of taurine, at that time available in the form of the proprietary tonic 'O-Due", in the management of heart disease[1].

Fig. 2. Photograph of Alberto Giotti (centre) and co-workers(centre) taken in 1992.

At 1986 Taurine Meeting, Professor Giotti presented a summary of the achievements of his research team at that stage[4], focussing on the mechanisms of the three most important effects of taurine in mammalian heart: its antiarrythmic action, its positive inotropism and its cardioprotective effect. His pioneering work in each of these aspects laid a solid foundation for further developments from research groups throughout the world as well as from his own laboratory. As just one example the study of Professor Giotti and his team published in 1992 in the Journal of Molecular and Cellular Cardiology[2] in which taurine was shown to protect the myocardium by reducing Ca^{2+} overload upon exposure to catecholamines has been particularly influential.

On a personal note, it is a great honour to me to present this appreciation since Alberto Giotti, who is now Emeritus Professor of Pharmacology at the University of Firenze, was my teacher of Pharmacology when I was student at the Medical School, and his subsequent advise and encouragement have been of great importance in the development of my own academic career.

There is much more that could be said about how Professor Giotti's insights, stimulating ideas and rigorous experimental approaches have fostered developments in our understanding of the functions and beneficial effects of taurine. However, perhaps the most fitting tribute is the many papers in this International Taurine Meeting that follow-up and attempt to provide more-detailed explanations for his observations on the pleiotypic effects of taurine on mammalian heart.

REFERENCES

1. Chapman R.A., Suleiman M.-S. and Earm Y.E., 1993, Taurine and the heart. *Cardiovasc. Res.*, 27:358-363.
2. Failli P., Fazzini P., Franconi F., Stendardi I. and Giotti A., 1992, Taurine antagonises the increase in intracellular calcium concentration induced by α-adrenergic stimulation in freshly isolated guinea-pig cardiomyocytes. *J. Mol. Cell Cardiol.*, 24:1253-1256.
3. Giotti A., 1956, Le basi fisiologiche della farmacologia sperimentale della diastole miocardica. *Arch. It. Sci. Farmacol.*, serie III, 6:9-156.
4. Giotti A., 1987, Cardiovascular pharmacology and experimental therapeutics of taurine and related compounds. *Adv. Exp. Med. Biol.*, 217:3-22.

MODE OF ACTION OF TAURINE AND REGULATION DYNAMICS OF ITS SYNTHESIS IN THE CNS

Jang-Yen Wu[1,4], Weiqing Chen[1], Xiao Wen Tang[1], Hong Jin[1], Todd Foos[1], John V. Schloss[2], Kathleen Davis[1], Morris D. Faiman[3] and Che-Chang Hsu[1]

[1]Depts. of Mol. Biosci., [2]Med. Chem.and [3]Pharmacol. & Toxicol., Univ. of Kansas, Lawrence, KS 66045, U.S.A. and [4]Inst. of Biol. Chem., Academia Sinica, Taipei, Taiwan

INTRODUCTION

Taurine is one of the most abundant amino acids in animals[18]. The brain is one of the organs containing the highest concentration of taurine, especially in the frontal and occipital lobes[22]. The retina has approximately 10-30 times more taurine than that in the brain, the concentrations ranging from 10 mM in frogs to 50 mM in mammals such as rats and rabbits[49,65].

Although the biological occurrence of taurine was reported over a hundred years ago, research focused on its biological significance and physiological roles has been conducted for only about a few decades[17,22]. Taurine was first considered to be the biochemically inert end product of methionine and cysteine metabolism and its main function was thought to be limited to bile salt synthesis[22,65]. The physiological role of taurine has received considerable attention since the report that cats fed a taurine deficient diet developed central retinal degeneration[15]. Now, taurine has been shown to be involved in many important physiological functions including neurodevelopment[10,15,48-50], membrane stabilization[12,16,19,28], detoxification[14,65,68], antiarrythmic action[9,63], regulation of calcium homeostasis[30,53,64], anticonvulsion[20,21,41,42,44,58,60,62], protein phosphorylation[32], antioxidation[18] and neuroprotection[55]. In addition, taurine has also been suggested as an inhibitory neurotransmitter or neuromodulator[29,34,37,45,52]. In this communication, the regulation dynamics of taurine biosynthesis and the mode of action of taurine in lowering intracellular level of free calcium, $[Ca^{2+}]_i$, are addressed.

Taurine 4, edited by Della Corte et al.
Kluwer Academic / Plenum Publishers, New York, 2000.

BIOSYNTHESIS OF TAURINE

Taurine may be formed in biological systems by five pathways[22]: pathway I: methionine→cysteine→cysteinesulfinic acid (CSA)→hypotaurine→ taurine; pathway II: methionine→cysteine→cysteinesulfinic acid→cysteic acid (CA)→taurine; pathway III: cysteamine→cystamine intermediates →hypotaurine→taurine; pathway IV: sulfate→sulfite intermediates→cysteic acid→taurine; pathway V: cystine→cystine disulfoxide→cystamine disulfoxide→hypotaurine→taurine. However, pathway I appears to be the main biosynthetic route to taurine and cysteine sulfinic acid decarboxylase (CSAD) appears to be the rate-limiting enzyme in taurine biosynthesis in mammalian tissues[22].

Although the final step of taurine biosynthesis along Pathway I is the oxidation of hypotaurine to taurine which is catalyzed by a NAD-dependent hypotaurine oxidase[51], the rate-limiting step is the decarboxylation of CSA to hypotaurine catalyzed by CSAD[22]. Davison[11] has established that the decarboxylation of CSA and CA in the liver is catalyzed by a single protein.

Blinderman *et al.*[6] proposed that the same enzyme that catalyzes the decarboxylation of L-glutamate and CSA was responsible for the biosynthesis of both GABA and taurine in the brain. However, this view was proven incorrect as we[66,67], as well as others[43,46], had identified and purified a specific taurine synthesizing enzyme, CSAD, from mammalian brain that is distinctly different from the GABA-synthesizing enzyme, L-glutamate decarboxylase (GAD). It is now well established that a specific and distinct brain enzyme, CSAD, which catalyzes the decarboxylation of both CSA and CA, but not glutamic acid, is responsible for the synthesis of taurine whereas GAD is responsible for GABA biosynthesis. There is less agreement regarding the cellular localizations of CSAD (i.e. neuronal versus glial localization) and the identity between brain and liver CSAD. Previously, we have shown that CSAD is localized in neurons in the cerebellum[7,8], retina[33-35] and hippocampus[52]. No glial localization was detected. More recently, CSAD and taurine were found to be co-localized in neurons in the cerebellum and hippocampus[38,39]. Recently, it was reported that the liver CSAD and the brain CSAD were indistinguishable based on their physical, chemical and immunological properties[2,47]. The same group further reported that brain CSAD was localized exclusively in glial cells (Bergmann fibers and oligodendrocytes) and not in neurons[1,57]. However, the majority of the reports as cited above support the notion that brain CSAD is distinctly different from the liver enzyme and is localized in neurons rather than in glial cells. The discrepancy could be due to heterogeneity of CSAD in the brain. It is possible that there are multiple forms of CSAD in the brain which show distinct cellular and subcellular localizations. Indeed, we have recently shown that there are multiple isoforms of CSAD in the brain which differ in their charge, size, immunological and enzymatic properties[55]. However, definitive proof of identity or non-identity of various forms of CSAD has to come from the sequence information.

REGULATION OF TAURINE BIOSYNTHESIS

Contrary to taurine transporters, which have been cloned and characterized[13,36,61], little is known regarding the molecular mechanisms in the regulation of CSAD activity. Jerkins and Steele[26] reported that the hepatic CSAD activity was decreased in female rats, adrenalectomized rats and rats fed L-methionine. The decrease of CSAD activity was due to reduction in CSAD protein. The same authors also reported that CSAD activity in liver of rats fed with sulfur amino acids was markedly reduced[24], suggesting that CSAD activity may be regulated by sulfur amino acids through the S-adenosylmethionine-dependent pathway. Furthermore, the hepatic and renal CSAD were modulated differently by thyroid hormone treatment. The hepatic CSAD was depressed by 65% whereas the renal CSAD activity was increased three-fold in rats treated with thyroid hormone[24,25].

Recently, Kaisakia *et al.*[27] reported cloning and characterization of rat liver CSAD and showed that in hyperthyroidism the decrease in liver CSAD activity was due to decrease in CSAD mRNA. CSAD activity in the liver was also reported to be reduced in rats fed a high protein diet[23]. Trenkner *et al.*[59] reported that in mouse cerebellum CSAD activity was developmentally regulated, reaching the lowest activity around postnatal day 14. Furthermore, CSAD activity was found to increase by: (i) depletion of intracellular taurine concentration (achieved by the taurine uptake inhibitors ß-alanine or GES); and (ii) through the excitotoxic effect of glutamate. There are at least two mechanisms that can account for the increase of total CSAD activity under the various conditions cited above. One is activation of CSAD by modification of the enzyme by protein phosphorylation/dephosphorylation and the other is increase in expression of CSAD gene resulting in increase in CSAD synthesis. Recently, we have found that the GABA-synthesizing enzyme, GAD, is activated by dephosphorylation and inhibited by protein phosphorylation whereas CSAD is activated by conditions favoring proteins in the phosphorylated state[3,4,56]. Furthermore, we have identified protein kinase C (PKC) and protein phosphatase type 2C to be responsible for phosphorylation and dephosphorylation of CSAD, respectively[56]. In addition, CSAD activity is increased with concomitant increase in CSAD phosphorylation when neurons are under depolarization conditions[56], presumably due to influx of Ca^{2+}. Also, taurine itself can regulate its own synthesis by inhibiting CSAD phosphorylation[56], presumably by its action in lowering the level of intracellular free calcium concentration, $[Ca^{2+}]_i$ (also see below).

EFFECT OF TAURINE ON CYTOSOLIC FREE CALCIUM LEVEL, [Ca^{2+}]$_i$

By using a dual-wavelength fluorescent spectroscopy with fura-2, a specific Ca^{2+} indicator, we were able to trace changes of [Ca^{2+}]$_i$ in cultured neurons in response to glutamate (Glu) treatment in the presence or absence of taurine. Changes of calcium level in response to Glu treatment were indicated as the ratio of the fluorescence at the wavelength of 346 nm to that at 386 nm. In some neurons, the Glu-induced elevation of [Ca^{2+}]$_i$ was transient and able to return to the basal level after removal of extracellular Glu (Fig. 1A) while in other neurons the elevation of [Ca^{2+}]$_i$ was sustained even after the Glu has been washed away from the media for over 10 min (Fig. 1B). When taurine was present in the media, the Glu-induced increase of [Ca^{2+}]$_i$ was suppressed and the level of [Ca^{2+}]$_i$ was returned to the basal level (Fig. B). Similar results were obtained using confocal microscopy (data not shown). These results suggest that taurine can effectively block elevation of [Ca^{2+}]$_i$ induced by excitatory transmitter such as Glu.

EFFECTS OF TAURINE ON REVERSE MODE OF Na$^+$-Ca^{2+} EXCHANGER ACTIVITY

The mechanisms underlying taurine inhibition of Glu-induced elevation of [Ca^{2+}]$_i$ are complicated, as many pathways can attribute to the elevation of [Ca^{2+}]$_i$ in response to Glu stimulation, including influx of Ca^{2+} through various Ca^{2+} channels, such as Glu receptor-associated channels, and the reverse mode of Na$^+$-Ca^{2+} exchangers. Previously, Matsuda *et al.*[40] have reported that taurine has an inhibitory effect on Na$^+$-Ca^{2+} exchanger activity which may account for the positive ionotropic effect of taurine at low Ca^{2+} concentration in guinea pig heart. It has also been shown that the Na$^+$-Ca^{2+} exchanger is bi-directional and can function in either a forward or reverse mode depending on the electrochemical gradient of Na$^+$ across the membrane[5]. Hence, when neurons are under depolarizing conditions such as stimulation with Glu, Na$^+$-Ca^{2+} exchanger activity is in a reverse mode because of a high intracellular Na$^+$ level induced by excitation. A reverse mode of Na$^+$-Ca^{2+} exchanger activity can also be triggered in cultures by application of ouabain, a Na$^+$-K$^+$ ATPase inhibitor, in combination with monensin, a Na$^+$ ionophore, and this process can be blocked by dichlorobenzamil, a specific blocker for the Na$^+$-Ca^{2+} exchanger[54]. Under these conditions, we found that the intracellular ^{45}Ca^{2+} in cultured neuron was significantly increased by treatment with either ouabain/monensin or Glu and this increase of [Ca^{2+}]$_i$ was partially inhibited in the presence of taurine or dichlorobenzamil (Fig. 2).

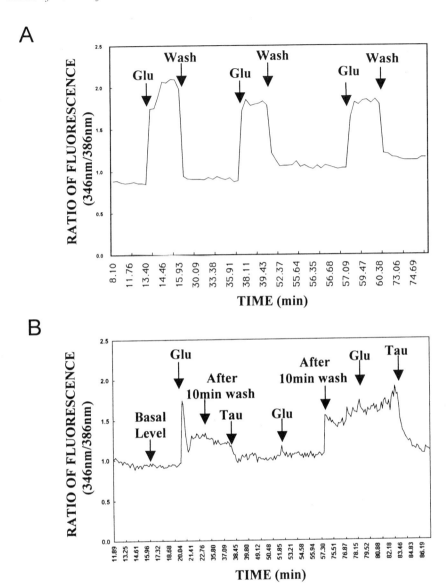

Figure 1. Effect of taurine on glutamate (Glu)-induced increase of $[Ca^{2+}]_i$. (A) Effect of Glu on $[Ca^{2+}]_i$. Cultured neurons were prepared as previously described[31] and plated on coverslips. After 14 days, the regular culture media were replaced with EBSS. Fura-2 AM was loaded after 1-h equilibration. Calcium level was detected using dual-wavelength fluorescence spectroscopy and analyzed as the ratio of the fluorescnece at 346 nm to 386 nm. Glutamate (50 mM) induces a significant increase in intracellular free calcium level (the value is the average of the ratio from six neurons in two trials). (B) Effect of taurine on Glu induced elevation of intracellular calcium level. Cultured neurons were labeled with Fura-2 AM and calcium level was detected as described in (A). Glu induced a sustained elevation of $[Ca^{2+}]_i$ which was suppressed or inhibited in the presence of 25 mM taurine (the value is the average of the ration from ten neurons in two trials).

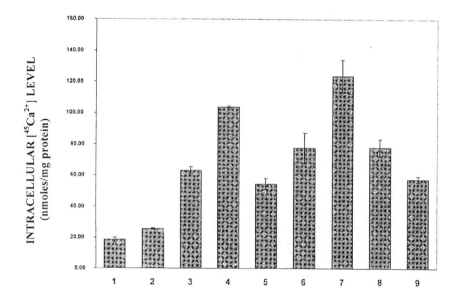

Figure 2. Effect of taurine on reverse mode of Na^+-Ca^{2+} exchanger activity. Neuronal cultures in EBSS medium as described in Fig. 1 were equilibrated for 1-h. The media were then switched to the EBSS containing 1.5 mCi/ml $^{45}CaCl_2$ in the presence of 1 mM ouabain, 20 mM monensin to trigger the reverse mode of Na^+-Ca^{2+} exchanger. The cultures were incubated for 10 min and then exposed to 50 mM glutamate in the presence or absence of 25 mM taurine or dichlorobenzamil, an inhibitor of Na^+-Ca^{2+} exchanger. The treatment was terminated by removal of the media, followed by 3x quick rinsing with 1 ml of ice-cold 4 mM EGTA/0.9% NaCl. The cells were solubilized in 700 ml of 0.3 M NaOH and aliquots were taken for determination of protein as well as radioactivity. The control groups without glutamate treatment were assayed to determine basal $[Ca^{2+}]_i$ level. 1: Control; 2: Taruine control; 3: Benzamil control; 4: + Ouabain/Monensin (O/M); 5: O/M + Taurine; 6: O/M + Benzamil; 7: Glu; 8: Glu + Tau; and 9: Glu + Benzamil.

CONCLUSION

The regulation of taurine biosynthesis can be summarized as following: (i) When neurons are stimulated, the arrival of action potential will open the voltage-dependent Ca^{2+}-channel, resulting in an increase of intracellular free Ca^{2+}, $[Ca^{2+}]_i$; (ii) Elevation of $[Ca^{2+}]_i$ will trigger release of taurine as well as activation of PKC, which in turn activates CSAD through protein phosphorylation; (iii) The activated CSAD then synthesizes more taurine to replenish that lost due to stimulation-mediated release; (iv) When intracellular taurine reach a certain level, it then inhibits the activation of PKC directly or indirectly (possibly through regulating Ca^{2+} availability), thus shutting down activation of CSAD through inhibition of CSAD phosphorylation by PKC; and (v) CSAD soon returns to its inactive state through the action of a protein phosphatase, most likely PrP-2C. The mode of action of taurine in lowering the level of $[Ca^{2+}]_i$ is at least partially due to its inhibition on the reverse mode of Na^+-Ca^{2+} exchanger activity

ACKNOWLEGMENTS

This work was supported in part by grants from National Science Foundation (IBN-9723079), Office of Naval Research (N00014-94-1-0457), The Research Development Fund, University of Kansas. The expert typing of the manuscript by Sharon Lee Green is greatly appreciated.

REFERENCES

1. Almarghini, K., Barbagli, B., and Tappaz, M., 1994, Production and characterization of a new specific antiserum against the taurine putative biosynthetic enzyme cysteine sulfinate decarboxylase. *J. Neurochem.* 62: 1604-1614.
2. Almarghini, K., Remy, A., and Tappaz, M., 1991, Immunocyto-chemistry of the taurine biosynthesis enzyme, cysteine sulfinate decarboxylase in the cerebellum. Evidence for a glial localization. *Neuroscience* 43: 111-119.
3. Bao, J., Cheung, W. Y., and Wu, J.-Y., 1995, Brain L-glutamate decarboxylase: Inhibition by phosphorylation and activation by dephosphorylation. *J. Biol. Chem.* 270: 6464-6467.
4. Bao, J., Nathan, B., and Wu, J.-Y., 1994, Role of protein phos-phorylation in the regulation of brain L-glutamate decarboxylase activity. *J. Biomed. Sci.* 1: 237-244.
5. Blaustein, M.P., 1989, Calcium transport and buffering in neurons. *Trends Neurosci.* 11: 438-443.
6. Blinderman, J. M., Maitre, M., Ossola, L., and Mandel, P., 1978, Purification and some properties of L-glutamate decarboxylase from human brain. *Eur. J. Biochem.* 86: 143-152.
7. Chan-Palay, V., Lin, C. T., Palay, S., Yamamoto, M., and Wu, J.-Y., 1982, Taurine in the mammalian cerebellum: Demonstration by autoradiography with [3H]taurine and immuno-cytochemistry with antibodies against the taurine-synthesizing enzyme, cysteine sulfinic acid decarboxylase. *Proc. Natl. Acad. Sci.* USA 79: 2695-2699.
8. Chan-Palay, V., Palay, S., Li, C., and Wu, J.-Y., 1982, Sagittal cerebellar micro-bands of taurine neurons: Immunocytochemical demonstration by using antibodies against the taurine synthesizing enzyme cysteine sulfinic acid decarboxylase. *Proc. Natl. Acad. Sci.* USA 79: 4221-4225.
9. Chazov, E.I., Malchikova, L.S., Lipina, N.V., Asafov, G.B., and Smirnov, V.N., 1974, Taurine and electrical activity of the heart. *Circ. Res.* 34/35 SIII: III-11-III-21.
10. Chen, X.-C., Pan, Z.-L., Liu, D.-S., and Han, X.B., 1998, Effect of taurine on human fetal neuron cells: Proliferation and differentiation. In *Adv. Exp. Med. Biol.*, "Taurine 3: Cellular and Regulatory Mechanisms", Schaffer, S., Lombardini, J.B. and Huxtable, R.J., eds., Plenum Press, New York, 442: 397-403.
11. Davison, A.V., 1956, Amino acid decarboxylase in rat brain and liver. *Biochem. Biophys. Acta.* 234: 107-108.
12. Green, P., Dawson, R., Wallace, D.R., and Owens, J., 1998, Treatment of rat brain membranes with taurine increases radioligand binding. In *Adv. Exp. Med. Biol.*, "Taurine3: Cellular and Regulatory Mechanisms", Schaffer, S., Lombardini, J.B. and Huxtable, R.J., eds., Plenum Press, New York, 442: 377-383.

13. Han, X., Budreau, A.M., and Chesney, R.W., 1998, Molecular cloning and functional expression of LLC-PK1 cell taurine transporter that is adaptively regulated by taurine. *Adv. In Expt. Med. & Biol.* 442: 261-268.

14. Hasal, S.J., Sun, Y., Yan, C.C., Brendel, K., and Huxtable, R.J., 1998, Effects of taurine in percision-cut liver slices exposed to the pyrrolizidine alkaloid, retrorsine. In *Adv. Exp. Med. Biol.*, "Taurine3: Cellular and Regulatory Mechanisms", Schaffer, S., Lombardini, J.B. and Huxtable, R.J., eds., Plenum Press, New York, 442: 7 9-83.

15. Hayes, K., Rabin, A., and Berson, E., 1975, An ultrastructural study of nutritionally induced and reversed retinal degeneration in cats. *Am. J. Pathol.* 78: 505.

16. Hayes, K., Carey, R.E., and Schmidt, S.Y., 1975, Retinal degeneration associated with taurine deficiency in the cat. *Science* 188: 949-951.

17. Huxtable, R.J., 1989, Taurine in the central nervous system and the mammalian actions of taurine. *Prog. Neurobiol.* 32: 471-533.

18. Huxtable, R.J., 1992, Physiological actions of taurine. *Physiol. Rev.* 72: 101-163.

19. Huxtable, R.J., and Bressler, R., 1973, Effect of taurine on a muscle intracellular membrane. *Biochem. Biophys. Acta.* 323: 573-583.

20. Izumi, K., Donaldson, J., Minnich, J.L., and Barbeau, A,. 1973, Ouabain-induced seizures in rats: suppression effects of taurine and g-aminobutyric acid. *Can. J. Physiol. Pharmacol.* 51: 885-889.

21. Izumi, K., Igisu, H., and Fukuda, T., 1974, Suppression of seisures by taurine specific or non-specific? *Brain Res.* 76: 171-173.

22. Jacobsen, J.G., and Smith, L.H., 1968, Biochemistry and physiology of taurine and taurine derivatives. *Physiol. Rev.* 48: 424-511.

23. Jerkins, A.A.; Bobroff, L.E., and Steele, R.D., 1989, Hepatic cysteine sulfinic acid decarboxylase activity in rat fed various levels of dietary casein. *J. Nutri.* 119: 1593-1597.

24. Jerkins, A.A., and Steele, R.D., 1991, Dietary sulfur amino acid modulation of cysteine sulfinic acid decarboxylase. *Am. J. Physiol.* 261: 551-555.

25. Jerkins, A.A., and Steele, R.D., 1991, Cysteine sulfinic acid decarboxylase activity in response to thyroid homone administration in rats. *Arch. Biochem. Biophys.* 286: 428-432.

26. Jerkins, A.A., and Steele, R.D., 1992, Quantification of cysteine sulfinic acid decarboxylase in male and female rats: Effect of adrenalectomy and methionine. *Arch. Biochim. Biophys.* 294: 534-538.

27. Kaisakia, P.J., Jerkins, A.A., Goodspeed, D.C., and Steele, R.D., 1995, Cloning and characterization of rat cysteine sulfinic acid decarboxylase. *Biochim. Biophys. Acta.* 1262: 79-82.

28. Kramer, J.H., Chivan, J.P., and Schaffer, S.W., 1981, Effect of taurine on calcium paradox and ischemic heart failure. *Am. J. Physiol.* 240: H238-246.

29. Kuriyama, K., 1980, Taurine as a neuromodulator. *Fed. Proc.* 39: 2680-2684.

30. Lazarewicz, J.W., Noremberg, K., Lehmann, A., and Hamberger, A., 1985, Effects of taurine on calcium binding and accumulation in rabbit hippocampal and cortical synaptosomes. *Neurochem. Int.* 7: 421-428.

31. Lee, Y.-H., Deupree, D.L., Chen, S.-C., Kao, L.-S., and Wu, J.-Y,. 1994, Role of Ca^{2+} in a-amino-3-hydroxy-5-methyl-4-isoxazolepropionic acid-mediated polyphosphoinositide turnover in primary neuronal cultures. *J. Neurochem.* 62: 2325-2332.

32. Li, Y.P., and Lombardini, J.B., 1991, Taurine inhibits protein kinase C catalyzed phosphorylation of specific proteins in a rat cortical P2 fraction. *J. Neurochem.* 56: 1747-1753.

33. Lin, C.-T., Li, H.-Z., and Wu, J.-Y., 1983, Immunocytochemical localization of L-glutamate decarboxylase, gamma aminobutyric acid transaminase, cysteine-sulfinic acid decarboxylase, aspartate aminotransferase and somatostatin in rat retina. *Brain Res.* 270: 273-283.

34. Lin, C.-T., Su, Y.Y.T., Song, G.-X., and Wu, J.-Y., 1985, Is taurine a neurotransmitter in rabbit retina? *Brain Res.* 337: 293-298.

35. Lin, C.-T., Song, G.-X., and Wu, J.-Y., 1985, Ultrastructural demonstration of L-glutamate decarboxylase and cysteine sulfinic acid decarboxylase in rat retina by immunocytochemistry. *Brain Res.* 331: 71-80 .

36. Liu, Q.R., Lopez-Corcuera, B., Nelson, H., Mandiyan, S., and Nelson, N., 1992, Cloning and expression of a cDNA encoding the transporter of taurine and beta-alanine in mouse brain. *Proc. Natl. Acad. Sci.* USA 89: 12145-12149.

37. Lombardini, J.B., Schaffer, S.W., and Azuma, J., (eds)., 1992, Taurine: Nutritional value and mechanisms of action. *Adv. Expt. Med. & Biol.* 315: 1-441.

38. Magnusson, K.R., Madl, J.E., Clements, J.R., Wu, J.-Y., Larson, A.A,. and Beitz, A.J., 1988, Co-localization of taurine- and cysteine sulfinic acid decarboxylase-like immunoreactivity in the cerebellum of the rat with the use of a novel monoclonal antibody against taurine. *J. Neurosci.* 8: 4551-4564.

39. Magnusson, K.R., Clements, J.R., Wu, J.-Y., and Beitz, A.J., 1989, Co-localization of taurine- and cysteine sulfinic acid decarboxylase-like immunoreactivity in the hippocampus of the rat. *Synapse* 4: 55-69.

40. Matsuda, T., Gemba, T., Baba, A., and Iwata, H., 1989, Inhibition by taurine of $Na+-Ca2+$ exchange in sarcolemmal membrane vesicles from bovine and guinea pig hearts. *Comp. Biochem. Physiol.* 94C(1): 335-339.

41. Mutani, R., Bergamini, L., Fariello, R., and Delsedime, M., 1974, Effect of taurine on accut epileptic foci. *Brain Res.* 70: 170-173.

42. Mutani, R., Monaco, F., Durelli, L., and Delsedime, M., 1975, Level of free amino acids in serum and cerebrospinal fluid after administration of taurine to epileptic and normal subjects. *Epilepsia* 16: 765-769.

43. Oertel, W.H., Schmechel, D.E., Weise, V.K., Ransom, D.H., Tappaz, M.L., Krutzsch, H.C., and Kopin, I.J., 1981, Comparison of cysteine sulfinic acid decarboxylase isoenzymes and glutamic acid decarboxylase in rat liver and brain. *Neuroscience* 6: 2701-2714.

44. Oja, S.S., and Kontro, P., 1983, *Taurine.* In Handbook of Neurochemistry Vol 3, 2nd edn, edited by Lajtha, A, Plenum press, New York, p. 501-533.

45. Okamoto, K., Kimura, H., and Sakai, Y., 1983, Evidence for taurine as an inhibitory neurotransmitter in cerebellar stellate interneurons: Selective antagonism by TAG (6-aminomethyl-3-methyl-4H,1, 2,4-benzothiadiazine-1,1-dioxide). *Brain Res.* 265: 163-168.

46. Reichert, P., and Urban, P.F., 1986, Purification and properties of rat brain cysteine sulfinate decarboxylase (E. C. 4.1.1.29). *Neurochem. Int.* 9: 315-321.

47. Remy, A, Henry, S., and Tappaz, M., 1990, Specific antiserum and monoclonal antibodies against the taurine biosynthesis-enzyme cysteine sulfinate decarboxylase: Identity of brain and liver enzyme. *J. Neurochem.* 54: 870-879.

48. Sturman, J.A., Wen, G.Y., Wisniewski, H.M., and Neuringer, M.D., 1984, Retinal degeneration in primates raised on a synthetic human infant formula. *Int. J. Dev. Neurosci.* 2: 121-130.

49. Sturman, J.A., 1993, Taurine in development. *Physiol. Rev.* 73: 119-147.

50. Sturman, J.A., and Chesney, R., 1995, Taurine in pediatric nutrition. *Pediatr. Clin. N. Amer.* 42(4): 879-897.

51. Sumizu, K., 1962, Oxidation of hypotaurine in rat liver. *Biochim. Biophys. Acta.* 63: 210-212.

52. Taber, T.C., Lin, C.-T., Song, G.-X., Thalman, R.H., and Wu, J.-Y., 1986, Taurine in the rat hippocampus-localization and postsynaptic action. *Brain Res.* 386: 113-121.

53. Takahashi, K., Harada, H., Schaffer, S.W., and Azuma, J., 1992, Effect of taurine on intracellular calcium dynamics of cultured myocardial cells during the calcium paradox. In *Adv. Exp. Med. Biol.*, "Taurine: Nutritional Value and Mechanisms of Action", Lombardini, J.B., Schaffer, S.W. and Azuma, J., eds., Plenum Press, New York, 315: 153-161.

54. Takuma, K., Matsuda, T., Hashimoto, H., Asano, S.,and Baba, A., 1994, Cultured rat astrocytes possess $Na+-Ca2+$ exchanger. *Glia* 12: 336-342.

55. Tang, X.-W., Deupree, D.L., Sun, Y., and Wu, J.-Y., 1996, Biphasic effect of taurine on excitatory amino acid-induced neurotoxicity. In *Adv. Exp. Med. Biol.,* "Taurine2: Basic and Clinical Aspects", Huxtable, R.J., Azuma, J., Kuriyama, K., Nakagawa, M., and Baba, A., eds., Plenum Press, New York, 403: 499-505.

56. Tang, X.W., Hsu, C.C., Schloss, J.V., Faiman, M.D., Wu, E., Yang, C.-Y., and Wu, J.-Y., 1997, Protein phosphorylation and taurine biosynthesis in vivo and in vitro. *J. Neuroscience* 17: 6947-6951.

57. Tappaz, M., Almarghini, K., Legay, F., and Remy, A., 1992, Taurine biosynthesis enzyme cysteine sulfinate decarboxylase (CSD) from brain: the long and tricky trail to identification. *Neurochem. Res.* 17: 849-859.

58. Thursby, M.H., and Nevis, A.H., 1974, Anticonvulsant activity of taurine in electrically and osmotically induced seizures in mice and rats. *Fed. Proc.* 33: 1494.

59. Trenkner, E., Gargano, A., Scala, P., and Sturman, J., 1992, Taurine synthesis in cat and mouse in vivo and in vitro. *Adv. Expt. Med. Biol.* 315: 7-14.

60. Tsukada, Y., Inoue, N., Donaldson, J., and Barbeau, A., 1974, Suppressive effects of various amino acids against ouabain-induced seizures in rats. *Can. J. Neurol. Sci.* 1: 214-221.

61. Uchida, S., Kwon, H.M., Yamauchi, A., Preston, A.S., Marumo, F., and Handler, J.S., 1992, The molecular cloning of the cDNA for an MDCK cell Na(+)- and Cl(-)-dependent taurine transporter that is regulated by hypertonicity. *Proc. Natl. Acad. Sci.* USA 89: 8230-8234.

62. van Gelder, N.M., and Courtois, A., 1972, Close correlation between changing content of specific amino acids in epileptogenic cortex of cats, and severity of epilepsy. *Brain Res.* 43: 477-484.

63. Wang, G.X., Duan, J., Zhou, S., Li, P., and Kang, Y., 1992, Antiarrhythmic action of taurine. In *Adv. Exp. Med. Biol.,* "Taurine: Nutritional Value and Mechanisms of Action", Lombardini, J.B., Schaffer, S.W. and Azuma, J., eds., Plenum Press, New York, 315: 187-192.

64. Welty, J.D., and McBroom, M.J., 1985, Effects of taurine on subcellular calcium dynamics in the normal and cardiomyopathic hamster heart. In *The Effect of Taurine on Excitable Tissues,* S.W. Schaffer, S.I. Baskin and J.J. Kocsis, eds., Spectrum Publications, New York, pp. 295-312.

65. Wright, C.E., Tallan, H.H., and Lin, Y.Y., 1986, Taurine: Biological update. *Ann. Rev. Biochem.* 55: 427-453.

66. Wu, J.-Y., Su, Y.Y.T., Brandon, C., Lam, D.M.K., Chen, M.S., and Huang, W.M., 1979, Purification and immunochemical studies of GABA-, acetylcholine- and taurine-synthesizing enzymes from bovine and fish brains. Seventh International Meeting of the ISN, p. 662.

67. Wu, J.-Y., 1982, Purification and characterization of cysteic/cysteinesulfinic acids decarboxylase and L-glutamate decarboxylase in bovine brain. *Proc. Natl. Acad. Sci.* USA 79: 4270-4274.

68. Yan, C.C., and Huxtable, R.J., 1998, Effect of taurine on biliary metabolites of glutathione in liver perfused with the pyrrolizidine alkaloid, monocrotaline. In Adv. Exp. Med. Biol., "Taurine3: Cellular and Regulatory Mechanisms", Schaffer, S., Lombardini, J.B. and Huxtable, R.J., eds., Plenum Press, New York, 442: 85-89.

TAURINE AND SKELETAL MUSCLE ION CHANNELS

Annamaria De Luca, Sabata Pierno, Domenico Tricarico, Jean-François Desaphy, Antonella Liantonio, Mariagrazia Barbieri, Claudia Camerino, Loredana Montanari and Diana Conte Camerino
Unit of Pharmacology, Department of Pharmacobiology, Faculty of Pharmacy, University of Bari, Via E. Orabona 4, 70125 Bari, Italy

INTRODUCTION

The sulfonic amino acid taurine is ubiquitously present in all tissues and exerts a wide variety of actions ranging from osmoregulation to neurotransmission[10]. Most of the actions of taurine are mediated by ion channels. For instance, during osmoregulation, taurine leaves cells exposed to hypotonicity through different channel-like molecules permeable to taurine and anions, among which swelling activated anion channels, phospholemman, chloride channels of the ClC family, etc.[15,17]. In excitable tissues, taurine has a recognized modulatory role on many different ion channels, thus controlling membrane excitability and consequently tissue function. In peripheral organs, the actions of taurine on voltage-gated ion channels have been studied electrophysiologically in heart as well as in skeletal muscle. The actions of taurine on cardiac ion channels have been extensively reviewed. It turned out that taurine modulates cardiac ion channel activity in a complex manner that strictly depends upon $[Ca^{2+}]_i$ and $[Ca^{2+}]_o$. In conditions of low Ca^{2+}, taurine stimulates Ca^{2+} influx through L-type channels and shortens action potential duration by activating fast delayed rectifier potassium currents (I_{Kr}), whereas the opposite is observed in condition of high Ca^{2+} [22,25]. Thus, the main action of taurine in the heart is to ensure a proper cardiac function through constant levels of intracellular Ca^{2+}. In fact, taurine exerts a positive inotropic effect in heart failure and an antiarrhythmic activity in conditions of Ca^{2+} overload, such as ischemia. The taurine-induced cardioprotection is further enhanced by the concomitant actions on other channels, i.e. taurine inhibits cardiac Na^+ currents, stimulates T-type Ca^{2+} current important for automaticity, inhibits ATP-dependent K^+ channels (K_{ATP}) and modulates the activity of ion exchangers[23,24,25,27].

The present report summarizes the results of electrophysiological experiments performed on skeletal muscle in our laboratory, which indicate that taurine also exerts a control of striated fiber function through its action on various ion channels.

Taurine and skeletal muscle chloride channels

Pharmacological actions

Skeletal muscle Cl^- channels are responsible for the large Cl^- conductance (gCl) of sarcolemma at rest. The large gCl stabilizes membrane excitability, since an abnormal hyperexcitability results from naturally occurring or pharmacologically induced lowering of gCl. Various evidences corroborate that the main muscle Cl^- channel is the ClC-1 protein, belonging to the large family of the ClC channels[13]. In fact, the expression of ClC-1 increases in parallel with the increase in gCl in rodent fast twitch muscle fibers during early post-natal development[1,28]. Also the spontaneous decrease in muscle gCl occurring during aging is correlated with a decrease in mRNA for ClC-1[20]. Several mutations in the gene coding for ClC-1 channels can account for the low gCl typical of Myotonia Congenita, an inherited myopathy characterized by membrane hyperexcitability and prolonged contractions[21]. We firstly demonstrated that *in vitro* application of mM concentrations of taurine to extensor digitorum longus (EDL) muscle decreases membrane excitability through a concentration-dependent increase of gCl, supporting its proposed beneficial effects in myotonic patients[3]. The action could be mediated by the interaction of taurine with a low affinity binding site strictly related to chloride channels. In fact taurine increases gCl lowered *in vivo* by 20,25-diazacholesterol that acts indirectly on the channel, but it is unable to antagonize the block of gCl brought about by direct channel blockers, such as the anthracene-9-carboxylic acid[2]. Nonetheless, the action of taurine is specific since its analogs, with an increased distance between the two charged heads and/or with a more delocalized positive charge for the replacement of the amino group with aza-cyclo moieties, show a decreased potency in increasing gCl[18].

The direct action of taurine on muscle Cl⁻ channel has been recently confirmed by evaluating its effects on Cl⁻ currents, measured by two microelectrode voltage-clamp technique, after heterologous expression of human ClC-1 channel in *Xenopous* oocytes. We found that the application of 20 mM taurine increased Cl⁻ currents in all the potential range (Fig. 1). At –80 mV, a membrane voltage close to that in resting condition, taurine increased Cl⁻ current from 1.2 ± 0.2 to 2.5 ± 0.5 µA (n = 5). Taurine was also able to shift by 8.2 ± 0.9 mV the channel activation toward a more negative potential, suggesting its ability to modulate the channel kinetic.

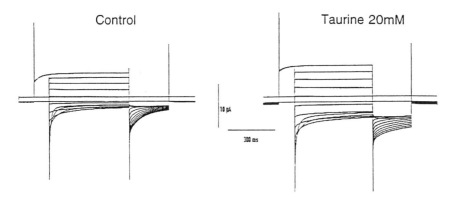

Figure 1. Sample of Cl⁻ currents measured by two microelectrode voltage clamp technique after injection of mRNA for human ClC-1 channel into *Xenopus* oocytes and enhancing effect of 20 mM taurine. Families of currents were obtained by test potentials from –140 to + 60 mV after full channel activation with a prepulse to + 60mV. A following pulse to –120 mV allowed to record tail currents from which activation parameters were calculated.

Physiological actions

In consideration of the effects of taurine on Cl⁻ channel and mostly the pivotal role of these channels for a proper tissue function, the possible physiological control of muscle Cl⁻ channels by intracellular taurine has been investigated. During post-natal development, the intracellular concentration of taurine increases in parallel with the increase in gCl. The chronic administration of taurine to female rats during both pregnancy and lactation is able to accelerate the post-natal development of gCl in skeletal muscle of the pups[4]. Accordingly, a 4 week chronic treatment of adult rats with guanidinoethane sulfonate (GES), a known inhibitor of the high affinity transporter, produces a significant 50% depletion of taurine content in EDL muscle and, in parallel, a significant decrease of gCl and a consequent increase in membrane excitability[7]. The alterations observed in

GES-treated, taurine depleted, muscles are similar to those naturally occurring in skeletal muscle of aged subjects. The HPLC determination revealed that indeed a significant decrease in taurine content occurs in skeletal muscle of aged rats, in parallel with the decrease in gCl. A 3-month chronic treatment with taurine to aged rats (1 g/kg in drinking water), significantly prevented both the loss of intracellular taurine and the decrease in gCl, supporting the strict link between appropriate tissue levels of the amino acid and the function of Cl⁻ channel[19]. In order to clarify the mechanism by which taurine exerts the long-term physiological control on Cl⁻ channels, it should be taken into account that the channel is controlled in a negative manner, by a phosphorylation pathway involving a Ca^{2+} and phospholipid dependent protein kinase C (PKC). An increased activity of this pathway can account for the low gCl recorded in skeletal muscle of aged rats as shown by the higher sensitivity of gCl to phorbol esters, specific PKC activators[5]. In rat synaptosomes, taurine has been described to inhibit the PKC mediated phosphorylation pathway, both by reducing the formation of diacylglycerol and by decreasing the level of available Ca^{2+}, through the stimulation of its uptake by intracellular stores[12]. This latter mechanism can be of importance in skeletal muscle, since taurine stimulates Ca^{2+} uptake by muscle sarcoplasmic reticulum[10,11].

In support of the view that taurine may control channel phosphorylation pathway, the chronic treatment of aged rats with taurine also restores the normal sensitivity of gCl to phorbol esters[19]. Finally, it should be taken into account that a strict relationship exists between intracellular taurine levels and the pharmacological ability of exogenous taurine to increase gCl. In fact, a higher than normal sensitivity to taurine has been found on gCl of taurine-depleted muscles (GES treated or aged)[6,7], whereas the *in vitro* application of taurine on the slow-twitch soleus muscle, which is characterized by higher levels of taurine *vs.* the fast EDL[31], exerts a small increase in gCl although this parameter is lower than in EDL muscle (Fig. 2). These observations, other than supporting the pharmacological use of taurine in conditions of tissue depletion, suggest that other yet unknown Cl⁻ channel-mediated action of taurine in skeletal muscle can occur in relation to different muscle phenotype-related metabolism.

Taurine and skeletal muscle sodium channels

It is well known that voltage-gated Na^+ channels play a pivotal role in excitable tissues for the generation and propagation of the action potential. Genetic defects of skeletal muscle Na^+ channels result in changes of excitation pattern and muscle contractility being responsible for recessive hereditary disorders ranging from myotonias to paralyses.

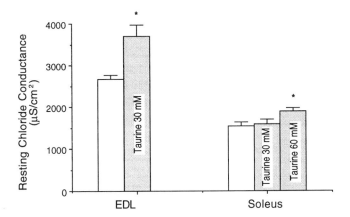

Figure 2. Effect of *in vitro* application of taurine on resting macroscopic chloride conductance of fast-twitch extensor digitorum longus (EDL) and slow-twitch soleus muscles. Each bar is the mean ± S.E. of gCl values recorded from 12-30 fibers. *Significantly different with respect to related control value in the absence of taurine.

In these disorders the mutated channels show biophysical defects of inactivation processes processes that commonly lead to an abnormal and disease-causing persistent Na^+ current. Drugs able to block voltage-gated Na^+ channels are clinically useful in these syndromes[21].

The effects of taurine on Na^+ channels of native muscle fibers have been investigated by patch clamp recordings in cell-attached configuration. Na^+ currents have been elicited with depolarizing test pulses to various membrane potentials from the holding potential of –110 mV. The *in vitro* application of 10 mM taurine, produced a peculiar effect. In fact taurine enhanced, up to twofold, the Na^+ transients elicited by threshold depolarizing test pulses (test pulse to –70/-50 mV), while it reduced the current at more depolarized test pulse potentials (Fig. 3). However, taurine did not produce any significant shift of the I-V curve and at 10 mM the maximal peak Na^+ current (elicited between –30/-20 mV) was reduced to about 50%. Taurine shifted the activation curve from –37.6 mV to –46.5 mV (n = 4), an effect that likely accounts for the increase of I_{Na} at negative potentials. In parallel taurine left-shifted the steady-state inactivation curve by 7.6 mV, indicating that the I_{Na} block is related to a reduction in channel availability, owing to the stabilization by taurine of the channel in the inactivated state. The double effect of both channel activation and inactivation is similar to what already observed in cardiac Na^+ currents[24,27]. Thus, taurine exerts an anesthetic-like action on skeletal muscle Na^+ channels, likely mediated by the amino group that is generally recognized as a pharmacophore moiety of Na^+ channel blockers. This action can account for a beneficial effect of exogenous taurine in Na$^+$ channel related disorders of skeletal muscle. In parallel, the ability to decrease the threshold for channel activation may represent a sort of safety factor to avoid excessive depression of membrane excitability.

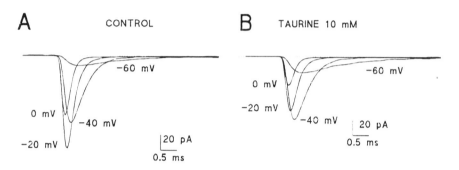

Figure 3. Effects of taurine on Na$^+$ currents recorded in cell-attached macropatches of striated fibers freshly dissociated from adult rat fast-twitch muscles. Panels A and B represent a family of Na$^+$ currents elicited from a holding potential of –110 mV to potentials indicated near the traces, before (A) and after (B) application of 10 mM taurine.

Taurine and skeletal muscle potassium channels

K$^+$ channels comprise a large family of genetically distinct entities with distinct biophysical properties and physiological roles. In heart taurine modulates the function of many types of K$^+$ channels, such as delayed rectifier K$^+$ channels, important for action potential duration, and ATP sensitive K$^+$ channels (K$_{ATP}$) that in excitable tissues couple cell metabolism to membrane electrical activity[22,23]. K$_{ATP}$ channels are abundantly present on sarcolemma of striated muscle fibers, where they exert a similar role since their opening allows membrane hyperpolarization, reduction of muscle work and saving of ATP content[30]. Similarly to the heart, in skeletal muscle this channel plays a pivotal role during ischemic conditions characterized by drop of ATP content[30]. A loss of taurine content occurs in ischemic heart and this in turn may affect potassium channel behavior. We previously found that a taurine depletion in GES-treated rat muscles significantly increases macroscopic resting K$^+$ conductance from 280 ± 27 (n = 15) to 508 ± 28 μS/cm^2 (n = 38)[7]. Thus, by means of patch clamp recordings on freshly isolated rat muscle fibers, we evaluated the effect of taurine on skeletal muscle K$_{ATP}$ channels. We found that taurine inhibited K$_{ATP}$ current with an half-maximal concentration of 19.3 mM, slightly higher than that found in the heart. Also taurine produced a left-shift of the concentration-response curve of glibenclamide, whereas it was ineffective on the dose-response curves of either calcium or ATP, suggesting that the amino acid can directly or indirectly interact with the SUR subunit of K_{ATP} channel, enhancing the effect of classical blockers. Differently from other channel modulators, the effects of taurine did not depend upon channel states, being not changed in condition of channel run-down. Thus, it can be proposed that the exit of taurine from the cell occurring during ischemia is an important mechanism to enhance the opening of K_{ATP} channels, and to help the protective effect of this current for tissue metabolism. Furthermore, we evaluated the effect of *in vitro* application of taurine on the calcium activated K^+ channels (K_{Ca}^{2+}), whose opening counteracts the membrane depolarization after trains of action potentials. We found that taurine exerted a block of K_{Ca}^{2+} channel, particularly at depolarized potentials. Since ischemic conditions are characterized by a large increase in the activity of K_{Ca}^{2+} channels (unpublished observations) paralleled by an hyperkalemic state, it can be proposed that the latter condition can be counteracted by taurine supplementation, through the block of K^+ channels.

Taurine and calcium ion availability in skeletal muscle

As already stated taurine is able to control excitation-contraction (e-c) coupling in cardiomyocytes by modulating Ca^{2+} channels and therefore the influx of Ca^{2+} that is pivotal for cardiac contraction[22]. As a consequence all the other Ca^{2+}-dependent mechanisms are affected[26].

The e-c coupling of skeletal muscle fiber is almost independent of external Ca^{2+}; nonetheless, taurine modulates intracellular Ca^{2+} levels. As proposed by Huxtable, intracellular taurine can bind, with a low affinity, to neutral phospholipids and such a binding can allosterically modulate the binding of Ca^{2+} ion to acid phopholipids[10]. Also, taurine has been found to increase the Ca^{2+} uptake by sarcoplasmic reticulum by increasing the activity of Ca^{2+} ATPase[11]. Conditions of taurine depletion are expected to increase the levels of free Ca^{2+} concentration and to affect e-c coupling of striated fibers. This hypothesis has been tested by recording, by point voltage clamp technique, the voltage threshold for striated fiber contraction in relation to test pulse duration (mechanical threshold, MT) up to reach the rheobase, i.e. the voltage at which the speed of Ca^{2+} release equals that of Ca^{2+} reuptake. Taurine depletion produced by chronic GES treatment, shifted (by about 10 mV) the rheobase voltage toward more negative potentials, as indicative of increased cytosolic Ca^{2+}. This shift was due to taurine depletion, since the *in vitro* application of 60 mM taurine, ineffective on control muscles, significantly counteracted the alteration of

MT in GES treated ones[7]. Similar observations have been made on skeletal muscles of aged rats, in which a natural decrease in taurine content occurs in parallel with a shift of the rheobase voltage toward more negative potentials. Chronic taurine administration, restored the intracellular levels of the amino acid and counteracted the change of MT typical of this condition[19].

Muscular dystrophies are hereditary disorders of skeletal muscle in which the progressive muscle fiber necrosis and degeneration have been related to increase in cytosolic Ca^{2+}. Duchenne muscular dystrophy (DMD) is the most common among these pathologies and is related to a disassembling of the sarcolemmal dystrophin-glycoprotein due to genetic lack of the cytoskeletal protein dystrophin[16]. As a consequence, the sarcolemma becomes mechanically weak, being unable to withstand the stress of contraction. This in turn leads to an increase in intracellular Ca^{2+} levels and to a loss of intracellular components. A muscle taurine loss may occur in this condition, since an increase of urinary excretion of taurine and other amino acids has been observed in DMD patients[9]. An alteration in the ability of dystrophin-missing fibers to retain appropriate amounts of taurine can contribute to muscle dysfunction. A recent study aimed to identify, by NMR analysis, markers of degeneration and regeneration events in dystrophin-missing fibers, has shown a marked decrease in the level of both taurine and creatine in muscles of dystrophic mdx mouse, the most widely used animal model for DMD[9], during the active degeneration period. These alterations are compensated during adulthood, when a spontaneous and successful regeneration takes place in this phenotype[8]. We have recently measured by HPLC the level of taurine in hindlimb muscles, heart and brain of 6-month old mdx mouse. In agreement with the data of McIntosch *et al.*[14], we found only slightly lower level of the amino acid in the tissues of the dystrophic animals; nonetheless, the plasma levels were significantly higher in mdx *vs.* control mice, suggesting that an alteration in taurine turnover does occur in this phenotype[8]. In agreement with the proposed higher level of Ca^{2+} in the muscle fibers of the mdx mouse, we found that the rheobase voltage of mdx EDL muscle fibers was significantly shifted toward more negative potentials of about 7 mV. The *in vitro* application of 60 mM taurine significantly shifted the altered MT towards normal values[8], suggesting a potential therapeutic use of this safe amino acid to improve muscle function in the dystrophic conditions.

Table 1. Summary of taurine effects on ion channels and e-c coupling of skeletal muscle

Channel and/or Ion Movement	Channel Physiological Role	Taurine Effect	Therapeutic Potential of Taurine
Chloride	Stabilization of membrane potential	Physiological: To keep the channel in a conductive state by reducing PKC-mediated phosphorylation Pharmacological: To increase channel conductance in relation to intracellular taurine content	To reduce hyperexcitability in myotonic syndromes. To counteract the decrease in gCl during aging and other conditions
Sodium	Generation and propagation of action potential	To block the channel with a local-anesthetic like mechanism To enhance the current at negative voltage range	To reduce hyperexcitability in sodium channel myotonias
K_{ATP}	Coupling cell metabolism to membrane potential	Physiological: To favor channel closed state	Its efflux during ischemia helps the opening of the channel and membrane hyperpolarization
K_{Ca}	Control of membrane potential after trains of action potentials	To block the channel	To reduce hyperkalemic state during ischemia
Calcium ions	Excitation-contraction coupling	To modulate calcium availability for contraction	To ameliorate muscle performance in degenerative conditions characterized by calcium overload

CONCLUSIONS

We have extensively investigated by means of macroscopic and single channel electrophysiological techniques, the effect of taurine on voltage-gated ion channels and on the e-c coupling of skeletal muscle fibers. We found that taurine exerts direct actions on many ion channels (Table 1). In particular, we found that taurine, is able both physiologically and pharmacologically to act on Cl⁻ channels, ensuring a proper large macroscopic resting gCl and sarcolemma excitability. The pharmacological effects of taurine on Cl⁻ channels are related to its intracellular levels, supporting its use in conditions of taurine depletion. Furthermore, taurine exerts a local-anesthetic like action on voltage-gated Na^+ channel. The concomitant action on Cl⁻ and Na^+ channels provides a mechanism for a therapeutic effect of taurine to solve hyperexcitability in myotonic syndromes. Also, taking into account the periodic occurrence of myotonic crises in patients[21], a preventing strategy could be that of maintaining constant the levels of taurine with taurine-rich diets. Physiologically taurine can maintain both K_{ATP} and K_{Ca}^{2+} channels in a closed state. The opening of the former following taurine loss, i.e. during ischemia, can facilitate hyperpolarization and cytoprotection to preserve muscle metabolism, whereas the ability of exogenous taurine to block K_{Ca}^{2+} channels can counteract the dangerous hyperkalemic state typical of ischemic conditions.

Finally, taurine acts on different channels being therefore a modulator, although non specific, of channel biophysic and allowing an harmonious function of channels that is pivotal for normal excitation and contraction of skeletal muscle. Thus, a generalized tissue dysfunction may result from alteration in taurine content that can occur in various pathophysiological conditions. This is further strengthened by the observation that exogenous taurine ameliorates e-c coupling of striated fibers in conditions of endogenous taurine depletion and/or Ca^{2+} overload, supporting its potential use as adjuvant drug in severe muscular dystrophies for which effective pharmacological agents are missing.

Taking into account the plethora of effects of taurine, it can be concluded that this amino acid may exert potential therapeutic effects in various skeletal muscle disorders characterized by alterations in ion channel function and e-c coupling mechanisms.

ACKNOWLEDGMENTS

Supported by Italian Cofin Murst 1998 and Telethon (projects # 1150 and 1208). The authors wish to thank Dr. Michael Pusch for collaboration to the experiments on taurine effects on expressed hClC-1 channels, and Prof. Alberto Giotti and Prof. Ryan Huxtable for continuous support to their work in the taurine field.

REFERENCES

1. Conte Camerino, D., De Luca, A., Mambrini, M., and Vrbovà G., 1989a, Membrane ionic conductances in normal and denervated skeletal muscle of the rat during development. *Pflügers Arch* **413**: 569-570.
2. Conte Camerino, D., De Luca, A., Mambrini, M., Ferrannini, E., Franconi, F., Giotti, S.H., and Bryant, S.H., 1989b, The effects of taurine on pharmacologically induced myotonia. *Muscle Nerve* **12**: 898-904.
3. Conte Camerino, D., Franconi, F., Mambrini, M., Bennardini, F., Failli, P., Bryant, S.H., and Giotti, A., 1987, The action of taurine on chloride conductance and excitability characteristics of rat striated fibers. *Pharmacol. Res. Commun.* **19**: 685-701.
4. De Luca, A., Conte Camerino, D., Failli, P., Franconi, F., Giotti, A., 1990, Effects of taurine on mammalian skeletal muscle fiber during development. *Prog. Clin. Biol. Res.* **351**: 163-173.
5. De Luca, A., Tricarico, D., Pierno, S., and Conte Camerino, D., 1994a, Aging and chloride channel regulation in rat fast-twitch muscle fibers. *Pflügers Arch* **427**: 80-85.
6. De Luca, A., Pierno, S., and Conte Camerino D., 1994b, Pharmacological interventions for the changes of chloride channel conductance of aging rat skeletal muscle. *Ann. N.Y. Acad. Sci.* **717**: 180-188.
7. De Luca, A., Pierno, S., and Conte Camerino, D., 1996, Effect of taurine depletion on excitation-contraction coupling and Cl- conductance of rat skeletal muscle. *Eur. J. Pharmacol.* **296**: 215-222.
8. De Luca, A., Pierno, S., Camerino, C., Huxtable, R.J., and Conte Camerino, D., 1998, Effect of taurine on excitation-contraction coupling of extensor digitorum longus muscle of dystrophic mdx mouse. *Adv. Exp. Med. Biol.* **442**: 115-119.
9. Engel, A.G., Yamamoto , M., and Fischbeck, K.H., 1994, Dystrophinopathies. In *Myology* (A.G. Engel and C. Franzini-Armstrong, eds), Mc Graw-Hill, Inc., New York, pp. 1133-1187.
10. Huxtable, R.J., 1992, The physiological actions of taurine. *Physiol. Rev.* **72**: 101-163.
11. Huxtable, R.J., and Bressler, R., 1973, Effect of taurine on a muscle intracellular membrane. *Biochim. Biophys. Acta* **323**: 573-583.
12. Li, Y-P., and Lombardini, J.B., 1991, Inhibition by taurine of the phosphorylation of specific synaptosomal proteins in the rat cortex: effects of taurine on the stimulation of calcium uptake in mitochondria and inhibition of phosphoinositide turnover. *Brain Res.* **553**: 89-96.
13. Jentsch T.J., Friedrich T., Scriever A., and Yamada H., 1999, The ClC chloride channel family. *Pflügers Arch* **437**: 783-795.
14. McIntosch, L., Granberg, K-E., Brière, K.M., Anderson, J.E., 1998, Nuclear magnetic resonance spectroscopy study of muscle growth, mdx dystrophy and glucocorticoid treatments: correlation with repair. *NMR Biomed.* **11**: 1-10.
15. Moorman, J.R., Ackerman, S.J., Kowdley, G.C., Griffin, M.P., Mounsey, J.P., Chen, Z., Cala, S.E., O'Brian, J.J., Szabo, G., and Jones, L.R., 1995, Unitary anion currents through phospholemman channel molecules, *Nature* **377**: 737-740.

16. Ozawa, E., Noguchi, S., Mizuno, Y., Hagiwara Y., and Yoshisa M., 1998, From dystrophinopathy to sarcoglycanopathy: evolution of a concept of muscular dystrophy. *Muscle Nerve* **21**: 421-438.

17. Pasantes-Morales, H., Quesada, O., Moràn, J., 1998, Taurine: an osmolyte in mammalian tissues. *Adv. Exp. Med. Biol.* **442**: 209-217.

18. Pierno, S., Tricarico, D., De Luca, A., Campagna, F., Carotti, A., Casini, G., Conte Camerino, D., 1994, Effects of taurine analogues on chloride channel conductance of rat skeletal muscle fibers: a structure-activity relationship investigation. *Naunyn-Schmied. Arch Pharmacol* **349**: 416-421.

19. Pierno, S., De Luca A., Camerino, C., Huxtable, R.J., and Conte Camerino, D., 1998, Chronic administration of taurine to aged rats improves the electrical and contractile properties of skeletal muscle fibers. *J. Pharmacol. Exp. Ther.* **286**: 1183-1190.

20. Pierno, S., De Luca, A., Beck, C.L., George, A.L., and Conte Camerino, D., 1999, Aging-associated down-regulation of ClC-1 expression in skeletal muscle: phenotypic-independent relation to the decrease of chloride conductance. *FEBS Lett* **449**: 12-16.

21. Ptacek, L., 1998, The familial periodic paralyses and nondystrophic myotonias. *Am. J. Med.* **104**: 58-70.

22. Satoh, H., and Sperelakis, N., 1998, Review of some actions of taurine on ion channels of cardiac muscle cells and others. *Gen Pharmacol.* **30**: 451-463.

23. Satoh, H., 1996, Direct inhibition by taurine of the ATP-sensitive K^+ channel in guinea pig ventricular cardiomyocytes. *Gen Pharmacol* **27**: 625-627.

24. Satoh, H., 1998, Inhibition of the fast Na^+ current by taurine in guinea pig ventricula myocytes. *Gen Pharmacol.* **31**: 155-157.

25. Satoh, H., 1999, Taurine modulates I_{Kr} but not I_{Ks} in guinea-pig ventricular cardiomyocytes. *Br. J. Pharmacol.* **126**: 87-92.

26. Schaffer, S.W., Punna, S., Duan, J., Harada, H., Hamaguchi, T., and Azuma, J., 1992, Mechanisms underlying physiological modulation of myocardial contraction by taurine. *Adv. Exp Biol. Med.* **315**: 193-198.

27. Schanne, O.F., and Dumaine, R., 1992, Interaction of taurine with the fast Na-current in isolated rabbit myocytes. *J. Pharmacol. Exp. Ther.* **263**: 1233-1240.

28. Steinmeyer, K., Ortland, C., and Jentsch, T.J., 1991, Primary structure and functional expression of a developmentally regulated skeletal muscle chloride channel. *Nature* **354**: 301-304.

29. Tricarico, D., and Conte Camerino D., 1994a, ATP-sensitive K^+ channels of skeletal muscle fibers from young adult and aged rats: possible involvement of thiol-dependent redox mechanisms in the age-dependent modifications of their biophysical and pharmacological properties. *Mol. Pharmacol.* **46**: 754-761.

30. Tricarico, D., and Conte Camerino, D., 1994b, Effects of ischaemia and post-ischaemic reperfusion on the passive and active electrical parameters of rat skeletal muscle fibres. *Pflügers Arch* **426**: 44-50.

31. Turinski, J., and Long, C.L., 1990, Free amino acids in muscle: effect of muscle fiber population and denervation. *Am. J. Physiol.* **258**: E485-E491.

TAURINE-DEFICIENT CARDIOMYOPATHY: ROLE OF PHOSPHOLIPIDS, CALCIUM AND OSMOTIC STRESS

Stephen Schaffer[1], Viktoriya Solodushko[1] and Junichi Azuma[2]
[1]*Department of Pharmacology, University of South Alabama, Mobile, AL, USA*
[2]*Department of Clinical Evaluation of Medicines and Therapeutics, Osaka University, Osaka, Japan*

INTRODUCTION

The concentration of taurine in excitable tissues is extremely high, usually in the mM range[17]. Although this amino acid is synthesized in the liver, the major source of taurine for the maintenance of the large intracellular taurine pools is the diet. In some species, such as infant monkeys, cats and the fox, taurine is an essential nutrient. Without an adequate dietary source of taurine, these animals develop numerous defects, including growth retardation[15], retinal degeneration[14,19,56], impaired neuronal development and function[26,34,37,50], poor reproductive outcome[12,51], abnormal immune response[47], altered platelet function[55] and myocardial failure[8,31,39]. Since the initial nutritional study by Hayes *et al.*[14] linking taurine deficiency to retinal degeneration, considerable progress has been made ascribing a physiological function to taurine. We now know that osmoregulation is perhaps its most important function, although taurine also exerts other actions, including calcium modulation, phospholipid regulation, antioxidation and detoxification[17]. The present review discusses the role of osmoregulation, calcium modulation and phospholipid regulation in the development of one of the taurine-deficient conditions, dilated cardiomyopathy.

The effects of taurine deficiency on cardiac function have been studied using both nutritional and drug-induced models of taurine depletion. As the degree of taurine depletion is greater in the nutritional models, the cardiac defects are also greater. However, differences also exist between the two

Taurine 4, edited by Della Corte *et al.*
Kluwer Academic / Plenum Publishers, New York, 2000.

drug-induced taurine deficient models. While the major abnormality found in the ß-alanine model is impaired myocardial relaxation (Figure 1), Eley *et al.*[8] found that the guanidinoethane sulfonate (GES) model of taurine deficiency exhibits significant systolic dysfunction. However, the systolic abnormality detected by Eley *et al.*[8] was not seen by Mozaffari *et al.*[30] despite a 56% decline in taurine content. One concern with the use of GES as a taurine depleting agent is its ability to lower creatine phosphate content[30]. In the heart creatine phosphate not only serves as an energy source but is also involved in a mitochondrial shuttle that balances the energy states of the mitochondria and cytoplasm[58]. Depletion of creatine phosphate using a compound related to GES, guanidinopropionic acid, induces heart failure[21]. ß-Alanine does not mediate the same decline in creatine phosphate levels, although it may mediate other adverse responses.

The nutritionally depleted heart exhibits defects in both systolic and diastolic function, with some evidence that diastolic dysfunction may precede systolic contractile failure[32]. Significant reductions in M-mode fractional shortening, left ventricular end systolic short axis diameter and velocity of circumferential fiber shortening are observed after onset of overt heart failure[31]. Novotny *et al.*[32] have also detected in taurine-depleted cats a significant rise in left ventricular diastolic chamber compliance, coupled with impaired systolic function, which are properties consistent with an eccentric form of hypertrophy. In this form of hypertrophy, the ventricle remodels, producing myocytes that are thicker and longer than normal[11]. However, in severe cases of the cardiomyopathy the ventricles also dilate and the wall thickness declines. Therefore, the taurine deficient cardiomyopathy is characterized initially by cellular hypertrophy, but as the condition worsens some of the cells die and the ventricular wall narrows.

A major cause of both systolic and diastolic dysfunction is defective calcium transport[2]. Although these calcium defects become exaggerated in dilated cardiomyopathy, recent studies suggest that the myofibrils may also contribute to the development of severe heart failure in some animal models[38]. In the case of the taurine deficient cardiomyopathy, there is evidence implicating both factors in the development of the disease. The present review discusses mechanisms underlying these defects in the taurine deficient heart.

INVOLVEMENT OF ALTERED CALCIUM MOVEMENT IN TAURINE-DEFICIENT CARDIOMYOPATHY

Numerous studies have established a link between taurine and calcium movement. However, many of these studies relate to the pharmacological actions of taurine (elevations in extracellular taurine). In this review, the pharmacological actions of taurine are largely ignored. Only the effects of reduced intracellular taurine content on calcium transport will be discussed.

According to the work of Satoh and Sperelakis[42], an increase in intracellular taurine content causes an increase in the L-type calcium current ($I_{Ca(L)}$) and slow inward sodium current ($I_{Na(s)}$). Moreover, Suleiman *et al.*[52] have established an interdependence between the intracellular content of taurine and sodium. The net effect is that an acute elevation in taurine can lead to an initial increase in both $[Ca^{2+}]_i$ and $[Na^+]_i$. However, this activates the taurine efflux mechanism to maintain a proper osmolyte balance[4]. Thus, a decrease in taurine could reduce $[Ca^{2+}]_i$ and thus contractile function.

The effect of taurine depletion on these transporters has not been determined. Interestingly, $[Na^+]_i$ is lower in the ß-alanine treated cardiomyocyte[44]. This effect may be related in part to the co-transport of taurine with sodium[4]. However, $[Ca^{2+}]_i$ is not depressed in the ß-alanine treated cell. Thus, further studies are required to clarify the direction and extent of ion movement through the various ion channels of the taurine depleted heart.

Role of Taurine-Phospholipid Interaction in Altered Calcium Transport

In a related study Earm *et al.*[7] suggested that taurine might also enhance flux through the Na^+/Ca^{2+} exchanger, although this effect appears to be secondary[22,46] to a rise in $[Ca^{2+}]_i$. The function of the Na^+/Ca^{2+} exchanger is to extrude calcium from the cell. In the normal functioning myocyte, the amount of calcium entering the cell via the calcium channel is balanced by the amount of calcium that leaves the cell via the Na^+/Ca^{2+} exchanger[3]. Therefore, maintenance of normal calcium homeostasis requires an active Na^+/Ca^{2+} exchange process. Taurine appears to regulate this efflux process by altering the amount of calcium available to be extruded. According to Hilgemann *et al.*[16] the K_d of the Na^+/Ca^{2+} exchanger for calcium is about 6 µM. Yet, peak systolic $[Ca^{2+}]_i$ usually approaches only 2 µM, and that only for a short period of time. Therefore, a dilemma develops over the mode of calcium efflux. A resolution of this dilemma was proposed by Langer[27], who felt that calcium must be compartmentalized within the cell. Based on evidence that the inner leaflet of the cell membrane is rich in anionic phospholipids capable of binding calcium, he hypothesized that the calcium concentration in the vicinity of the anionic phospholipids reaches 600 µM. Because this special pool of calcium interacts with the Na^+/Ca^{2+} exchanger, a continuous supply of calcium is provided for efflux. Significantly, taurine affects the size of this calcium pool, an effect which should facilitate the removal of calcium from the myocyte during diastole. Indeed, one of the defects noted in the taurine deficient heart is an impairment in the relaxation phase of the contraction cycle (Figure 1).

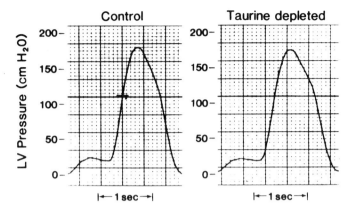

Figure 1. Effect of taurine depletion on cardiac cycle. Hearts from normal (Control) and ß-alanine treated (Taurine-depleted) rats were perfused with Krebs Henseleit buffer. Shown is a representative left ventricular pressure tracing obtained for the Control and Taurine depleted hearts. The major effect of taurine depletion is to prolong the relaxation phase of the cardiac cycle.

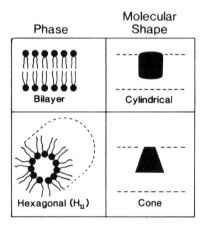

Figure 2. Schematic representation of aqueous phase preference and molecular shape of bilayer formers, such as phosphatidylcholine, and hexagonal formers, such as phosphatidylethanolamine.

The mechanism by which taurine enhances the binding of calcium to the cell membrane has been the focus of several studies by Schaffer and co-workers[5,6,24] and Huxtable and coworkers[18,28,48]. In 1979, Chovan *et al.*[5] reported that incubation of isolated heart sarcolemma with 20 mM taurine increased calcium binding to La^{3+} sensitive, low affinity membrane binding sites. As the increase in calcium binding induced by taurine was not associated with alterations in the rotational properties of a membrane associated electron spin resonance probe, the authors concluded that taurine does not cause gross structural changes. However, taurine appears to cause localized membrane changes related to its interaction with anionic phospholipids[18,24]. In 1978, Kulakowski *et al.*[24] found that taurine binds to two different sites on the membrane. Based on reconstitution studies, the high affinity sites belong to the taurine transporter[45] while the low affinity sites appear to represent the binding of taurine to anionic phospholipids[43,18]. The low affinity binding sites are of interest because they exhibit positive cooperativity indicative of a conformational change[24]. Moreover, molecular modeling studies indicate that the ionic interaction between taurine and the anionic phospholipids require complete exposure of the phospholipid headgroup to the aqueous medium[43]. When taurine binds to one head group, it is possible that localized structural changes could be induced in a neighboring headgroup, thereby increasing its calcium binding capacity.

While the direct effects of the taurine-phospholipid interaction probably affect only a small number of membrane transporters, of which the Na^{+}/Ca^{2+} exchanger is one, the indirect effects of the interaction can be dramatic. One important indirect effect is the modulation of phospholipid metabolism, of which the conversion of phosphatidylethanolamine to phosphatidylcholine is an example. This three-step N-methylation reaction plays an important cellular role by altering both membrane function and structure. It has been established that phosphatidylethanolamine preferentially assumes a hexagonal conformation in aqueous solution while phosphatidylcholine organizes into a bilayer configuration (Figure 2). As a result, phosphatidylcholine is considered a stabilizer of the biological membrane while excessive levels of phosphatidylethanolamine can be disruptive. The proper mix of bilayer formers and hexagonal formers is required for normal cellular events, such as membrane transport. Phospholipid N-methylation also causes a redistribution of phospholipids within the membrane. This occurs because phosphatidylethanolamine preferentially localizes to the inner leaflet of the cell membane while phosphatidylcholine is preferentially found in the outer leaflet[40]. As a result of these effects, the conversion of phosphatidylethanolamine to phosphatidylcholine in the heart is associated with changes in the activity of a number of membrane transporters[10,13,35,36,41].

Taurine influences membrane structure and function by inhibiting phospholipid N-methylation[13,41]. Several ion transporters are sensitive to phospholipid N-methylation and respond to taurine. In 1991, Hamaguchi *et al.*[13]

found that perfusion of the isolated heart with buffer containing L-methionine promoted phospholipid N-methylation, causing a decrease in Na^+/Ca^{2+} activity and mediating a negative inotropic effect. Inclusion of 10 mM taurine in the buffer partially prevented these effects of L-methionine. While the reversal of the negative inotropic effect may involve an effect on Na^+/Ca^{2+} exchanger activity, it more likely relates to the fact that calcium release from the junctional sarcoplasmic reticulum is inhibited by phospholipid N-methylation but enhanced by taurine[41]. These effects of taurine on phospholipid metabolism and calcium binding may contribute to the depressed contractile state of the taurine deficient heart.

Role of Osmotic Stress in Altered Calcium Transport

Another action of taurine that may contribute to altered calcium transport in the taurine depleted heart involves an osmotic linked mechanism. Cardiomyocytes exposed to medium containing 5 mM ß-alanine lose 50% of their intracellular taurine pool, a large enough change to affect the osmotic balance across the cell membrane and trigger a volume regulatory mechanism[44]. This compensatory mechanism is important to the cell, not only because it restores cell volume and osmotic balance, but because it alters $[Ca^{2+}]_i$ and contractile function. It has been established that acute, moderate hyperosmotic stress elevates $[Ca^{2+}]_i$ and enhances contractile function while acute, severe hyperosmotic stress has the opposite effect[33].

However, the major defect developing in the heart following chronic exposure to a mild form of hyperosmotic stress is diastolic dysfunction. This is seen in cardiomyocytes incubated for 3 days with medium containing 30 mM mannitol (Figure 3). The dominant defect in these cells, as well as in the ß-alanine-treated myocyte, is a prolongation of the calcium transient (Figure 3). This suggests that osmotic stress is the likely cause of impaired contractile function in the ß-alanine-treated heart (Figure 1). However, as the nutritionally depleted heart loses significantly more taurine than the ß-alanine-treated heart, it experiences a much larger osmotic shock, which appears capable of causing a more dramatic alteration in cardiac function[32]. Clearly, further studies are required to clarify the contribution of osmotic stress towards the development of the cardiomyopathy in the nutritionally depleted heart.

ALTERED MUSCLE PROTEIN FUNCTION IN THE TAURINE-DEPLETED HEART

Very little information is available on the status of the muscle proteins in the taurine depleted heart. In one study, Lake[25] reported that the GES

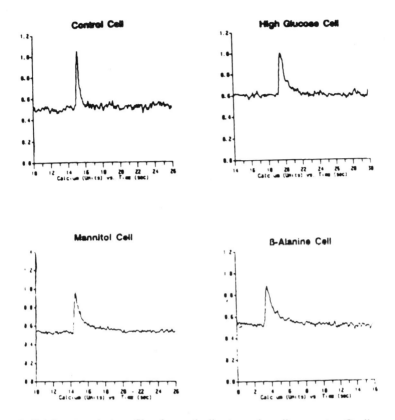

Figure 3. Calcium transient profile of osmotically stressed cardiomyocytes. Cardiomyocytes were incubated for 3 days in serum-substituted medium containing either 5 mM glucose (Control Cell), 30 mM glucose (High Glucose Cell), 30 mM mannitol (Mannitol Cell) or 5 mM ß-alanine (ß-alanine Cell). The cells were loaded with 2.0 µM Indo-1, a fluorescent calcium indicator. Shown are the calcium transients monitored using a point scan of an ACAS confocal microscope. The data are expressed as calcium units versus time in seconds. Osmotic stress prolongs the relaxation phase of the calcium transient.

taurine depleted heart contained less contractile protein than the normal myocardium. However, in a follow-up study using the same GES model, no difference in the relative protein levels of myosin heavy chain, myosin light chain, actin, troponin and tropomyosin were observed[8]. In the latter study, there was a tendency for total myocardial tissue protein content to decrease following GES treatment. However, the effect was not significant. According to these authors, the largest effect of GES-induced taurine depletion was a reduction in strontium-mediated activation of the myofibrils, which the authors attributed to impaired myofibril function. The GES taurine-depleted heart also exhibited increased compliance, raising the possibility that the density of the actin filaments might be reduced.

Particularly intriguing from the work of Eley et al.[8] is the observation that the calcium sensitivity of the myofibrils is reduced in the GES taurine-depleted heart. This finding is consistent with the reports of Steele et al.[49] and Galler et al.[9] who showed that taurine increased the calcium sensitivity of the force generating myofilaments of skinned muscle fibers. However, as the concentration dependency of the taurine effect has not been studied, it is unclear if the effect contributes to the reduction in systolic function seen in the heart severely depleted of taurine.

INTERACTION OF TAURINE AND ANGIOTENSIN II - ROLE IN TRIGGERING CONGESTIVE HEART FAILURE

Two mechanisms, hemodynamic overload and neurohumoral stimulation, serve as the primary initiators of cardiac hypertrophy in the failing heart[29,57]. Based on the beneficial clinical effects of the ACE inhibitors, it is clear that one of the most important neurohumoral factors is angiotensin II. The effects of angiotensin II on the heart are multiple and include such opposite effects as the stimulation of protein synthesis and the promotion of apoptosis[20,23]. Despite this wide array of activity, taurine is able to block many of the actions of angiotensin II. By inference, it is likely that reductions in the cellular content of taurine might be expected to potentiate the actions of angiotensin II. In fact, as the effects of taurine on systolic function may not be that large, the angiotensin II-taurine interaction might play a central role in driving the taurine-deficient heart into failure.

Two effects of angiotensin II deserve special mention because of their relationship to cardiac hypertrophy and heart failure. First, angiotensin II induces early response genes, such as c-fos and c-jun[53]. These genes are characterized by their rapid and transient induction following exposure to various growth factors. They influence transcription through their interaction with the consensus AP-1 site. Therefore, it is logical to assume that they contribute to angiotensin II-induced stimulation of protein synthesis[23]. As expected, these actions of angiotensin II are blocked by AT_1 receptor antagonists[54]. However, they are also prevented by preincubating cardiomyocytes with medium containing 20 mM taurine[53]. Although not verified experimentally, indirect evidence suggests that taurine depletion may potentiate the protein stimulating effects of angiotensin II. First, the taurine deficient cardiomyopathy is associated with an eccentric form of hypertrophy, the development of which requires the stimulation of protein synthesis[32]. Second, taurine deficiency activates a protein kinase cascade, with protein kinase C being one of the enzymes activated (Figure 4). This is significant because the activation of protein kinase C with phorbol esters accelerates protein synthesis in cultured cardiomyocytes[23]. Moreover, the induction of c-fos and the stimulation of protein synthesis by angiotensin II are blocked by inhibitors of protein kinase C[23]. However, the obvious experiment examining the hypertrophic effects of angiotensin II in the taurine deficient cell has not been performed.

Protein Kinase C Activity

1   2   3   4

Figure 4. Effect of taurine depletion on protein kinase C activity. The activity of protein kinase C in taurine depleted cardiomyocytes was assessed by monitoring the phosphorylation status of an *in vivo* 29 kD protein kinase C substrate using Western blot analysis. Lanes 1 and 2 represent the phosphorylation status of the substrate in control cells before and following downregulation of protein kinase C using phorbol myristate acetate (100 nM). Lanes 3 and 4 represent taurine depleted cells before and following the downregulation of protein kinase C. Taurine depletion significantly increased the phosphorylation of the 29 kD protein. This effect is largely reversed by downregulation of protein kinase C.

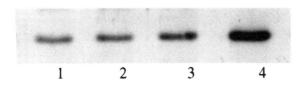

1   2   3   4

Figure 5. Elevation of cellular Bax content by taurine depletion and angiotensin II. Taurine depleted and normal cells were incubated with medium containing 1 nM angiotensin II for a period of 24 h. The cells were harvested and Western blots obtained for Bax. Lanes 1 and 2 represent normal cells incubated in the absence and presence of angiotensin II, respectively. Lanes 3 and 4 are taken from taurine depleted cells after exposure to 0 or 1 nM angiotensin II, respectively. The upregulation of Bax by angiotensin II was significantly potentiated by taurine depletion.

While increased protein synthesis is required for cardiomyocytes to undergo hypertrophy, the transition from compensated to decompensated hypertrophy involves cell loss[1]. Recently, Kajstura *et al.*[20] have shown that angiotensin II initiates apoptosis in some cardiomyocytes after 24 h in culture. Since angiotensin II appears to promote apoptosis in part by increasing the levels of the pro-apoptotic factor, Bax, it is significant that taurine depletion alone has no effect on either the levels of Bax or the number of apoptotic cells in culture (Figure 5). However, ß-alanine treatment dramatically potentiates the upregulation of Bax by angiotensin II (Figure 5). As a result, the taurine-depleted myocyte is significantly more sensitive to angiotensin II-induced apoptosis than the cell containing normal taurine levels, as evidenced by the greater intensity of its DNA ladder (Figure 6).

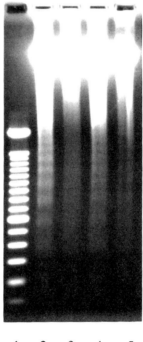

1   2   3   4   5

Figure 6. Effect of taurine depletion on angiotensin II-induced apoptosis. Taurine depleted and normal cardiomyocytes were exposed to medium containing 1 nM angiotensin II for a period of 24 h. The cells were harvested, DNA isolated and the samples subjected to agarose gel electrophoresis in the presence of ethidium bromide. DNA was visualized by UV light. Lanes 1 represent a standard DNA ladder. Lanes 2 and 3 represent taurine-depleted cells treated with 1 nM and 0 angiotensin II, respectively. Lanes 4 and 5 were taken from normal cells after treatment with either 1 nM or 0 angiotensin II, respectively. Taurine depletion significantly increased the intensity of the DNA ladder induced by angiotensin II treatment.

Normally, the series of hemodynamic events that culminate in the onset of congestive heart failure are initiated by a drop in cardiac output. One of the important events in this sequence is an elevation in angiotensin II levels. The taurine-deficient heart exhibits impaired contractile function. However, it is unclear if the initial systolic defect is sufficient to cause an activation of the renin-angiotensin system. Yet, it is clear that the effects of angiotensin II are potentiated by taurine deficiency. Therefore, the angiotensin II-taurine interaction probably plays a central role in the development of the taurine deficient cardiomyopathy.

REFERENCES

1. Anversa, P., Kajstura, J., and Olivetti, G., 1996, Myocyte cell death in heart failure. *Curr. Opin. Cardiol.* 11: 245-251.
2. Arai, M., Matsui, H., and Periasamy, M., 1994, Sarcoplasmic reticulum gene expression in cardiac hypertrophy and heart failure. *Circ Res* 74: 555-564.
3. Bridge, J.H.B., Smolley, J.R., Spitzer, K.W., 1990, The relationship between charge movements associated with I_{Ca} and I_{Na-Ca} in cardiac myocytes. *Science* 248: 376-378.
4. Chapman, R.A., Suleiman, M-S, and Earm, Y.E., 1993, Taurine and the heart. *Cardiovasc. Res.* 27: 358-363.
5. Chovan, J.P., Kulakokwski, E.C., Benson, B.W., and Schaffer, S.W., 1979, Taurine enhancement of calcium binding to rat heart sarcolemma. *Biochim. Biophys. Acta* 551: 129-136.
6. Chovan, J.P., Kulakowski, E.C., Sheakowski, S., and Schaffer, S.W., 1980, Calcium regulation by the low-affinity taurine binding sites of cardiac sarcolemma. *Mol. Pharmacol.* 17: 295-300.
7. Earm, Y.E., Ho, W.K., and So, I.S., 1991, Inward current generated by Na-Ca exchange during the action potential in single atrial cells of the rabbit. *Proc. R. Soc. Lond.* 240: 61-81.
8. Eley, D.W., Lake, N., and ter Keurs, H.E.D.J., 1994, Taurine depletion and excitation-contraction coupling in rat myocardium, *Circ. Res.* 74: 1210-1219.
9. Galler, S., Hutzler, C., Haller, T., 1990, Effects of taurine on Ca^{2+}-dependent force development of skinned muscle fibre preparations. *J. exp. Biol.* 152: 255-264.
10. Ganguly, P.K., Panagia, V., Okumura, K., and Dhalla, N.S., 1985, Activation of Ca^{2+} stimulated ATPase by phospholipid N-methylation in cardiac sarcoplasmic reticulum. *Biochem. Biophys. Res. Commun.* 130: 472-478.
11. Gerdes, A.M., 1992, The use of isolated myocytes to evaluate myocardial remodeling. *Trends Cardiovasc. Med.* 2: 152-155.
12. Gottschall-Pass, K.T., Gorecki, D.K.J., and Paterson, P.G., 1995, Effect of taurine deficiency on tissue taurine concentrations and pregnancy outcome in the rat. *Can. J. Physiol. Pharmacol.* 73: 1130-1135.
13. Hamaguchi, T., Azuma, J., and Schaffer, S., 1991, Interaction of taurine with methionine: inhibition of myocardial phospholipid methyltransferase. *J. Cardiovasc. Pharmacol.* 18: 224-230.
14. Hayes, K.C., Carey, R.E., and Schmidt, S.Y., 1975, Retinal degeneration associated with taurine deficiency in the cat. *Science* 188: 949-951.
15. Hayes, K.C., Stephan, Z.F., and Sturman, J.A., 1980, Growth depression in taurine-depleted infant monkeys. *J. Nutr.* 110: 2058-2064.
16. Hilgemann, D.W., Collins, A., and Matsuoka, S., 1992, Steady-state and dynamic properties of cardiac sodium-calcium exchange: secondary modulation by cytoplasmic calcium. *J. Gen. Physiol.* 109: 933-961.
17. Huxtable, R.J., 1992, Physiological actions of taurine. *Physiol. Rev.* 72: 101-163.
18. Huxtable, R.J. and Sebring, L., 1986, Towards a unifying theory for the action of taurine. *Trends Pharmacol. Sci.* 7: 481-485.

19. Imaki, H., Jacobson, S.G., Kemp, C.M., Knighton, R.W., Neuringer, M., and Sturman, J., 1993, Retinal morphology and visual pigment levels in 6- and 12-month-old rhesus monkeys fed a taurine-free human infant formula. *J. Neurosci. Res.* 36: 290-304.
20. Kajstura, J., Cigola, E., Malhotra, A. Li, P., Cheng, W., Meggs, L.G., and Anversa, P., 1997, Angiotensin II induces apoptosis of adult ventricular myocytes in vitro. *J. Mol. Cell. Cardiol.* 29: 859-870.
21. Kapelko, V.I., Kupriyanov, V.V., Novikova, N.A., Lakomkin, V.L., Steinscheneider, A.Y., Severina, M.Y., Veksler, V.I., and Saks, V.A., 1988, The cardiac contractile failure induced by chronic creatine and phosphocreatine deficiency. *J. Mol. Cell. Cardiol.* 20: 465-479.
22. Katsube, Y., and Sperelakis, N., 1996, Na^+/Ca^{2+} exchange current: lack of effect of taurine. *European J. Pharmacol.* 316: 97-103.
23. Kinugawa, K, Takahashi, T., Kohmoto, O., Yao, A., Ikenouchi, H., and Serizawa, T., 1995, Ca^{2+}-growth coupling in angiotensin II-induced hypertrophy in cultured rat cardiac cells. *Cardiovasc. Res.* 30: 419-431.
24. Kulakowski, E.C., Maturo, J., and Schaffer, S.W., 1978, The identification of taurine receptors from rat heart sarcolemma. *Biochem. Biophys. Res. Commun.* 80: 936-941.
25. Lake, N., 1993, Loss of cardiac myofibrils: mechanism of contractile deficits induced by taurine deficiency. *Am. J. Physiol.* 264: H1323-H1326.
26. Lake, N., Malik, N., and DeMarte, L., 1988, Taurine depletion leads to loss of rat optic nerve axons. *Vision Res.* 28: 1071-1076.
27. Langer, G.A., 1994, Myocardial calcium compartmentation. *Trends Cardiovasc. Med.* 4: 103-109.
28. Lleu, P.-L., and Huxtable, R.J., 1992, Phospholipid methylation and taurine content of synaptosomes from cerebral cortex of developing rat. *Neurochem. Int.* 21: 109-118.
29. Morgan, H.E., and Baker, K.M., 1991, Cardiac hypertrophy: mechanical, neural and endocrine dependence. *Circulation* 83: 13-25.
30. Mozaffari, M.S., Tan, B.H., Lucia, M.A., and Schaffer, S.W., 1986, Effect of drug-induced taurine depletion on cardiac contractility and metabolism. *Biochem. Pharmacol.* 35: 985-989.
31. Novotny, M.J., Hogan, P.M., and Flannigan, G., 1994, Echocardiographic evidence for myocardial failure induced by taurine deficiency in domestic cats. *Can. J. Vet. Res.* 58: 6-12.
32. Novotny, M.J., Hogan, P.M., Paley, D.M., and Adams, H.R., 1991, Systolic and diastolic dysfunction of the left ventricle induced by dietary taurine deficiency in cats. *Am. J. Physiol.* 261: H121-H127.
33. Ogura, T., You, Y., and McDonald, T.F., 1997, Membrane currents underlying the modified electrical activity of guinea-pig ventricular myocytes exposed to hyperosmotic solution. *J. Physiol.* 504: 135-151.
34. Palackal, T., Moretz, R., Wisniewski, H., and Sturman, J., 1986, Abnormal visual cortex development in the kitten associated with maternal dietary taurine deprivation. *J. Neurosci. Res.* 15: 223-239.
35. Panagia, V., Makino, N. Ganguly, P.K., and Dhalla, N.S., 1987, Inhibition of Na^+-Ca^{2+} exchange in heart sarcolemmal vesicles by phosphatidylethanolamine N-methylation. *Eur. J. Biochem.* 166: 597-603.
36. Panagia, V., Okumura, K., Makino, N., and Dhalla, N.S., 1986, Stimulation of Ca^{2+}-pump in rat heart sarcolemma by phosphatidylethanolamine N-methylation. *Biochim. Biophys. Acta* 856: 383-387.
37. Pasantes-Morales, H., Arzate, M.E., Quesada, O., and Huxtable, R.J., 1987, Higher susceptibility of taurine-deficient rats to seizures induced by 4-aminopyridine. *Neuropharmacology* 26: 1721-1725.
38. Perez, N.G., Hashimoto, K., McCune, S., Altschuld, R.A., and Marban, E., 1999, Origin of contractile dysfunction in heart failure. *Circulation* 99: 1077-1083.

39. Pion, P.D., Kittleson, M.D., Roger, Q.R., and Morris, J.G., 1987, Myocardial failure in cats associated with low plasma taurine: a reversible cardiomyopathy. *Science* 237: 764-768.
40. Post, J.A., Langer, G.A., Op den Kamp, J.A.F., Verkleij, A.J., 1988, Phospholipid asymmetry in cardiac sarcolemma: analysis of intact cells and gas-dissected membranes. *Biochim Biophys Acta* 943: 256-266.
41. Punna, S., Ballard, C., Hamaguchi, T., Azuma, J., and Schaffer, S., 1994, Effect of taurine and methionine on sarcoplasmic reticular Ca^{2+} transport and phospholipid methyltransferase activity. *J. Cardiovasc. Pharmacol.* 24: 286-292.
42. Satoh, H., and Sperelakis, N., 1998, Review of some actions of taurine on ion channels of cardiac muscle cells and others. *Gen. Pharmacol.* 30: 451-463.
43. Schaffer, S.W., Azuma, J., and Madura, J.D., 1995, Mechanisms underlying taurine-mediated alterations in membrane function. *Amino Acids* 8: 231-246.
44. Schaffer, S.W., Ballard-Croft, C., Azuma, J., Takahashi, K., Kakhniashvili, D.G., and Jenkins, T.E., 1999, Shape and size changes induced by taurine depletion in neonatal cardiomyocytes. *Amino Acids* 15: 135-142.
45. Schaffer, S.W., Kulakowski, E.C., and Kramer, J.H., 1982, Taurine transport by reconstituted membrane vesicles. In *Taurine in Nutrition and Neurology* (R.J. Huxtable and H. Pasantes-Morales, eds.) Plenum Press, New York, pp. 143-160.
46. Schaffer, S.W., Punna, S., Duan, J., Harada, H., Hamaguchi, T., and Azuma, J., 1992, Mechanism underlying physiological modulation of myocardial contraction by taurine. In *Taurine: Nutritional Value and Mechanisms of Action* (J.B. Lombardini, S.W. Schaffer and J. Azuma, eds.), Plenum Press, New York, pp. 193-198.
47. Schuller-Levis, G., Mehta, P.D., Rudelli, R., and Sturman, J., 1990, Immunologic consequences of taurine deficiency in cats. *J. Leukocyte Biol.* 47: 321-331.
48. Sebring, L.A., and Huxtable, R.J., 1985, Taurine modulation of calcium binding to cardiac sarcolemma. *J. Pharmacol. Expt. Therap.* 232: 445-451.
49. Steele, D.S., Smith, G.L., and Miller, D.J., 1990, The effects of taurine on calcium uptake by sarcoplasmic reticulum and calcium sensitivity of chemically skinned rat heart. *J. Physiol.* 422: 499-511.
50. Sturman, J.A., and Lu, P., 1997, Role of feline maternal taurine nutrition in fetal cerebellar development: an immunohistochemical study. *Amino Acids* 13: 369-377.
51. Sturman, J.A., Moretz, R.C., French, J.H., and Wisniewski, H.M., 1985, Taurine deficiency in the developing cat: persistence of the cerebellar external granule cell layer. *J. Neurosci. Res.* 13: 405-416.
52. Suleiman, M-S, Rodrigo, G.C., and Chapman, R.A., 1992, Interdependence of intracellular taurine and sodium in guinea pig heart. *Cardiovasc. Res.* 26: 897-905.
53. Takahashi, K., Azuma, M., Taira, K., Baba, A., Yamamoto, I., Schaffer, S.W., and Azuma, J., 1997, Effect of taurine on angiotensin II-induced hypertrophy of neonatal rat cardiac cells. *J. Cardiovasc. Pharmacol.* 30: 725-730.
54. Thienelt, C.D., Weinberg, E.O., Bartunek, J., and Lorell, B.H., 1997, Load-induced growth responses in isolated adult rat hearts: role of the AT_1 receptor. *Circulation* 95: 2677-2683.
55. Welles, E.G., Boudreaux, M.K., and Tyler, J.W., 1993, Platelet, antithrombin and fibrinolytic activities in taurine-deficient and taurine-replete cats. *Am. J. Vet. Res.* 54: 1235-1243.
56. Wen, G.Y., Sturman, J.A., Wisniewski, H.M., Lidsky, A.A., Cornwell, A.C. and Hayes, K.C., 1979, Tapetum disorganization in taurine-depleted cats. *Invest. Ophthalmol. Visual Sci.* 18: 1200-1206.
57. Yamazaki, T., Komuro, I., Yazaki, Y., 1995, Molecular mechanism of cardiac hypertrophy by mechanical stress. *J. Mol. Cell. Cardiol.* 27: 133-140.
58. Zweier, J.L., Jacobus, W.E., Korecky, B., Brandejs-Barry, Y., 1991, Bioenergetic consequences of cardiac phosphocreatine depletion induced by creatine analogue feeding. *J. Biol. Chem.* 266: 20296-20304.

POST-TRANSCRIPTIONAL REGULATION OF CYSTEINE DIOXYGENASE IN RAT LIVER

Deborah L. Bella, Young-Hye Kwon, Lawrence L. Hirschberger, and
Martha H. Stipanuk
Division of Nutritional Sciences, Cornell University, Ithaca, NY 14853

Abstract: Changes in hepatic cysteine dioxygenase (CDO) activity in response to diet
play a dominant role in regulation of cysteine catabolism and taurine synthesis.
We have conducted several studies of the molecular regulation of CDO activity in rat liver and rat hepatocytes. Compared to levels observed in liver of rats
fed a basal 10% casein diet, up to 180-fold higher levels of CDO activity and
protein were observed in liver of rats fed diets that contained additional protein, complete amino acid mixture, methionine, or cystine[5,6]. Neither CDO activity nor CDO protein was induced by excess non-sulfur amino acids alone.
Excess sulfur amino acids or protein did not significantly increase the concentration of hepatic CDO mRNA. Preliminary studies indicate that the polysome
profile for association of CDO mRNA with polysomes is not altered by an increase in dietary protein level, suggesting that regulation may be posttransla-
tional and possibly involve a decrease in the rate of CDO degradation. In primary cultures of rat hepatocytes, CDO mRNA, protein, and activity all virtually disappeared by 12 to 24 h of culture in standard medium[14] whereas CDO
protein, but not CDO mRNA, accumulated markedly between 12 and 24 h in
hepatocytes cultured in medium with excess methionine or cyst(e)ine. These
observations are also consistent with a limited role of transcriptional or translational regulation of CDO in response to diet.

INTRODUCTION

Cysteine, one of the two sulfur-containing amino acids, is used in the synthesis of protein, and for the synthesis of several essential nonprotein compounds, including taurine, sulfate, inorganic sulfur, and glutathione. In nutri-

Taurine 4, edited by Della Corte *et al.*
Kluwer Academic / Plenum Publishers, New York, 2000.   71

tional regulation studies of cysteine metabolism, we have consistently found
that the activities of the key regulatory enzymes of cysteine metabolism
[cysteine dioxygenase (CDO), EC 1.13.11.20; cysteinesulfinate decarboxylase (CSDC), EC 4.1.1.29; γ-glutamylcysteine synthetase (GCS), EC 6.3.2.2]
change in liver of rats fed different levels of dietary protein[1,2,4]. In comparing
the magnitude of changes in these three enzymes in response to changes in
dietary protein levels, hepatic CDO activity is barely detectable in rats fed
10% casein diets but is markedly increased to 50-times as much activity
when protein levels are increased to 20%[3]. In contrast to the increases in
CDO, both GCS and CSDC undergo decreases in activity with increases in
protein intake. Within the same range of protein intake, decreases in GCS
and CSDC activity are on the order of 50 to 75%; fold changes of approximately 1-3 versus 50 for CDO. Thus, changes in level of CDO activity, in
addition to changes in cysteine availability, are a key factor in determining
the flux of cysteine between cysteine catabolism/taurine synthesis and glutathione synthesis. The purpose of these studies was to (i) determine if CDO
activity is regulated at the level of mRNA or protein and (ii) to determine if
the response of CDO to dietary protein is dependent on the presence of non-
sulfur amino acids or of a particular sulfur amino acid.

Table 1. Composition of semipurified diets containing various levels of protein and methionine for feeding study # 1

Ingredients	Diet				
	LP	MP	HP	LP+MM	LP+HM
	g per kg diet				
Vitamin-free casein	100	200	400	100	100
L-Methionine	--	--	--	3.0	10.0
Cornstarch	376.5	326.5	226.5	375.0	371.5
Sucrose	376.5	326.5	226.5	375.0	371.5
Cellulose	50	50	50	50	50
Corn oil	50	50	50	50	50
Vitamin mix (AIN 76)	10	10	10	10	10
Mineral mix (AIN 76)	35	35	35	35	35
Choline bitartrate	2	2	2	2	2

Diets were prepared by Dyets, Inc. (Bethlehem, PA) in powdered form. LP, low protein; MP,
moderate protein; HP, high protein; LP+MM, moderate methionine; LP+HM, high methionine. The basal LP diet provided 3.2 g sulfur amino acids (2.8 g met + 0.4 g cys) per kg diet.
Both the MP and the LP + MM diets provided ~ 6 g sulfur amino acids and both the HP and
the LP+HM diets provided ~13 g sulfur amino acids.

MATERIALS AND METHODS

Animal Feeding Studies

Animals and dietary treatments. Male Sprague-Dawley rats were purchased from Harlan Sprague-Dawley (Indianapolis, IN). Rats were housed individually in stainless steel mesh cages in a room maintained at 20°C and 60-70% humidity with light from 20:00 to 08:00 h. Rats had free access to diet and water for the duration of each experiment. Animals were blocked into groups by initial body weight. In each feeding study, rats within each block were randomly assigned to receive one of five semi-purified experimental diets. The composition of the experimental diets for feeding studies 1 and 2 are shown in Tables 1, 2, and 3.

The care and use of animals for all studies was approved by the Cornell University Institutional Animal Care and Use Committee. Rats were fed the experimental diets for 3 weeks. Rats were killed using CO_2 anaesthesia + decapitation. The liver was removed, rinsed with ice-cold saline, blotted, and weighed. Approximately 100 mg of liver from each animal was homogenized in denaturation solution (ToTALLY™ RNA kit, Ambion, Inc., Austin, TX) and then stored at -70°C for later measurement of CDO mRNA as described below. The liver was then minced, homogenized in 50 mM Mes [2-(N-morpholino)ethanesulfonic acid], pH 6.0, and immediately used for the CDO enzyme assay[2] or to obtain the soluble fraction, which was stored at -70°C for subsequent determination of the concentration of CDO (western blot analysis) and hepatic protein level[3].

Western and Northern Blot Methods

Sources of antibodies and cDNA. The purified IgG fraction from rabbit anti-CDO serum and the cDNA for CDO were gifts from Dr. Yu Hosokawa (National Institute of Health and Nutrition, Tokyo, Japan)[12]. The cDNA fragment was cloned into pBluescript SK$^+$. A partial CDO cDNA was prepared using complementary DNA probes corresponding to bp 459-717 of rat liver CDO[13]. DECAprobe™ template-actin-mouse and DECAprobe™ template-18S-mouse (Ambion, Inc., Austin, TX) were used as internal standards for Northern analyses. DECAtemplate-cyclophilin-mouse and the complete heavy chain ferritin cDNA in pBluescript SK+ (a gift of Dr. Patrick Stover, Division of Nutritional Sciences, Cornell University) were used as internal standards for polysome profile analysis. The 258 bp CDO PCR product and the Northern blot and polysome profile analysis internal standard probes were labelled with [^{32}P]dCTP using the Prime-It RmT® random primer labelling kit (Stratagene, La Jolla, CA).

Table 2. Composition of semipurified diets supplemented with excess sulfur amino acids, nonsulfur amino acids, or both for feeding study # 2

Ingredients	Diet				
	B	B+AA	B+M	B+AA+M	B+AA+C
	g per kg diet				
Vitamin-free casein	100	100	100	100	100
Amino acid mixture					
(Table 3)	--	300	--	300	300
L-Methionine	--	--	9.6	9.6	--
L-Cystine	--	--	--	--	7.8
Cornstarch	282.5	282.5	282.5	282.5	282.5
Dextrose	94	94	94	94	94
Sucrose	376.5	76.5	366.9	66.9	68.7
Cellulose	50	50	50	50	50
Corn oil	50	50	50	50	50
Vitamin mix (AIN 76A)	10	10	10	10	10
Mineral mix (AIN 76)	35	35	35	35	35
Choline chloride	2	2	2	2	2

Diets were prepared by Dyets, Inc. (Bethlehem, PA) in pelleted form. B, basal; B + AA, basal + amino acids; B + M, basal + methionine; B + AA + M, basal + amino acids + methionine; B + AA + C, basal + amino acids + cystine.

Table 3. Composition of sulfur amino acid-free amino acid mixture*

Amino acid	g per 300 g mixture
L-Asparagine.H$_2$O	20
L-Arginine	16
L-Histidine	6
L-Lysine.HCl	26
L-Tyrosine	8
L-Tryptophan	4
L-Phenylalanine	16
L-Threonine	16
L-Leucine	22
L-Isoleucine	16
L-Valine	16
Glycine	20
L-Proline	16
L-Glutamate	20
L-Alanine	20
L-Aspartate	5
L-Serine	20
L-Glutamine	20
Sodium bicarbonate	13

* Modified from Rogers and Harper[16].

Western blot analysis. Western blot analyses were conducted as previously described[5]. Briefly, 15 µg of total liver supernatant protein from each experimental diet group was separated by one-dimensional SDS-PAGE, and the proteins were electroblotted onto Immobilon-P membranes (Millipore Corporation, Medford, MA). Immunoreactive protein was detected by chemiluminescence with exposure to Kodak X-OMAT XRP film. For quantitative analysis, 5-210 µg of total soluble protein was loaded, with the amount dependent on the treatment group because the range of CDO protein among dietary groups was much greater than the linear range of the standard curve. The film images were scanned using a desktop scanner (Hewlett Packard Scanjet 3c, Hewlett Packard, Camas, WA). Two-dimensional quantitative densitometric analysis of the regions of interest was performed using Molecular Analyst ™ software (Bio-Rad Laboratories, Hercules, CA). The apparent molecular weight for CDO was consistent with previously published values[19].

Isolation of total RNA and northern blot analysis. Northern blot analyses were performed as previously described[5]. Briefly, total RNA was isolated from liver using the Totally RNA™ kit (Ambion, Inc., Austin, TX) based on the method of Chomczynski & Sacchi[10], and Northern blot analysis was conducted using a ^{32}P-labeled probe as described by Brown[8]. Results were quantified using Bio-Rad GS 363 Phosphorescence Imaging System (Bio-Rad Laboratories, Inc, Melville, NY) and Molecular Analyst™ program (Bio-Rad Laboratories, Inc, Hercules, CA).

Polysome Profile Analysis

For one week, rats (4 per experimental diet) were fed either the low protein or high protein diet used in feeding study #1. Polysome profile analysis was carried out as described by Chambers and Ness[9]. A portion of the liver from each rat was homogenized in PMS buffer (250 mM sucrose, 20 mM HEPES, pH 7.5, 250 mM KCl, 5mM MgCl, 2 mM DTT, 150 µg/ml cycloheximide, 1 mg/ml sodium heparin), and nuclei were collected by centrifugation. The supernatant was layered over a sucrose gradient of 15-40% in PMS buffer, and the gradient was centrifuged at 100,000 x g for 3 h. The gradients were pumped through a flow cell with monitoring of absorbance at 254 nm and collection of the eluant as fractions. RNA was purified from each fraction and a Northern blot conducted as outlined above. Cyclophilin mRNA and ferritin mRNA distribution were determined for comparison; cyclophilin is not translationally regulated whereas ferritin is translationally regulated, but only in the presence of supplemental iron. When rats consume a diet without supplemental iron (i.e., both the LP and HP diets), ferritin mRNA is not associated with ribosomes, will not be heavy enough to move into the sucrose gradient and thus, should be observed at the

move into the sucrose gradient and thus, should be observed at the beginning of the gradient.

Primary Cultures of Rat Hepatocytes

Hepatocytes from Sprague-Dawley rats were isolated aseptically by collagenase perfusion as described by Berry *et al.*[7]. The initial viability of isolated hepatocytes was more than 85% as determined by 0.2% (w/v) Trypan blue exclusion. The freshly isolated hepatocytes were resuspended in Williams E (WE) medium to give a final cell concentration of 7.5×10^5 cells per ml. The basal WE medium contained 1 µg/ml insulin, 50 ng/ml EGF, 50 nM dexamethasone, 3 nM Na_2SeO_3, 100 units/ml penicillin G, 100 µg/ml streptomycin sulfate, 0.25 µg/ml amphotericin B. Five ml of the diluted cell suspension (0.18 ml per cm^2) were plated on each 60 mm diameter collagen-coated dish[14]. Cells were allowed to attach in basal medium over for a 4-h period. At 4 h, the basal medium was replaced with either fresh basal medium or the designated treatment medium. Cells were cultured with either standard WE with or without 2 mM methionine plus 0.05 mM bathocuproine disulfonate (BCS) for a total culture time of 48 h.

Table 4. Effects of experimental diets on food intake and weight gain of rats and on hepatic protein concentrations for Feeding Study # 1

	Feeding Study #1				
	LP	**MP**	**HP**	**LP+MM**	**LP+HM**
	n = 6	n = 8	n = 5	n = 5	n = 6
Average daily diet consumption, g	25.8 ± 2.0	22.6 ± 0.6	22.0 ± 0.5	23.4 ± 0.5	25.8 ± 1.5
Average daily weight gain, g	4.1 ± 0.6^a	6.1 ± 0.3^b	6.0 ± 0.4^b	5.0 ± 0.3^{ab}	5.4 ± 0.2^{ab}
Body weight at end of feeding period, g	281 ± 9.1^a	321 ± 4.3^b	321 ± 5.8^b	313 ± 5.1^b	312 ± 2.7^b
Liver protein, mg/g liver	210.2 ± 3.9	215.2 ± 4.2	223.2 ± 1.9	204.7 ± 4.2	200.5 ± 5.1

Values are means ± SE. Within a row, values with different superscripts are significantly different ($p \leq 0.05$) by ANOVA and Tukey-Kramer's ω-procedure.

For measurement of CDO activity and CDO concentration (western blot analysis), cultured hepatocytes were disrupted by sonication for three 15-s intervals using a High Intensity Ultrasonic Processor (Sonics and Materials Inc., Dansbury, CT) in 50 mM Mes [2-(N-morpholino)ethanesulfonic acid], pH 6.0. The supernatant fraction of cell homogenates was obtained by centrifugation at 20,000 × g for 30 min at 4°C. Cells were stored in denaturation buffer (described above) for further analysis of mRNA via northern blot analyses. Each experiment was conducted in triplicate.

Statistics

Data were analyzed either by paired t-test or by analysis of variance (Minitab 10.5., Minitab Inc., State College, PA) and Tukey's or Tukey-Kramer's ω-procedure[17]. Correlation coefficients were calculated using Microsoft Excel 5.0 (Microsoft Corp., Cambridge, MA). Differences were accepted as $p \leq 0.05$.

RESULTS

Animal Study #1

The body weight at the end of the feeding period for rats fed the LP diet was significantly lower ($p < 0.05$) than that for the rats fed the other four diets (Table 4). The LP or basal diet contained (wt/wt) 10% casein and ~0.3% sulfur amino acids, which provides just below the National Research Council (NRC, 1978) recommended levels of protein and sulfur amino acid intakes for growing rats of 12% protein and 0.6% sulfur amino acids. De-

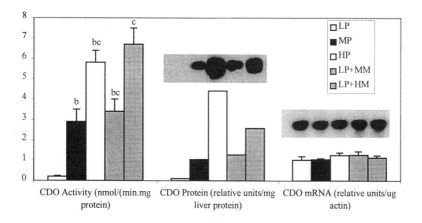

Figure 1. Changes in CDO activity, CDO Protein, and CDO mRNA levels in rats fed diets that varied in protein and methionine levels. Values are means SE for 4-8 rats. For each parameter, units are listed below the label on the horizontal axis. For CDO activity, values designated by different letters are significantly different ($p \leq 0.05$) by Tukey-Kramer's ω-procedure.

spite the lower body weight at the end of the feeding period, rats fed the LP diet did have adequate rates of weight gain. All results except for mRNA levels are expressed on the basis of liver protein, which was similar (p > 0.05) for rats fed all diets.

Table 5. Effects of experimental diets on food intake and weight gain of rats and on hepatic protein concentrations for Feeding Study # 2

| | **Feeding Study #1** | | | | |
	B	**B+AA**	**B+M**	**B+AA+M**	**B+AA+C**
	n = 6	n = 6	n = 7	n = 7	n = 6
Average daily diet consumption, g	18.6 ± 0.7^b	16.5 ± 0.4^a	20.1 ± 0.6^b	19.5 ± 0.2^b	19.4 ± 0.4^b
Average daily weight gain, g	3.7 ± 0.3^a	2.8 ± 0.1^a	5.3 ± 0.2^b	5.9 ± 0.2^b	5.8 ± 0.3^b
Body weight at end of feeding period, g	234 ± 8.2^a	211 ± 4.9^a	275 ± 8.7^b	293 ± 3.0^b	279 ± 12^b
Liver protein, mg/g liver	208.0 ± 8.4	214.6 ± 5.2	221.8 ± 7.1	230.8 ± 4.8	221.9 ± 12.4

Values are means ± SE. Within a row, values with different superscripts are significantly different ($p \le 0.05$) by ANOVA and Tukey-Kramer's ω-procedure.

Figure 1 shows changes in the level of CDO activity, CDO protein concentration, and CDO mRNA levels in response to diets with varying levels of protein and/or sulfur amino acids. Both CDO activity and relative CDO protein levels increased in a dose-dependent manner in response to increases in dietary protein or sulfur amino acid levels. Hepatic CDO activity in rats fed the MP, HP, LP+MM, and LP+HM diets was 14, 29, 18, and 35 times, respectively, the level of activity in rats fed the LP diet. Of particular interest, the magnitude of increase in CDO activity was greater for equisulfur levels of protein than for sulfur amino acid supplementation alone (MP vs. LP+MM and HP vs. LP+HM). In addition, the fold increase in CDO activity was greater for the step between LP vs. MP or LP vs. LP+MM than for the step from the moderate diets to those with the highest levels of protein or methionine. The relative CDO protein levels in liver of rats fed the MP, LP+MM, HP, and LP+HM diets were 10.5, 13, 44 and 26 times, respectively, the level of CDO protein observed in rats fed the LP diet. In contrast to enzyme activity and CDO protein levels, no marked differences were observed in the steady-state hepatic CDO mRNA levels among the five dietary groups.

Animal Feeding Study #2

Daily weight gain and final body weight of rats fed both the B and B+AA diets were significantly lower ($p < 0.05$) than that of rats fed the sulfur amino acid-supplemented diets (Table 5). These lower values were anticipated due to the protein/sulfur amino acid content of the diets as reviewed above. The B diet did support a reasonable rate of weight gain, and food intake was similar ($p > 0.05$) to that of rats fed the sulfur amino acid-supplemented diets. The daily intake of rats fed the B+AA diet was significantly lower ($p < 0.05$) than that of rats fed the other four diets. The imbalanced amino acid content of this diet may have contributed to the lower intake. Results are expressed on the basis of liver protein, which was similar ($p > 0.05$) for rats fed all diets.

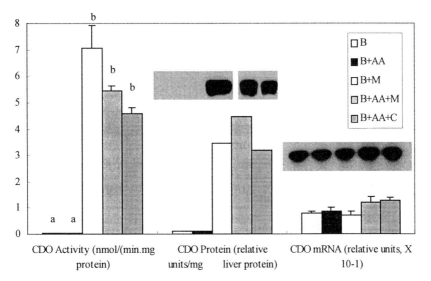

Figure 2. Changes in CDO activity, CDO protein, and CDO mRNA levels in rats fed diets supplemented with excess sulfur amino acids, nonsulfur amino acids, or both. Values are means ± SE for 5-7 rats. For each parameter, units are listed below the label on the horizontal axis. For CDO activity, values designated by different letters are significantly different ($p \leq 0.05$) by Tukey-Kramer's ω-procedure.

Figure 2 shows changes in the level of CDO activity, CDO protein concentration, and CDO mRNA levels in rats fed diets supplemented with excess sulfur amino acids, nonsulfur amino acids, or both. Hepatic CDO activity and protein were similar ($p > 0.05$) in rats fed the B vs. B+AA diets. CDO activity was significantly higher ($p < 0.001$) with addition of sulfur amino acids to the basal diet. The levels of CDO activity in rats fed the B+M, B+AA+M, and B+AA+C diets were 178-, 138-, and 115-fold, respectively, the activity observed in rats fed the B diet. The relative CDO protein levels in the liver of rats fed the B+M, B+AA+M, and B+AA+C diets were

34, 45, and 32 times, respectively, the level observed in rats fed the basal diet. Again, in contrast to enzyme activity and CDO protein levels, no marked differences were observed in the steady-state hepatic CDO mRNA levels among the five dietary groups.

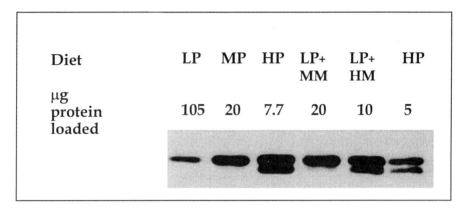

Figure 3. Western blot of hepatic CDO in rats fed diets with graded levels of protein or sulfur amino acids. Dietary treatment group from which pooled sample was obtained and the amount of soluble protein loaded are indicated above the bands

For western blot analysis of CDO, two distinct bands were detected by the anti-CDO IgG in liver samples from rats fed diets with graded levels of protein and methionine (Figure 3). A CDO band with an estimated molecular mass of 25.5 kDa was observed in all dietary treatments for both feeding studies. However, an additional band with estimated molecular mass of 23.5 kDa was observed in liver of rats fed the HP and LP+HM diets in animal study #1, and the B+M, B+AA+M, and B+AA+C diets in animal study #2 (data not shown). The molecular mass of the lower band is in agreement with the molecular mass calculated from the amino acid sequence (23 kDa) and the molecular mass estimated (22 kDa) by SDS-PAGE for purified CDO as reported by Yamaguchi *et al.*[19]

Polysome Profile

Figure 4 shows the distribution of CDO mRNA along the liver polysome profile. The distribution of CDO mRNA with polysomes of various sizes was similar in liver of rats fed the low or high protein diets. For cyclophilin mRNA, there was a clear increase in association of the mRNA with polysomes of higher ribosome number, which paralleled the distribution observed for CDO. Ferritin mRNA, whose translation is regulated in response to cellular iron status, showed markedly less association with polysomes of higher ribosome number.

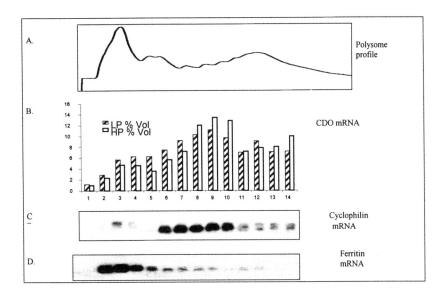

Figure 4. Distribution of CDO mRNA along the liver polysome profile as compared to cyclophilin and ferritin controls. (A) Representative liver polysome profile. (B) Effect of dietary protein on the distribution of hepatic CDO mRNA in polysomes. Values are means ± SE for 4 rats per diet group. (C) Distribution of cyclophilin mRNA across polysome gradient. (D) Distribution of ferritin mRNA across polysome gradient.

Cell Culture Study

Figure 5 shows the effect of cell culture and methionine on the expression of CDO in rat hepatocytes. Lane 1 shows the initial cell culture levels of CDO protein and mRNA. Lanes 3, 5, 7, and 9 show the effect of culture time and lanes 2, 4, 6, and 8 show the effect of methionine over time on CDO protein and mRNA levels. CDO protein levels decreases markedly over the first 10 h in standard WE medium. However, when the medium was supplemented with 2 mM methionine, CDO protein increases at all time points with a marked elevation noted at 10 h. In contrast, CDO mRNA levels decrease overtime with no apparent differences with or without methionine supplementation. As in the western blot analyses for the feeding studies, a second band was also observed at 10 and 48 h in cells cultured in medium supplemented with methionine.

DISCUSSION

From the animal feeding studies, it is clear that CDO activity showed a dose-dependent increase in response to the addition of either casein, methionine, or cyst(e)ine in approximately equisulfur amounts. However, no changes in CDO activity or protein concentration were observed with supplementation of non-sulfur amino acids. In both studies, the changes in CDO protein concentrations paralleled the changes in CDO activity. CDO mRNA abundance was essentially unaffected by dietary treatment, but CDO mRNA transcript was abundant in rats fed the LP, B, or B+AA diets despite extremely low levels of both CDO activity and protein. The results of these studies suggest that the regulation of hepatic CDO activity was largely in response to dietary sulfur amino acids and occurred by a posttranscriptional mechanism. This mechanism involved changes in CDO protein and may reflect either increased efficiency of translation of CDO mRNA or decreased degradation of CDO protein.

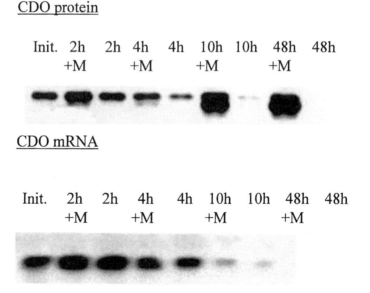

CDO protein

Init. 2h   2h   4h   4h   10h 10h   48h   48h
       +M   +M   +M   +M

CDO mRNA

Init. 2h   2h   4h   4h   10h 10h   48h   48h
       +M   +M   +M   +M

Figure 5. The effect of culture time and methionine supplementation on the levels of CDO protein and mRNA in primary hepatocytes. The culture time at which each sample was collected and whether the medium was supplemented with methionine (+ or -) are indicated above each band.

No changes were observed in the distribution of CDO mRNA across the liver polysome profile, which suggests that regulation of CDO does not occur by changes in the rate of CDO mRNA translation. Therefore, increases in CDO protein and activity in response to sulfur amino acids appears to be due to stabilization of the CDO protein. Although the specific mechanism has not been identified, several lines of evidence support a role for cysteine, or a closely related metabolite, in stabilization of CDO. In cultured hepatocytes, increases in CDO protein were observed when the medium was supplemented with methionine (Figure 5) versus no methionine supplementation. In addition, increases in CDO protein and activity were observed when cultured hepatocytes were supplemented with methionine or cyst(e)ine, but not when methionine was co-incubated with propargylgylcine, an inhibitor of transsulfuration[14]. Yamaguchi *et al.*[19] and Hosokawa *et al.*[11] also demonstrated that the half-life of hepatic CDO increased in rats injected with cysteine. Thus, it appears that the upregulation of CDO activity requires the presence of cysteine or a closely related metabolite and may be due to a lower rate of degradation of the enzyme.

The nature of the differences between the two CDO protein bands detected on the western blot analyses has not been fully explored. However, it is clear that the most marked changes in CDO activity occur over the range of recommended protein and sulfur amino acid intakes for growing rats (10 to 20% protein and 0.3 to 0.6% sulfur amino acid, wt/wt). Thus, these changes in activity clearly occur at the level of enzyme concentration and do not appear to involve the second CDO species.

CONCLUSIONS

Changes in CDO activity appear to be predominantly due to changes in CDO protein, which most likely occur via changes in the rate of protein degradation. Cysteine, or a closely related metabolite, appears to be necessary for the upregulation of CDO activity and protein and may play a role in CDO stabilization. CDO clearly plays an important role in determining the flux of cysteine between cysteine catabolism/taurine synthesis and glutathione synthesis. Indeed, the large fold changes in CDO activity provide a robust system for the removal of sulfur amino acids when methionine and cysteine are in excess and for the conservation of cysteine for glutathione and protein synthesis when sulfur amino acid availability is low.

ACKNOWLEDGEMENTS

We gratefully acknowledge the guidance and advice of Dr. Patrick Stover. We also thank Dr. Yu Hosokawa for the anti-CDO IgG and Dr. Hosokawa and Nubuyo Tsuboyama for the *Eco*R I cut cDNA for CDO. This research was supported in part by National Research Initiative Competitive Grants Program/US Department of Agriculture (USDA) Grant 92-37200-7583, USDA/Cooperative State Research, Education, and Extension Service Grant 94-34324-0987, and by the President's Council for Cornell. Women. D. L. Bella was supported by a National Institute of Diabetes and Digestive and Kidney Diseases training grant (T32-DK-07158).

REFERENCES

1. Bagley, P.J., and Stipanuk, M.H., 1994. The activities of rat hepatic cysteine dioxygenase and cysteinesulfinate decarboxylase are regulated in a reciprocal manner in response to dietary casein level. *J. Nutr.* 124: 2410-2421.
2. Bella, D.L., and Stipanuk, M.H., 1995, Effects of protein, methionine, or chloride on acid-base balance and on cysteine catabolism. *Am. J. Physiol.* 269: E910-E917.
3. Bella, D. L., Kwon, Y. H., and Stipanuk, M.H., 1996, Variations in dietary protein but not in dietary fat plus cellulose or carbohydrate levels affect cysteine metabolism in rat isolated hepatocytes. *J. Nutr.* 126: 2179-2187.
4. Bella, D.L., and Stipanuk, M.H., 1996, High levels of dietary protein or methionine have different effects on cysteine metabolism in rat hepatocytes. In: *Taurine 2: Basic and Clinical Aspects*, edited by R. Huxtable, J. Azuma, M. Nakagawa, K. Kuriyama, and A. Baba. New York, Plenum, p. 73-84.
5. Bella, D.L., Hirschberger, L.L. , Hosokawa, Y., and Stipanuk, M.H., 1999, The mechanisms involved in the regulation of key enzymes of cysteine metabolism in rat liver in vivo. *Am. J. Physiol.* 276 (*Endocrinol. Metab. 39*): E326-E335.
6. Bella, D.L., Hahn, C. and Stipanuk, M.H., 1999, Effects of nonsulfur and sulfur amino acids on the regulation of hepatic enzymes of cysteine metabolism. *Am. J. Physiol.* 277 (*Endocrinol. Metab. 40*): E144-E153.
7. Berry, M.N., Edwards, A.M., and Borritt G.J., 1991, *Isolated Hepatocytes: Preparation, Properties, and Applications.* Elsevier, New York, p. 16-32 and 56-57.
8. Brown, C., 1993, Analysis of RNA by Northern and slot blot hybridization. In: *Current Protocols in Molecular Biology*, edited by F.M. Ausubel, R. Brent, R.E. Kingston, D.D. Moore, J.G. Seidman, J.A. Smith, and K. Strunl. New York: Wiley- Interscience, p. 4.9.1-4.9.14.
9. Chambers, C.M., and Ness, G.C., 1997, Translational regulation of hepatic HMG-CoA reductase by dietary cholesterol. *BBRC* 232: 278-281.
10. Chomczynski, P., and Sacchi, N., 1987, Single-step method of RNA isolation by acid guanidinium thiocyanate-phenol-chloroform extraction. *Anal. Biochem.* 162: 156-159.
11. Hosokawa, Y., Yamaguchi, K., Kohashi, N., Kori, Y., and Ueda, I., 1978, Decrease of rat liver cysteine dioxygenase (cysteine oxidase) activity mediated by glucagon. *J. Biochem.* 84: 419-424.

12. Hosokawa, Y., Yamaguchi, K., Kohashi, N., Kori, Y., Fujii, O., and Ueda, I., 1980, Immaturity of the enzyme activity and the response to inducers of rat liver cysteine dioxygenase during development. *J.Biochem.* 88: 389-394.

13. Hosokawa, Y., Matsumoto, A., Oka, J., Itakura, H., and Yamaguchi, K., 1990, Isolation and characterization of a complementary DNA for rat liver cysteine dioxygenase. *Biochem. Biophys. Res. Commun.* 168: 473-478.

14. Kwon, Y.-H., Regulation of cysteine dioxygenase and γ-glutamylcysteine synthetase in response to sulfur amino acids in primary cultures of rat hepatocytes. Ph.D. thesis, Cornell University, Ithaca, NY.

15. National Research Council., 1978, *Nutrient Requirements of Laboratory Animals.* Washington, DC: National Academy of Sciences, p. 13-16.

16. Rogers, Q.R., and Harper, A.E., 1965, Amino acid diets and maximal growth in the rat. *J. Nutr.* 87: 267-73.

17. Steel, R.G.D., and Torrie,J.H., 1960, *Principles and Procedures of Statistics.* New York: McGraw-Hill, p. 99-160.

18. Yamaguchi, K., Sakakibara, S., Koga, K., and Ueda, I., 1971, Induction and activation of cysteine oxidase of rat liver. I. The effects of cysteine, hydrocortisone, and nicotinamide injection on hepatic cysteine oxidase and tyrosine transaminase activities of intact and adrenalectomized rats. *Biochem. Biophys. Acta* 237: 502-512.

19. Yamaguchi, K., Hosokawa, Y., Kohashi, N., Kori, Y., Sakakibara, S., and Ueda, I., 1978, Rat liver cysteine dioxygenase (cysteine oxidase): further purification, characterization, and analysis of the activation and inactivation. *J. Biochem.* 83: 479- 491.

THE MITOCHONDRIAL PERMEABILITY TRANSITION AND TAURINE

[1]Mitri Palmi, [1]Gisèle Tchuisseu Youmbi, [1]Giampietro Sgaragli, [1]Antonella Meini, [1]Alberto Benocci, [1]Fabio Fusi, [1]Maria Frosini, [2]Laura Della Corte, [3]Gavin Davey and [3]Keith Francis Tipton
[1]*Istituto di Scienze Farmacologiche, Università di Siena, Siena, Italy*
[2]*Dipartimento di Farmacologia Preclinica e Clinica, Università di Firenze, Firenze, Italy*
[3]*Department of Biochemistry, Trinity College, Dublin, Ireland.*

Abstract: Perturbed cellular calcium homeostasis has been implicated in both apoptosis and necrosis., but the role of altered mitochondrial calcium handling in the cell death process is unclear. Recently we found that taurine, a naturally occurring amino acid potentiates Ca^{2+} sequestration by rat liver mitochondria. These data, which accounted for the taurine antagonism on Ca^{2+} release induced by the neurotoxins 1-methyl-4-phenylpyridinium plus 6-hydroxy dopamine previously reported, prompted us to investigate the effects of taurine on the permeability transition (PT) induced experimentally by high Ca^{2+} plus phosphate concentrations. The parameters used to measure the PT were, mitochondrial swelling, cytochrome *c* release and membrane potential changes. The results showed that, whereas taurine failed to reverse changes of these parameters, cyclosporin A completely reversed them. Even though these results exclude a role in PT regulation under such gross insult conditions, they cannot exclude an important role for taurine in controlling pore-opening under milder more physiological PT-inducing conditions.

Taurine 4, edited by Della Corte *et al.*
Kluwer Academic / Plenum Publishers, New York, 2000.

INTRODUCTION

Mitochondria from mammalian tissues possess an elaborate system for transporting Ca^{2+} across their inner membrane which consists of Ca^{2+} import, via the Ca^{2+} uniporter, in response to the mitochondrial membrane potential ($\Delta\Psi$), and of Ca^{2+} release by an antiport system in exchange for H^+ or Na^+ (see Fig. 1)[9,23]. Because the uniporter is dependent upon the external Ca^{2+} concentration ($[Ca^{2+}]_o$), mitochondria accumulate Ca^{2+} until the $[Ca^{2+}]_o$ decreases to the level at which the uniporter activity balances the Ca^{2+} efflux. The $[Ca^{2+}]_o$ at which the uniporter and efflux activities are equal is defined as the "set point" and corresponds to values between 0.3-3 µM.

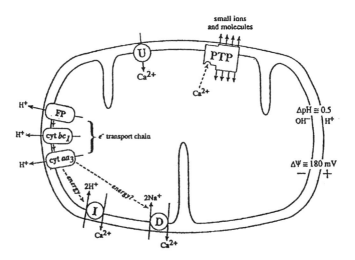

Figure 1. The Ca^{2+} transport system of the inner membrane of mammalian mitochondria. U, uniporter. I, Na^+-independent efflux mechanism or $Ca^{2+}/2H^+$ exchanger. D, Na^+-dependent efflux mechanism or $Ca^{2+}/2Na^+$ exchanger. PTP, permeability transition pore. FP, flavoprotein. $\Delta\Psi$, membrane potential. ΔpH, pH gradient. Adapted from[11].

Ca^{2+} transport and storage play important roles under both physiological and pathological conditions. Whereas under physiological conditions, Ca^{2+} transport has been proposed to function as a metabolic signal for energy production[11,12,20,24], under pathological conditions that are characterised by elevated cytosolic $[Ca^{2+}]$ levels, the same process seems to contribute to Ca^{2+} sequestration and protection of the cytosol against Ca^{2+} overload[10,17,29]. This latter event is considered to play a critical role in the development of damage in those cells that are destined to undergo either the apoptotic or the necrotic form of cell death[1]. Although not fully characterised, the mechanism involved in apoptosis seem to include a pore-mediated increase in the permeability of the mitochondrial inner membrane (permeability transition, PT) to small ions and molecules with the consequent complete collapse of the $\Delta\Psi$, colloidosmotic swelling of the matrix and release of caspase activators such as cytochrome c[3,14,15,18,30]. Recent findings showed that potentiation of the capacity of mitochondria to accumulate Ca^{2+} is associated with either a protection[21] against or inhibition[19] of apoptosis in neural cell mitochondria. Furthermore spermine and other aliphatic polyamines which enhance Ca^{2+} uptake by liver or heart mitochondria, have been reported to inhibit PT[28]. Taken together all these data indicate that Ca^{2+} accumulation by mitochondria plays a pivotal role in the cell death process. Since we recently demonstrated that taurine potentiates mitochondrial Ca^{2+} sequestration[26], in the present study, we investigated the effect of this amino acid on experimentally-induced PT, to establish whether taurine might have a role in the regulation of this process.

MATERIALS AND METHODS

Mitochondrial preparation

Rat liver mitochondria were prepared by differential centrifugation, following the method of Chappell and Hansford[5]. Briefly, male Wistar rats (200-250 g) were anaesthetised with a mixture of Ketavet and Rompun and rapidly exsanguinated. Livers were rapidly removed and homogenised in a medium containing 250 mM sucrose, 5 mM Tris and 1 mM EGTA, pH 7.4, and mitochondria were separated by differential centrifugation. After centrifugation, the mitochondrial pellet was resuspended in 1-2 ml of a medium (KSH) containing 150 mM KCl, 25 mM sucrose, 5 mM Mg^{2+} and 2.7 mM HEPES, pH 7.2. The mitochondrial protein content was determined by the biuret method[8] using bovine serum albumin as standard.

Mitochondrial swelling: swelling induction and measurement

Following preparation in KSH, the mitochondria (0.5 mg/ml) were suspended in a spectrophotometer cuvette, 1 cm path-length at 25°C, in a medium (195 mM mannitol, 65 mM sucrose, 3 mM HEPES, pH 7.2) containing 10 mM succinate (K^+ salt) plus 13 μM rotenone. After the absorbance at 540 nm stabilised, swelling was induced by Ca^{2+} (50, 25, or 10 μM) plus potassium phosphate (P_i) (5 mM)[7]. After addition of P_i, the mitochondrial swelling was measured according the method of Broekemeir[4], by monitoring the decrease in light-scattering at 540 nm.

Detection of cytochrome *c* by western blotting

Sample preparation

Mitochondria (5 mg/ml) were incubated under high Ca^{2+} (100 nmol/mg of mitochondrial protein) plus P_i (50 mM) conditions. After 15 min incubation and centrifugation at 15000g for 2 min, 20 µl from each sample was removed, supplemented with 5 µl of sample buffer, dithiothreitol (DTT) and sodium dodecyl suphate (SDS) (10:1 vol/vol) and boiled for 2 min at 100 °C.

Electrophoresis

Samples together with molecular weight markers and horse heart cytochrome *c* as a positive control, were loaded onto SDS-PAGE gels consisting of 5% stacking and 1.5 % resolving gel. The gel was run (Atto electrophoresis unit, model AE 6450) at 80 V in the stacking and at 120 V in the resolving gel for a total of about 1.5 h. Following electrophoresis, the gel was transferred to nitrocellulose membrane (Atto apparatus) at 110 mA/5 V for a period of 30 min. Following this, the membrane was placed in blocking buffer (5% Marvel dry milk in phosphate-buffered saline, PBS) for 24 h at 4°C. After washing in PBS containing Temex (PBST), it was incubated for 60 min with diluted (1/250 in PBST containing 1% dry milk) monoclonal antibody (mouse anti-cytochrome *c*) and then incubated in diluted (1/1000) secondary antibodies (goat anti-mouse IgG - peroxidase conjugate) for 45 min. After washing in PBST, the secondary antibody marker was revealed by using the ECL detection system (Amersham).

Measurement of mitochondrial membrane potential (ΔΨ)

Mitochondria (1mg/ml) were incubated in 195 mM mannitol, 65 mM sucrose and 3 mM HEPES, pH 7.2 ,containing 13.3 µM rotenone. After energising mitochondria with 10 mM succinate (K^+ salt), 50 µM Ca^{2+} was added, followed, at steady-state Ca^{2+} accumulation, by addition of 5 mM P_i The mitochondrial $\Delta\Psi$ was measured by using a methyltriphenylphosphonium ($TPMP^+$)-selective electrode to monitor the distribution of the permeant lipophilic cation $TPMP^+$ across the mitochondrial inner membrane[24]. The electrode was calibrated before use by sequential additions of TPMP$^+$ up to a final concentration of 5 μM.

RESULTS

Fig. 2 shows the effects of taurine on mitochondrial swelling induced by 5 mM P_i plus Ca^{2+} at different (50, 25 and 10 μM) concentrations in mitochondria energised by 10 mM succinate. The addition of P_i in the presence of Ca^{2+} was followed by a fast and extensive Ca^{2+}–dependent decrease of absorbance, indicative of mitochondrial swelling. The absorbance change correlated with the concentration of Ca^{2+} and was completely reversed by 0.5 μM cyclosporin A (CsA), but was not affected by 10 mM taurine at any of the Ca^{2+} concentrations used.

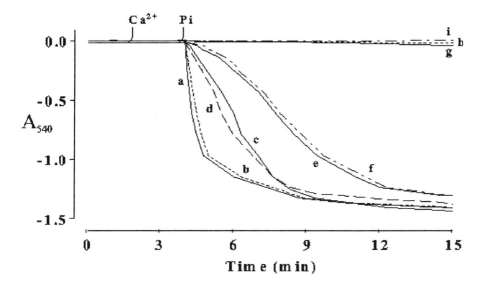

Figure 2. Effect of taurine on mitochondrial swelling induced by Pi and different concentrations of Ca^{2+} in the presence of succinate as a respiratory substrate. Mitochondria (0.5 mg/ml) were incubated, in medium (195 mM mannitol, 65 mM sucrose, 3 mM HEPES, pH 7.2) containing 10 mM succinate (K^+ salt) plus 13.3 μM rotenone, in the absence or in the presence of taurine. After the absorption stabilised, 10, 25 or 50 μM Ca^{2+} was added, followed 2 min later by 5 mM Pi. Changes in volume of mitochondrial suspensions were monitored with time by measuring the decrease in absorbance at 540 nm. **Trace a,** control 1 (Ca^{2+} 50 μM + 5 mM Pi); **trace b,** control 1 + 10 mM taurine; **trace c,** control 2 (Ca^{2+} 25 _M + 5 mM Pi); **trace d,** control 2 + 10 mM taurine; **trace e,** control 3 (Ca^{2+} 10 μM+ 5 mM Pi); **trace f,** control 3 + 10 mM taurine; **trace g,** control 3 + of 0.5 μM cyclosporin A (CsA); **trace h,** control 2 + 0.5 μM CsA; **trace i,** control 1 + 0.5 μM CsA. All traces are representative of 3-5 separate experiments.

These results indicate that mitochondrial swelling, induced by a combination of Ca^{2+} plus P_i, is mediated by changes of the permeability of the mitochondrial inner membrane, but that taurine does not affect this process, at least under these experimental conditions.

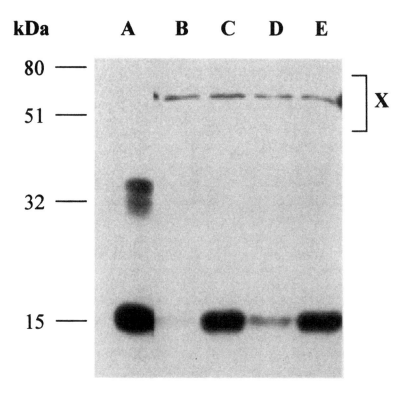

Figure 3. Effect of taurine on cytochrome *c* release induced by Ca^{2+} and Pi, in the presence of succinate as a respiratory substrate. Mitochondria (5 mg/ml) were incubated in medium (195 mM mannitol, 65 mM sucrose, 3 mM HEPES, pH 7.2) containing 10 mM succinate (K^+ salt), 13.3 μM rotenone, 100 nmol Ca^{2+}/mg of mitochondrial protein and 50 mM Pi, in the absence or in the presence of taurine. Mitochondria were pelleted after 15 min and supernatants subjected to SDS-PAGE electrophoresis. The presence of PT was assessed by western blotting analysis of cytochrome *c* release, as described in the methods section. **Lane A,** standard cytochrome *c*, appearing as a monomer at 15 kD and a dimer at 32 kD; **lane B,** control 1 (in the absence of Ca^{2+} and Pi); **lane C,** control 2 (in the presence of Ca^{2+} and Pi); **lane D,** control 2 + 0.5 μM cyclosporin A; **lane E,** control 2 + 10 mM taurine. The [X] denotes protein bands that cross-reacted with the primary antibody.

Since release of cytochrome *c* from mitochondria is known to characterise the PT, immunoblotting analysis was carried out to investigate its release from a mitochondrial suspension submitted to combined Ca^{2+} plus P_i treatment. The results reported in Fig. 3, showed that treatment of mitochondria (5 mg/ml) with Ca^{2+} (100 nmol/mg of mitochondrial protein) and P_i (50 mM) induced release of cytochrome c that was reversed by 0.5 μM CsA but not by 10 mM taurine.

A further validation of these findings was obtained by measures of mitochondrial ΔΨ, known to collapse during PT, under high Ca^{2+} plus P_i conditions. As shown in Fig. 4, addition of Ca^{2+} (50 μM) to mitochondria (1 mg/ml) induced a transient depolarisation followed by a rapid repolarisation to a lower ΔΨ, as a result of the steady-state Ca^{2+} cycling. Subsequent addition of P_i (5 mM) was followed by a short period of repolarisation before a complete collapse of the mitochondrial potential. Again CsA, but not taurine reversed this effect.

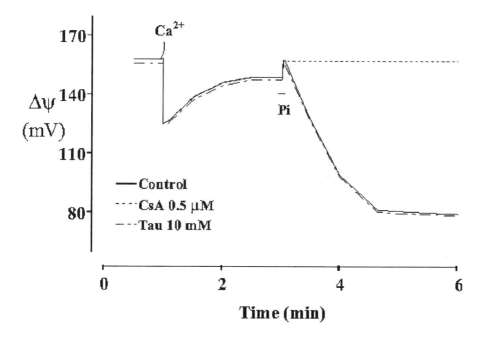

Figure 4. Effect of taurine on mitochondrial membrane depolarisation induced by Ca^{2+} and Pi. Mitochondria (1mg/ml) were incubated in medium (195 mM mannitol, 65 mM sucrose, 3 mM HEPES, pH 7.2) containing 13.3 μM rotenone, in the presence or absence of 10 mM taurine. ΔΨ was measured by using a $TPMP^+$-ion selective electrode. After the calibration, 10 mM succinate (K^+ salt) was added, followed by addition of 50 μM Ca^{2+}. At steady-state Ca^{2+} accumulation, addition of 5 mM Pi was made. All traces are representative of 3-5 separate experiments.

DISCUSSION

Previous findings by our group showed hat taurine stimulates Ca^{2+} uptake by rat liver mitochondria by a mechanism envisaging direct stimulation of the Ca^{2+} uptake system (uniporter) rather than inhibition of the Ca^{2+} efflux[26]. The increased Ca^{2+} uptake caused by taurine counteracted mitochondrial Ca^{2+} release induced by a combined treatment with the neurotoxins 1-methyl-4-phenylpyridinium and 6-hydroxy dopamine[25]. Whereas the effect of taurine on Ca^{2+} uptake is potentially protective towards moderate elevation of cytosolic $[Ca^{2+}]_o$, under the far-from physiological conditions used here and routinely used by others to investigate PT, this effect of taurine might exacerbate the, already extensive, increase in intra- matrix Ca^{2+}. Thus, under these extreme conditions taurine might not be expected to be protective but to enhance the probability of pore-opening. This hypothesis, while explaining the failure of taurine to inhibit PT induced by the gross insults used in our experiments, may also account for the contradictory data described in literature, where a potentiation of mitochondrial Ca^{2+} uptake by bcl-2 over-expressing neural cells, has been alternatively reported to prevent or potentiate, cell death induced by apoptotic insults[19,21]. Similarly spermine and other aliphatic polyamines, that enhance Ca^{2+} uptake by liver or heart mitochondria, inhibit, but do not abolish, the PT-mediated Ca^{2+} release induced by an acute increase in Ca^{2+} concentration in a cytosol-adapted incubation medium (Ca^{2+} pulse)[28]. In keeping with the structural analogy between taurine and the charged headgroups of the neutral phospholipid, taurine was found to interact with the neutral phospholipids in biological membranes thereby altering membrane architecture, fluidity and properties such as ion channel functions, membrane–bound enzyme activities and several Ca^{2+}-dependent cellular functions, including Ca^{2+} transport[16,22].

Taurine has also been proposed to act as an osmolyte to control cell volume (for review see[27]). This role may indirectly affect phospholipid behaviour given the experimental evidence showing that the phospholipid bilayer membranes fuse under osmotic swelling[6]. Particularly affected may be the highly specialised tissues, such as brain, muscle, or photoreceptors where an uncontrolled changes in intracellular water, accompanying ion fluxes across the membrane during physiological activity, may result in disruption of their highly organised cytoarchitecture and cell functions. It is interesting that in the photoreceptor outer segment, which contains no other osmolyte but taurine, there are dramatic consequences of taurine deficiency, including membrane disorganisation and cell dysfunction[13].

These properties, while provide a mechanistic basis for the, much discussed "membrane stabilising" effect of taurine[16], may have profound influence on the stability of the mitochondrial inner membrane and consequently the PT, nevertheless this effect can not be observed at the gross insult level used in the present study.

Further experiments under more mild conditions that are known to increase the probability of opening the PT pore will be necessary to determine whether taurine exerts anti-apoptotic effects at this level.

ACKNOWLEDGEMENTS

This work was supported by grants from the European Community (BIOMED 1 CT-94-1402) and from MURST, Rome, Cofin. '98.

REFERENCES

1. Ankarcrona, M., Dypbukt, J.M., Orrenius, S., and Nicotera, P., 1996, Calcineurin and mitochondrial function in glutamate-induced neuronal cell death. *FEBS Lett.* **394**: 321-324.
2. Azzone, G.F., Pietrobon, D., and Zoratti, M., 1984, Determination of the proton electrochemical gradient across the biological membranes. *Curr. Topics Bioenergetics* **13**: 1-77.
3. Bernardi, P., 1992, Modulation of the mitochondrial cyclosporin A-sensitive permeability transition pore by the electron electrochemical gradient. Evidence that the pore can be opened by membrane depolarisation. *J. Biol. Chem.* **267**: 8834-8839.
4. Broekemeier, K.M., Dempsey, M.E., and Pfeiffer, D.R., 1989, Cyclosporin A is a potent inhibitor of the inner membrane permeability transition in liver mitochondria. *J. Biol. Chem.* **264**: 7826-7830.
5. Chappell, J.B., and Hansford, R.G., 1972, Preparation of mitochondria from animal tissues and yeast. In: *Subcellular Components: Preparation and fractionation* (Eds. Birmie GD), pp. 77-91. Butterworths, London.
6. Cohen, F.S., Akabas, M.H., and Finkelstain, A., 1982, Osmotic swelling of phospholipid vescicles causes them to fuse with a planar phospholipid bilayer membrane. *Science (Wash. D.C.)* **217**: 458-460.
7. Crompton, M., and Costi, A., 1998, Kinetic evidence for a heart mitochondrial pore activated by Ca^{2+}, inorganic phosphate and oxidative stress. A potential mechanism for cellular dysfunction during mitochondrial Ca^{2+} overload. *Eur. J. Biochem.* **178**: 489-501.
8. Gornall, A.G., Bardawill, C.J., and David, M.M., 1949, Determination of serum protein by means of the biuret reaction. *J. Biol. Chem.* **177**: 751-766.
9. Gunter, K.K., and Gunter, T.E., 1994, Transport of calcium by mitochondria. *J. Bioenerg. Biomembr.* **26**: 471-485.
10. Gunter, T.E., Gunter, K.K., Sheu, S.S., and Gavin, C.E., 1994, Mitochondrial calcium transport: physiological and pathological relevance. *Am. J. Physiol.* **267**: C313-C339.
11. Hansford, R.G., 1985, Relation between mitochondrial calcium transport and control of energy metabolism. *Rev. Physiol. Biochem. Pharmacol.* **102**: 1-72.
12. Hansford, R.G., 1994, Physiological role of mitochondrial Ca^{2+} transport. *J. Bioenerg. Biomembr.* **26**: 495-508.

13.   Hayes, K.C., Carey, R.E., and Schmidt, S.Y., 1975, Retinal degeneration associated with taurine deficiency in the cat. *Science (Wash. D.C.)* **188**: 949-951.

14.   Hunter, D.R., and Haworth, R.A., 1979, The Ca^{2+}-induced membrane transition in mitochondria. *Arch. Biochem. Biophys.* **195**: 453-459.

15.   Hunter, D.R., Haworth, R.A., and Southard, J.H., 1976, Relationship between configuration, function and permeability in calcium-treated mitochondria. *J. Biol. Chem.* **251**:5069-5077.

16.   Huxtable, R.J., 1992, Physiological actions of taurine, *Physiol. Rev.*. **72**: 101-159.

17.   Kiedrowski, L., and Costa, E., 1995, Glutamate-induced destabilisation of intracellular calcium concentration homeostasis in cultured cerebellar granule cells: role of mitochondria in calcium buffering. *Mol. Pharmacol.* **47**: 140-147.

18.   Kluck, R.M., Bossy-Wetzel, E., Green, D.R., and Newmeyer, D.D., 1997, The release of cytochrome *c* from mitochondria: a primary site for Bcl-2 regulation of apoptosis. *Science (Wash. D.C.)* **275**: 1132-1136.

19.   Kruman, I., and Mattson, M.P., 1999, Pivotal role of mitochondrial calcium uptake in neural cell apoptosis and necrosis. *J. Neurochem.* **72**: 529-540.

20.   Mc Cormack, J.G., Halestrap, A.P., and Denton, R.M., 1990, Role of calcium ions in the regulation of mammalian intramitochondrial metabolism. *Physiol. Rev.* **70**: 391-425.

21.   Murphy, A.N., Bredesen, D.E., Cortopassi, G., Wang, E., and Fiskum, G., 1996, Bcl-2 potentiates the maximal Ca^{2+} uptake capacity of neural cell mitochondria. *Proc. Natl. Acad. Sci. USA.* **93**: 9893-9898.

22.   Nakashima, T., Shima, T., Sakai, M., Yama, H., Mitsuyoshi, H., Inaba, K., Matsumoto N., Sakamoto, Y., Kashima, K., and Nishikawa, H., 1996, Evidence of a direct action of taurine and calcium on biological membranes. A combined study of ^{31}P-nuclear magnetic resonance and electron spin resonance. *Biochem. Pharmacol.* **52**: 173-176.

23.   Nicholls, D.G., 1978, The regulation of extramitochondrial free calcium ion concentration by rat liver mitochondria. *Biochem. J.* **176**: 463-474.

24.   Nicholls, D.G., and Ferguson, S.J., 1992, Secondary transport. In: *Bioenergetics 2* (Eds. Nicholls DG and Ferguson S.J.), pp. 207-233. Academic Press, London.

25.   Palmi, M., Tchuisseu-Youmbi, G., Fusi, F., Frosini, M., Sgaragli, G.P., Della Corte, L., Bianchi.L., and Tipton, K.F., 1998, Antagonism by taurine on ruthenium red-induced and 6-hydroxy-dopamine plus 1-methyl-4-phenylpyridinium-induced Ca^{2+} release from rat liver mitochondria. *Adv. Exp. Med. Biol.* **442**: 91-98.

26.   Palmi, M., Tchuisseu-Youmbi, G., Fusi, F., Sgaragli, G.P., Dixon, H.B.F., Frosini, M., and Tipton, K.F., 1999, Potentiation of mitochondrial Ca^{2+} sequestration by taurine. *Biochem. Pharmacol.* **58**: 1123-1131.

27.   Pasantes-Morales, H., Quesada, O., and Moràn, J., 1998, Taurine: an osmolyte in mammalian tissues. *Adv. Exp. Med. Biol.* **442**: 209-217.

28.   Rustenbeck, I., Löptien, D., Fricke, K., Lenzen, S., and Reiter, H., 1998, Polyamine modulation of mitochondrial calcium transport. Inhibition of permeability transition by aliphatic polyamines but not by aminoglucosides. *Biochem. Pharmacol.* **56**: 977-985.

29.   White, R.J., and Reynolds, I.J., 1997, Mitochondrial accumulate Ca^{2+} following intense glutamate stimulation of cultured rat forbrain neurons. *J. Physiol. (London)* **498**: 31-47.

30.   Zoratti, M., and Szabò, I., 1994, Electrophysiology of the inner mitochondrial membrane. *J. Bioenerg. Biomembr.* **26**: 543-553.

CLONING AND CHARACTERIZATION OF THE PROMOTER REGION OF THE RAT TAURINE TRANSPORTER (TauT) GENE

Xiaobin Han, Andrea M. Budreau and Russell W. Chesney
Department of Pediatrics, University of Tennessee, and the Crippled Children's Foundation Research Center at Le Bonheur Children's Medical Center, Memphis, TN 38103

INTRODUCTION

The maintenance of adequate tissue levels of taurine is essential to the normal development of the retina and the central nervous system. It functions as the principal organic osmolyte protecting the brain against massive cell volume changes induced by hypo- or hyperosmolar states[13]. The renal adaptive response to dietary taurine intake has been localized in part to the proximal tubule brush border membrane, where NaCl-dependent taurine transporter activity is shared with other ß-amino acids and GABA[1]. Our studies show that the taurine transporter gene itself appears to be the primary target for adaptive regulation by dietary taurine availability; e.g. the increase (or decrease) in transcription is primarily responsible for the increase (or decrease) in taurine transporter mRNA, which in turn is responsible for the change in activity of the transporter after manipulation of taurine concentration[4]. We have precisely localized the adaptive regulation of the rat kidney taurine transporter gene *in vivo*, which is limited to the S3 segment of the proximal tubule[8]. Characterization of the taurine transporter gene will provide valuable insights about how expression of an amino acid transporter is regulated by its substrate, a nutrient-gene interaction.

Taurine 4, edited by Della Corte *et al.*
Kluwer Academic / Plenum Publishers, New York, 2000.

Table 1. Sense and anti-sense primers used in 5'-deletion constructs

---

Sense: K0 (-124/+48) 5'-GGGGTACCCGGCCAAGCTGGTATT-3'
           K1 (p-177/+48) 5'-GGGGTACCTGTGTGTGGGCGT-3'
           K2 (p-269/+48) 5'-GGGGTACCCGGGTTCTTTGTG-3'
           K3 (-574/+48) 5'-GGGGTACCGAGTTGGGGAGGGA-3'
           K4 (-963/+48) 5'-GGGGTACCTTACTGAAGGTCACACAG-3'
           K5 (-1532/+48) 5'-GGGGTACCTTCCCAGGTTTCCGAT-3'

Antisense: 5'-AAGATCTTGGCACGGGAGTTCA-3'

METHODS AND MATERIALS

Cloning of the 5'-Flanking Region of the Rat Taurine Transporter Gene

The rat P1 genomic DNA library was screened (Genome System, Inc., St. Louis, MO) by polymerase chain reaction (PCR) using three sets of oligonucleotide primers based on the DNA sequence of rat brain taurine transporter cDNA (rB16a)[12]. The primers used for generating the PCR probes were sense primer 5'-GCCAACGCCGCGATCGCCGCCAA-3'/antisense primer 5'-CTCCTCGTTTTGC-TTGAGAGGC-3'; sense primer 5'-ATGGCCACC AAGGAGA-AGCTTCAA-3'/antisense primer 5'-GAGAAATGCACCTCC ACCAT-3'; and sense primer 5'-TCAGAGGGAGAAGTGGTCCAGCAA GA-3'/antisense primer 5'-AGCCAGACACAAAACTGGTACCA-3'. A full length (~28 kb) rat taurine transporter gene (TauT) was isolated from the rat P1 library, as determined by Southern blot analysis using 5', 3', and internal probes of rB16a. The 28 kb DNA was digested using Hind III restriction enzyme, subcloned into pBluescript vector (Promega, Madison, WI), and transferred into JM109 bacteria by transformation. The DNA was isolated and analyzed by Southern blot using a 5' probe (5'-AGCCAGGTCCCGGAGT ACGA-3') of rB16a generated by 5'-rapid amplification of rat taurine transporter cDNA. A 7.2 kb Hind III fragment containing the 5'-flanking region of the gene was isolated and sequenced using an automatic DNA sequencer. Plasmids containing nested deletions of the proximal 5'-flanking region were generated by PCR using the 7.2 kb 5'-flanking region DNA as the template. The conditions used were 40 cycles of 1 min of denaturation at 94°C, 1 min of annealing at 58°C, and 1 min of elongation at 72°C. The sense primer designed for PCR contains a unique site for kpnI and the antisense primer contains a unique site for BagI II. The sense primers are listed in Table 1. PCR products were digested with kpnI and BagI II and re-ligated into the kpnI and BagI II unique sites of pGL3-Basic. This generates plasmids containing segments of the taurine transporter gene extending from the position at the

+48 nucleotide corresponding to the transcriptional start site. The constructs were verified by DNA sequencing.

Cell Culture and Transient Transfection

MDCK (Madin-Darby canine kidney) and LLC-PK1 (porcine kidney) cells were grown as confluent monolayers in 10 cm diameter tissue culture plates in Dulbecco's mininum essential medium (DMEM)/F_{12} (1:1) with 10% fetal calf serum (FCS) at 37°C in the presence of 5% CO_2 in a humidified incubator. Cells were plated 18 h before transfection and fed with fresh medium 4 h before transfection. Luciferase reporter plasmids were introduced into the cultured mammalian cells using cationic liposomes (LipofectAMINE, Life Technologies, Grand Island, NY). The transfection was carried out for 16-18 h. Cells were then washed twice with phosphate-buffered saline (PBS) and incubated in fresh medium for 24-48 h before harvesting. pGL3-control, which contains a luciferase gene driven by the SV40 early region promoter/enhancer, and empty pGL3-Basic vectors were used as positive and negative controls, respectively. To standardize the transfection efficiency, 0.1 μg of pRL-CMV vector (pRL Renilla Luciferase control reporter vector, Promega) was co-transfected in all experiments. Cells were harvested 48 h after transfection and lysed in 200 μl of reporter lysis buffer. A luciferase assay was performed using a dual-luciferase assay kit. Luciferase activity was measured with an Optocomp 1 luminometer (MGM Instruments, Inc., Hamden, CT). Protein concentrations of the cell extracts were determined using the Bradford method (BioRad protein assay, BioRad, Hercules, CA).

Primer Extension Analysis and S1 Nuclease Protection

Primer extension analysis was done following the instructions of the manufacturer (Promega). A 20-base antisense oligonucleotide primer (5'-GGGTGGCTAAGCCCACG CG-3') was end-labelled with [γ-^{32}P]ATP using T4 polynucleotide kinase. Ten μg of mRNA isolated from normal rat kidney (NRK-52E) cells were annealed to 1 x 106 cpm of the primer and extended with 200 units of Moloney murine leukemia virus reverse transcriptase (Life Technologies, Inc.). Yeast tRNA was used as a negative control. The primer-extended products were analyzed on 7 M urea, 8% polyacrylamide gels in parallel with sequencing reactions. An S1 nuclease protection assay was performed using an S1-Assay kit (Ambion, Austin, TX). A single-stranded DNA probe (317 bp) was prepared and hybridized with 20 μg of total RNA isolated from rat kidney. S1 nuclease-digested fragments were analyzed on an 8% acrylamide DNA sequencing gel with sequence ladders.

Statistical Analysis

The mean data from independent experiments using different cell preparations are reported. Error bars are ± SE. Data were analyzed using Student's *t* test for paired data. Statistical significance was defined as p<0.05.

RESULTS

A 7.2 kb Hind III fragment containing the 5'-flanking region of the gene was isolated and sequenced using an automatic DNA sequencer. This 7.2 kb DNA contains the exon 1 and exon 2 coding regions of the rat taurine transporter gene. The exon-intron boundaries were determined by "GT-AG" consensus rules for splicing. The 7.2 kb fragment was further digested using SacI restriction enzyme and subcloned into pBluescript vector. A 1.6 kb SacI fragment containing the 5'-flanking region of the gene was isolated using the same 5'-primer probe for the 7.2 kb fragment isolation. The sequence of this 1.6 kb 5'-flanking region (Fig. 1) was found to contain typical eukaryotic promoter elements, including a TATA box and four consensus binding sites for SP1, one overlapping site for WT-1/EGR-1/Sp1, two consensus p53 half-sites, two estrogen receptor half-sites and a simple sequence repeat consisting of $(T-G)_{22}/(A-C)_{22}$.

Transcriptional Start Site

Primer extension was used to determine the transcriptional start site using mRNA from rat normal kidney (NRK-52E) cells . As shown in Fig. 2, one major band (120 bp) was identified by the primer extension assay. The transcription start site was mapped to 25 nucleotides downstream from the TATA box. This product was not detected when the assay was carried out using tRNA. S1 nuclease protection assay confirmed this transcriptional start site.

Deletion Analysis of the Proximal 5'-Flanking Region of the Rat TauT Gene

To determine the basal promoter sequence of TauT, luciferase reporter plasmids were constructed that contained a nested set of deletions extending from nucleotide +48 within the first exon to -124, -269, -574, -963 and -1532 within the proximal 5'-flanking region. Plasmids were transfected into LLC-PK1 and MDCK cells, and luciferase activity was measured. As shown in Fig. 3, luciferase activity was high and quite similar in cells transfected with the constructs having deletions from -1532 to -269. However, p-124 reporter

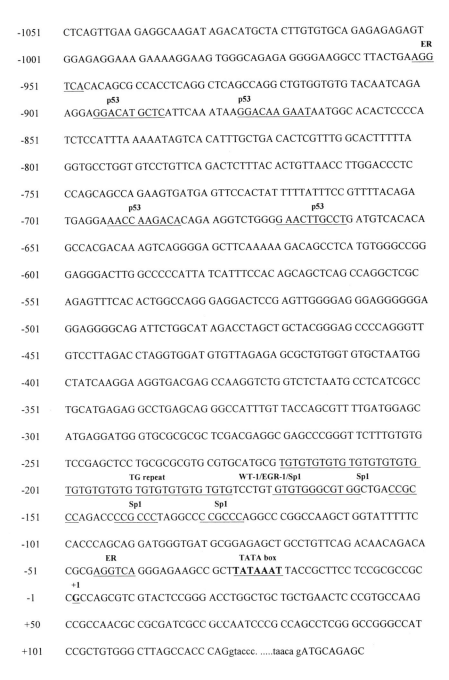

-1051	CTCAGTTGAA GAGGCAAGAT AGACATGCTA CTTGTGTGCA GAGAGAGAGT
	ER
-1001	GGAGAGGAAA GAAAAGGAAG TGGGCAGAGA GGGGAAGGCC TTACTGA<u>AGG</u>
-951	<u>TCA</u>CACAGCG CCACCTCAGG CTCAGCCAGG CTGTGGTGTG TACAATCAGA
	p53 p53
-901	AGGA<u>GGACAT GCTC</u>ATTCAA ATAA<u>GGACAA GAAT</u>AATGGC ACACTCCCCA
-851	TCTCCATTTA AAAATAGTCA CATTTGCTGA CACTCGTTTG GCACTTTTTA
-801	GGTGCCTGGT GTCCTGTTCA GACTCTTTAC ACTGTTAACC TTGGACCCTC
-751	CCAGCAGCCA GAAGTGATGA GTTCCACTAT TTTTATTTCC GTTTTACAGA
	p53 p53
-701	TGAGGA<u>AACC AAGACA</u>CAGA AGGTCTGGGG <u>AACTTGCCT</u>G ATGTCACACA
-651	GCCACGACAA AGTCAGGGGA GCTTCAAAAA GACAGCCTCA TGTGGGCCGG
-601	GAGGGACTTG GCCCCCATTA TCATTTCCAC AGCAGCTCAG CCAGGCTCGC
-551	AGAGTTTCAC ACTGGCCAGG GAGGACTCCG AGTTGGGGAG GGAGGGGGGA
-501	GGAGGGGCAG ATTCTGGCAT AGACCTAGCT GCTACGGGAG CCCCAGGGTT
-451	GTCCTTAGAC CTAGGTGGAT GTGTTAGAGA GCGCTGTGGT GTGCTAATGG
-401	CTATCAAGGA AGGTGACGAG CCAAGGTCTG GTCTCTAATG CCTCATCGCC
-351	TGCATGAGAG GCCTGAGCAG GGCCATTTGT TACCAGCGTT TTGATGGAGC
-301	ATGAGGATGG GTGCGCGCGC TCGACGAGGC GAGCCCGGGT TCTTTGTGTG
-251	TCCGAGCTCC TGCGCGCGTG CGTGCATGCG <u>TGTGTGTGTG TGTGTGTGTG</u>
	TG repeat WT-1/EGR-1/Sp1 Sp1
-201	<u>TGTGTGTGTG TGTGTGTGTG TGTGTCCTGT GTGTGGGCGT GGCTGA</u>CCGC
	Sp1 Sp1
-151	CCAGACC<u>CCG CCC</u>TAGGCCC <u>CGCCC</u>AGGCC CGGCCAAGCT GGTATTTTTC
-101	CACCCAGCAG GATGGGTGAT GCGGAGAGCT GCCTGTTCAG ACAACAGACA
	ER TATA box
-51	CGCG<u>AGGTCA</u> GGGAGAAGCC GCT**TATAAAT** TACCGCTTCC TCCGCGCCGC
	+1
-1	C**G**CCAGCGTC GTACTCCGGG ACCTGGCTGC TGCTGAACTC CCGTGCCAAG
+50	CCGCCAACGC CGCGATCGCC GCCAATCCCG CCAGCCTCGG GCCGGGCCAT
+101	CCGCTGTGGG CTTAGCCACC CAGgtaccc. .....taaca gATGCAGAGC

Figure 1. DNA sequence of 5'-flanking promoter region of rat taurine transporter gene. The putative transcription start site is indicated by a bold **G** (+1) in the nucleotide sequence. The DNA consensus binding sites for transcrition factors p53 , WT-1/EGR-1/Sp1, Sp1, and ER are indicated. The TG repeat and TATA motif are also indicated. Positions are given above each sequence according to the transcription start site.

b

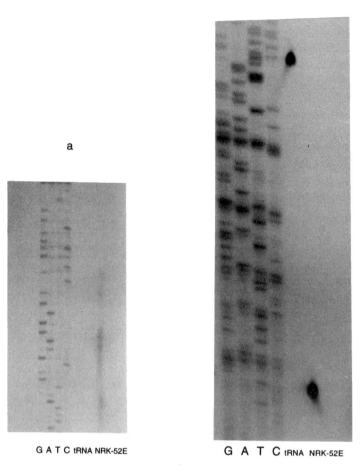

G A T C tRNA NRK-52E   G A T C tRNA NRK-52E

Figure 2. Identification of the transcription start site for the rat taurine transporter gene. a) Ten μg of mRNA isolated from normal rat kidney (NRK-52E) cells were annealed to 1×10^6 cpm of the primer and extended with 200 units of Moloney murine leukemia virus reverse transcriptase. Yeast tRNA was used as a negative control. The primer-extended products were analyzed on 7M urea, 8% polyacrylamide gels, in parallel with sequencing reactions. b) S1 nuclease protection assay was performed by using a single-stranded DNA probe (317 bp). The probe was hybridized with 20 μg of total RNA isolated from rat kidney, and S1 nuclease-digested fragments were analyzed on an 8% acrylamide DNA sequencing gel with sequence ladders.

failed to demonstrate any promoter activity. These results indicated that positive regulatory elements that controlled TauT promoter activity in renal cells were likely to be located from -269 to -124. This region contains a TG repeat, one overlapping consensus site for WT-1/EGR-1/Sp1, and three other putative sites for Sp1.

Transcription Factor Sp1 is Critical for the Basal Promoter Function of TauT

The TauT basal promoter sequence contains four potential Sp1 sites, located at -162 to -168, -155 to -150, -144 to -139, and -133 to -128 in relationship to the transcription start site. To test if Sp1 transcription factor is

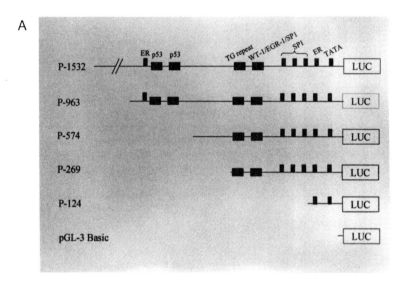

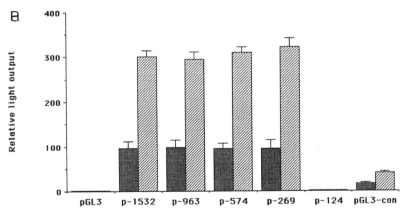

Figure 3. Deletion analysis of the proximal 5'-flanking region of the TauT gene. Recombinant plasmids containing nested deletions of the proximal 5'-flanking region of TauT (in pGL3-Basic) were generated by PCR. A (top): Constructs are numbered according to the position of their 5'-end from the putative transcription start site. B (Bottom): Plasmids were transfected into MDCK and LLC-PK1 cells and luciferase activity was measured in cell lysates after 24 h. To control for transfection efficiency, cells were co-transfected with pRL-CMV vector and luciferase activity was measured in the same cell lysates. The promoter activity (mean ± SE of four samples in relative light units) of each construct is represented by relative light output normalized to pRL-CMV control.

essential for the promoter function of TauT, p-177 reporter, which contains the basal promoter of TauT, was transiently co-transfected with or without a CMV-driven Sp1 expression vector into Sp1- deficient Drosophilia SL2 cells. The pRL-CMV vector was also co-transfected and used as an internal control in transient transfection studies. Our results demonstrate that Sp1 is obligatory for basal activation of the TauT promoter (Fig. 4).

TG Polymorphism is a Strong Enhancer for the Promoter Function of the Rat Taurine Transporter Gene

To determine the role of the TG repeat (TG_{22}) that is found upstream of the basal promoter DNA sequence (from -191 to -235), a p-177 reporter with TG repeat deletion was prepared by PCR. As shown in Fig. 5, transient transfection of the p-269 reporter (which contains the TG repeat) into MDCK and LLC-PK cells showed a 100- to 200-fold increase in luciferase reporter gene expression, as compared with control (pGL3-Basic vector). However, transient transfection of the p-177 reporter, which does not contain

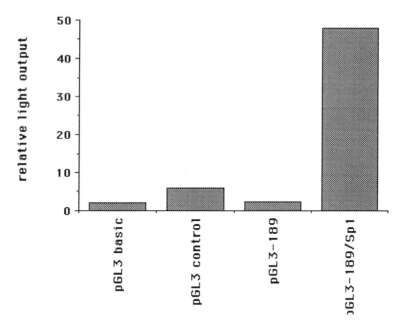

Figure 4. Effect of Sp1 transcription factor on TauT promoter activity. p-177 reporter, containing the basal TauT promoter DNA sequence, was transiently transfected into Sp1- deficient *Drosophila* SL2 cells, with or without co-transfection with Sp1 expression vector, for 72 h. To control for transfection efficiency, cells were co-transfected with pRL-CMV vector, and luciferase activity was measured in the same cell lysates. The promoter activity (mean ± SE of four samples in relative light units) of each construct is represented by relative light output normalized to pRL-CMV control.

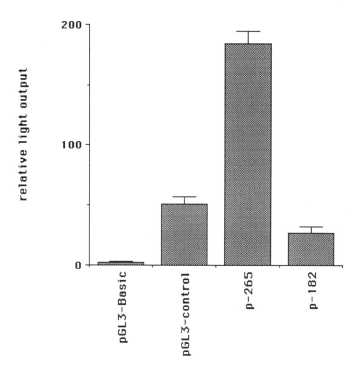

Figure 5. The role of TG polymorphism in rat TauT promoter function: Reporter p-269 or p-177 was transfected into MDCK and LLC-PK1 cells and luciferase activity was measured in cell lysates after 24 h. To control for transfection efficiency, cells were co-transfected with pRL-CMV vector and luciferase activity was measured in the same cell lysates. The promoter activity (mean ± S.E. of four samples in relative light units) of each construct is represented by relative light output normalized to pRL-CMV control.

the TG repeat, resulted in an 85% decrease in luciferase reporter gene expression as compared to the p-269 reporter, which contains both the TG repeat and the putative TauT promoter sequence. These results indicate that the TG repeat is critical for full expression of the rat taurine transporter gene

DISCUSSION

Taurine serves many important cellular functions, including membrane stabilization, antioxidation, detoxification, and cell volume regulation. Arguably, the most important biologic function of taurine is osmoregulation. Osmoregulation is critical to both normal brain development and ongoing neural function. Moreover, taurine, by its chemical nature a free intracellular organic solute, is important in maintenance of cell shape, and apoptosis[6,7].

Hence, factors that impair the uptake of taurine into renal cells could potentially result in the death of a significant fraction of undifferentiated cells within the nephrogenic zone of developing kidneys, both because of the role of taurine in cell volume maintainence and its role in the regulation of cell calcium content[11].

To understand better the importance of the renal adaptive regulation of taurine transporter, we cloned and characterized the promoter region of the rat taurine transporter gene (TauT). We have found that the 5'-flanking region of TauT contains three consensus sites for the Sp1 transcription factor, an overlapping site for WT-1/EGR-1/Sp1, two consensus p53 half-sites, two estrogen receptor half-sites and a simple sequence repeat consisting of (T-G)$_{22}$/(A-C)$_{22}$. To determine if Sp1 is necessary for transactivation of the TauT gene, a p-189 reporter was transiently cotransfected with or without Sp1 expression vector into SL2 cells (an Sp1-deficient cell line). Our results demonstrate that Sp1 is obligatory for basal activation of the TauT promoter. Clearly this observation has to be substantiated using more specific methods such as electrophoretic mobility shift assays (EMSA); nevertheless, our preliminary results suggest that Sp1 is a critical regulator of the rat taurine transporter gene. Studies have shown that Sp1 expression is temporally and spatially regulated during nephrogenesis, suggesting that Sp1 itself, as well as Sp1-regulated genes, may be important in kidney development[2]. We have previously demonstrated that expression of the rat renal taurine transporter is developmentally regulated[5]. Therefore, it would be interesting to precisely determine the role of Sp1 in regulation of the TauT gene during renal development.

Deletion of the TG repeat decreased TauT promoter activity by more than 85%, suggesting that the TG repeat is a strong enhancer of the TauT promoter, and that it is critical for full expression of the TauT gene. Simple repetitive DNA sequences are abundantly interspersed in eukaryote genomes and therefore useful in genome research and genetic fingerprinting in animals, including man, regarding their influences on the DNA structure, gene expression, genomic (in)stability and their development on an evolutionary time scale. Other studies have demonstrated that the human taurine transporter gene also contains the dinucleotide TG repeat[3]. We have cloned a full length human taurine transporter gene which may provide a tool for studying the function of the TG polymorphism in human taurine transporter gene expression. The TG microsatellite repeat could be used as a probe to explore the potential linkage of certain genetic disorders, such as 3p-syndrome, that may be associated with this gene.

The genes encoding the taurine transporter of different species and tissues share a high degree of homology. TauT gene is located on the central region of mouse chromosome 6, and on human chromosome 3 p21-25, where a conserved linkage group of genes has been found between mouse and man[10]. It has been demonstrated that in patients with 3p-syndrome, deletion of

3p25-pter is associated with profound growth failure, characteristic facial features, retinal changes, and mental retardation[9], suggesting that deletion of TauT might contribute to some phenotypic features of the 3p-syndrome. Furthermore, the F1 generation of inbred, taurine-deficient kits showed a characteristic facial appearance, blindness, ataxia, cerebellar abnormalities and hyphosis, as well as severe neonatal renal damage, which progresses to renal scarring. These observations led us to postulate that the TauT gene may play an important role in mammalian brain and renal development and differentiation.

In summary, we have cloned the promoter region of the TauT gene that contains typical eukaryotic promoter elements, including two consensus p53 binding sites and an overlapping binding site for WT-1/EGR-1/Sp1, located up-stream of the TATA box. WT-1 is required for normal kidney development. Overexpression of p53 in transgenic mice causes abnormal kidney development and progressive renal failure. It would be of interest to study if and how the TauT gene is regulated by the tumor suppressors p53 and WT-1, and the importance of TauT in renal development.

REFERENCES

1. Chesney, R.W., Gusowski, N., and Friedman, A.L., 1983, Renal adaptation to altered amino acid intake occurs at the luminal brush border membrane, *Kidney Int.,* 24:588-594.
2. Cohen, H.T., Bossone, S.A., Zhu, G., McDonald, G.A., and Sukhatme, V.P., 1997, Sp1 is a critical regulator of the Wilm's tumor-1 gene, *J. Biol. Chem.,* 272:2901-2913.
3. Gregor, P., Hoff, M., Holik, J., Hadley, D., Fang, N., Coon, H., and Byerly, W., 1994, Dinucleotide repeat polymorphism in the human taurine transporter gene (TAUT), Hum. Mol. Genet, 3:2263.
4. Han, X., Budreau, A.M., and Chesney, R.W., 1996, Adaptive regulation of MDCK cell taurine transporter (pNCT) mRNA: Transcription of pNCT gene is regulated by external taurine concentration, *Biochim. Biophys. Acta,* in press:
5. Han, X. and Chesney, R.W., 1993, Developmental expression of rat kidney taurine transporter and its regulation by diet in *Xenopus laevis* oocytes, *J. Am. Soc. Nephrol.,* 1:114.
6. Law, R.O., 1998, The role of taurine in the regulation of brain cell volume in chronically hyponatremic rats, *Neurochem. Int.,* 33:467-472.
7. Law, R.O., 1991, Amino acids as volume-regulatory osmolytes in mammalian cells, *Comp. Biochem. Physiol.,* 99A:263-277.
8. Matsell, D.G., Bennett, T., Han, X., Budreau, A.M., and Chesney, R.W., 1997, Regulation of the taurine transporter gene in the S3 segment of the proximal tubule, *Kidney Int.,* 52:748-754.
9. Mowrey, P.N., Chorney, M.J., Venditti, C.P., Latif, F., Modi, W.S., Lerman, M.I., Zbar, B., Robins, D.B., Rogan, P.K., and Ladda, R.L., 1993, Clinical and molecular analyses of deletion 3p25-pter syndrome, *Am. J. Med. Genet.,* 46:623-629.
10. Patel, A., Rochelle, J.M., Jones, J.M., Sumegi, G., Uhl, G.R., Seldin, M.F., Meisler, M.H., and Gregor, P., 1995, Mapping of the taurine transporter gene to mouse chromosome 6 and to the short arm of human chromosome 3., *Genomics,* 1:314-317.
11. Seabra, V., Stachlewitz, R.F., and Thurman, R.G., 1998, Taurine blunts LPS-induced increases in intracellular calcium and TNF-alpha production by Kupffer cells, *J. Leukocyte. Biol.,* 64:615-621.

12. Smith, K.E., Borden, L.A., Wang, C.D., Hartig, P.R., Branchek, T.A., and Weinshank, R.L., 1992, Cloning and expression of a high affinity taurine transporter from rat brain, *Mol. Pharmacol.,* 42:563-569.
13. Sturman, J. and Chesney, R., 1995, Taurine in pediatric nutrition, *Pediatr. Clin. N. Amer.,* 42:879-897.

TAURINE PREVENTS ISCHEMIA DAMAGE IN CULTURED NEONATAL RAT CARDIOMYOCYTES

Kyoko Takahashi, Yuko Ohyabu, [1]Stephen W. Schaffer and Junichi Azuma

Dept. of Clinical Evaluation of Medicines and Therapeutics, Graduate School of Pharmaceutical Sciences, Osaka University, Osaka, Japan: [1]Dept. of Pharmacology, University of South Alabama, School of Medicine, Mobile, AL, USA

INTRODUCTION

Myocardial ischemia leads to cardiac cell loss and scar formation. This results in reduced pumping capacity which eventually leads to congestive heart failure and death[1]. Cell loss is due predominantly to the death of cardiomyocytes, with the loss occurring as a result of both apoptosis and necrosis[1,2]. Previously, we have suggested that taurine plays an important role in myocardial remodeling and in myocyte dysfunction arising from altered calcium homeostasis[3,4].

We have recently developed a simulated model of ischemia which involves placing cardiomyocytes in sealed culture flasks[5]. This model mimics ischemia by combining the stresses of hypoxia, acidosis and inadequate perfusion. In the present study we have used this new simulated-ischemia model to test the idea that taurine improves the outcome of an ischemic event.

Taurine 4, edited by Della Corte *et al.*
Kluwer Academic / Plenum Publishers, New York, 2000.

MATERIALS AND METHODS

Preparation of cultured cardiac myocytes and the new simulated ischemia model

Primary cultures of cardiac myocytes and non-myocyte cells were prepared from 1-day old Wistar rats by the procedure of Sadoshima et al.[6] This yielded cultures containing 90-95% myocytes, as assessed by microscopic observation of cell beating. The myocytes were kept in serum-containing culture medium for 48 h and then transferred to serum-free medium. Fig 1 shows the procedure used to simulate ischemia[5]. An incubation flask was filled with 50 ml of phosphate-buffered saline (PBS) and then tightly sealed so that no gas could enter the flask. The control cells were incubated in 2.5 ml medium that was continuously equilibrated in an atmosphere of 5% CO_2-95% air.

Evaluation of morphology and beating status

The morphological status of the cardiomyocytes was monitored with an inverted phase-contrast microscope and videomonitor[3]. The shape and location of each myocyte was recorded before initiating the experiment and then after 72 h of simulated ischemia. Changes in morphology and beating status were estimated for each cell and expressed as percent of the total number of cells observed.

Determination of creatine phosphokinase (CPK) content

The activity of CPK in the myocyte culture was measured using a commercially-available kit, (CPK-test Wako kit from Wako Chem., Osaka, Japan). The protein concentration was determined by the method of Lowry et al.[7], using bovine serum albumin as a standard.

Apoptosis detection

Genomic DNA was isolated and detected as described by Cigola et al.[8]. DNA samples were separated electrophoretically on 1.5 % agarose gels and stained with ethidium bromide. For visualizing apoptotic nuclei, cells were fixed in 1% paraformaldehyde for 30 min at room temperature and then stained for 15 min with the fluorescent dye Hoechst 33258 (Sigma, MO, USA)[9,10], before examination using an Olympus fluorescence microscope system (Tokyo, Japan).

Intracellular taurine content

The intracellular taurine content of myocytes was measured by the procedure of Jones & Gilligan[11] using a high-performance liquid chromatography (HPLC) system (JASCO880-PU) equipped with a Hitachi F1000 fluorescence detector and Hitachi D-2500 integrator. An example of the elution profile of taurine is shown in Figure 5.

Statistics

Statistical significance was determined by Student's t-test or by analysis of variance (ANOVA), with either the Bonferroni's method or the χ^2 test being used to compare individual data points when a significant F value was obtained. Each value was expressed as the mean $\pm$ S.E.M. Differences were considered statistically significant when the calculated p value was less than 0.05.

RESULTS AND DISCUSSION

The effect of taurine on ischemia-induced injury of isolated cardiomyocytes

A well-characterized animal model is a prerequisite for the development and assessment of molecular strategies for the treatment of ischemia. We have established a new *in vitro* model of cardiac ischemia (Fig 1) which mimics distinct features of the human disease. In this model neonatal rat heart cells were cultured in a sealed flask for a period of 24-72 h. In this environment, the cells were exposed to the stresses of hypoxia, acidosis and a stagnant incubation medium (no reflow). The pO_2 and pH of the medium gradually decreased during the period to values of 14 mm Hg and 6.8, respectively, after 72 h. This treatment resulted in morphological degeneration of the cells, CPK release, beating impairment and ATP depletion. Apoptotic nuclei and nucleosomal ladders of DNA fragments were apparent in the ischemic myocytes. This model should be useful in in studies of ischemic injury and myocardial apoptosis[5].

We have shown taurine to play an important role in myocardial remodeling through inhibition of angiotensin II action[4]. Since, myocardial ischemia causes angiotensin II activation[2], we examined the effects of including 20 mM taurine in the serum-free medium used in this model system. In the absence of taurine, the simulated ischemia led to significant

CPK loss, morphological degeneration, beating cessation and ATP depletion within 24-72 h (data not shown).

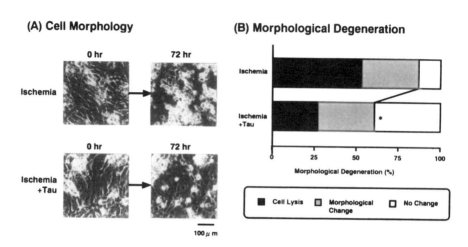

Figure 1. A cell-culture model of simulated ischemia.

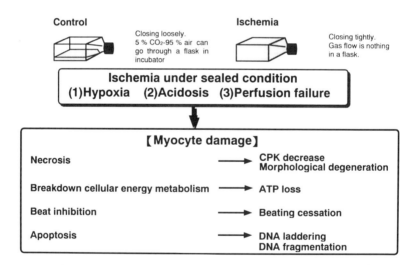

Figure 2. The effects of taurine on the ischemia-induced morphological changes in cultured neonatal rat cardiomyocytes. Cells (n = 71-110) were incubated under ischemic conditions for 72 h. (A) Cell morphology; (B) Morphological degeneration. *p < 0.01 versus the ischemic group to which taurine had not been added.

Typical photographs of the cells before and after 72 h incubation under the ischemic conditions are shown in Figure 2A. The unsupplemented ischemic myocytes showed pronounced injury, including ballooning and cellular lysis. The presence of 20 mM taurine reduced this ischemia-induced morphological degeneration from 90% to 65% (Figure 2B).

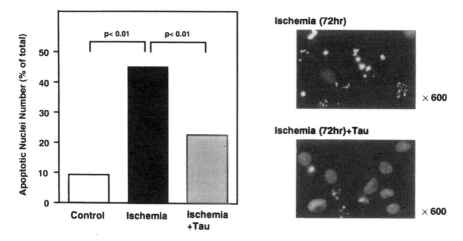

Figure 3. Effects of taurine on cell damage during exposure of cultured neonatal rat cardiomyocytes to a 72 h ischemic insult. Data shown represent the mean values ± S.E.M. (A) CPK loss (n = 6); (B) Beating cells (n = 34-40).

The effects of taurine on ischemia-induced CPK loss and impaired beating activity are shown in Figure 3. After 72 h under the ischemic conditions the CPK activity of untreated myocytes was reduced 10-fold. In the presence of taurine the reduction was only 3-fold. Similarly, taurine treatment reduced the loss of beating function in many of the cells. Whereas 90% of the untreated cells stopped beating during the 72 h hypoxic insult, only 52% of the taurine-treated cells ceased their beating activity under the same conditions. In contrast, taurine treatment failed to alter the degree of ischemia-induced ATP depletion (data not shown).

Apoptosis plays an important role in myocardial ischemic injury [12-14]. Hypoxia has also been found to induce apoptosis in cultured neonatal cardiomyocytes[15]. Our simulated ischemic model causes apoptosis to occur in cardiomyocytes. Fragmentation of DNA into integral multiples of the internucleosomal DNA length was observed after 72 h of hypoxia. Interestingly, the non-myocyte cells in the preparation did not show fragmentation of DNA, even after 72 h of hypoxia (data not shown).

Figure 4 shows a typical fluorescent microphotograph of control and taurine-treated cells after the 72 h ischemic insult. In the absence of taurine, a significant number of cells (48%) became apoptotic during this time. The myocytes treated with 20 mM taurine were more resistant to ischemia-induced apoptosis, as only 24% of the cells underwent apoptosis during the 72 h period.

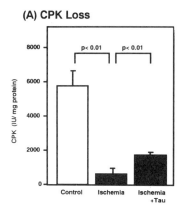
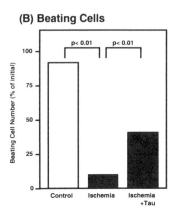

Figure 4.   Effect of taurine on ischemia-induced apoptosis in cultured neonatal rat cardiomyocytes. The cells were stained with Hoechst 33342 after the 72 h ischemic treatment in the presence and absence of 20 mM taurine. The data are expressed as % apoptotic nuclei before and after the 72 h period of ischemia.

        Ischemia is associated with multiple alterations in the extracellular and intracellular milieu, many of which could act to induce apoptosis. Cardiomyocytes are terminally differentiated, and as such lose their ability to duplicate soon after birth. Thus the regulation of apoptosis is of major importance in determining the cellular content of the adult heart.

        The mechanisms underlying this attenuation of ischemia-induced injury by taurine require further investigation. Very high concentrations of calcium ions can lead to necrosis through the activation of Ca^{2+}-sensitive proteases and the disruption of mitochondrial function[16]. Elevated concentrations of free cytoplasmic Ca^{2+} have also been implicated in apoptosis[17]. Furthermore, myocardial ischemia is known to promote the upregulation of the systemic renin-angiotensin system, which in turn influences the outcome of a myocardial ischemic event through changes in hemodynamic and hemostatic activity[2]. The benefit of taurine to patients suffering from congestive heart failure may be related to its suppression of angiotensin II-mediated cellular responses.

The relationship between intracellular taurine content and taurine cardioprotection against ischemic injury

        The myocardial taurine content is elevated in response to the chronic stresses that trigger the development of heart failure[18,19]. Conversely, in acute episodes of stress, myocardial taurine levels decline[20,21]. There is no evidence that elevated levels of taurine are harmful to humans. As shown in Figure 5, exposure of the cells to 20 mM taurine for 24 h elevated the intracellular taurine content from 240 pmol to 1600 pmol/mg protein.

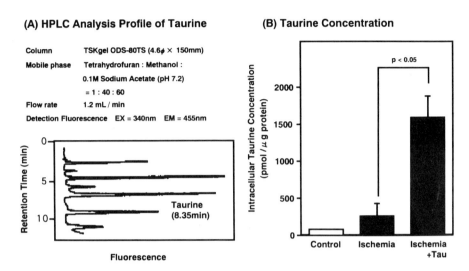

Figure 5. The intracellular taurine concentration after the ischemic insult. (A) HPLC elution profile showing the resolution of taurine; (B) Intracellular taurine concentrations in the control, untreated ischemic and the taurine-treated, ischemic cardiomyocyte. Incubation time: 24 h, *p < 0.05.

In summary, the most important finding of this study is that taurine renders the cell resistant to ischemia-induced injury, including morphological degeneration, CPK loss, beating abnormalities, DNA fragmentation and apoptosis. These effects appear to be related to an elevation in intracellular taurine levels.

REFERENCES

1. Buja, L.M., 1998, Modulation of the myocardial response to ischemia. *Lab. Invest.* **78**: 1345-1373.
2. Swynghedauw, B., 1999, Molecular mechanisms of myocardial remodeling. *Am. J. Phys. Soc.* **79**: 215-262.
3. Takahashi, K., Schaffer S.W. and Azuma, J., 1997, Taurine prevents intracellular calcium overload during calcium paradox of cultured cardiomyocytes. *Amino acids* **13**: 1-11.
4. Takahashi, K., Azuma, M., Taira, K., Baba, A., Yamamoto I., Schaffer S.W. and Azuma, J., 1997, Effect of taurine on angiotensin II-induced hypertrophy of neonatal rat cardiac cells. *J. Cardiovasc. Pharmacol.* **30**: 725-730.

5. Ohyabu, y., Takahashi, K., Yamamoto, I. and Azuma, J., 1999, An in vitro model of ischemia under sealed condition in cultured neonatal rat cardiomyocytes. In *Cardiac Structure and Metabolism* -1998 (M. Nagano, ed.), Rupoh Press, Tokyo, pp. 451-457.

6. Sadoshima, j., Jahn, L., Takahashi, T., Kulik, T. and Izumo, S., 1992, Molecular characterization of the stretch-induced adaptation of cultured cardiac cells : an in vitro model of load-induced cardiac hypertrophy. *J. Biol. Chem.* **267**: 10551-10560.

7. Lowry, O.H., Rosebrough, N.J., Farr, A.L. and Randall, R.J., 1951, Protein measurement with the Folin phenol regent. *J. Biol. Chem.* **193**: 265-275.

8. Cigola, E., Kajstura, J., Li, B., Meggs, L.G. and Anversa, P., 1997, Angiotensin II activates programmed myocyte cell death in vitro. *Exp. Cell Res.* **231**: 363-371.

9. Kajstura, J., Mansukhani, M., Cheng, W., Reiss, K., Krajewski, S., Reed, J.C., Quaini, F., Sonnenblick, E.H. and Anversa, P., 1995, Programmed cell death and expression of the protooncogene bcl-2 in myocytes during postnatal maturation of the heart. *Exp. Cell Res.* **219**: 110-121.

10. Kajstura, J., Cheng, W., Reiss, K., Clark, W.A., Sonnenblick, E.H., Krajewski, S., Reed, J.C., Olivetti, G. and Anversa, P., 1996, Apoptotic and necrotic myocyte cell death are independent contributing variables of infarct size in rats. *Lab. Invest.* **74**: 86-107.

11. Jones, B.N. and Gilligan, J.P., 1983, o-Phthaldialdehyde precolumn derivatization and reversed-phase high-performance liquid chromatography of polypeptide hydrolysates and physiological fluids. *J. Chromatogr.* **266**: 471-482.

12. Olivetti, G., Abbi, R., Quaini F., Kajstura, J., Cheng, W., Nitahara, J.A., Quaini, E., Loreto, C.D., Beltrami, C.A., Krajewski, S., Reed, J.C. and Anversa, P., 1997, Apoptosis in the failing human heart. *N. Engl. J. Med.* **336**: 1131-1141.

13. Haunstetter, A. and Izumo, S., 1998, Apoptosis: Basic mechanisms and implications for cardiovascular disease. *Circ. Res.* **82**: 1111-1129.

14. Fliss, H. and Gattinger, D., 1996, Apoptosis in ischemic and reperfused rat myocardium. *Circ. Res.* **79**: 949-956.

15. Tanaka, M., Ito H., Adachi, S., Akimoto H., Nishikawa, T., Kasajima, T., Marumo, F. and Hiroe, M., 1994, Hypoxia induces apoptosis with enhanced expression of Fas antigen messenger RNA in cultured neonatal rat cardiomyocytes. *Circ. Res.* **75**: 426-433.

16. Berridge, M.J., Bootman, M.D. and Lipp, P., 1998, Calcium-a life and death signal. *Nature* **395**: 645-648.

17. McConkey, D.J. and Orrenius, S., 1997, The role of calcium in the regulation of apoptosis. *Biochem. Biophys. Res. Commun.* **239**: 357-366.

18. Huxtable, R. and Bressler, R., 1974a, Taurine concentration in congestive heart failure. *Science* **184**: 1187-1188.

19. Huxtable, R. and Bressler, R., 1974b, Elevation of taurine in human congestive heart failure. *Life Sci.* **14**: 1353-1359.

20. Crass, M.F. and Lombardini, J.B., 1977, Loss of cardiac muscle taurine after acute left ventricular ischemia. *Life Sci.* **21**: 951-958.

21. Lombardini, J.B., 1980, Effect of ischemia on taurine level. In *Nature Sulfur Compounds* (D. Cavallini, et al, eds.), Plenum Press, New York, pp.255-306.

EFFECTS OF TAURINE AND THIOSULFATE ON THE SPECTROPHOTOMETRIC PROPERTIES OF RNA AND DNA

S.I. Baskin, T.L. Barnhouse, K.T. Filipiak, J.C. Bloom, and M.J. Novak

Pharmacology Division, United States Army Medical Research Institute of Chemical Defense, 3100 Ricketts Point Road, Aberdeen Proving Ground, EA, MD 21010-5400

INTRODUCTION

It has been proposed that multiple factors can signal cells to differentiate or proliferate[14]. Many mechanisms, involving proteins and enzymes[5], have been suggested to signal nucleic acids to aid in this cellular process. Baskin *et al.*[3] suggested the action of taurine on nucleic acids may be mediated by either an interaction through histones or by affecting RNA synthesis[10,12]. The intracellular concentration of taurine and its uptake into cells changes early in cellular growth[11] and proliferation[3]. Marked differences in taurine concentration are seen at various stages of development, both *in vivo* and in isolated cells[6,17,19]. Thiosulfate shows similar differences[7]. Taurine[3], thiosulfate, 2-mercaptoethane sulfonic acid[4] and other sulfur-containing compounds also show effects on cellular kinetics in comparing normal versus malignant cells. Some of these changes in concentration could be due to changes in uptake[27,28] or synthesis[8] or both[20].

As early as 1904 Kohler and Reimer[13] suggested that compounds such as guanosine might interact with certain sulfur substances, thereby modifying nucleic acid structure. Hayatsu *et al.*[9], Wakayama *et al.*[26] and Novak *et al.*[16] have shown that sulfites, thiosulfate and sulfonic acids interact with RNA[16] or DNA[2]. The protective effects of these agents, against radiation, alkylating agents and other carcinogens, suggest that these small anions may serve to regulate the conformational state of nucleic acid and thus reduce the toxic effects of chemical mutagens.

Taurine 4, edited by Della Corte *et al.*
Kluwer Academic / Plenum Publishers, New York, 2000.

Our studies were designed, in part, to examine the interaction (using ab-sorbance differences in this study) of sulfur-containing anionic substances with RNA or DNA in order to find if these compounds may protect nucleic acids against alkylating agents. Our experiments suggest that compounds, such as thiosulfate, which produce a hypochromic effect on RNA or DNA may protect against the alkylating agents by supercoiling the nucleic acid.

Name of Compound	Structure	Molecular Weight
2- Aminoethane Thiosulfate (sodium salt)	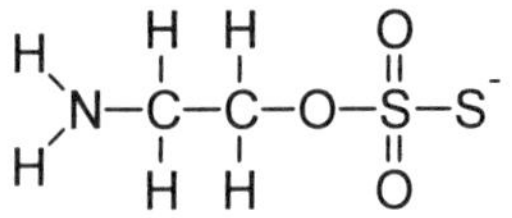 H H H O / H–N–C–C–O–S–S⁻ / H H H O	156.2
2- Hydroxyethane Thiosulfonate (Na) (sodium salt)	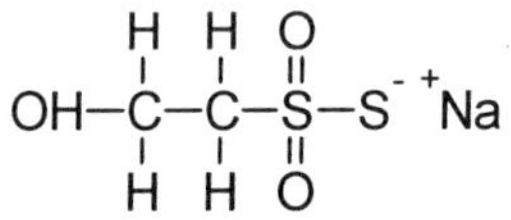 H H O / OH–C–C–S–S⁻ ⁺Na / H H O	164.2
2- Aminoethanesulfonic Acid (Taurine)	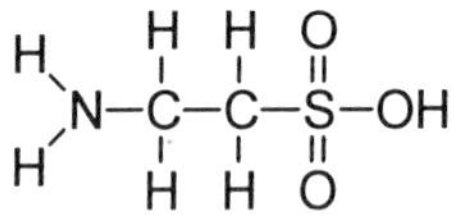 H H H O / H–N–C–C–S–OH / H H H O	125.1
2- Mercaptoethanesulfonic acid (sodium salt)	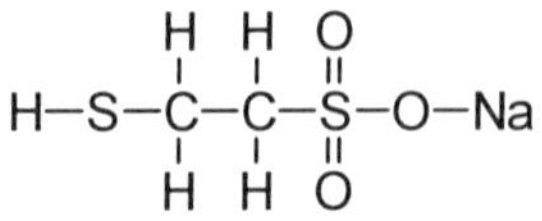 H H O / H–S–C–C–S–O–Na / H H O	164.1
Sodium Thiosulfate (sodium salt)	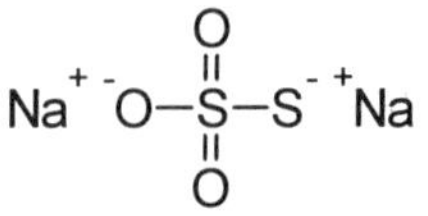 O / Na⁺ ⁻O–S–S⁻ ⁺Na / O	248.2

Figure 1. Chemical names, chemical structures and molecular weights of compounds used in this study

MATERIALS AND METHODS

RNA (~ 3-8 kD from yeast), (DNA (a highly polymerized form from calf thymus; 10-15 x 10^3 kD), Tris and sodium thiosulfate (Na_2SO_3) were pur-chased from Sigma (St. Louis MO, USA). Figure 1 shows the structures, chemical names and molecular weights of the sulfur-containing chemicals used in this study. Sulfur-containing salts were recrystallized within one month of experimentation. All other compounds employed were analytical–grade commercial products.

Spectral assays were performed on a Beckman DU7400 spectrophotometer (Fullerton, CA). Samples were prepared in 0.1 M Tris-HCl buffer, pH 7.4, and measured[22] at 37°C in 1.0 cm Starna quartz cuvettes between 205 and 360 nm. The pH of the buffer was adjusted with 0.1 M NaOH. RNA (0.005% w/v) in 0.1 M Tris-HCl buffer at 37°C was monitored over a period of 20 h to verify its stability in solution under the above conditions. An appropriate blank for each sulfur-containing compound employed in this study was performed over the concentration range used. In those cases where an individual sulfur-containing compound produced an absorbency response, the absorbency of the compound itself was subtracted from the response observed from the nucleic acid. Concentration-response curves were established for RNA or DNA alone (0.053-0.17 mg/ml) (using similar concentration ranges) and in the presence of sodium thiosulfate, sodium sulfate, sodium sulfite, or sodium metabisulfite (0.0526 mM, 0.105 mM, 0.210 mM, 0.419 mM and 0.836 mM, respectively)[16]. Spectral assays for each compound, in the presence of RNA or DNA, (0.021 mg/ ml) were performed. The total volume of the reaction mixture was 3.0 ml.

Statistical evaluation of the group data was performed using a one-way ANOVA followed by a post hoc Dunnett's test. This was done to compare all compounds and concentrations to RNA alone (control) or DNA alone (control) in order to employ statistical significance of $p < 0.05$[30]. Additional trend tests (covariance) were employed for the MESNA data to show that the slope of increasing the concentration of MESNA differed from that of control for peak 1.[30]

RESULTS

Figure 2 shows the effect of increasing concentrations of thiosulfate on the maximal absorbance (λ_{max}) of RNA and DNA at peak 1. The absorbance maxima for RNA alone were between at 210-212 nm (peak one) and 257-259 nm (peak 2). The absorbance maxima for DNA, alone, were 212 nm (peak one) and 257-259 nm (peak 2). Thiosulfate at low concentrations (0.053-0.21 mM) increased the absorbency of RNA or DNA. Greater concentrations of thiosulfate decreased the absorbency below that of the control.

The wavelength of maximal absorbency of both RNA and DNA shifted upwards with increasing concentrations of thiosulfate (Fig. 3). The λ_{max} represented by the line suggests that the effect is greater for RNA than for DNA. However, this may not be the case as the molecular weight of the DNA used was much larger than that of the RNA. This phenomenon was

observed to a greater degree in the presence of 2-hydroxyethane thiosulfate
(Fig. 4).

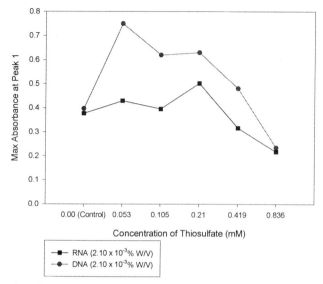

Figure 2. The effect of sodium thiosulfate on the maximum absorbance of peak 1 for RNA
and DNA. The assays were performed in 0.1 M Tris-HCl buffer, pH 7.4, at 37° C. The y-axis
shows max absorbance at peak 1 (λ_{max}) and the x-axis shows the concentration of thiosulfate
(mM). Each data circle represents the average of five values.

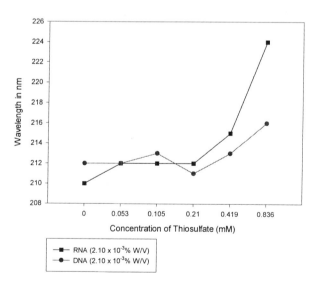

Figure 3. The effect of sodium thiosulfate on the maximum absorbance of peak 1 for RNA
and DNA. The assays were performed in 0.1 M Tris-HCl buffer, pH 7.4, at 37° C. The y-axis
shows wavelength (nm) location of peak 1 and the x-axis shows the concentration of thiosulfate (mM). Each data circle represents the average of five values.

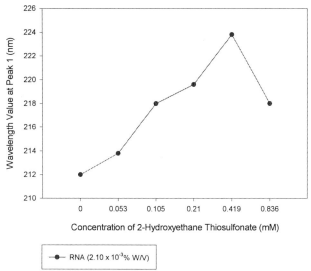

RNA (2.10 x 10⁻³% W/V)

Figure 4. The effect of sodium thiosulfate and 2-hydroxyethane thiosulfonate on the maximum absorbance of peak 1 for RNA and DNA. The assays were performed in 0.1 M Tris-HCl buffer, pH 7.4, at 37° C. The y-axis shows wavelength (nm) location of peak 1 and the x-axis shows the concentration of 2-hydroxyethane thiosulfonate (mM). The concentration of sodium thiosulfate was held constant at 0.836 mM for all assays. Each data circle represents the average of five values.

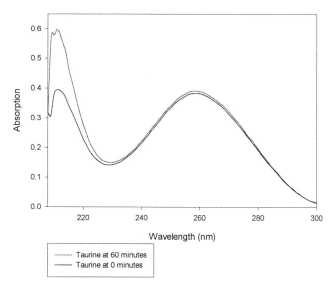

Taurine at 60 minutes
Taurine at 0 minutes

Figure 5. Superimposed spectra of RNA in the presence of taurine with sodium thiosulfate at 0 min and at 60 min. The assays were performed in 0.1 M Tris-HCl buffer, pH 7.4, at 37°C. The y-axis shows absorption and the x-axis shows wavelength (nm). The individual spectra represent the means $\pm$ SE of means of five experiments.

The addition of taurine (28 mM) directly with RNA in the Tris buffer caused an increase absorption at 211-212 nm of the RNA peak (Fig. 5). At 60 min, the reaction between taurine and RNA has reached its maximum response. The absorbency of the peak at 212 nm for DNA was similarly increased. Adding additional thiosulfate to the taurine and RNA, for example, decreased the absorbency compared to that of taurine and RNA alone.

Figure 6 illustrates the effects of increasing concentration of 2-mercaptoethane sulfonic acid sodium salt on the absorption of RNA. At lower concentrations (i.e., 0.053 mM) the spectrum of RNA showed a hyperchromic effect. This effect was not present at higher concentrations (up to 0.836 mM). A significant but small hyperchromic effect is observed for peak 1. 2-Hydroxyethane thiosulfonate also showed moderate hyperchromic effects at lower concentrations such as 0.053 mM, however the absorbance changes were not as significant as taurine.

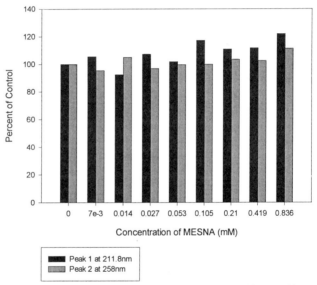

Figure 6. The effects of sodium thiosulfate and 2-mercaptoethane sulfonic acid on the maximum absorbance of peak 1 for RNA. The assays were performed in 0.1 M Tris-HCl buffer, pH 7.4, at 37°C. The y-axis shows percent control and the x-axis shows concentration of 2-mercaptoethane sulfonic acid (mM). The individual bars represent the means ± SE of five experiments.

DISCUSSION

Our data and the literature[9,24] suggested that certain small anionic sulfur containing compounds appear to regulate activity (possibly the coiling or uncoiling) of RNA or DNA. This study using the methods of Szinicz et al.[22] allows for the measurement of changes in the UV-visible spectrum that provides an *in vitro* method for determining the interaction of compounds that may protect against alkylating agents. It is assumed that uncoiled nucleic acids will block more light in the cuvette because the conformation of the DNA and RNA will cover more area or there is a greater molar extinction coefficient for the uncoiled species. Further molecular experiments need to be performed to establish the precise mechanism of action. Changing the wavelength of λ_{max} may reflect a conformational change in the nucleic acid and thus a change in the coiling or uncoiling of the macromolecule. It is also possible that the interaction of the anionic sulfane sulfur moiety with guanosine, for example, could alter the chromophore density of the resulting species. Small structural changes, such as changing from sulfate to a sulfonate or from one sulfur to a persulfide or a sulfane sulfur, appear to produce hyper- or hypo-chromic responses that suggest conformational changes in RNA or DNA. This shift in λ_{max} observed with thiosulfate and 2-hydroxyethane thiosulfonate may reflect a supercoiling, which would be consistent with a therapeutic action of thiosulfate against alkylating agents. It has been proposed that the alkylating agent sulfur mustard (using half mustard) primarily (~72.3-97.6%) interacts at the guanosine site[15,29]. As thiosulfate, a mustard therapeutic agent, is thought to interact with guanosine, we proposed that certain sulfur containing anionics could act as an antidote that would protect against the toxicity of sulfur mustard by acting at the guanosine site. On the other hand, topoisomerase is thought to be involved by interaction at an adenosine/thymidine site[18]. Therefore, it is proposed that at least two signal mechanisms for nucleic acids may exist for folding/unfolding nucleic acids, one acting through adenosine/thymidine (thought to relate to topoisomerase activity) and another which acts through sulfane sulfur/ sulfonic acids.

The interaction of low molecular weight thiols and other sulfur compounds with DNA[21,31] could be a molecular mechanism to protect nucleic acid information. Our study suggests that certain thiols or sulfane sulfur compounds that endogenously exist perform this task of coiling and uncoiling RNA or DNA. Another possible role for these low molecular weight sulfur compounds is to serve as initiators or inhibitors of RNA or DNA coding activity. These initiators or inhibitors could coil or uncoil the nucleic acids[23]. It would be speculative to imagine if a chemical code exist using sulfonate, thiosulfate, persulfide sulfur and other chemical sulfur forms to regulate this process. Much further evidence is needed to prove that those compounds that produce the folding of RNA or DNA or result in cellular expression in proliferation or differentiation of cells[23]. Additional studies need to be performed *in vitro* and at the cellular and *in vivo* levels to establish the exact roles of these compounds in cellular metabolism and as therapies for exposure to toxic substances.

All of these data suggest that if a sulfone sulfur is substituted by a thiosulfonate, for example, the absorbency is decreased (for RNA) and therefore

RNA is uncoiling. At high concentrations of these compounds, there are further evidences that suggest that the nucleic acids are being supercoiled. These small changes seem to regulate the cell cycle for certain nucleic acids, particularly guanosine. The second structural alteration occurs at the other end of the molecule, with the hydroxy, amino, mercapto groups, which appear to modify the quantitative effects but not the fundamental properties.

CONCLUSIONS

Thiosulfate, a treatment for compounds that alkylate RNA or DNA, produces a hyperchromic action at lower concentrations and a hypochromic action at higher concentrations on RNA or DNA. A similar effect occurs with 2-aminoethane thiosulfate and 2-hydroxyethane thiosulfonate. On the other hand, taurine and sodium 2-mercaptoethane sulfonate produce only a hyperchromic effect. The effect of 2-mercaptoethane sulfonate was less than 20% over the concentrations employed. Taurine's hyperchromic effect appeared to antagonize the hypochromic effect of thiosulfate. The effects of thiosulfate at high concentrations (the shift of the λ_{max} of RNA or DNA to a greater wavelength) are indicative of increased coiling of RNA or DNA. A greater wavelength shift was observed with 2-hydroxyethane thiosulfonate than for thiosulfate.

The data suggest that small sulfur-containing anionic compounds may affect the conformational status of larger RNA or DNA species. Small structural differences among sulfate, sulfonate and sulfane sulfur functional groups appear to affect the conformation of the nucleic acid species. It is also possible that ionic strength or osmolar effects may cause these changes. It is known that changes in ionic strength or osmolar effects may also cause conformational alteration of nucleic acids. However, many believe that several of these sulfur containing anionic substances such as taurine and isethionic acid can be, in part, responsible for these actions[1]. Other functional groups may provide quantitative differences to affect RNA or DNA folding.

ACKNOWLEDGMENTS

The authors thank Mr. Eric Neally for his expertise in computer graphics. The help of Mr. Steven Miller for the maintenance of the lab equipment is also appreciated. Finally, the authors thank Barbara A. Schultz and Leanna E. Bush for their help with literary references.

REFERENCES

1. Baskin, S.I., and Finney, C.M., 1979, Effects of taurine and taurine analogs on the cardio-vascular system. *Sulfur-Containing Amino Acids* 2:1-18.
2. Baskin, S.I., Prabhakaran, V., Bowman, J.D., and Novak, M.J, 1998, The in vitro effects of anionic sulfur compounds with DNA. *Toxicologist* 42: 392.
3. Baskin, S.I., Wakayama, K., Knight, T., Jepson, J.H., and Besa, E.C., 1982, The sulfur-containing amino acid pathway in normal and malignant cell growth. In *Taurine in nutrition and neurology* (R. J. Huxtable and H. Pasantes-Morales, eds.), Plenum Press, New York, pp.127-141.
4. Blomgren, H., and Hallstrom, M., 1991, Inhibition of tumor cell growth in vitro by 2-mercaptoethanesulfonate (Mesna) and other thiols. *Meth. Find. Exp. Clin. Pharmacol.* 13: 579-582.
5. Chambon, P., Ramuz, M., Mandel, P., and Doly, J., 1968, The influence of ionic strength and a polyanion on transcription in vivo. I. Stimulation of the aggregate RNA polymerase from rat liver nuclei. *Biochim. Biophys. Acta.* 157: 504-519.
6. Christensen, H.N., Hess, B., and Riggs, T.R., 1954, Concentration of taurine, β-alanine and tri-iodothyronine by ascites tumor cells. *Cancer Res.* 14: 124-127.
7. Fasth, A., and Sorbo, B., 1973, Protective effect of thiosulfate and metabolic thiosulfate precursors against toxicity of nitrogen mustard. *Biochem. Pharmacol.* 22: 1337-1351.
8. Green, J.P., and Day, M., 1963, Biosynthetic pathways in mastocytoma cells in culture and in vivo. *Ann. N.Y. Acad. Sci.* 103: 334-350.
9. Hayatsu, H., Wataya, Y., Kai, K., and Iida, S., 1970, Reaction of sodium bisufite with uracil, cytosine, and their derivatives. *Biochemistry* 9: 2858-2865.
10. Holubek, V., Fanshier, L., and Crocker, T.T., 1966, The inhibition of nuclear RNA synthesis by added RNA. *Exp. Cell. Res.*, 44: 362-368.
11. Jacobsen, G., and Smith, L.H. Jr., 1968, Biochemistry and physiology of taurine and taurine derivatives. *Physiol. Rev.* 48: 424-511.
12. Kinoshita, S., 1971, Heparin as a possible initiator of genomic RNA synthesis in the early development of sea urchin embryos. *Exp. Cell Res.* 64: 403-410.
13. Kohler, E.P., and Reimer, M., 1904, Some addition-reactions of sulfinic acids. *Amer. Chem. J.* 31: 163-184.
14. Leffert, H.L., 1980, Growth regulation by ion fluxes. *Ann. N.Y. Acad. Sci.* 339: 1-335.
15. Ludlum, D.B., Kent, S., and Mitra, J.R., 1986, Formation of O^6- ethyl thioethyl-guanine in DNA by reaction with the sulfur mustard, chloroethyl ethylsulfide and its apparent lack of repair by O^6 - alkylguanine-DNA-alkyltransferase. *Carcinogenesis* 7: 1203-1206.
16. Novak, M.J., and Baskin, S.I., 1997, The effects of inorganic sulfur species on the spectrophotometric properties of RNA. *Toxicol. Methods* 7: 193-206.
17. Piez, K.A., and Eagle, H., 1958, The free amino acid pool of cultured human cells. *J. Biol. Chem.* 231: 533-545.
18. Pu, Q.Q., and Bezwooda, W.R., 1999, Induction of alkylator (Melphlan) resistance in HL60 cells is accompanied by increased levels of topoisomerse II expression and function. *J. Pharmacol. Expt. Therap.* 56: 147-153.
19. Rouser, G., Samuels, A.L., Kinugasa, K., Jelinek, B., and Heller, D., 1962, Free amino acids in the blood of man and animals. In *Amino Acid Pools* (J.T. Holder, ed.), Elsevier, New York.
20. Schrier, B.K., and Thompson, 1974, E.J., On the role of glial cells in the mammalian nervous system: uptake, excretion and metabolism of putative neurotransmitters by cultured tumor cells. *J. Biol. Chem.* 249: 1769-1780.
21. Smoluk, G.D., Fahey, I.M., and Ward, J.F., 1988, Interaction of glutathione and other low-molecular-weight thiols with DNA. Evidence for counterion condensation and colon depletion near DNA. *Radiation Research* 114: 3-10.

22. Szinicz, L., Allbrecht, G.L., and Weger, N., 1981, Effect of various compounds on the reactions of tris-(2-chloroethyl) amine with ribonucleic acid in vitro and on its toxicity in mice. *Arzneimittel-Forsch.* 31: 1713-1717.
23. Toohey, J.I., 1989, Sulphane sulphur in biological systems: a possible regulatory role. *Biochem. J.* 264: 625-632.
24. Wakayama, K., Baskin, S.I., and Besa, E.C., 1983, Changes in intracellular taurine content of human leukemic cells. *Nagoya J. Medical Sci.* 45: 89-96.
25. Wakayama, K., Besa, E. C., and Baskin, S.I., 1984, Two components in L1210 cells and their growth characterization. *Nagoya J. Medical Sci.* 46: 125-142.
26. Weisberger, A.S., and Levine, B., 1954, Incorporation of radioactive L-cystine by normal and leukemic leukocytes in vivo. *Blood* 9: 1082- 1094.
27. Weisberger, A.S., Suhrland, L.G., and Griggs, R.C., 1954, Incorporation of radioactive L-cystine and L-methionine by leukemic leukocytes in vitro. *Blood* 9: 1095-1104.
28. Whitfield, D., 1987, literature review upon the toxicology, mechanism of action and treatment of sulphur and nitrogen mustard poisoning. Chemical Establishment Porton Down Salisbury Wilts, Technical Note No. 840, March.
29. Zar, J.H., 1996, Biostatistical Analysis, 3rd ed. Prentice Hall, Englewood Cliffs, N.J.
30. Zheng, S., Newton, Gonick, G., Fahey, R.C., and Ward, J.F., 1988, Radioprotection of DNA by thiols: relationship between the net charge on a thiol and its ability to protect DNA. *Radiation Research.* 114: 11-27.

IONOMYCIN RESTORES TAURINE TRANSPORTER ACTIVITY IN CYCLOSPORIN-A TREATED MACROPHAGES

[1]Ha Won Kim, [2]Eun Jin Lee, [3]Won Bae Kim and [2]Byong Kak Kim

[1]*Department of Life Science, University of Seoul, Seoul 130-743,* [2]*College of Pharmacy, Seoul National University, Seoul 151-742, and* [3]*Dong-A Pharmaceutical Company, Ltd., Kyunggi-Do 449-900, Korea*

Abstract: Taurine is accumulated at high concentrations in various tissues. The taurine transporter (TAUT) is responsible for the transportation of taurine in the cell. The transporter is affected by various stimuli to maintain cell volume. Macrophage cell volume varies in its activated states. In our experiment, it was found that the murine macrophage cell line, RAW264.7, expressed TAUT protein in its membrane. Its transporting activities could be blocked by β-amino acid such as β-alanine, but not by α-amino acids in this cell line. when assessed in RAW264.7 cells under the influence of immunosuppressive reagents, the activity of the TAUT was decreased by treatment with rapamycin (RM) or cyclosporin A (CSA). However, when ionomycin (IM) was added to this system, TAUT activity was recovered only in CSA-treated cells, in a concentration-dependent manner. in order to inhibit voltage gated Ca^{2+} channels, calmidazolium was added to the RAW264.7 cell line. Treatment of the cells with calmidazolium completely blocked TAUT. Furthermore, addition of IM to this system resulted in recovery the activity of TAUT again. When we added phorbol myristyl acetate (PMA) to the cell line, secretion of nitric oxide (NO) was increased 4-fold and the TAUT activity was decreased 5-fold. However, the addition of *N*-nitro L-arginine methyl ester (L-NAME), an inducible NO synthase (iNOS) inhibitor, to the PMA-treated cells induced recovery of TAUT activity. These results showed that the activity of TAUT was sensitive to both the intracellular concentrations of Ca^{2+} and NO.

Taurine 4, edited by Della Corte *et al.*
Kluwer Academic / Plenum Publishers, New York, 2000.

INTRODUCTION

Taurine (2-aminoethanesulfonic acid, $^+NH_3CH_2CH_2SO_3^-$) is a β-amino acid which is biosynthesized from cysteine in mammalian animals. It is an abundant intracellular free amino acid in mammalian tissues, and an energy-dependent, cell membrane-associated transport mechanism is responsible for the maintenance of high intracellular taurine concentration in most tissues. Such high concentrations of taurine in various tissues suggest that it exerts important functions, such as neuromodulation[1], antioxidation[2,3], and osmoregulation[4]. A taurine transporter (TAUT) has been cloned from human[5], mouse[6], rat[7], dog[8], and pig[9]. Murine TAUT was cloned from a mouse brain cDNA library. The deduced amino acid sequence was 590 amino acids with typical characteristics of sodium-dependent neurotransmitter transporters. TAUT is abundant in proximal tubules at which β-amino acids such as β-alanine, β-aminobutyric acid and gamma-aminobutyric acid (GABA) are actively reabsorbed through this transporter[1-13]. TAUT was found to be a Na^+/Cl^- co-transport system where the ratio of transportation of taurine:Na^+:Cl^- was $1:2:1$[13,14].

Protein kinase C plays central roles in the signal transduction and the regulation of transportation/absorption of nutrients. Inhibition of TAUT by stimulation of PKC was first reported by Kulanthaivel[15] in a placental choriocarcinoma cell line. Inhibition of TAUT due to the phosphorylation by PKC were reported in the kidney[16], colon carcinoma[17] and murine macrophage[18] cell lines. The blocking of TAUT could be affected by calmodulin[19]. Inhibition of TAUT by the immunosuppressive agent cyclosporin A could be reversed by the calmodulin antagonist W-7 or calmidazolium. However another immunosuppressive agent, FK506, does not affect TAUT activity[20]. Blocking of TAUT by PMA could be prevented by addition of the PKC inhibitor staurosporine,. However, this recovery was not affected by protein synthesis inhibitors, such as cycloheximide, actinomycin D, colchicine and cytochalasin D[17], suggesting that the TAUT was not induced, but expressed constitutively on the cell.

In this report we investigated the effect of immunosuppressive agents such as cyclosporin A and rapamycin on the TAUT activity, and found these agents blocked TAUT activity of murine macrophage. And we also found that ionomycin could reverse the blockade in cyclosporin-A, but not in rapamycin, treated cells.

MATERIAL AND METHODS

Cell Line and Chemicals

Macrophage RAW 264.7 cell line, an Abelson leukemia virus-transformed murine macrophage cell line, was obtained from the American Type Culture Collection (Rockville, MD, USA). Dulbecco's modified Eagle's medium (DMEM) and [2-^{3}H(N)]-Taurine (21.9 Ci/mmol) were bought from Gibco-BRL (Grand Island, NY) and NEN (Boston, MA), respectively. Fetal bovine serum (FBS), penicillin and streptomycin, dexamethasone (Dex), PMA, taurine, β-alanine, L-serine, L-leucine, rapamycin, EDTA, calmidazolium, staurosporine, ionomycin (IM) and N$^\omega$-nitro-L-arginine methyl ester (L-NAME) were obtained from Sigma Chemical (St. Louis, MO). Interferon-γ was obtained from R&D Systems (Minneapolis, MN, USA). Cyclosporin A (CsA, Sandoz) was kindly provided by Han-mi Pharmaceutical Co., Korea.

Cell Culture and Taurine Transport

RAW 264.7 cells were cultured and maintained in 90 mm culture dishes in DMEM medium containing 10% heat-inactivated fetal bovine serum (FBS) and penicillin (100 U/ml) /streptomycin (100 μg/ml), and in 6-well plates for taurine uptake experiments. Cultures were maintained at 37 °C in 5% CO_2 incubator until they reached confluence. Taurine transportation study was performed according to the previous report[21], i.e., when the cells were confluent on the dish, the medium was removed from the monolayer culture and replaced with 2 ml transporter buffer, which contained 25 mM-HEPES/Tris (pH 7.5), 140 mM NaCl, 5.4 mM KCl, 1.8 mM $CaCl_2$, 0.8 mM $MgSO_4$, 5 mM D-glucose. After stabilizing the cells for 30 min at 37 °C, 1 μCi of [^{3}H]-taurine was added to each culture and incubated for 30 min. Then the cells were washed thrice with ice-cold Na$^+$ deficient buffer and lyzed with 1 ml of 0.2 N NaOH/0.5% SDS. The lysate was mixed with 2 ml cocktail solution and its radioactivity was measured by a liquid scintillation counter.

Cell stimulation

For the stimulation of the RAW264.7 cells, confluent monolayer cells in DMEM containing 10% fetal bovine serum in transportation buffer were stimulated with each reagents for 30 min in 6-well plates. Then the cells were treated with 1 μCi of [^{3}H]-taurine for another 30 min. The cells were washed thrice with cold buffer followed by the lysis buffer. The transported radioactive taurine in the lysate was measured by γ-counter. For the analysis of nitric oxide, RAW cells were stimulated for 48 h and the resulting supernatants were obtained by centrifugation at 2,000 rpm. One hundred μl of the supernatant were transferred to ELISA titer plate and mixed with 100 μl of Griess reagent for 10 min at room temperature. An ELISA reader at 540 nm was used for the measurement of absorbance, together with standard sodium nitrite solutions.

RESULTS AND DISCUSSION

Expression of taurine transporter on raw cells

Murine macrophage cell line, RAW264.7, was used for the measurement of TAUT activity. Confluent cells on 6-well plate was pretreated with unlabelled α- or β-amino acids at a concentration of 1 mM for 10 min and followed by the addition of 1 μCi [³H]-taurine for another 30 min. Pretreatment of the cell with unlabelled taurine for 30 min led to blocking of radioactive taurine transport, to 4.8% of the control. Pretreatment with β-alanine also blocked taurine transport, to 12.9% of the control. However pretreatment of the cell with α-amino acids such as L-serine or L-serine did not affect taurine transportation (Table I). From these experiments murine macrophages were confirmed to express β-amino acid specific transporter on the cell membrane. As unlabelled taurine prevented radioactive taurine transport, this transporter seems to be TAUT. Furthermore, TAUT in the RAW cells was confirmed by the dependence of uptake on the presence of Na⁺ ions.

Table I. Substrate specificity of the carrier system for taurine uptake.
RAW264.7 cell line was stabilized with transporter buffer for 30 min in 6-well plates and then pretreated with the reagent for 10 min. Then 1 μCi [³H]-taurine was added for 30 min. After lysis, intracellular [³H]-taurine was measured by liquid scintillation counting

Pretreated reagent	[³H]-Taurine transportation in cpm and	%
PBS	391,412 ± 5,810	100.0%
1 mM Taurine	18,943 ± 83	4.8%
1 mM β-Alanine	50,543 ± 1,697	12.9%
1 mM L-Serine	400,575 ± 2,367	102.3%
1 mM L-Leucine	438,381 ± 1,125	111.9%

Inhibition of taurine transporter activity by rapamycin

RAW264.7 cell line was exposed to immunosuppressive agents, such as rapamycin and CsA, for 5 min and activity of the TAUT was then measured the. When we used 1, 10, 100 and 1,000 ng/ml rapamycin in this system, a concentration of 1 ng/ml reduced the TAUT activity to 11.9%, when compared to that of the control (Fig. 1). The blocking of the transporter could be induced within 5 min when the cell was cultured with rapamycin for 5, 30, 60 and 300 min.

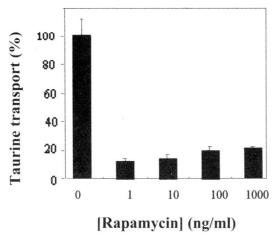

Fig 1. Effects of rapamycin on taurine transpoter activity in RAW264.7 cell line

When the same experiment was done on the CsA, the minimal concentration and time for the blocking of the TAUT were 5 nM and 5 min, respectively (data not shown). The blocking pattern of CsA was quite similar to that of the rapamycin suggesting that high intake of taurine would be required to maintain normal concentration of taurine during the period of taking immunosuppressive agents. This blocking phenomenon may be correlated with the inhibition of ATP-dependent taurocholate transportation by CsA, which results in the cholestasis as a side effect of CsA therapy[22]. Immunosuppressants, FK506 (tacrolimus) and CsA, were reported to inhibit the multidrug efflux transport in isolated rat hepatocytes[23] and HeLa cells[24].

Recovery of taurine transporter activity by ionomycin

As the intracellular concentration of Ca^{2+} is important for the activation of macrophages, RAW264.7 cells were co-treated with rapamycin and ionomycin (IM) simultaneously. The concentration of rapamycin was fixed at 1 ng/ml and the concentration of IM was increased 10-fold from 20 to

2,000 ng/ml. When the concentration of IM was increased in the rapamycin-treated cells, the recovery of the activity of TAUT was not altered by the addition of IM (Table II). However in the case of co-treatment with CsA and IM, increasing the concentration of IM in the CsA-treated cells resulted in recovery of TAUT activity (Fig. 2). Treatment of the RAW cell with 5 nM CsA reduced the taurine transport activity (17,268 ± 1,763 cpm) to 12.6% of that of the control (136,598 ± 3,112 cpm). When IM was added to this system at the concentrations of 20, 200 and 2,000 ng/ml, the TAUT activities were increased to 21.7% (29750 ± 9288 cpm), 59.4% (81,235 ± 134 cpm) and 74.1% (101,252 ± 1,357 cpm), respectively. This result indicates that increasing intracellular concentration of Ca^{2+} activates the TAUT.

To prove the involvement of intracellular Ca^{2+} on the recovery of TAUT activity, EDTA, a Ca^{+2} chelator, was added to the above system. When 800 nM EDTA was added to the cell systems that were treated with 5 nM CsA plus 20 ng/ml IM, 5 nM CsA plus 200 ng/ml IM, and 5 nM CsA plus 2,000 ng/ml IM, the corresponding TAUT activities were 12.8% (17,786 ± 1,244 cpm), 13.2% (18,333 ± 2,447 cpm) and 13.9% (19,294 ± 5,224 cpm) of the control (138,600 ± 2,510 cpm), respectively (Fig. 2). This result indicates that, after removal of free Ca^{2+} from the medium by EDTA, no recovery was observed by various concentrations of IM. Therefore intracellular Ca^{2+} was essential for the recovery of the TAUT activity that had been blocked by CsA. Our conclusion seems to be in conflict with the inactivation of taurine transporter by extracellular calcium reported by Kulanthaivel *et al.*[25]. The discrepancy may be due to the two reasons. First, in our experiment we measured the recovery of activity. So it may be different from the activity of the transporter itself. The second reason may be due to the difference of the cell lines used. The TAUT of murine macrophages may be different from that of the human placental brush border membrane.

Table II. Effects of ionomycin on taurine transporter activity in the rapamycin-treated RAW264.7 cell line.

RAW264.7 cell was stabilized with transportation buffer for 30 min and stimulated for 10 min before 1 μCi [^{3}H]-taurine was added for 30 min.

Stimulator	[^{3}H]-Taurine transportation in cpm and	%
PBS	200,431 ± 1,323	100.0%
RM (1 ng/ml)	21,465 ± 4,764	10.7%
RM (1 ng/ml) + IM (20 ng/ml)	21,281 ± 588	10.6%
RM (1 ng/ml) + IM (200 ng/ml)	23,469 ± 2,198	11.7%
RM (1 ng/ml) + IM (2,000 ng/ml)	25,356 ± 1,741	12.6%

To prove the involvement of calmodulin on the modulation of the TAUT activity, calmidazolium, an inhibitor of calmodulin, was added to the RAW cell. When 5 µg/ml of calmidazolium was added to the cell, taurine transportation activity was 3.9% of that of the control, indicating that calmodulin is responsible for the activity of TAUT. However when 20 ng/ml of IM was added to this system, TAUT activity was recovered to 64% of that of the control, suggesting involvement of intracellular Ca^{2+} in the recovery of taurine transporter activity (Table III).

Table III. Effects of calmodulin antagonists on taurine transporter activity in the RAW264.7 cell line.
RAW264.7 cells were stabilized with transportation buffer for 30 min and stimulated for 12 h. Then 1 µCi [^{3}H]-taurine was added for30 min. Final concentrations of CsA, calmidazolium and IM were 5 nM, 5 µg/ml and 20 ng/ml, respectively.

Stimulator	[^{3}H]-Taurine transportation in cpm and %	
PBS	$300,693 \pm 3,124$	100.0%
CsA	$11,379 \pm 978$	3.7%
Calmidazolium	$11,786 \pm 513$	3.9%
CsA + Calmidazolium	$11,542 \pm 1,003$	3.8%
Calmidazolium + IM	$192,543 \pm 2,668$`	64.0%

Nitric oxide blocks taurine transporter activity

Activation of protein kinase C (PKC) was known to inhibit TAUT activity. Addition of 10 ng/ml of PMA to the RAW cell reduced TAUT activity to 19.4% of that of the control. When L-NAME, inhibitor of NOS, was added to this culture system, TAUT activity recovered to 60.7% of that of the control, indicating that blocking of the production of NO could induce recovery of TAUT activity.

ACKNOWLEDGMENTS

This research was supported in part by the grant of University of Seoul (year of 1998 entitled of studies on the function of taurine transporter) and a grant of Korea Science and Engineering Foundation.

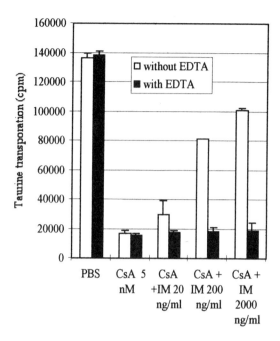

Fig 2. Effects of pretreatment with the Ca^{2+} chelator, EDTA on taurine transporter activity in CsA and IM treated cells.

REFERENCES

1. Oja ,S.S., and Kontro, P., 1990, Neuromodulatory and trophic actions of taurine. *Prog. Clin. Biol. Res.* **351**: 69-76.
2. Sawamura, A., Azuma, J., Awata, N., Harada, H. and Kishimoto, S., 1990, Modulation of cardiac Ca^{++} current by taurine. *Prog. Clin. Biol. Res.* **351**: 207-215.
3. Gordon, R.E., Heller, R.F., and Heller, R.F., 1992, Taurine protection of lungs in hamster models of oxidant injury: a morphologic time study of paraquat and bleomycin treatment. *Adv. Exp. Med. Biol.* **315**: 319-328.
4. Koyama, I., Nakamura, T., Ogasawara, M., Nemoto, M., and Yoshida, T., 1992, The protective effect of taurine on the biomembrane against damage produced by the oxygen radical. *Adv. Exp. Med. Biol.* **315**: 355-359.
5. Jhiang, S.M., Fithian, L., Smanik, P., McGill, J., Tong, Q., and Mazzaferri, E.L., 1993, Cloning of the human taurine transporter and characterization of taurine uptake in thyroid cells. *FEBS Lett.* **318**: 139-144.
6. Liu, Q.R., Lopez-Corcuera, B., Nelson, H., Mandiyan, S., and Nelson, N., 1992, Cloning and expression of a cDNA encoding the transporter of taurine and beta-alanine in mouse brain. *Proc. Natl. Acad. Sci. USA* **89**: 12145-12149.
7. Smith, K.E., Borden, L.A., Wang, C.H., Hartig, P.R., Branchek, T.A. and Weinshank, R.L., 1992, Cloning and expression of a high affinity taurine transporter from rat brain. *Mol. Pharmacol.* **42**: 563-569.
8. Uchida, S., Kwon, H.M., Yamauchi, A., Preston, A.S., Marumo, F., and Handler, J.S., 1992, Molecular cloning of the cDNA for an MDCK cell Na(+)- and Cl(-)-dependent

taurine transporter that is regulated by hypertonicity. *Proc. Natl. Acad. Sci. USA* **89**: 8230-8234.

9. Han, X., Budreau, A.M., and Chesney, R.W., 1998, Molecular cloning and functional expression of an LLC-PK1 cell taurine transporter that is adaptively regulated by taurine. *Adv. Exp. Med. Biol.* **442**: 261-268.

10. Chesney, R.W., Gusowski, N., Dabbagh, S., Theissen, M., Padilla, M., and Diehl, A., 1985, Factors affecting the transport of beta-amino acids in rat renal brush-border membrane vesicles. The role of external chloride. *Biochim. Biophys. Acta.* **812**: 702-712.

11. Chesney, R.W., Gusowski, N., Zeilkovic, I., and Padilla, M., 1986, Developmental aspects of renal beta-amino acid transport. V: Brush border membrane transport in nursing animals-effect of age and diet. *Pediatr. Res.* **20**: 890-894.

12. Chesney, R.W., Gusowski, N., and Zelikovic, I., 1987, Developmental aspects of renal beta-amino acid transport. VI. The role of membrane fluidity and phospholipid composition in the renal adaptive response in nursing animals. *Pediatr. Res.* **22**: 163-167.

13. Chesney, R.W., Zelikovic, I., Friedman, A.L., Dabbagh, S., Lippincott, S., Gusowski, N., and Stjeskal-Lorenz, E., 1987, Renal taurine transport-recent developments. *Adv. Exp. Med. Biol.* **217**: 49-59.

14. Zelikovic, I., Stejskal-Lorenz, E., Lohstroh, P., Budreau, A., and Chesney, R.W., 1989, Anion dependence of taurine transport by rat renal brush-border membrane vesicles. *Am. J. Physiol.* **256** (4 Pt 2): F646-655.

15. Kulanthaivel, P., Cool, D.R., Ramamoorthy, S., Mahesh, V.B., Leibach, F.H., and Ganapathy, V., 1991, Transport of taurine and its regulation by protein kinase C in the JAR human placental choriocarcinoma cell line. *Biochem. J.* **277**: 53-58.

16. Jones, D.P., Miller, L.A., Dowling, C., Chesney, R.W., 1991, Regulation of taurine transporter activity in LLC-PK1 cells: role of protein synthesis and protein kinase C activation. *J. Am. Soc. Nephrol.* **2**:1021-1029.

17. Brandsch, M., Miyamoto, Y., Ganapathy, V. and Leibach, F.H., 1993, Regulation of taurine transport in human colon carcinoma cell lines (HT-29 and Caco-2) by protein kinase C. *Am. J. Physiol.* **264** (5 Pt 1): G939-946.

18. Kim, H. W., Shim, M. J., Kim, W. B., and Kim, B. K., 1995, Regulation of taurine transporter activity by glucocorticoid hormone. *J. Biochem. Mol. Biol.* **28**: 527-532.

19. Ramamoorthy, S., Del Monte, M.A., Leibach, F.H., and Ganapathy, V., 1994, Molecular identity and calmodulin-mediated regulation of the taurine transporter in a human retinal pigment epithelial cell line. *Curr. Eye Res.* **13**: 523-529.

20. Ramamoorthy, S., Leibach, F.H., Mahesh, V.B., and Ganapathy, V., 1992, Selective impairment of taurine transport by cyclosporin A in a human placental cell line. *Pediatr. Res.* **32**: 125-127.

21. Kim, H.W., Lee, E.J., Shim, M.J., and Kim, B.K., 1998, Effects of steroid hormones and cyclosporine A on taurine-transporter activity in the RAW264.7 cell line. *Adv. Exp. Med. Biol.* **442**: 247-254.

22. Bohme, M., Jedlitschky, G., Leier, I., Buchler, M. and Keppler, D., (1994), ATP-dependent export pumps and their inhibition by cyclosporins. *Adv. Enzyme Regul.* **34**: 371-380.

23. Takeguchi, N., Ichimura, K., Koike, M., Matsui, W., Kashiwagura, T. and Kawahara, K., 1993, Inhibition of the multidrug efflux pump in isolated hepatocyte couplets by immunosuppressants FK506 and cyclosporine. *Transplantation* **55**: 646-650.

24. Kirk, J. and Kirk, K., 1994, Inhibition of volume-activated I- and taurine efflux from HeLa cells by P-glycoprotein blockers correlates with calmodulin inhibition. *J. Biol. Chem.* **269**: 29389-29394.

25. Kulanthaivel, P., Miyamoto, Y., Mahesh, V.B., Leibach, F.H. and Ganapathy, V., 1991, Inactivation of taurine transporter by calcium in purified human placental brush border membrane vesicles. *Placenta* **12**: 327-340.

DAILY DIETARY TAURINE INTAKE IN JAPAN

E. Kibayashi[1,2], H. Yokogoshi[3], H. Mizue[2], K. Miura[4], K. Yoshita[4], H. Nakagawa[4], Y. Naruse[5], S. Sokejima[1] and S. Kagamimori[1]

[1]Department of Life and Culture, Sonoda Woman's College, Amagasaki, Japan, [2]Department of Welfare Promotion and Epidemiology, Toyama Medical and Pharmaceutical University, Toyama, Japan, [3]School of Food and Nutritional Sciences, University of Shizuoka, Shizuoka, Japan, [4]Department of Public Health, Kanazawa Medical University, Kanazawa, Japan, [5]Department of Community Health and Gerontological Nursing, Toyama Medical and Pharmaceutical University, Toyama, Japan

INTRODUCTION

Although fish and shellfish intake has gradually decreased, the Japanese people still have the highest intake of fish and shellfish in the world. Japanese people have the lowest incidence of ischemic heart disease in the world, and the urinary taurine excretion was the highest in the world in the CARDIAC Study. In that study, there was a significant correlation between urinary taurine excretion and fish and shellfish intakes. Taurine can be biosynthesized from sulfur-containing amino acids such as methionine or cysteine, and these taurines are excreted in the urine. Thus, the urinary taurine excretion may not reflect dietary taurine intake. It is difficult to calculate dietary taurine intakes from dietary research, due to the lack of inclusion of taurine in standard tables of food composition in Japan.

The aim of this study was to develop taurine composition tables of foods, and to calculate according to these tables the daily dietary taurine intake using a 24-hour dietary recall method. The final aim was to evaluate the validity and reliability of the developed taurine composition tables of food.

Taurine 4, edited by Della Corte et al.
Kluwer Academic / Plenum Publishers, New York, 2000.

SUBJECTS AND METHODS

Subjects

The subjects were 163 males and 161 females aged between 20-59 in Toyama, Japan. Japan is divided into 9 districts, Toyama City where the subjects are living is situated in the Hokuriku district. On the Japanese map, Toyama City is marked with a star (Fig 1). According to the Ministry of Health and Welfare in Japan, fish and shellfish intake in the Hokuriku district is the third highest among the 9 districts in Japan. The highest fish and shellfish intake is in the Hokkaido district, and the second highest is the Tohoku district. The bar graph shows the difference from average of dietary fish and shellfish intake and dietary meat intake in Japan. The centerline shows the average in Japan. The 3 districts with the highest fish and shellfish intake are just above the average.   However, the dietary meat intakes of the highest 3 districts are below the average. On the other hand, in the Kyusyu district with the lowest fish and shellfish intake, dietary meat intake was the highest in Japan. However, fish and shellfish intake in this district was higher than meat intake.

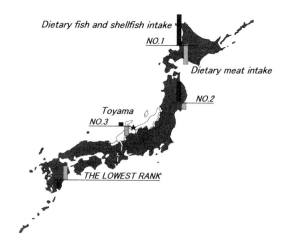

Figure 1. Dietary fish and shellfish intake for one day in Yoyama, Japan (Ministry of Health and Welfare, 1998)

Dietary Survey Methods

The dietary data were collected from one 24-hour dietary recall. The dietary interviewers were dieticians.

Taurine Composition Tables

In preparing taurine composition tables of animal food and algae, data on 91 foods containing taurine were collected from reference papers. Also, 51 processed foods and others were analyzed for taurine content (mg /100g).

In this analysis, food samples were first homogenized with 3% sulfosalicylic acid solution, then each homogenate was centrifuged to obtain the supernatant for an amino acid analysis. An automatic amino acid analyser determined the concentrations of the amino acid.

Propriety of the Taurine Composition Tables

The 24-hour dietary taurine intake was analyzed from 30 dietary recall data by a random sampling method in order to confirm the propriety of the taurine composition tables. The relationship between the results of analysis and the results of calculation by taurine composition tables of animal food and algae were compared in a random sample of 30 dietary recall data.

All analyses were carried out by the SPSS for Windows statistical package.

CORRELATION BETWEEN DIETARY TAURINE INTAKE, CALCULATED OR MEASURED IN 30 RECALLED DIETS

The correlation diagram of Figure 2 shows the relation between the calculated values of dietary taurine intake and the measured ones in order to evaluate the validity and reliability and confirm the propriety of the taurine composition tables. There was a significant correlation between both values of dietary taurine intake. However, we need to elucidate why intake values of calculated taurine intake were lower than values of measured taurine.

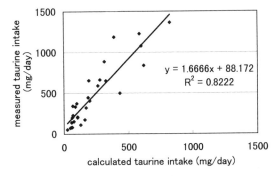

Figure 2. Correlation between values of dietary taurine intake, calculated or measured in 30 recalled diets.

PHYSICAL AND NUTRITIONAL STATUS OF SUBJECTS

Table 1 shows the mean of body mass index of the subjects, and for nutritional status of the subjects shows mean of energy intake, protein intake, animal protein intake, fish and shellfish intake and meat intake, respectively. Animal protein intake is about half the total protein intake, and fish and shellfish intake is higher than meat intake. This is standard in Japan. Our survey had a similar result.

Table 1. Physical and nutritional status of subjects

	Total	Males	Females
Sample size	324	163	161
BMI	22.8 ± 3.17	23.1 ± 3.09	22.4 ± 3.21
Energy intake (kcal)	2042 ± 606	2225 ± 629	1857 ± 522
Protein intake (g)	75.1 ± 24.3	82.8 ± 25.7	67.4 ± 20.2
Animal protein intake (g)	36.5 ± 18.3	42.5 ± 20.3	30.5 ± 13.7
Fish and shellfish intake (g)	89.7 ± 75.0	107.7 ± 87.8	71.6 ± 53.6
Meat intake (g)	60.6 ± 56.1	69.9 ± 61.0	51.2± 49.0

Data are means ± SD

DIETARY TAURINE INTAKE FOR ONE DAY

The histogram of Figure 3 shows dietary taurine intake for one day, calculated using the taurine composition tables of foods. A large number of the subjects were concentrated around the 100-200mg/day range for daily taurine intake. There were also many subjects with a higher taurine intake distributed between 300 and 600mg/day. The highest intakes were over 1000mg/day. The mean of all subjects was 194.2mg/day. Males were 225.5, and females were 162.6mg/day.

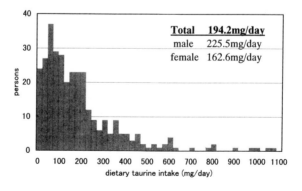

Figure 3. Dietary taurine intake for one day

CORRELATION BETWEEN DIETARY TAURINE INTAKE AND FISH AND SHELLFISH INTAKE

Fish and shellfish are rich in taurine. Therefore, the relation between fish and shellfish intake and taurine intake is shown in this correlation diagram of Figure 4. There was a significant correlation between fish and shellfish intake and taurine intake.

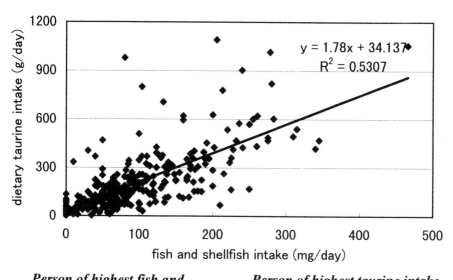

$$y = 1.78x + 34.137$$
$$R^2 = 0.5307$$

Person of highest fish and shellfish intake		*Person of highest taurine intake*	
ID 124	male	ID 144	male
BMI	23.0	BMI	17.6
Taurine intake	1055 mg/day	Taurine intake	1089 mg/day
fish and shellfish intake	465 g	fish and shellfish intake	205 g
meat intake	0 g	meat intake	0 g
energy intake	3366 kcal	energy intake	1560 kcal
protein intake	132 g	protein intake	54.9 g

Figure 4. Correlation between dietary taurine intake and fish and shellfish intake

Below the correlation diagram in Figure 4 are data for the person with the highest fish and shellfish intake and the person with the highest taurine intake. Dietary fish and shellfish intake for the latter person is about half of that of the person with the highest fish and shellfish intake, and energy intake is also lower. The reason may be due to the latter eating more oysters, in which taurine concentration is high.

CORRELATION BETWEEN DIETARY TAURINE INTAKE AND MEAT INTAKE

The relation between dietary meat intake and dietary taurine intake is shown in Figure 5. There was no significant correlation between meat intake and taurine intake. The data of the subject with highest meat intake is also shown in Figure 5. This female had higher energy intake and protein intake than others.

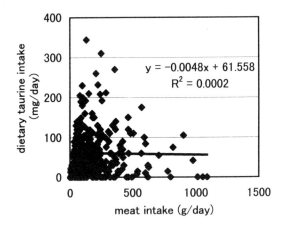

Person of highest meat intake

ID 309	female
BMI	26.6
Taurine intake	133 mg/day
fish and shellfish intake	0 g
meat intake	344 g
energy intake	4262 kcal
protein intake	143 g

Figure 5. Correlation between dietary taurine intake and meat intake

CONCLUSION

Our data indicate that dietary taurine intake calculated from 24-hour dietary recalls using the taurine composition tables of foods and using dietary fish and shellfish intake calculated from 24-hour dietary recalls may be useful indexes of dietary taurine intake for nutritional epidemiological studies. Further examination will be required.

COMPARATIVE STUDIES ON SERUM TAURINE AND PLASMA FATTY ACIDS IN HUMANS BETWEEN THE SEA SIDE AREA IN TOYAMA, JAPAN AND THE MOUNTAIN AREAS IN INNER MONGOLIA, CHINA

E. Kibayashi, M. Zhang, Z.Y. Liu, M. Sekine, S. Sokejima, and S. Kagamimori
Department of Welfare Promotion & Epidemiology, Toyama Medical & Pharmaceutical University, Toyama, Japan

INTRODUCTION

The incidence of ischemic heart disease in Japan is the lowest of the developed countries. The urinary taurine excretion was the highest in the world in the CARDIAC Study[1]. In that study, there was a significant correlation between urinary taurine excretion and fish and shellfish intake. Urinary taurine excretion level was also related to mortality from ischemic heart disease.

The purpose of this study was to investigate the relationship between taurine level and biochemical markers related to atherosclerosis. The relationship between serum taurine level and urinary taurine excretion was compared between the sea side area in Toyama, Japan, where people often eat fish and shellfish and three mountain districts in Inner Mongolia, China, where the people almost never eat seafood.

METHODS

The subjects were 100 females aged 35-49 years (average age of 40.1±3.6) in Japan and China. Twenty-five females were selected in Toyama City, Japan, which is located near the sea. The remainders of subjects were

Taurine 4, edited by Della Corte *et al.*
Kluwer Academic / Plenum Publishers, New York, 2000.

selected in Inner Mongolia, China, which is located in a mountain district. They are 25 females in Fuhhot City (city area), 25 females in Tuzhuoqi (agricultural area) and 25 females in the periphery of Tamaoqi (grassland area).

BMI (weight kg/height, m^2) and blood pressure were measured in the people in Inner Mongolia. Blood samples and morning urine samples were collected from July to August, 1996. In the mean while, they were questioned about their lifestyle and dietary habits. The participants of Toyama City were examined in the same way in September, 1996.

The serum, plasma and urine were quickly frozen at $-30°C$ until analysis. The urinary samples were analyzed for taurine and glycine. Urinary taurine concentration was adjusted by both urinary glycine and urinary creatinine. The serum samples were analyzed for taurine, glycine, lipoperoxide (thiobarbituric acid method) and cholesterol. The plasma samples were analysed for fatty acids. Serum taurine concentration was also adjusted by serum glycine.

All analyses were carried out using SPSS statistical package (Norussis, 1993). Biochemical data among the four districts were compared using one-way analysis of variance (ANOVA) for continuous variables. Bonferroni t-test was used for multiple comparisons. Pearson correlation coefficient was calculated to evaluate the relationship between serum or urinary taurine to glycine ratios and biochemical markers related to atherosclerosis.

COMPARISON OF TAURINE IN SERUM AND URINE

Comparison of taurine (Tau)/glycine (Gly) ratios in serum and urine among Toyama City and three areas in Inner Mongolia is shown in Table 1. In Toyama city, serum Tau/Gly ratio was significantly higher than that of city and grassland areas in Inner Mongolia ($p<0.05$ for both). There was no significant difference in serum Tau/Gly ratio between Toyama City and the agricultural area in Inner Mongolia. Urinary Tau/Gly ratio was not significantly different among the four areas.

COMPARISON OF PLASMA N-6 FATTY ACIDS AND PLASMA N-3 FATTY ACIDS

Comparison of plasma n-6 fatty acids concentration and plasma n-3 fatty acids concentration between Toyama City and three areas in Inner Mongolia are shown in Table 2. Plasma n-6 fatty acids concentration in both Toyama City and City area in Inner Mongolia was significantly higher than that of agricultural area and grassland area in Inner Mongolia. Plasma n-3 fatty acids concentration in Toyama City was significantly higher than that of the three areas in Inner Mongolia. Plasma concentration of eicosapentanenoic acid (EPA) and docosahexaenoic acid (DHA)[2] in Toyama City were significantly higher than those of the three areas in Inner Mongolia.

Table 1. Comparison of serum taurine and urinary taurine between Toyama City and three areas in Inner Mongolia

	Toyama City	Inner Mongolia			ANOVA
		City area	Agricultural area	Grassland area	
Serum Taurine/Glycine ratio	0.36 ± 0.46	0.11 ± 0.05	0.18 ± 0.08	0.15 ± 0.11	p<0.01
Urinary Taurine/Glycine ratio	1.09 ± 1.00	0.66 ± 0.37	0.63 ± 0.35	1.02 ± 0.85	ns

Each value represents the mean $\pm$ S.D.   Bonferroni t-test *p<0.05.

COMPARISON OF BIOCHEMICAL MARKERS RELATED TO ATHEROSCLEROSIS

The serum lipoperoxide in Toyama City was significantly higher than that of the three areas in Inner Mongolia.

Serum total cholesterol concentration and blood pressure were not significantly different among 4 areas (Table 2).

CORRELATION BETWEEN TAURINE AND BIOCHEMICAL MARKERS RELATED TO ATHEROSCLEROSIS

Correlations between taurine and biochemical markers related to atherosclerosis in Toyama City and Inner Mongolia are showed in Table 3. Biochemical data obtained from four areas were combined in the analysis of correlation between taurine and biochemical markers for atherosclerosis.

Table 2. The comparison of related factors of atherosclerosis among Toyama City and three areas in Inner Mongolia

	Toyama City (n=25)	Inner Mongolia City area (n=25)	Agricultural area (n=25)	Grass land area (n=25)	ANOVA
Plasma Fatty Acids (µg/ml)					
n-6	1049±210	1145±314	892.5±175	828.8±193	p<0.001
n-3	199±64.6	120±38.7	151±68.1	104±33.7	p<0.001
EPA	47.9±23.2	14.2±8.49	28.4±19.8	21.4±8.70	p<0.001
DHA	114±35.2	40.8±8.13	42.4±13.9	35.5±8.56	p<0.001
n-6 / n-3 ratio	5.63±1.52	10.2±3.47	6.49±1.69	8.23±1.49	p<0.001
Serum Cholesterol (mg/dl)					
Total cholesterol	174±29.1	177±22.5	157±36.7	177±26.9	ns
HDL cholesterol	57.2±14.5	51.2±8.71	43.5±7.96	55.2±9.32	p<0.001
Total / HDL ratio	2.22±1.00	3.57±0.72	3.70±1.02	3.22±0.63	ns
Serum Lipoperoxide (nmol/ml)	3.16±0.93	2.15±0.70	1.78±0.79	2.22±1.09	p<0.001
Blood Pressure (mmHg)					
DBP	74.0±9.98	75.8±7.66	83.0±12.2	75.2±11.1	ns
SBP	113±10.6	111±15.9	124±18.3	107±15.1	ns

Each value represents the mean±SD. Bonferroni t-test *p<0.05.

There was a significant correlation between serum Tau/Gly ratio and plasma n-3 fatty acids (r=0.30, p<0.01). There was also a significant correlation between serum Tau/Gly ratio and plasma EPA (r=0.23, p<0.05) and DHA (r=0.45, p<0.001). For urinary Tau/Gly ratio, there was no correlation with plasma n-3 fatty acids, EPA and DHA.

Table 3. Correlation coefficients between taurine and biochemical markers related to atherosclerosis

		Serum Taurine/Glycine ratio (n=100)	Urine Taurine/Glycine ratio (n=100)
Taurine/Glycine ratio			
	Serum		0.03
	Urine	0.03	
Plasma Fatty Acids (μg/ml)			
	n-6 / n-3 ratio	- 0.19	-0.03
	n-6	0.13	0.05
	n-3	0.30**	0.06
	EPA	0.23*	0.15
	DHA	0.45***	0.13
Serum Cholesterol (mg/dl)			
	Total cholesterol	0.05	0.14
	HDL cholesterol	0.24*	0.17
	Total / HDL ratio	- 0.14	-0.07
Serum Lipoperoxide (nmol/ml)		0.19	0.10
Blood Pressure (mmHg)			
	DBP	0.08	-0.08
	SBP	0.08	-0.21*

*p<0.05; ** p<0.01; *** p<0.001

For serum HDL cholesterol, there was a significant correlation with serum Tau/Gly ratio (r=0.24, p<0.05).

There was no correlation between serum Tau/Gly ratio and diastolic or systolic blood pressure. However, there was a significant negative correlation between urinary Tau/Gly ratio and systolic blood pressure (r=-0.21, p<0.05). Diastolic blood pressure showed no correlation with urinary Tau /Gly ratio.

CONCLUSION

Serum taurine/glycine ratio was associated with biochemical markers related to atherosclerosis. Serum taurine/glycine ratio was highest in Toyama City. Therefore, there is a possibility that daily dietary intake of taurine derived from dietary intake of fish and shellfish might contribute to the low incidence of ischemic heart disease in Japan, through the change of serum lipid metabolism, including the increase in n-3 fatty acids and HDL cholesterol.

REFERENCES

1. Yamori Y., Nara Y., Mizushima S., Mano M., Sawamura M., Kihara M., and Horie R., 1992, *Nutrition and Health,* 8: 77-90.
2. Wlliam E.M.L., 1986, *Fish and Human Health,* Academic Press, pp 96-117.

BIOCHEMICAL AND ULTRASTRUCTURAL ALTERATIONS IN RAT AFTER HYPEROXIC TREATMENT: EFFECT OF TAURINE AND HYPOTAURINE

Giuseppina Pitari[1], Silvestro Duprè[2], Anna Maria Ragnelli[1], Pierpaolo Aimola[1], Camillo Di Giulio[3] and Fernanda Amicarelli[1]

[1]Dipartimento di Biologia di Base ed Applicata, Università di L'Aquila, L'Aquila, Italy; [2]Dipartimento di Scienze Biochimiche e Centro di Biologia molecolare del CNR, Università di Roma "La Sapienza", Roma, Italy; [3]Dipartimento di Scienze Biomediche, Università "G. D'Annunzio", Chieti, Italy.

Abstract: The cell ultrastructure and some detoxifying enzyme activities were studied in skeletal muscles of young rats kept for 84 h under normobaric hyperoxia (95% O_2) or normoxia as control. Rat were injected ip. Every 12 h either with 1ml saline, 1 ml saline+30 mg hypotaurine or 1 ml saline+30 mg taurine. Ultrastructural observation revealed an highly protective effect on tissue damages due to hyperoxia in taurine-treated rats and, at less extent, in hypotaurine-treated ones. Enzymatic assays suggest a different mechanism of the two molecules in their protective action.

INTRODUCTION

We studied under normobaric hyperoxic conditions the effect of hypotaurine and taurine on young rats. Taurine exerts high protective effects on biological structures by specifically reacting with hypochlorite[11]. Taurine has protective effects in many other systems and its action can be ascribed to a membrane stabilizing action[11]. Hypotaurine is known to protect liposome, isolated lipoprotein[19] and cells[8] from oxidative damages.

Hyperoxic normobaric conditions are model systems to study oxidative stress: it is well documented[1,10,16] that the generation of reactive oxygen species (ROS) causes tissue damage. Hyperoxic normobaric treatment damages animals, plants and aerobic bacteria[4,9]. Powerful scavenging enzyme systems participate to protect the organism: superoxide dismutase (SOD, EC

Taurine 4, edited by Della Corte et al.
Kluwer Academic / Plenum Publishers, New York, 2000.

1.15.1.1.), catalase (EC 1.11.1.6), glutathione reductase (GSSG Rx, EC 1.6.4.27), glutathione peroxidase (GSH Px, EC 1.11.1.9), glutathione trans- ferase (GST, EC 2.5.1.18) and others. The response to oxidative stress of some protective enzymes has been studied[1,2,13] in rats and different behav- iours were found depending on the age of animals.

In this paper we studied the protective effect of hypotaurine and taurine in skeletal muscles of young rats under normobaric normoxic and hyperoxic conditions. The specific activities of some detoxifying enzymes were deter- mined and the ultrastructure of the tissue was observed.

MATERIALS AND METHODS

Young female rats (200-230 g) were divided into 6 groups of three rats each, treated for 84 h, and fed *ad libitum*. The three control groups were kept in normobaric normoxic conditions and received: 1 ml saline (control group); 30 mg of hypotaurine in 1 ml saline; or 30 mg of taurine in 1 ml sa- line i.p. every 12 h. The other groups were kept in normobaric hyperoxic conditions (95% O_2) in a large plexiglass chamber and received the same above treatments. After 84 h rats were killed and skeletal muscles rapidly excised and frozen at once in liquid nitrogen. Some pieces of the tissue were fixed for ultrastructural studies.

Enzyme activities were determined on cytosol of all tissues. Homoge- nates were obtained by using 10 mM phosphate buffer, pH 7.0, containing 1 mM dithiothreitol for GSH peroxidase and GSSG reductase assays and 1 mM Triton X-100 for catalase and SOD assays.

Superoxide dismutase activity was determined as described by Sun and Zigman[18]. The inhibitory effect of superoxide dismutase in samples on the autoxidation of epinephrine (0.1 M) in 50 mM carbonate buffer, pH 10.0 was assayed spectrophotometrically at 480 nm. One unit was defined as the amount of enzyme required to halve the rate of autoxidation.

Catalase activity was measured according to Luck[12]. The reduction of H_2O_2 was followed spectrophotometrically at 240 nm. One unit was defined as 1 μmol of H_2O_2 reduced/min.

Glutathione peroxidase activity was measured as described by Paglia and Valentine[14] as modified by Di Ilio *et al.*[7] The activity of Se-dependent GSH- Px was measured in presence of H_2O_2 (0.25 mM) and glutathione; the oxida- tion of NADPH was followed at 340 nm. One unit was defined as 1 μmol of glutathione oxidized/min.

Glutathione reductase was measured as described by Di Ilio *et al.*[6] The oxidation of NADPH (0.16 mM) in phosphate buffer, pH 7.4 in the presence of 1 mM oxidized glutathione, 1 mM EDTA was followed at 340 nm. One unit was defined as 1 μmol of NADPH oxidized/min.

For electron microscopy, pieces of muscle were fixed with 3% glutaraldehyde in 0.1 M cacodylate buffer, pH 7.2 and postfixed in 1% osmium tetroxide. Dehydration was performed in ethanol series and embedding in Durcupan ACM resin. Ultrathin sections were stained with 5% uranyl acetate in 70% ethanol and lead citrate and observed by a JEOL JEM 100 C electron microscope.

RESULTS

Tissue Ultrastructure

After treatment normoxic rats survived well, hyperoxic rats suffered and one of the saline-treated group died. Hyperoxic treatment caused tissue damage in the untreated group, with diffused hematoma in liver and lungs and edema in the thorax. Hypotaurinetreated hyperoxic rats and, to a lesser extent, taurine-treated ones appeared normal from the anatomic point of view.

It is well known that hyperoxic treatment causes severe damage to mitochondria and membrane derangement. In our study, hyperoxic conditions gave the same extensive damage previously observed in rat skeletal muscle[2]. Both taurine and hypotaurine seem to protect against hyperoxic damage: taurine- and hypotaurine-treated hyperoxic skeletal muscles appear ultrastructurally similar to the normoxic one; the taurine protective effect is greater (Fig. 1) compared to the protection effected by hypotaurine were some mitochondria appeared damaged.

Figure 1. Electron micrographs of longitudinal sections of rat skeletal muscle after hyperoxic treatment without (A) and with (B) taurine. m, mitochondria; my, myofibrillar elements. Bar: 1 μm.

Detoxifying Enzymes

Table 1 shows the levels of catalase, superoxide dismutase, Se-dependent GSH peroxidase and GSSG reductase, together with catalase/SOD, GSH peroxidase/SOD and GSSG reductase/GSH peroxidase ratios in all the considered conditions.

SOD assay on control muscles evidenced a large increase of this specific activity in hyperoxic conditions not paralleled by an increase in catalase activity. Taurine treatment significantly decreases SOD specific activity in hyperoxic conditions and catalase levels both in normoxia and hyperoxia compared with the control values. After hyperoxic treatment, catalase/SOD ratio results lowered in the control and hypotaurine groups. The same behaviour cannot be observed in taurine groups in which this ratio results three times higher in hyperoxia respect to normoxia.

Table 1. Enzymatic specific activities of some detoxifying enzymes (mU/mg)

Enzyme specific activities mU/mg	Normoxia	Hyperoxia	Normoxia + taurine	Hyperoxia + taurine	Normoxia + hypo-taurine	Hyperoxia + hypo-taurine
SOD	1523 ± 66	3919 ± 142	2304 ± 103	1025 ± 66	2053 ± 99	2412 ± 56
Catalase	3517 ± 196	2453 ± 80	1938 ± 260	2250 ± 35	3834 ± 31	1657 ± 79
Catalase / SOD	2.31	0.62	0.84	2.19	1.87	0.68
GSH Px	15.2 ± 0.7	7.7 ± 0.4	11.1 ± 0.3	8.8 ± 0.9	18.3 ± 0.6	10.5 ± 0.6
GSH Px/ SOD	0.010	0.002	0.048	0.085	0.009	0.004
GSSG Rx	7.7 ± 0.5	6.1 ± 0.3	6.4 ± 0.3	6.2 ± 0.2	6.2 ± 0.1	< 0.6
GSSG Rx/Px	0.50	0.79	0.58	0.71	0.34	< 0.06

Each value is the mean of at least 5 replicates ± SEM.

The levels of GSH Px decrease in hyperoxic conditions in all the considered treatments. In hypotaurine groups, GSH Px activity is higher with respect to the control. It is also higher when compared to the corresponding taurine groups, both in normoxic and hyperoxic conditions. GSH peroxidase/SOD ratio after hyperoxia markedly decreases in the control and hypotaurine treated rats, while it doubles in taurine groups.

The levels of GSSG reductase activity do not significantly change under any condition but hypotaurine-treated hyperoxic rats, where this activity is undetectable. As a consequence, GSSG reductase/GSH peroxidase ratio increases in hyperoxia in the control and taurine-treated rats but is negligible in hypotaurine-treated hyperoxic rats.

DISCUSSION

Tissue Ultrastructure

The presence of high concentration of oxygen is toxic for animals, plants and aerobic bacteria[49]. This toxicity is not due only to oxygen *per se* but more to the production of highly reactive reduced products from oxygen. In hyperoxic conditions the production of ROS may be so high as to exceed the natural scavenging defences of the organism and a large variety of damage can be observed.

In our experiments the damages caused by hyperoxia can be easily observed at the anatomic level. All organs appear damaged: liver and lung show large hematomas and there is edema in the thoracic chest. In hyperoxic rats treated with taurine and hypotaurine, however, all tissues appear normal and well conserved from the anatomic point of view. These effects are more evident in hypotaurine-treated rats. These findings suggest that both taurine and hypotaurine have important roles in protecting tissues from ROS damages This is not surprising, considering the well-studied scavenging activity of taurine[11] and hypotaurine[8,15,19].

In hyperoxic conditions (60 h at 95% oxygen) rat skeletal muscles appear severely damaged. In particular, ultrastructural studies revealed a massive mitochondria degeneration, larger in old than young rats[2]. Figure 1 shows the micrographs of rat hyperoxic skeletal muscle without and with taurine. Muscle ultrastructure similar to the normal one with well-preserved mitochondria is observed. Hypotaurine-treated muscle appears less protected as some mitochondria are damaged (data not shown). Thus, both molecules exert a protective effect on tissue damage, probably by maintaining membrane integrity. It is known that taurine is a stabilizer for cellular membranes[11]; for hypotaurine this property has not been demonstrated, but its scavenging properties towards ROS, in particular singlet oxygen and hydroxyl radicals, has been described[3,8,15].

Detoxifying Enzymes

Under hyperoxic conditions some detoxifying enzymes activities change (SOD, GSH Px)[17]. Our results show that in the saline-treated group hyperoxia produces an increase of SOD specific activity. These data are consistent with the finding that SOD biosynthesis is sensitive to oxygenation and is high in rats subjected to strong oxygen tensions[5]. This increase is not observed in taurine- and hypotaurine-treated groups. On the other hand the other specific activities decrease in hyperoxia in almost all the considered conditions.

The detoxifying power of a tissue can be better described by monitoring some enzymatic ratios. In particular, catalase/SOD and GSH Px/SOD ratios can be considered as an index of the tissue ability to counteract oxidative stress injury. When these ratios increase, this might indicate an activation of antioxidant enzymes system. When these ratios decrease, this may be indicative of a lower scavenging efficiency that might result in an increase of ROS levels and in a higher risk of oxidative damage[17]. In our experiments, in the control and hypotaurine-treated groups both the above ratios decrease, while in the taurine-treated group they increase after hyperoxia. These results might indicate that the presence of taurine enhances enzymatic scavenging efficiency when the tissue undergoes oxidative stress risk. This increased efficiency protects the tissue as shown by the ultrastructural analysis. In the untreated muscles, the scavenging efficiency is lowered under hyperoxia with the result the muscles are severely damaged. Hypotaurine does not induce an enhanced enzymatic scavenging efficiency under hyperoxic conditions, although the tissue is not damaged and therefore protected from ROS. It might be hypothesized that hypotaurine and taurine protect against oxidative damage through different mechanisms. This hypothesis is supported by the GSSG Rx specific activity behaviour: no significant change in activities can be observed in any group apart from the hyperoxic hypotaurine-treated one. In this last case the activity is undetectable, therefore the value of the ratio GSSG Rx/GSH Px becomes negligible. This ratio shows how well the system is able to recycle glutathione[17]. An increase in this ratio, as observed in hyperoxic control and taurine-treated groups with respect to normoxic ones, indicates that tissues are mantaining high GSH levels, thus providing protection to the cells. A decrease, on the contrary, is indicative of oxidative damage. In hypotaurine-treated rats, however, the observed decrease is not connected to oxidative damage. In conclusion, in the case of hypotaurine we observe after hyperoxic stress a well preserved ultrastructural tissue organization not imputable to an increased enzymatic scavenging efficiency. On the other hand, the taurine protection mechanism involves the induction of antioxidant enzymatic defence.

ACKNOWLEDGMENTS

Financial support by MURST (40% and 60%) is acknowledged. Useful suggestions and facilities by Prof. C. Di Ilio are also gratefully acknowledged.

REFERENCES

1. Amicarelli, F., Di Ilio, C., Masciocco, L., Bonfigli, A., Zarivi, O., D'Andrea, M.R., Di Giulio, C. and Miranda, M., 1997: Aging and detoxifying enzymes responses to hypoxic and hyperoxic treatment. *Mech. of Ageing and Dev.*, **97**, 215-226.
2. Amicarelli, F., Ragnelli, A.M., Aimola, P., Bonfigli, A., Colafarina, S., Di Ilio, C. and Miranda, M., 1999, Age dependent ultrastructural alterations and biochemical response of rat skeletal muscle after hypoxic and hyperoxic treatments, *Biochim. Biophys. Acta*, **1453**, 105-114.
3. Aruoma, O.I., Hallivell, B., Hoey, B.M. and Butler, J., 1988, The antioxidant action of taurine, hypotaurine and their metabolic precursors, *Biochem. J.*, **256**, 251-255.
4. Balentine, J.D., (ed.) 1982, Pathology and oxygen toxicity, Academic Press, New York.
5. Capro, J.D. and Tierney, D.F., 1974, Superoxide dismutase and pulmonary oxygen toxicity, *Am. J. Physiol.*, **226**, 1401-1407.
6. Di Ilio, C., Polidoro, G., Arduini, A., Muccini, A. and Federici, G., 1983, Glutathione peroxidase, glutathione reductase, glutathione S-transferase and γ-glutamyl-transpeptidase activities in human early pregnancy placenta, *Biochem. Med.*, **29**, 143-148.
7. Di Ilio, C., Sacchetta, P., Lo Bello, M., Caccuri, G. and Federici, G., 1986, Selenium dependent glutathione peroxidase activity associated with cationic forms of glutathione transferase in human hearts, *J. Mol. Cell. Cardiol.*, 1986, 18, 983.
8. Duprè, S., Costa, M., Spirito, A., Pitari, G., Rossi, P. and Amicarelli, F., 1998, Hypotaurine protection on cell damage by H_2O_2 and on protein oxidation by Cu^{++} and H_2O_2 in *Taurine 3*, Schaffer *et al.*, Ed. Plenum Press, New York, pagg. 17-23.
9. Forman, H.J., 1986, Oxidant production and bactericidal activity of fagocytes, *Annu. Rev. Physiol.*, **48**, 669-680.
10. Halliwell, B. and Gutteridge, M.C., 1984, Oxygen toxicity, oxygen radicals, transition metals and disease, *Biochem. J.*, **219**, 1-14.
11. Huxtable, R.J., 1992, Physiological action of taurine, *Physiol. Rev.*, **72**, 101-163.
12. Luck, H., 1965, in "Methods in enzymatic analysis" Bergmeyer H.U. (Ed), Verlag Chemie, Weinheim, pp. 865-894.
13. Pacifici, R.E. and Davies, K.J.A., 1990, Protein degradation as an index of oxidative stress. *Methods Enzymol.*, **186**, 485-512.
14. Paglia, D. E. and Valentine, W. N., 1967, Studies on qualitative and quantitative characterization of erythrocytes glutathione peroxidase, *J. Lab. Clin. Med.*, **70**, 158-169.
15. Pecci, L., Costa, M., Montefoschi, G., Antonucci, A. and Cavallini, D., 1999, Oxidation of hypotaurine to taurine with photochemically generated singlet oxygen: the effect of azide, *Biochem. Biophys. Res. Commun.*, **254**, 661-664.
16. Pryor, W.A., 1986, Oxy-radicals and related species: their formation, lifetimes and reactions. *Annu. Rev. Physiol.*, **48**, 657-667.
17. Somani, S.M., Husain, K. and Schlorff, E.C., 1996, Response of antioxidant system to physical and chemical stress, in I.S. Baskins, H. Salem (eds), Oxidants, Antioxidants and Free Radicals, Taylor and Francis, London, pp. 125-141.
19. Sun, M. and Zigman, S., 1978, An improved spectrophotometric assay for superoxide dismutase based on epinephrine autoxidation, *Anal. Biochem.*, **90**, 81-89.

20. Tadolini, B., Pintus, G., Pinna, G.G., Bennardini, F. and Franconi, F., 1995, Effect of taurine and hypotaurine on lipid peroxidation, *Biochem. Biophys. Res. Commun.*, **213**, 820-826.

HYPOTAURINE PROTECTION ON CELL DAMAGE BY SINGLET OXYGEN

Giuseppina Pitari[1], Silvestro Duprè[2], Alessandra Spirito[2], Giovanni Antonini[1] and Fernanda Amicarelli[1]

[1]*Dipartimento di Biologia di Base ed Applicata, Università di L'Aquila, L'Aquila, Italy;* [2]*Dipartimento di Scienze Biochimiche e Centro di Biologia Molecolare del C.N.R.., Università di Roma "La Sapienza", Roma, Italy*

Abstract: Singlet oxygen ($^1O_2$), generated by irradiating methylene blue, is toxic to melanoma cell cultures. Hypotaurine is known to scavenge efficiently singlet oxygen; the addition of hypotaurine (800 μM) to the medium during irradiation of the dye produces a greater protective effect on cells than taurine added at the same concentration. The assay of some detoxifying enzymatic activities indicate a different mechanism of protection of the two molecules: taurine induces an efficient detoxifying enzymatic action with respect to the control; hypotaurine exerts its effect greatly by specifically scavenging singlet oxygen.

INTRODUCTION

Hypotaurine is rapidly oxidized to taurine by singlet oxygen ($^1O_2$) generated by irradiating methylene blue[12]. Singlet oxygen is one of most reactive oxygen products, inducing oxidative damage and cell death in cultured cells[10]. The antioxidant activity of hypotaurine on cell structure has been studied using artificial liposomes and isolated lipoprotein[15]. Most studies on whole cells have been performed on sperm cells: cell function was studied under oxidative conditions in the presence of hypotaurine[7,9]. Studies on melanoma cell cultures[5] showed that hypotaurine protects cells from oxidative damage induced by H_2O_2.

In this paper we describe the effects of photochemically generated singlet oxygen on cultured melanoma cells and the protective effect of hypotaurine on cell death. The levels of superoxide dismutase (SOD, EC 1.15.1.1), catalase (EC 1.11.1.6), glutathione reductase (GSSG, Rx EC 1.6.4.27) and glu-

Taurine 4, edited by Della Corte *et al.*
Kluwer Academic / Plenum Publishers, New York, 2000.

tathione peroxidase (GSH Px, EC 1.11.1.9) were also measured to establish the efficiency of antioxidant enzyme activity in these conditions.

MATERIAL AND METHODS

Carlini human melanoma cells were cultured in RPMI 1640 supplemented with 10% fetal calf serum, at 37°C, 5% CO_2. Cells were treated with 10 µM methylene blue in serum-free RPMI for different times, washed with PBS and incubated for 24, 48 or 72 h in complete medium. Treatment with light in the presence or absence of methylene blue (MB) (10 µM) with or without hypotaurine (800 µM), was performed by irradiating cells with an alogen lamp (100 watt) at a distance of 15 cm for 10 min. A Petri dish containing water was placed between lamp and cells to avoid medium heating. Viability test was performed by trypan blue exclusion.

Enzymatic activities were performed on superrnatants of water treated cells (30 x 10^6). Phosphate buffer, pH 7.0 (final concentration 10 mM) containing respectively 1 mM dithiothreitol, for GSH Px and GSSG Rx assay, or 1 mM Triton X-100 for catalase and SOD detection, was added to water-treated cells.

Catalase activity was measured according to Luck[8]. The reduction of H_2O_2 was followed spectrophotometrically at 240 nm. One unit was defined as 1 µmol of H_2O_2 reduced/min.

Table 1. Cell treatments with methylene blue: viability after 24 h

Sample	Viability %
Control	100 ± 13.9
MB 10 min.	97.5 ± 8.8
MB 30 min.	96.2 ± 2.5
MB 40 min.	65.8 ± 3.4
MB 50 min.	60.8 ± 2.5
MB 60 min.	54.4 ± 8.8

Superoxide dismutase activity was determined as described[14]. The inhibitory effect of superoxide dismutase in samples on the autoxidation of epinephrine (0.1 M) in 50 mM carbonate buffer, pH 10.0, was assayed spectrophotometrically at 480 nm. One unit was defined as the amount of enzyme required to halve the rate of autoxidation.

Glutathione peroxidase activity was measured as described by Paglia and Valentine[11] as modified by Di Ilio *et al.*[4]. The activity of Se-dependent GSH-Px was measured in presence of H_2O_2 (0.25 mM) and glutathione. The oxidation of NADPH was followed at 340 nm. One unit was defined as 1 μmole of glutathione oxidized/min.

Glutathione reductase was measured as described by Di Ilio *et al.*[3]. The oxidation of NADPH (0.16 mM) in phosphate buffer, pH 7.4 in the presence of 1 mM oxidized glutathione and 1mM EDTA was followed at 340 nm. One unit was defined as 1 μmol of NADPH oxidized/min.

Protein concentration was determined by Bradford's assay[2].

Table 2. Cell treatment with methylene blue (10 μM) and light: cell viability after 24, 48, 72 h

Sample	Viability % 24 h	Viability % 48 h	Viability % 72 h
Control	100	206.4	405.4
MB + light 5 min	59.2	49.6	109.4
MB + light 10 min	45.0	42.0	61.9
MB + light 15 min	28.2	20.3	42.1

RESULTS AND DISCUSSION

Carlini melanoma cells were tested for resistance to methylene blue (MB), irradiation, or both (Tables 1 and 2). These cells are resistant to 10 μM MB for times as long as 60 min (Table 1) and do not suffer when irradiated in the absence of the dye (data not shown). For protection experiments an incubation time of 10 min was chosen to ensure good viability and negligible toxic effects of that vital dye. The presence of MB during irradiation strongly decreases viability (Table 2). This treatment produces in solution singlet oxygen, one of the most reactive ROS. Viability tests after 24, 48, 72 h confirm the high toxicity of singlet oxygen towards cells.

Cells were irradiated in the presence or in absence of hypotaurine or taurine (800 μM) in combination with methylene blue for 10 min. Table 3 shows the effect on cell viability of taurine and hypotaurine at the same concentrations, when cell cultures were stressed with irradiated methylene blue.

The presence of hypotaurine markedly decreases cell death respect to the not treated cells. After 24 h viability in the presence of hypotaurine was 72% and 58% in the presence of taurine. These results indicate that, in the presence of both taurine and hypotaurine, cells are protected against oxidative stress induced by singlet oxygen and that hypotaurine ensures a more efficient protection in this system. It is known that hypotaurine specifically "traps" singlet oxygen *in vitro*[12] whereas taurine apparently does not react with it [Pecci, unpublished result]. Hypotaurine is also a much more efficient scavenger of hydroxyl radicals than taurine[1]. We could hypothesize that the less efficient protective effect of taurine involves a different mechanism.

Table 3. Cell viability (evaluated by trypan blue exclusion) in melanoma cells after different treatments: dye irradiation was 10 min in all conditions. Hyp: hypotaurine; Tau: taurine; MB: methylene blue

Sample	Control	Hyp 800 μM	Tau 800 μM	MB 10 μM + light	MB 10 μM + light +Hyp 800 μM	MB 10 μM + light + Tau 800 μM
Viability % 24 h	100	107	85	42	72	58

Table 4. Ratios of enzymatic specific activities in cytosolic fractions of cultured melanoma cells 24 h after treatment

SAMPLES Ratio of enzymatic activities	Control	MB + light	Control + hypotaurine	MB + light + hypotaurine
Catalase/ SOD	13.2	28.6	18.1	17.1
GSH Px/ SOD	0.030	0.11	< 0.006	< 0.006
GSSG Rx/GSH Px	2.13	2.29	> 13.4	> 20

methylene blue was 10 μM, hypotaurine 800 μM and the irradiation time 10 min.

After an irradiation time of 10 min the effect of hypotaurine was high and cell viability allowed measurement of enzymatic activities, recorded 24 h after treatment on the cytosol of cultured cells. Ratios of specific activities of some detoxifying enzymes are reported in Table 4. Absolute activity values are not shown.

The ratios catalase/SOD and GSH Px/SOD are indicative of the efficiency of enzymatic systems to scavenge ROS: an increase indicates a positive response of the cells, a decrease may be indicative of the risk of oxidative damage[13]. The ratio GSSG Rx/GSH Px is indicative of the capability of the cells to recycle glutathione: a decrease indicates a possible oxidative damage, because the cell cannot produce enough GSH to get rid of hydrogen peroxide, organic hydroperoxides or ROS.

The efficiency of ROS scavenging measured as a ratio between specific activities of catalase and SOD show some important modifications. Catalase specific activity markedly increased in methylene blue + light treated cells. SOD specific activity does not strongly change in any of the considered conditions, even if there is a little decrease when cells are stressed in the presence or in the absence of hypotaurine. The ratio catalase/SOD doubles with respect to the control cells, indicating a positive response of the cells to oxidative stress. The ROS scavenging efficiency is increased even if it is not sufficient to avoid cell death. In the presence of hypotaurine this ratio does not change when cells are stressed and it is a little over the control. These results suggests that the protection by hypotaurine is not imputable to an induction of antioxidant enzymatic response.

The activity of GSH Px parallels that of catalase, as it increases in stressed conditions. In the presence of hypotaurine, however, both in the control and in stressed cells, GSH Px activity is undetectable, and this explains the very low GSH Px/SOD ratio values.

In cells without hypotaurine, the activity of GSSG Rx parallels that of peroxidase, suggesting a relatively enhanced efficiency of glutathione recycling in stressed cells respect to the control. The observed high increase in the GSSG Rx/GSH Px ratio in hypotaurine-treated cells is due mainly to the above reported very low GSH Px activity.

It is well known that GSH Px is responsible for the detoxification of hydrogen peroxide and organic hydroperoxides. Its activity increases in those conditions where peroxide concentration increases in cells. Hypotaurine reacts very slowly[6] with H_2O_2 and is not a valuable candidate for peroxide scavenging.

The unexpected disappearance of GSH Px activity in the presence of hypotaurine may indicate that peroxide production is lowered by the scavenging activity of hypotaurine towards singlet oxygen and/or hydroxyl radicals and that recycling of glutathione as a peroxide-scavenging system is not very important in these conditions.

In conclusion, in cells treated with methylene blue + light the activation of enzymatic defences is not sufficient to counteract ROS attack and we observe cell death. In hypotaurine treated cells viability is higher respect to the control and this would indicate that the enzymatic scavenging activation is not necessary for cell surviving. The protective effect of hypotaurine does not involve detoxifying enzymes; probably hypotaurine decreases the amount of ROS, in particular singlet oxygen, thus avoiding modification in enzyme activities.

ACKNOWLEDGMENTS

Financial support from MURST (40% and 60%) is acknowledged.

REFERENCES

1. Aruoma, O.I., Halliwell, B., Hoey, B.M. and Butler, J., 1988, The antioxidant action of taurine, Hypotaurine and their metabolic precursors, *Biochem. J.,* **256**, 251-256.
2. Bradford, M.M., 1976, A rapid and sensitive method for the quantitation of micrograms quantities of protein utilizing the principle of protein-dye binding. *Anal. Biochem.*, **72**, 248-254.
3. Di Ilio, C., Polidoro, G., Arduini, A., Muccini, A. and Federici, G., 1983, Glutathione peroxidase, glutathione reductase, glutathione S-transferase and γ-glutamyl-transpeptidase activities in human early pregnancy placenta, *Biochem. Med.*, **29**, 143-148.
4. Di Ilio, C., Sacchetta, P., Lo Bello, M., Caccuri, G. and Federici, G., 1986, Selenium dependent glutathione peroxidase activity associated with cationic forms of glutathione transferase in human hearts, *J. Mol. Cell. Cardiol.*, 1986, 18, 983.
5. Duprè, S., Costa, M., Spirito, A., Pitari, G., Rossi, P. and Amicarelli, F., 1998, Hypotaurine protection on cell damage by H_2O_2 and on protein oxidation by Cu^{++} and H_2O_2 in *Taurine 3*, Schaffer *et al.* ed., Plenum Press, New York, pagg. 17-23.
6. Fiori, A. and Costa, M., 1969, Ossidazione dell'ipotaurina con H_2O_2. *Acta Vitaminol. Enzymol.* **26**, 204-207.
7. Huxtable, R. J., 1992, Physiological action of taurine, *Biol. Rew.*, **72**, 101-163.
8. Luck, H., 1965, in " Methods in enzymatic analysis" Bergmeyer H.U. (ed), Verlag Chemie, Weinheim, pp. 865-894.
9. Meizel, S., Lui, C.W., Working, P.K. and Mrsny, M.J., 1980, Taurine and hypotaurine: their effects on motility, capacitation and the acrosome reaction of hamster sperm in vitro and their presence in sperm and reproductive tracts fluids of several mammals, *Dev. Growth Differ.*, **22**, 483-494.
10. Noodt, B.B., Rodal, G.H., Wainwright, M., Peng, Q., Nesland, J.M. and Berg, K., 1998, Apoptosis induction by different pathways with methylene blue derivative and light from mitochondrial sites in V79 cells. *Int. J. Cancer*, **75**, 941-948.
11. Paglia, D. E. and Valentine, W. N., 1967, Studies on qualitative and quantitative characterization of erythrocytes glutathione peroxidase, *J. Lab. Clin. Med.*, **70**, 158-169.
12. Pecci, L., Costa, M., Montefoschi, G., Antonucci, A. and Cavallini, D., Oxidation of hypotaurine to taurine with photochemically generated singlet oxygen: the effect of azide. 1999, *Bioch. Biophys. Res. Comm.*, **254**, 661-664.
13. Somani, S.M., Husain, K. and Schlorff, E.C., 1996, Response of antioxidant system to physical and chemical stress, in I.S. Baskins, H. Salem (eds), Oxidants, Antioxidants and Free Radicals, Taylor and Francis, London, pp. 125-141.
14. Sun, M. and Zigman, S., 1978, An improved spectrophotometric assay for superoxide dismutase based on epinephrine autoxidation, *Anal. Biochem.*, **90**, 81-89.
15. Tadolini, B., Pintus, G., Pinna, G.G., Bennardini, F. and Franconi, F., 1995, Effect of taurine and hypotaurine on lipid peroxidation, *Biochem. Biophys. Res. Commun.*, **213**, 820-826.

HYPOTAURINE AND SUPEROXIDE DISMUTASE

Protection of the Enzyme against Inactivation by Hydrogen Peroxide and Peroxidation to Taurine

Laura Pecci, Gabriella Montefoschi, Mario Fontana, Silvestro Duprè, Mara Costa, and Doriano Cavallini

Dipartimento di Scienze Biochimiche, Università di Roma "La Sapienza", Roma; Centro di Biologia Molecolare, CNR, Roma, Italy

Abstract: Hypotaurine is able to prevent the inactivation of SOD by H_2O_2. The protection is concentration-dependent: at 20 mM hypotaurine the inactivation of SOD is completely prevented. It is likely that hypotaurine exerts this effect by reacting with hydroxyl radicals, generated during the inactivation process, in competition with the sensitive group on the active site of the enzyme. According to this, spectral studies indicate that in presence of hypotaurine the integrity of the active site of SOD is preserved by the disruptive action of H_2O_2. An interesting outcome of the SOD/H_2O_2/hypotaurine interaction is that SOD catalyzes the peroxidation of hypotaurine to taurine. Indeed, the formation of taurine increases with the reaction time and with the enzyme concentration. Although the peroxidase activity of SOD is not specific and relatively slow compared to the dismutation of superoxide, it might represent another valuable mechanism of production of taurine.

INTRODUCTION

Hypotaurine, a metabolic precursors of taurine, is a potent scavenger of some reactive oxygen species (ROS) *in vitro* like hydroxyl radicals[2] and singlet oxygen[12]. This property is related to the proposed role of hypotaurine as an antioxidant and free radical trapping agent *in vivo*. According to this, it has been reported that hypotaurine quenches oxidants released by human neutrophils[5], inhibits lipid peroxidation[14] and acts as an antioxidant to protect sperm cells against infertility[1]. This antioxidant function of hypotaurine has been related to the high level of taurine found in tissues or cells typically subjected to oxidative stress (sperm, neutrophils, liver and retinal tissue)[9]. Among the biological sources of ROS, it has been reported that hydroxyl radicals are generated during the inactivation of Cu,Zn-superoxide dismutase (SOD) with H_2O_2[7,15]. Moreover it has been shown that some small anionic radical scavengers, such as formate, azide and urate, which gain access to the active site, protect the enzyme against the inactivation[7].

We report that hypotaurine is able to intercept hydroxyl radicals generated by reaction of SOD with H_2O_2. As result of this interaction hypotaurine protects the enzyme against inactivation and the enzyme catalyzes the peroxidation of hypotaurine to taurine.

MATERIALS AND METHODS

Superoxide dismutase (SOD) from bovine erythrocytes, xanthine oxidase (XO), and cytochrome c (cyt c) were purchased from Sigma. Other reagents were purchased from Sigma, Fluka and Merck. H_2O_2 concentration was measured at 230 nm using a molar extinction coefficient of 72.4[11]. The amino acid analyzer was a Carlo Erba 3A30. SOD activity was determined by the XO/cyt c-based assay[10]. In brief, the reaction mixture contained 10 µM cyt c, 50 µM xanthine in 3 ml of 50 mM carbonate buffer containing 0.1 mM EDTA, pH 10 at 25°C. XO was added to start the reaction, the exact amount of XO being adjusted so to give $\Delta A_{550nm} = 0.06\text{-}0.09$ min^{-1}. Sufficient SOD was added to the reaction mixture to produce about 75% inhibition of the cyt c reduction rate.

RESULTS

Protection of SOD by Hypotaurine

It has been shown that the inactivation of Cu, Zn superoxide dismutase by H_2O_2 can be followed by the spectral changes of the enzyme[7]. Inactivation is preceded by a bleaching of the absorbance in the visible at 680 nm, due to the rapid reduction of Cu^{2+} on the enzyme and, following this, there is a gradual appearance of a new absorption at 450 nm which is coincident with the loss of catalytic activity. This new absorbance is indicative of a modification of the active site where one residue of histidine undergoes oxidative attack by the hydroxyl radicals generated by the enzyme-H_2O_2 interaction.

As shown in Fig. 1, in the presence of hypotaurine, SOD is still rapidly reduced by H_2O_2 but the absorbance at 450 nm does not appear indicating that in these conditions the integrity of the active site is preserved by the disruptive action of H_2O_2.

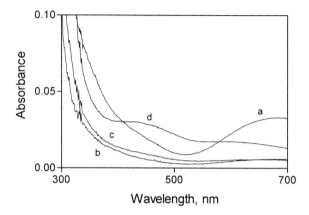

Figure 1. The effect of H_2O_2 upon the absorption spectrum of SOD in the presence of hypotaurine. Enzyme (5.6 mg/ml) in 20 mM sodium pyrophosphate, pH 9 at 25°C was treated with 10 mM hypotaurine (line a) and then with 1 mM H_2O_2 (line b: 1 min after H_2O_2 addition; line c: 30 min later). Enzyme treated with 1 mM H_2O_2 for 30 min in the absence of hypotaurine is reported for comparison (line d).

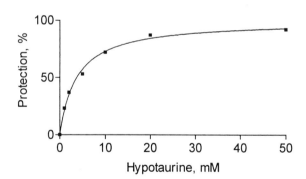

Figure 2. Concentration-dependence of the protective effect of hypotaurine against the inactivation of SOD by H_2O_2. Hypotaurine, at the indicated concentration, was mixed with SOD (12 μg/ml) in 50 mM carbonate buffer, pH 10, and 0.1 mM EDTA, before H_2O_2 addition (0.15 mM). After 15 min incubation at 25°C, 10 μl of the sample were withdrawn for assay of SOD activity (xanthine oxidase/cit c-based assay). Appropriate controls of untreated enzyme and of enzyme treated with H_2O_2 in the absence of hypotaurine were carried out in parallel to estimate the level of protection.

According to the spectral studies, the results reported in Fig. 2 show that hypotaurine is able to protect SOD by H_2O_2 inactivation. The protection is concentration-dependent: at 20 mM hypotaurine the inactivation of SOD is completely prevented. Hypotaurine is not protective if added to the enzyme preincubated with H_2O_2 for 15 min, indicating that the compound does not reverse but prevents SOD inactivation. Taurine is without effect.

Because hypotaurine does not noticeably react with H_2O_2 ($k = 10^{-2}$ $M^{-1}sec^{-1}$)[4], it is likely that the compound exerts its protective effect by reacting with hydroxyl radicals[2] ($k = 10^{10}$ $M^{-1}s^{-1}$), postulated as the reactive species that causes the inactivation.

Peroxidation of Hypotaurine

Hodgson and Fridovich[8] reported that compounds which protect SOD against H_2O_2 inactivation are peroxidized in competition with the sensitive group on the enzyme. Therefore SOD was expected to catalyze also the peroxidation of hypotaurine. According to this expectation, Fig. 3 shows that SOD in the presence of H_2O_2 acts upon hypotaurine with progressive formation of taurine.

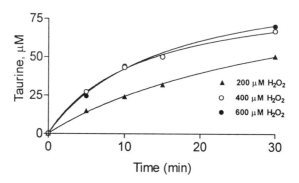

Figure 3. Time-course production of taurine. Reaction mixture contained SOD (200 μg/ml), 1 mM hypotaurine, 0.2-0.6 mM H_2O_2 and 0.1 mM EDTA in 50 mM carbonate buffer, pH 10, 25°C. At the indicated time intervals, aliquots of 100 μl were withdrawn, diluted 1:4 (v/v) with citrate buffer, pH 2.2 and analyzed for taurine formation by amino acid analyser.

Taurine formation increases with H_2O_2 concentration, indicating an optimum value of 400 μM; beyond this value the reaction rate becomes constant. In control experiments, there is no formation of taurine if either H_2O_2 or SOD is omitted. The rate of peroxidation is proportional to SOD concentration while the boiled enzyme does not carry out the reaction (Fig. 4).

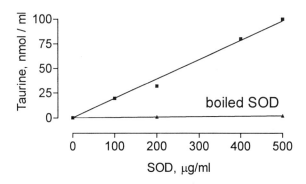

Figure 4. Taurine production as a function of SOD concentration. Reaction mixture contained 2.5 mM hypotaurine, 0.25 mM H_2O_2 and the indicated concentration of enzyme in 50 mM carbonate buffer, pH 10, containing 0.1 mM EDTA. After 5 min incubation at 25°C, taurine formation was determined as described in the legend of fig. 3.

DISCUSSION

The results reported in this study clearly demonstrate that hypotaurine is a good protective agent against inactivation of SOD by H_2O_2. As hydroxyl radicals are postulated as the reactive species that cause the inactivation[7] it is concluded that hypotaurine gains access to the active site of the enzyme and exerts its effect by scavenging these radicals. In this way hypotaurine not only prevents the inactivation of SOD, which is an essential antioxidant enzyme for superoxide dismutation *in vivo*, but also intercepts hydroxyl radicals which could escape from the active site and cause oxidative damage to biologically important molecules. An interesting outcome of the SOD/H_2O_2/hypotaurine interaction is that SOD catalyzes the peroxidation of hypotaurine to taurine. Although, *in vivo*, hypotaurine is readily converted to taurine, the search for a specific enzymatic system which carries out this oxidation has given, instead, negative results[3]. Only not specific oxidants, such as singlet oxygen[12], hydroxyl radicals[2], hypochlorite[6] and UV radiations[13] have been reported to accomplish this oxidation in good yield. Therefore the peroxidation of hypotaurine by SOD may represent another valuable, although not specific, mechanism that can account for the last metabolic step in the production of taurine.

REFERENCES

1. Alvarez, J.G., and Storey, B.T., 1983, Taurine, hypotaurine, epinephrine and albumin inhibit lipid peroxidation in rabbit spermatozoa and protect against loss of motility. *Biol. Reprod.* 29: 548-555.

2. Aruoma, O.I., Halliwell, B., Hoey, B.M., and Butler, J., 1988, The antioxidant action of taurine, hypotaurine and their metabolic precursors. *Biochem. J.* 256: 251-255.

3. Cavallini, D., De Marco, C., Mondovì, B., and Stirpe, F., 1954, The biological oxidation of hypotaurine. *Biochim. Biophys. Acta* 15: 301-302.

4. Fiori, A., and Costa, M., 1969, Ossidazione dell'ipotaurina con H_2O_2. *Acta Vitaminol. Enzymol.* 26: 204-207.

5. Green, T.R., Fellman, J.H., and Eicher, A.L., 1985, Myeloperoxidase oxidation of sulfur-centered and benzoic acid hydroxyl radical scavengers. *FEBS Lett.* 192: 33-36.

6. Green, T.R., Fellman, J.H., Eicher, A.L., and Pratt, K.L., 1991, Antioxidant role and subcellular location of hypotaurine and taurine in human neutrophils. *Biochim. Biophys. Acta* 1073: 91-97.

7. Hodgson, E.K., and Fridovich, I., 1975a, The interaction of bovine erythrocyte superoxide dismutase with hydrogen peroxide: inactivation of the enzyme. *Biochemistry* 14: 5294-5299.

8. Hodgson, E.K., and Fridovich, I., 1975b, The interaction of bovine erythrocyte superoxide dismutase with hydrogen peroxide: chemiluminescence and peroxidation. *Biochemistry* 14: 5299-5303.

9. Huxtable, R.J., 1992, Physiological actions of taurine. *Physiol. Rev.* 72: 101-163.

10. McCord, J.M., and Fridovich, I., 1969, Superoxide dismutase. *J. Biol. Chem.* 244: 6049-6055.

11. Nelson, D.P., and Kiesow, L.A., 1972, Enthalpy of decomposition of hydrogen peroxide by catalase at 25°C (with molar extinction coefficients of H_2O_2 solutions in the UV). *Anal. Biochem.* 49, 474-478.

12. Pecci, L., Costa, M., Montefoschi, G., Antonucci. A., and Cavallini, D., 1999, Oxidation of hypotaurine to taurine with photochemically generated singlet oxygen: the effect of azide. *Biochem. Biophys. Res. Commun.* 254, 661-665.

13. Ricci, G., Duprè, S., Federici, G., Spoto, G., Matarese, R.M., and Cavallini, D., 1978, Oxidation of hypotaurine to taurine by ultraviolet irradiation. *Physiol. Chem. Phys.* 10: 435-441.

14. Tadolini, B., Pintus, G., Pinna, G.G., Bennardini, F., and Franconi, F., 1995, Effect of taurine and hypotaurine on lipid peroxidation. *Biochem. Biophys. Res. Commun.* 213: 820-826.

15. Yim, M.B., Chock, P.B., and Stadtman, E.R., 1990, Copper, zinc superoxide dismutase catalyzes hydroxyl radical production from hydrogen peroxide. *Proc. Natl. Acad. Sci. USA* 87: 5006-5010.

DIETARY TAURINE CHANGES ASCORBIC ACID METABOLISM AND CHOLESTEROL METABOLISM IN RATS FED DIETS CONTAINING POLYCHLORINATED BIPHENYLS

[1]Hidehiko Yokogoshi, [1]Hideki Mochizuki, and [2]Hiroaki Oda

[1]*School of Food and Nutritional Sciences, The University of Shizuoka, 52-1 Shizuoka 422-8526, Japan;* [2]*Department of Applied Biological Sciences, Nagoya University, Nagoya 464-8601, Japan*

INTRODUCTION

Xenobiotics such as polychlorinated biphenyls (PCB), 1,1,1-trichloro-2,2-bis(*p*-chloro phenyl)ethane (DDT), 2,6-di-*tert*-2,2-butyl-*p*-cresol (BHT), and barbital derivatives are widely distributed in the environment as harmful contaminants. Administration of xenobiotics to animals cause metabolic and pathologic changes, includes: (i) induction of hepatic drug metabolizing enzymes[16]; (ii) elevation of serum levels of high density lipoprotein (HDL) cholesterol and apolipoprotein A-I[11,12]; (iii) accumulation of liver lipids[5,13]; and (iv) enhancement of ascorbic acid in urine and tissues[2,6].

These parameters are also influenced by dietary nutrients such as protein and sulfur-containing amino acids (SAA). However, it has not been investigated whether taurine influences the enhancement of these parameters caused by PCB. Taurine affects various biological and physiological functions, including cell membrane stabilization[15], antioxidation[9], detoxification, osmoregulation[4], neuromodulation[7], and brain and retinal development[18]. Our previous studies showed that taurine enhanced the serum HDL-cholesterol concentration in normal rats[8], and that taurine exerted a hypocholesterolemic action in the hypercholesterolemia caused by high-cholesterol diet[10,19].

In the present study, we have investigated the effect of taurine on the hepatic drug metabolizing enzymes and the concentrations of ascorbic acid in tissues and urine in the hypercholesterolemia caused by PCB.

Taurine 4, edited by Della Corte *et al.*
Kluwer Academic / Plenum Publishers, New York, 2000.

MATERIALS AND METHODS

Animals and Diets

Young male rats of the Wistar strain weighing about 60 g (Japan SLC, Hamamatsu, Japan) were maintained at 24°C with a 12-h light (7:00-19:00 h) and dark cycle. To accustom the rats to the experimental conditions, they were initially allowed free access to a 20% casein diet (normal diet, 200 g/kg diet) for 2 days, and divided into four groups. The compositions of normal diet are (in weight percentage): casein, 20; mineral mixture (AIN-93), 5.0; corn oil, 5.0; vitamin mixture, 1.0; choline-chloride, 0.15; and a mixture of sucrose and corn starch (1:2; w/w) to 100%. Animals were given free access to the experimental diets and water for 14 days. The rats were killed by decapitation at 10:00 a.m. after 16 h of fasting on the last day in the experiment, and blood was collected from the cervical wound. The experimental procedures used in this study met the guidelines of the Animal Care and Use Committee of the University of Shizuoka.

Biochemical Analyses

The serum lipids (total cholesterol and HDL-cholesterol) were enzymatically measured by commercial kits (cholesterol C-test and HDL-cholesterol-test; Wako Pure Chemical, Osaka, Japan, respectively). Urine was collected with 15ml of 10% (100 g/L) metaphosphoric acid solution to determine the concentrations of ascorbic acid. The concentrations of ascorbic acid in the metaphosphoric acid supernatant of urine were determined by the 2,4-dinitrophenylhydrazine method[17]. The hepatic postmitochondrial supernatant was further centrifuged for 60 min at $105,000 \times g$ at 4°C to obtain the microsomal pellets. Their pellets were suspended in 1.15% KCl, and the activities of cytochrome P-450 were determined[14].

Statistics

Experimental data were statistically analyzed by one-way analysis of variance (ANOVA), and the differences between means were analyzed by Duncan's multiple-range test[1].

RESULTS

Body weight gains of rats fed the normal diet supplemented with taurine and/or PCB did not differ from those of rats fed the normal diet. The concentrations of serum total cholesterol and HDL-cholesterol were significantly higher in rats fed PCB containing diets than those in rats fed PCB-untreated diets (Fig. 1). In the case of PCB-treated groups, the addition of taurine to the diet significantly amplified the increase of total- and HDL-cholesterol. The concentrations of hepatic cholesterol and lipids were significantly higher in rats fed PCB containing diets, and the simultaneous supplementation of taurine amplified these increases (Fig. 2).

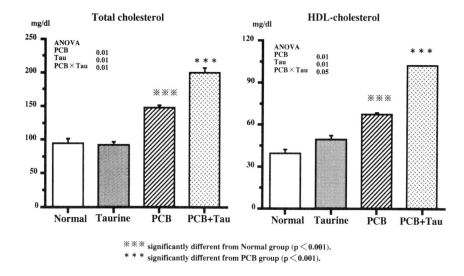

※※※ significantly different from Normal group (p＜0.001).
＊＊＊ significantly different from PCB group (p＜0.001).

Figure 1. Serum cholesterol concentrations in rats fed 0.02% PCB containing diets with or without 3% taurine for 15 days. Vertical bars indicate SEM, and the different superscript symbols are significantly different from each control (p<0.05).

The activities of drug-metabolizing enzymes such as cytochromes p-450 in the liver were significantly higher in rats fed PCB-containing diets than those in rats fed control diets (Fig. 3). In the case of PCB-treated groups, dietary taurine significantly amplified the enhancement of cytochrome P-450 caused by PCB. The levels of various mRNAs of CYP7A1, CYP1A2 and CYP2B1/2B2 were significantly higher in rats fed PCB-containing diets, and taurine significantly amplified the enhancement of each mRNAs (Fig. 4).

Figure 5 shows the time (day) dependent changes in urinary excretion of ascorbic acid. In the case of control groups, dietary taurine did not affect the excretion of ascorbic acid, whereas in the PCB-treated groups, dietary taurine significantly amplified the PCB-induced enhancement of the urinary ascorbic acid excretion during the entire experimental period.

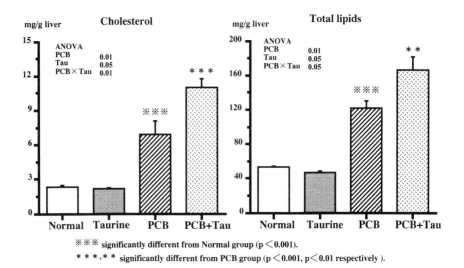

※※※ significantly different from Normal group (p ＜0.001).

* * *,* * significantly different from PCB group (p ＜0.001, p＜0.01 respectively).

Figure 2. Liver lipids concentrations in rats fed 0.02% PCB containing diets with or without 3% taurine for 15 days. Vertical bars indicate SEM, and the different superscript symbols are significantly different from each control (p<0.05).

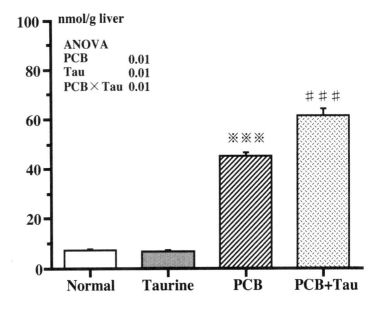

Figure 3. Cytochrome P-450 contents in the liver in rats fed 0.02% PCB containing diets with or without 3% taurine for 14 days. Vertical bars indicate SEM, and the different superscript symbols are significantly different from each control (p<0.05).

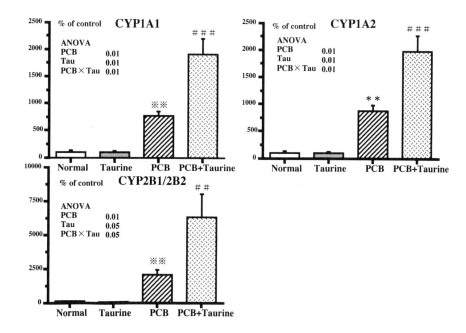

Figure 4. Induction of various mRNAs concerning drug metabolizing enzymes. Vertical bars indicate SEM, and the different superscript symbols are significantly different from each control (p<0.05).

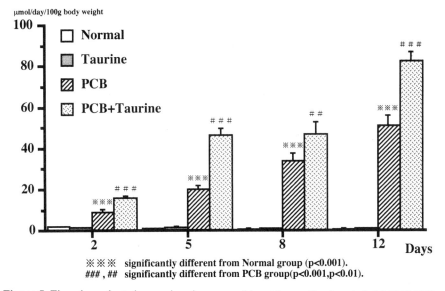

※※※ significantly different from Normal group (p<0.001).
###, ## significantly different from PCB group(p<0.001,p<0.01).

Figure 5. Time dependent changes in urinary ascorbic acid excretion in rats fed 0.02% PCB containing diet with or without 3% taurine. Vertical bars indicate SEM, and the different superscript symbols are significantly different from each control (p<0.05).

DISCUSSION

The administration of various types of xenobiotics causes marked increases in urinary excretion and tissue concentrations of ascorbic acid in rats[2]. Previous reports suggested that this phenomenon was caused by the enhancement of ascorbic acid synthesis in the liver, accompanying the high activities of the hepatic drug metabolizing enzymes[3]. Because the administration of large amounts of ascorbic acid in the diet ameliorated the toxicity of PCB in guinea pigs, it was postulated that the enhancement of ascorbic acid synthesis might be an active adaptive response to the exposure to PCB. In this study, we examined the effect of dietary taurine on the concentrations of ascorbic acid and drug metabolizing enzymes in the liver. Dietary PCB significantly enhanced not only the concentrations of ascorbic acid in urine, but also the activities of cytochromes P-450, which were always accompanied by increases of serum total cholesterol and HDL-cholesterol. On the other hand, the simultaneous supplementation of taurine with PCB significantly amplified the urinary excretion of ascorbic acid and the concentrations of hepatic microsomal cytochrome P-450 in rats. These results suggested that, in the case of PCB-treated groups, dietary taurine might enhance the biosynthesis of ascorbic acid, presumably through the acceleration of the hepatic drug metabolizing systems. Also taurine might have influenced the activity of ascorbic acid biosynthesis and the gene expression of cytochrome P-450 under the PCB-treated condition. Further studies are required to explore the mechanism of the acceleration of ascorbic acid metabolism induced by taurine in rats fed PCB-containing diet.

CONCLUSION

The effect of dietary taurine on ascorbic acid metabolism and hepatic drug-metabolizing enzymes was investigated in rats fed diets containing 0.02% PCB. The rats fed the PCB containing diets showed greater liver weight, and greater activities of drug metabolizing enzymes in the liver than control rats. In PCB-fed rats, urinary ascorbic acid excretion was enhanced, and serum cholesterol concentration (especially HDL-cholesterol) was significantly elevated compared with those in normal rats. Dietary taurine potentiated the increases of urinary excretion of ascorbic acid and the rise in the levels of cytochrome P-450 which were induced by PCB . On the other hand, the supplementation of taurine to normal diet did not alter these parameters. These results suggested that taurine might enhance the hepatic drug-metabolizing systems, leading consequently to the stimulation of the ascorbic acid metabolism in rats fed PCB-containing diet.

REFERENCES

1. Duncan, D. B., 1955. Multiple range and multiple F test. *Biometrics* 11: 1-42.
2. Horio, F. and Yoshida, A., 1982. Effect of some xenobiotics on ascorbic acid metabolism in rats. *J. Nutr.* 112: 416-425.
3. Horio, F., Ozaki, K. Yoshida, A. Makino, S. & Hayashi, Y., 1986. Ascorbic acid requirement for the induction of microsomal drug-metabolizing enzymes in a rat mutant unable to synthesize ascorbic acid. *J. Nutr.* 116: 2278-2289.
4. Huxtable, R. J., 1992. Physiological actions of taurine. *Physiol. Rev.* 72: 101-163.
5. Kato, N., Tani, T., and Yoshida, A., 1980. Effect of dietary PCB on hepatic cholesterogenesis in rats. *Nutr. Rep. Inter.* 21: 107-112.
6. Kato, N., Tani, T. and Yoshida, A., 1980. Effect of dietary level of protein on liver microsomal drug-metabolizing enzymes, urinary ascorbic acid and lipid metabolism in rats fed PCB-containing diets. *J. Nutr.* 110: 1686-1694.
7. Kuriyama, K., 1980. Taurine as a neuromodulator. *Fed. Proc.* 39: 2680-2684
8. Mochizuki, H., Oda, H. & Yokogoshi, H., 1998. Increasing effect of dietary taurine on the serum HDL-cholesterol concentration in rats. *Biosci. Biotechnol. Biochem.* 62: 578-579.
9. Nakamura, T., Ogasawara, M., Nemoto, M. & Yoshida, T., 1993. The protective effect of taurine on the biomembrane against damage produced by oxygen radicals. *Biol. Pharm. Bull.* 16: 970-972.
10. Nanami, K., Oda, H. & Yokogoshi, H., 1996. Antihypercholesterolemic action of taurine on streptozotocin-diabetic rats or on rats fed a high cholesterol diet. *Adv. Exp. Med. Biol.* 403: 561-568.
11. Oda, H., and Yoshida, A., 1994. Effect of feeding xenobiotics on serum high density lipoprotein and apolipoprotein A-I in rats. *Biosci. Biotech. Biochem.* 58: 1646-1651.
12. Oda, H., Matsushita, N., Hirabayashi, A., and Yoshida, A., 1990. Hyperlipoproteinemia in rats fed polychlorinated biphenyls. *J. Nutr. Sci. Vitaminol.* 36: 117-122.
13. Oda, H., Matsushita, N., Hirabayashi, A., and Yoshida, A., 1994. Cholesterol-rich very low density lipoproteins and fatty liver in rats fed polychlorinated biphenyls. *Biosci. Biotech. Biochem.* 58: 2152-2158.
14. Omura, T. & Sato, R., 1964. The carbon monooxide binding pigment of liver microsomes. *J. Biol. Chem.* 239: 2370- 2378.
15. Pasantes-Morales, H., Wright, C. E. & Gaull, G. E., 1985. Taurine protection of lymphoblastoid cells from iron- ascorbate-induced damage. *Biochem. Pharmacol.* 34: 2205-2207.
16. Poland, A., and Kunttson, J. C., 1982. 2, 3, 7, 8- Tetrachlorodibenzo-p-dioxin and related halogenated aromatic hydrocarbons. *Ann. Rev. Pharmacol. Toxicol.* 22: 517-554.
17. Roe, J. H. & Kuether, C. A., 1943. The determination of ascorbic acid in whole blood and urine through the 2, 4- dinitrophenylhydrazine derivative of dehydroascorbic acid. J. Biol. Chem. 147: 399-407.
18. Sturman, J. A., 1986. Nutritional taurine and central nervous system development. *Ann. NY Acad. Sci.* 477: 196-213.
19. Yokogoshi, H., Mochizuki, H., Nanami, K., Hida, Y. Miyachi, F. & Oda, H., 1999. Dietary taurine enhances cholesterol degradation and reduces serum and liver cholesterol concentrations in rats fed a high-cholesterol diet. *J. Nutr.* 129: 1705-1712.

EFFECTS OF LONG-TERM TREATMENT WITH TAURINE IN MICE FED A HIGH-FAT DIET

Improvement in cholesterol metabolism and vascular lipid accumulation by taurine

Shigeru Murakami, Yukiko Kondo, and Takatoshi Nagate
Medicinal Research Laboratories, Taisho Pharmaceutical Co. Ltd., Ohmiya, Japan

Abstract: Hypocholesterolemic effects of taurine in rats fed a high-fat and high-cholesterol diet are well established. However, there are few studies on long-term effects of taurine on cholesterol metabolism. In the present study, taurine was dissolved in drinking water and given to C57BL/6J mice during 6 months-feeding of a high fat diet. Taurine treatment significantly decreased serum LDL and VLDL cholesterol, while it significantly increased serum HDL cholesterol. In the liver, taurine decreased cholesteryl ester contents, accompanied by decrease in acyl Co-A:cholesterol acyltransferase (ACAT) activity. Hepatic activity of cholesterol 7α-hydroxylase, a rate-limiting enzyme for bile acid synthesis, was doubled with taurine. Taurine reduced by 20% the high-fat diet-induced arterial lipid accumulation. Thus, taurine prevented elevation of serum and liver cholesterol levels, as possibly related to accelerated cholesterol elimination from the body through the stimulation of bile acid synthesis. Long-term treatment with taurine is beneficial for prevention of hypercholesterolemia and atherosclerosis.

INTRODUCTION

Taurine is widely distributed in animal tissues and has a role in maintaining physiological functions[11,37]. The world-wide epidemiological study, CARDIAC Study, suggested beneficial effects of taurine for preventing cardiovascular disease[39]. Cholesterol-lowering actions of taurine have been studied in experimental animal[10,12,18,33,38]. Taurine decreases plasma cholesterol levels in animals fed a high-fat and/or high-cholesterol diets, while taurine has no significant effect in animals fed regular chow. Thus, taurine attenuates abnormal cholesterol metabolism, and normalizes plasma and tissues cholesterol levels. Although the mechanism responsible for

Taurine 4, edited by Della Corte *et al.*
Kluwer Academic / Plenum Publishers, New York, 2000.

hypocholesterolemic action of taurine was not well defined, recent studies revealed that taurine stimulates cholesterol catabolism to bile acid by enhancing cholesterol 7α-hydroxylase activity, a rate-limiting enzyme for bile acid synthesis [13,18,21,34]. This event result in an increase in receptor-mediated LDL uptake and cholesterol synthesis in the liver.

Less attention has focused on the anti-atherosclerotic effects of taurine[27], although the beneficial effects of taurine on plasma cholesterol is expected to prevent the development of vascular disease. The C57BL/6J mouse serves as a model of diet-induced atherosclerosis, since this strain develops atherosclerotic lesions when maintained on a high-fat diet[25,26]. In the present study, we evaluated effects of taurine on cholesterol metabolism and progression of atherosclerosis in C57BL/6J mice fed a high-fat diet.

MATERIALS AND METHODS

Materials

Autosera CHO-2 was purchased from Daiichi Kagaku Yakuhin (Tokyo, Japan). Radiochemicals for the enzymatic assay were purchased from New England Nuclear (Boston, MA). All other chemicals were purchased from Sigma Chemical Co. (St. Louis, MO). Taurine was synthesized at Taisho Pharmaceutical Co. (Ohmiya, Japan).

Animals and diets

C57BL/6J female mice were obtained from Charles River Japan (Atsugi, Japan). The regular chow and high-fat diet were purchased from the Oriental Yeast Co. (Tokyo, Japan). The high-fat diet was prepared by mixing a diet containing 30% cocoa butter, 5% cholesterol, 2% sodium cholate, 30% casein, 5% cellulose powder, 4% vitamin mixture, 4% mineral mixture, 6.5% sucrose, 6.5% glucose, &.5% dextran, and 0.5% choline chloride with standard chow in a ratio of 1:3 parts, respectively, and then pelleting it. This diet has a 15% total fat content, and the cholesterol content was 1.25%. Taurine was dissolved in drinking water at 1% (w/v) and was freely available to the mice. This concentration of taurine is equivalent to 1,700 mg/kg/day, calculated based on the consumption of drinking water.

Experimental design

Seven week-old mice were grouped into 4 (n=9 per group): Group 1 (control), in which the mice consumed a regular chow; Group 2 (control plus taurine), in which the mice consumed a regular chow and 1% taurine; Group 3 (high fat), in which the mice consumed a high-fat diet; and Group 4 (high-hat diet plus taurine), in which the mice consumed a high-fat diet and 1% taurine. At the end of experiment, blood samples were collected from femoral arteries and hearts and aortas were excised. The livers were also excised and used to measure cholesterol contents and to determine enzymatic activities.

Measurement of serum and liver cholesterol

Serum was prepared by centrifugation at 3,000 rpm for 10 min. Serum cholesterol was measured using an enzymatic method and commercial kits, Autosera CHO-2. HDL cholesterol was measured in the same manner after precipitating out the LDL and VLDL cholesterol with sulfate-Mg^{2+} [36]. LDL plus VLDL cholesterol was calculated by subtracting HDL cholesterol from the total cholesterol. The cholesterol content of the liver was determined enzymatically after extraction with isopropanol [20].

Measurement of enzyme activities

Livers were excised, and microsomes were prepared by ultracentrifugation[23]. Activities of 3-hydroxy-3-mehtylglutaryl coenzyme (HMG-Co) A reductase, acyl-coenzyme A:cholesterol acyltransferase (ACAT), and cholesterol 7α-hydroxylase in the liver microsomes were determined using methods described by Brown *et al.*[1], Lichtenstein and Brecher[15], and Nicolau *et al.*[22], respectively. Protein was measured by the method of Lowry *et al.*[16].

Quantitative assessment of lipid accumulation

The ether anesthetized mice were exanguinated. Fixation and tissue preparation for *en face* Oil red-O staining were done by perfusing the aorta *via* the left ventricle, first with phosphate buffered saline and then with 10% buffered formalin as described [35]. The heart and aorta were immersed in 10% buffered formalin. Under a dissecting microscope, the adventitia was separated from the intima and media using fine forceps and scissors. The aortic valve was excised from the aorta, and stained for lipid with Oil red-O. The segment of the aortic valve was laid out flat on a glass slide with the endothelium face up. The areas of Oil red-O positive lesions were analyzed quantitatively using an image analyzer LEICA Q 500 MC (Cambridge, England).

RESULTS

Body weight

The body weight of the mice on the high-fat diet was 1.5 times heavier than that of the control mice at the end of the experiment (Table 1).

Serum cholesterol levels

Ingestion of the high-fat diet led to a marked elevation of serum LDL and VLDL cholesterol from 10.9 to 93.8 mg/dl (Table 1). Taurine treatment decreased serum LDL and VLDL cholesterol levels by 49%. Although serum HDL cholesterol tended to decrease by ingestion of the high-fat diet, taurine significantly increased serum HDL cholesterol.

Table 1. Effect of taurine on body weight and serum cholesterol levels

	Body weight	Total cholesterol	LDL+VLDL cholesterol	HDL cholesterol
	(g)		(mg/dl)	
Control	23.4 ± 0.6	84.1 ± 2.8	10.9 ± 3.4	73.2 ± 2.8
Control + Taurine	24.1 ± 0.5	83.0 ± 1.6	3.4 ± 1.6	78.8 ± 1.3
High-fat	35.3 ± 1.38 ***	160.3 ± 11.7***	93.8 ± 9.7***	64.3 ± 3.2
High-fat + Taurine	35.8 ± 0.7	151.0 ± 2.6	$52.9 \pm 3.0^{\#}$	$93.5 \pm 2.0^{\#\#}$

Each value represents the mean ± SEM for data obtained from 9 mice. Significant difference; ***$p<0.001$ (*vs* Control), $^{\#\#}p<0.01$ (*vs* High-fat). Student's *t*-test was used to determine the statistical significance.

Hepatic cholesterol levels

Liver cholesterol was markedly increased by ingesting of a high-fat diet for 6 months (Fig. 1). Taurine significantly decreased total cholesterol contents, attributed to a decrease in cholesteryl ester. In the control mice, taurine had no apparent effect.

Hepatic enzyme activity

Activities of enzymes in the liver responsible for cholesterol synthesis and metabolism were determined (Fig. 2). HMG-CoA reductase activity was markedly suppressed in mice fed a high-fat diet, and taurine tended to increase the enzymatic activity. Acyl-CoA:cholesterol acyltransferase (ACAT) activity was measured to determine conversion of free cholesterol to cholesteryl ester. Feeding of high-fat diet markedly increased the ACAT activity, while taurine significantly decreased it. In agreement with reported data[21], cholesterol 7α-hydroxylase activity decreased in case of high-fat diet supplemented with cholesterol and cholic acid. Taurine doubled the enzymatic activity up to the control level in mice fed a high-fat diet. The stimulatory effect of taurine on cholesterol 7α-hydroxylase activity was also seen in mice fed regular chow.

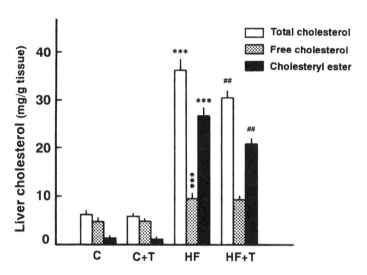

Figure 1. Effect of taurine on liver cholesterol content. C57BL/6J mice consumed a high-fat diet for 6 months (HF) and the control group (C) consumed regular chow. Taurine was dissolved in drinking water at 1% (w/v), and provided during ingestion of regular chow (C+T) or the high-fat diet (HF+T). Cholesterol levels were measured enzymatically using commercial kits after extraction with isopropanol. Each value represents the mean ± SEM for data obtained from 9 mice. Significant difference; ***p<0.001 (*vs* control group), ##p<0.01 (*vs* high-fat group).

Arterial lipid accumulation

Lipid accumulation was evaluated quantitatively as the Oil red-O stained area, using an image analyzer. Ingestion of a high-fat diet led to lipid accumulation in the aortic valve. Treatment of mice on the high-fat diet with taurine resulted in a decrease in lipid accumulation by 20% (Fig. 3).

DISCUSSION

We investigated the effect of long-term treatment with taurine on cholesterol metabolism and development of atherosclerosis in mice on a high-fat diet. Ingestion of a high-fat diet by C57BL/6J mice for 6 months led to development of arterial lesions, accompanied by elevation of LDL and VLDL. Taurine treatment decreased serum LDL and VLDL, and increased HDL. Reduction in elevated levels of atherogenic LDL and VLDL cholesterol results in a reduction in risks for coronary artery disease[9]. Moreover, a number of epidemiological studies have clearly demonstrated that elevated plasma concentrations of HDL cholesterol exert a protective effect against the development of coronary heart disease[8,17]. Therefore, alteration of the serum cholesterol profile by taurine may prevent development of atherosclerosis. In fact, taurine reduced aortic lipid accumulation by 20% in C57BL/6J mice.

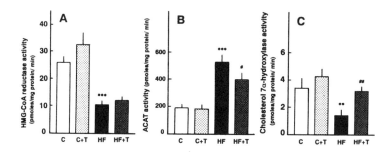

Figure 2. Effect of taurine on enzymatic activities responsible for cholesterol metabolism. C57BL/6J mice consumed a high-fat diet for 6 months (HF). The control group (C) consumed regular chow. Taurine was dissolved in drinking water at 1% (w/v), and provided during ingestion of regular chow (C+T) or the high-fat diet (HF+T). Microsomes were prepared from the liver of each mouse, and used for assays of HMG-CoA reductase (panel A), acyl-coenzyme A:cholesterol acyltransferase (ACAT) (panel B), and cholesterol 7α-hydroxylase (panel C). Each value represents the mean ± SEM for data obtained from 9 mice. Significant difference; **$p<0.01$, ***$p<0.001$ (*vs* control group), #$p<0.05$, ##$p<0.01$ (*vs* high-fat group).

Cholesterol-lowering effects of taurine were noted in various experimental animals[10,12,33,38]. We have also shown that taurine decreased serum cholesterol and prevented lipid accumulation in mesenteric arteries in stroke-prone spontaneously hypertensive rats fed a high-fat diet[18]. In addition, taurine accelerates the regression of pre-established hypercholesterolemia[19]. The mechanism by which taurine decreases serum LDL and VLDL cholesterol, and increases HDL cholesterol remains elusive. Several investigations showed that taurine stimulates conversion of cholesterol to bile acid *via* elevation of cholesterol 7α-hydroxylase activity[13,18,21,34]. Conversion of cholesterol to bile acid represents the major regulated pathway whereby cholesterol is eliminated from the body[30]. Hepatic cholesterol 7α-hydroxylase is the initial and rate-limiting enzyme in the synthesis of bile acid. There are data that indicate a relation between cholesterol 7α-hydroxylase activity and serum cholesterol levels. A rabbits strain resistant to diet induced-hypercholesterolemia has a higher level of cholesterol 7α-hydroxylase[28]. Hamsters infected with a cholesterol 7α-hydroxylase adenoviral construct exhibit a higher level of hepatic cholesterol 7α-hydroxylase mRNA, and the plasma level of LDL decreases more than 50%[31]. The bile acid sequestrant, cholestyramine, increases cholesterol catabolism by binding bile acid in the intestine, which in turn increases activities of cholesterol 7α-hydroxylase and HMG-CoA reductase, and the plasma cholesterol level is reduced[7]. Loss of bile acid from the entrahepatic circulation results in elevation of cholesterol 7α-hydroxylase activity an increase in the rate of bile acid synthesis. The liver compensates for the loss of cholesterol by increasing the rate of de novo cholesterol biosynthesis and, in many cases, by up-regulating hepatic LDL receptor activity. The increase in LDL receptor activity elevates LDL uptake and thereby decreases plasma LDL level[14,29]. Thus, it is likely that stimulation of bile acid synthesis through elevation of cholesterol 7α-hydroxylase activity may be related to beneficial effect of taurine on serum and liver cholesterol levels.

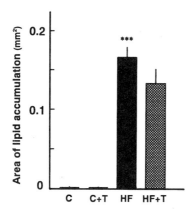

Figure 3. Effect of taurine on arterial lipid accumulation. C57BL/6J mice consumed a high-fat diet for 6 months (HF), and the control group (C) consumed regular chow. Taurine was dissolved in drinking water at 1% (w/v), and provided during ingestion of regular chow (C+T) or the high-fat diet (HF+T). The lipid in the aortic valve was stained with Oil red-O, and the area of lipid accumulation was determined using an image analyzer. Each value represents the mean ± SEM for data obtained from 9 mice. Significant difference; ***p<0.001 (*vs* control group).

A marked increase in hepatic cholesterol content and acyl-coenzyme A:cholesterol acyltransferase (ACAT) activity in mice fed a high-fat diet was significantly reduced by taurine. In the liver, cholesterol exists as free cholesterol and as cholesteryl esters. The amount of free cholesterol within the cell is tightly regulated by several cellular mechanisms. One important mechanism is the conversion of free cholesterol to cholesteryl ester, which is catalyzed by ACAT[32]. Cholesteryl ester is used for assembly of VLDL particles in the liver which are secreted into the blood stream[4]. The decrease in liver ACAT activity and cholesterol content suggests that stimulation of bile acid synthesis leads to decreased hepatic cholesteryl esters and a reduction in hepatic VLDL secretion and serum VLDL and LDL levels.

Taurine significantly increased serum HDL cholesterol levels and elevated hepatic cholesterol 7α-hydroxylase (7αCH) activity. Sulfur amino acid, dietary proteins, carbohydrates, fibe, and xenobiotic have been reported to affect serum HDL levels by altering the expression of apolipoprotein (apo) A-1 gene[3,5,24]. Dueland *et al.* showedhepatic expression of apoA-1 and 7αCH activity varied in parallel in C57BL/6J mice and that there was a significant correlation between HDL cholesterol and the expression of 7αCH activity[6], although the mechanism by which expression of cholesterol 7α-hydroxylase influences the expression of apoA-1is unknown. Similar observations have been made in humans[2]. These findings suggest that 7αCH may influence serum HDL cholesterol level in taurine-treated mice.

Despite many studies on hypocholesterolemic ef fects of taurine, less attention has focused on its anti-atherosclerotic effects. Petty *et al.* noted that taurine prevented the progression of atherosclerosis without affecting plasma cholesterol levels in rabbits fed a high-fat diet and considered the possible involvement of anti-oxidative actions of taurine[27]. Based on our present data, it is likely that improvement in serum lipoprotein profiles is the main mechanism by which taurine reduced arterial lipid accumulation. However, our study on apoE-deficient mice model revealed that taurine prevents progression of atherosclerosis, despite an increase in atherogenic lipoproteins, and we suggest that anti-oxidative actions of taurine may be related to its anti-atherosclerotic effect. Thus, anti-atherosclerotic effects of taurine may attributed to both a decrease in atherogenic lipoproteins and an increase in HDL cholesterol, and anti-oxidative action.

REFERENCES

1. Brown, M.S., Goldstein, J.L. and Dietschy, J.M., 1979, Active and inactive forms of 3-hydroxy-3-methylglutaryl coenzyme A reductase in the liver of the rat, *J. Biol. Chem.*, **254**:5144-5149.
2. Buchwald, H., Varco, R.L., Matts, J.P., Long, J.M., Fitch, L.L., Campbell, G.S., Pearce, M.B., Yellin, A.E.A. and Smink, R.D., 1990, Effect of partial ileal bypass surgery on mortality and morbidity from coronary heart disease in patients with hypercholesterolemia, Report of the program on the surgical control of the hyperlipidemias, *N. Engl. J. Med.*, **323**:946-955.
3. Carr, T.P. and Lei, K.Y., 1989, In vivo apoprotein catabolism of high density lipoproteins

in copper-deficient, hypercholesterolemic rats, *Proc. Soc. Exp. Biol. Med.*, **191**:370-376.
4. Carr, T.P., hamilton, R.L. and Ludel, L.L., 1995, ACAT inhibitors decreases secretion of cholesteryl esters and apolipoprotein B by perfused livers of African green monkeys, *J. Lipid Res.*, 36:25-36.
5. Chong, K.S., Nicolosi, R.J., Rodger, R.F., Arrigo, D.A., Yuan, R.W., MacKey, J.J., Georas, S., and Herbert, P.N., 1987, Effect of dietary fat saturation on plasma lipoproteis and high density lipoprotein metabolism of the rhesus monkey, *J. Clin. Invest.* 79:675-683.
6. Dueland, S., France, D., Wang, S.L., Trawick, J.D. and Davis, R.A., 1997, Cholesterol 7α-hydroxylase influences the expression of hepatic apoA-1 in two inbred mouse strains displaying different susceptibilities to atherosclerosis and in hepatoma cells, *J. Lipid Res.*, **38**:1445-1453.
7. Einarsson, K., Ericsson, S., Ewerth, S., Reihner, E., Rudling, M., Stahlberg, D., and Angelin, B., 1991, Bile acid sequestrants: mechanisms of action on bile acid and cholesterol metabolism, *Eur. J. Clin. Pharmacol.*, **40**:S53-S58.
8. Gordon, T., Kannel, W.B., Castelli, W.P. and Dawber, T.R., 1981, Lipoproteins,cardiovascular disease, and death. The Framingham Study, *Arch. Intern. Med.*, **141**:1128-1131.
9. Gould, A.L., Rossouw, J.E., Santanello, N.C., Heyse, J.F., and Furberg, C.D., 1995, Cholesterol reduction yields clinical benefit: A new look at old data, *Circulation*, **91**:2274-2282.
10. Herrman, R.G., 1959, Effect of taurine, glycine and β-sitosterol on serum and tissues cholesterol in the rabbit, *Cir. Res.*, **7**:224-227.
11. Huxtable, R.J., 1992, Physiological actions of taurine, *Physiol. Rev.*, 72:101-163.
12. Kamata, K., Sugiura, M., Kojima, S. and Kasuya, Y., 1996, Restoration of endothelium-dependent relaxation in both hypercholesterolemia and diabetes by chronic taurine, *Eur. J. Pharmacol.*, **303**:47-53.
13. Kibe, A., Wake, C., Kuramoto, T. and Hoshita, T., 1980, Effect of dietary taurine on bile acid metabolism in guinea pigs, *Lipids*, **15**:224-229.
14. Kovanen, P.T., Bilheimer, D.W., Goldstein, J.L., Jaramillo, J.J. and Brown, M.S., 1985, Regulatory role of low density lipoprotein receptors in vivo in the dog, *Proc. Natl. Acad. Sci. USA*, **78**:1194-1198.
15. Lichtenstein, A.H. and Brecher, P., 1980, properties of acyl-CoA:cholesterol acyltransferase in rat liver microsomes, *J. Biol. Chem.*, **255**: 9098-9104.
16. Lowry, O.H., Rosebrough, N.J., Farr, A.L. and Randall, R.J., 1951, Protein measurement with the Folin phenol reagent, *J. Biol. Chem.*, **193**: 265-275.
17. Miller, N.E., Forde, O.H., Thelle, D.S. and Mjos O.D., 1977, The Tromso heart-study. High density lipoprotein and coronary heart-disease: a prospective case-control study, *Lancet*, **1**: 965-968.
18. Murakami, S., Yamagishi, I., Asami, Y., Ohta, Y., Toda, Y., Nara. Y. and Yamori, Y., 1996, Hypolipidemic effect of taurine in stroke-prone spontaneously hypertensive rats, *Pharmacology*, **52**:303-313.
19. Murakami, S., Nara, Y. and Yamori, Y., 1996, Taurine accelerates the regression of hypercholesterolemia in stroke-prone spontaneously hypertensive rats, *Life Sci.*, **58**:1643-1651.
20. Murakami, S., Yamagishi, I., Sato, M., Tomisawa, K., Nara, Y. and Yamori, Y., 1997, ACAT inhibitor HL-004 accelerates the regression of hyper-cholesterolemia in stroke-prone spontaneously hypertensive rats (SHRSP): stimulation of bile acid production by HL-004, *Atherosclerosis*, **133**: 97-104.
21. Nakamura-Yamanaka, Y., Tsuji, K. and Ichikawa, T., 1987, Effect of dietary taurine on cholesterol 7α-hydroxylase activity in the liver of mice fed a lithogenic diet, *J. Nutr. Sci. Vitaminol.*, **33**:239-243.

22. Nicolau, G., Shefer, S., Salen, G., and Mosbach E.H., 1974, Determination of hepatic cholesterol 7α-hydroxylase activity in man, *J. Lipid Res.*, **15**:146-151.
23. Nordstrom, J.L., Rodwell, V.W. and Mitscelen, J.J., 1977, Interconversion of active and inactive forms of rat liver hydroxymethylglutaryl-CoA reductase, *J. Biol. Chem.*, **252**: 8924-8934.
24. Oda, H. and Yoshida, A., 1994, Effect of feeding xenobiotics on serum high density lipoprotein and apolipoprotein A-1 in rats, *Biosci. Biotech. Biochem.*, **58**:1646-1651.
25. Paigen, B., Morrow, A., Brandon, C., Mitchell, D. and Holmes, P.A., 1985, Variation in susceptibility to atherosclerosis among inbred strains of mice, *Atherosclerosis*, **57**: 65-73.
26. Paigen, B., Ishida, B.Y., Verdtuyft, J., Winters, R.B. and Albee, D., 1990, Atherosclerosis susceptibility differences among progenitors of recombinant inbred strain of mice, *Arteriosclerosis*, **10**: 316-323.
27. Petty, M.A., Kintz, J. and DiFrancesco, G.F., 1990, The effects of taurine on atherosclerosis in cholesterol-fed rabbits, *Eur. J. Pharmacol.*, **180**:119-127.
28. Poorman, J.A., Buck, R.A., Smith, S.A., Overturf, M.L. and Loose-Mitchell, D.S., 1993, Bile acid excretion and cholesterol 7α-hydroxylase expression in hypercholesterolemia-resistant rabbits, *J. Lipid Res.*, **34**:1675-1685.
29. Reihner, E., Angelin, B., Rudling, M., Ewerth, S. and Einarsson, K., 1990, Regulation of hepatic cholesterol metabolism in humans: stimulatory effects of cholestyramine on HMG-CoA reductase activity and low density lipoprotein receptor expression in gallstone patients, *J. Lipid Res.*, **26**:465-472.
30. Russel, D.W., 1992, Bile acid biosynthesis. *Biochemistry*, **31**:4737-4749.
31. Spady, D.K., Cuthbert, J.A., Willard, M.N. and Meidell, R.S., 1995, Adenovirus-mediated transfer of a gene encoding cholesterol 7α-hydroxylase into hamsters increases hepatic enzyme activity and reduces plasma total and low density lipoprotein cholesterol, *J. Clin. Invest.*, **96**:700-709.
32. Suckling, K.E., and Stange, E.F., 1985, Role of acyl-CoA:cholesterol acyltransferase in cellular cholesterol metabolism, *J. Lipid Res.*, **26**:647-671.
33. Sugiyama, K., Kushima, Y., and Muramatsu, K., 1984, Effect of methionine, cystein, and taurine on plasma cholesterol level in rtas fed a high cholesterol diet, *Agric. Biol. Chem.*, **48**:2897-2899.
34. Sugiyama, K., Ohishi, A., Ohnuma, Y., and Muramatsu, K., 1989, Comparison between the plasma cholesterol-lowering effects of glycine and taurine in rats fed on high cholesterol diets, *Agric. Biol. Chem.*, **53**:1647-1652.
35. Sugiyama, F., Haraoka, S., Watanabe, T., Shiota, N., Taniguchi, K., Ueno, Y., Tanimoto, K., Murakami, K. and Fukamizu, A., 1997, Acceleration of atherosclerosis lesions in transgenic mice with hypertension by the activated renin-angiotensin system, *Lab. Invest.*, 76:835-842.
36. Warnic, G.R., Benderson, J. and Albers, J.J., 1982, Dextran sulfate-Mg^{2+} precipitation procedure for quantification of high-density lipoprotein cholesterol, *Clin. Chem.*, **28**:1379-1388.
37. Wright, C.E., Tallan, H.H. and Lin, Y.Y., 1986, Taurine: Biological update, *Ann. Rev. Biochem.*, **55**:427-453.
38. Yamanaka, K., Tsuji, K. and Ichikawa, T., 1986, Stimulation of chenodeoxycholic acid excretion in hypercholesterolemic mice by dietary taurine, *J. Nutr. Sci. Vitaminol.*, 32:287-296.
39. Yamori, Y., Horie, R., Nara, Y., Tagami, M., Kihara, M., Mano, M. and Ishino, H., Pathogenesis and dietary prevention of cerebrovascular diseases in animal models and epidemiological evidence for the applicability in man, In *Prevention of cardiovascular diseases: An approach to active long life* (Y. Yamori and C. Lenfant, eds), Elsevier, Amsterdam.

HYPOLIPIDEMIC EFFECT OF TAURINE IN GOLDEN SYRIAN HAMSTERS

Takaaki Takenaga, Keisuke Imada and Susumu Otomo
Pharmacological Evaluation Laboratory, Taisho Pharmaceutical Co., Ltd., Ohmiya, Japan

INTRODUCTION

Hyperlipidemia is implicated in development of atherosclerosis and coronary heart disease[1, 2]. Taurine, as well as glycine, plays a critical role in cholesterol catabolism by subserving its conjugation to bile acids, and has well known blood cholesterol-lowering effects[3-6]. Most of the experiments concerned with the cholesterol lowering effect of taurine, however, have been carried out on rats and mice[3-6]. In these species, the majority of plasma lipoproteins is HDL, while in humans that is represented by LDL. Furthermore, in rats or mice, unlike humans, conjugation of bile acids occurs preferentially with taurine. Thus, in rats and humans lipid metabolism follows different pathways. In hamster and guinea pig, on the contrary, the ratio of glycine/taurine (G:T) in bile acid conjugation is similar to that found in humans (G:T=3:1)[7]; consequently, hamster seems to be an appropriate animal model to evaluate the efficacy of taurine against hyperlipidemia. Indeed, the hyperlipidemic hamster has been frequently used as a model system for the research and development of anti-hyperlipidemic drugs[8, 9]. In this study we examined the blood lipid-lowering effect of taurine in golden syrian hamsters.

Taurine 4, edited by Della Corte *et al.*
Kluwer Academic / Plenum Publishers, New York, 2000.

MATERIALS AND METHODS

Animals and diets

Male golden syrian hamsters weighing 70-80 g obtained from Charles River Inc. (Atsugi, Japan), were housed in groups of six for 1 week prior to the beginning of the experiment. The animals were exposed to a 12 hr light cycle (7:00 to 19:00), and received for two weeks the experimental diet (6 hamsters/each diet) *ad libitum* shown in Table.1. At the end of feeding period the animals were fasted for 24 hr, and then bled from vena cava under diethyl ether anesthesia. Livers were excised after perfusion with PBS through the portal vein. Blood was added with EDTA to avoid coagulation and plasma was obtained after centrifugation at 3,000 r.p.m. for 15 min. Livers were washed with PBS and kept at –80 °C until use.

Table 1. Compositions of experimental diets (%)

	Normal	1%T	HF/HC	HF/HC/0.5%T	HF/HC/1%T
Casein	25	25	25	25	25
Corn starch	40	36	36	35.8	35.4
Sucrose	20	18	18	17.7	17.6
Corn oil	5	5			
Coconuts oil			10	10	10
Vitamin Mixture (AIN-93)	1	1	1	1	1
Mineral Mixture (AIN-93G)	3.5	3.5	3.5	3.5	3.5
Choline chloride	0.5	0.5	0.5	0.5	0.5
Cellurose	5	5	5	5	5
Cholesterol			1	1	1
Taurine		1		0.5	1

HF/HC: High fat/high cholesterol diet
T:Taurine

Plasma lipids concentration

Plasma concentration of total cholesterol, HDL-cholesterol, triglycerides, NEFA and phospholipids were determined with use of commercial assay systems (Wako Pure Chemical Industries, Ltd.).

Hepatic cholesterol and triglycerides

A 0.5 g portion of liver was minced and suspended into a chloroform/methanol mixture (3:1) and homogenized for 3 min at room temperature for 24 hr. Lipids extracted into the organic phase were subsequently analysed for cholesterol and triglyceride contents (Wako Pure Chemical Industries, Ltd.).

Agarose gel electrophoresis of plasma lipoproteins

One ml of plasma was applied onto 1% agarose gel and run in barbital buffer (pH 8.8) at 90 V for 25 min. Cholesterol was stained using a commercial reagent (Titan gel S-cholesterol, Helena Laboratories) and the relative visible band was analyzed by densitometric scanning.

Statistical analyses

Students' t-test, Dunnett's multiple test and Williams' multiple test were performed with use of a SAS program, version 6.12 (SAS Institute, Inc.).

RESULTS AND DISCUSSION

The various diets listed in table 1 did not affect body weight gain nor did they modify food consumption (Table 2). However, liver weight was significantly increased by high fat/high cholesterol diet, either alone or supplemented with taurine (Table 2).

After feeding these diets for 14 days, plasma was taken from vena cava, and lipids were examined by enzymatic methods. Total cholesterol was markedly increased in animals fed on high fat/high cholesterol diet. Taurine, however, significantly and dose-dependently, counteracted this effect, although it did not (Table 3) modify total cholesterol level in animals fed on a normal diets. HDL-cholesterol level was also increased by high fat/high cholesterol diet, but taurine had no influence upon it (Table 3). Taurine was also able to counteract drastically in a dose dependent fashion the effect of high fat/high cholesterol diet on triglycerides which resulted markedly elevated as compared to values found in animals fed on a normal diet (Table 3). High fat/high cholesterol diet also caused a marked increase in plasma levels of NEFA as well as phospholipids. When taurine was added to this diet, the increase in plasma levels of NEFA and phospholipids were not so pronounced; thus indicating that taurine counteracted, at least in part, the increase in NEFA/phospholipid promoted by high fat/high cholesterol diet.

Table 2. Body weight gain, Food consumption and Liver weight of hamsters fed on a high fat/high cholesterol-diet supplemented with taurine *,**:significantly different from Normal-diet animals (Dunnett's test, P<0.05 and P<0.01, respectively).

	Normal	1%T	HF/HC	HF/HC/0.5%T	HF/HC/1%T
Body weight gain (g/14 days)	14.2 ±0.84	13.3 ±1.07	17.2 ±1.15	18.7 ±0.83**	17.7 ±0.80
Food consumption (g/day)	6.43 ±0.48	5.78 ±0.45	6.48 ±0.54	6.32 ±0.41	6.28 ±0.47
Liver weight (g)	3.41 ±0.23	3.09 ±0.11	4.82 ±0.17**	4.76 ±0.30**	4.35 ±0.14*

*,** : significantly different from Normal group (Dunnett's test, p<0.05 and p<0.01, respectively)

Table 3. Effect of taurine on plasma and liver lipids concentrations in hamsters fed on a high fat/high cholesterol-diet. #:significantly different from Normal-diet animals (Student's t-test, P<0.01) *:significantly different from HF/HC-diet animals (William's test, P<0.05).

	Normal	1%T	HF/HC	HF/HC/0.5%T	HF/HC/1%T
Plasma					
Total Cholesterol (mg/dl)	92.5 ± 7.00	90.1 ± 15.8	307 ± 6.98 [#]	275 ± 7.39 [*]	224 ± 15.4 [*]
HDL-Cholesterol (mg/dl)	53.8 ± 2.70	54.9 ± 6.31	101 ± 3.31 [#]	102 ± 15.5	90.6 ± 3.60
Triglyceride (mg/dl)	256 ± 34.9	227 ± 53.5	795 ± 96.1 [#]	507 ± 46.2 [*]	332 ± 65.0 [*]
NEFA (mEq/l)	1.06 ± 0.07	1.23 ± 0.13	2.01 ± 0.11 [#]	1.49 ± 0.21 [*]	1.37 ± 0.15 [*]
Phospholipids (mg/dl)	192 ± 6.64	190 ± 25.1	384 ± 10.6 [#]	353 ± 13.4	300 ± 18.7 [*]
Liver					
Total Cholesterol (mg/g)	1.39 ± 0.10	1.55 ± 0.09	10.7 ± 1.06 [#]	13.2 ± 1.17	12.2 ± 0.84
Triglyceride (mg/g)	9.38 ± 0.90	11.7 ± 0.57	16.1 ± 0.77 [#]	16.5 ± 0.60	15.5 ± 0.75

#:significantly different from Normal group (Student's t-test, p<0.01)
*:significantly different from HF/HC group (William's test, P<0.05)

As shown in Fig. 1, high fat/high cholesterol diet caused a marked increased of all the serum lipoprotein fractions, especially LDL/VLDL fraction. Taurine added to this diet, counteracted selectively the increase of LDL/VLDL fraction. Although taurine was able to reduce the increase in serum of lipids promoted by high fat/high cholesterol diet, it did not influence the liver accumulation promoted by this diet of cholesterol and triglycerides (Table 3), thus suggesting that taurine increases the serum clearance of both cholesterol and triglycerides. Hamster has been used in a number of studies for the assessment of the hypolipidemic action of drugs, owing to the similarities in lipid metabolism with humans where the increase in LDL account for most of the increase in serum total cholesterol[8, 9]. Moreover, in hamster bile acid conjugation ratio (Glycine/Taurine ratio) is similar to that observed in humans[7]. Thus, hamster is thought to be more

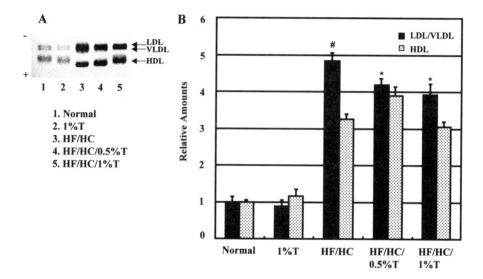

Figure 1. Agarose gel electrophoresis of plasma lipoproteins. A; Electrophoresis pattern of lipoproteins. B; Densitometric analysis of the agarose gel. Plasma lipoproteins were analyzed by agarose gel electrophoresis followed by cholesterol staining as described in the «Materials and Methods» section. Relative amounts of the visible bands on the gel were analyzed by densitometric scanning, taking values observation in Normal-diet animals, as 1. #:significantly different from Normal-diet animals (Student's t-test, $P<0.001$) *:significantly different from HF/HC-diet animals (Williams test, $P<0.05$).

appropriate as an animal model for the evaluation of lipid lowering drugs as compared to rat and mouse. In the present study, we demonstrated that taurine lowers plasma total cholesterol and lipids in hamster fed on high fat/high cholesterol diet by modifying preferentially LDL/VLDL-cholesterol fraction, while leaving almost unaltered HDL-cholesterol.

These observations suggest a possible anti-atherogenic effect of taurine. It is noticeable that taurine was able to lower plasma concentrations of triglycerides and NEFA which were raised by high fat/high cholesterol diet; thus, taurine might be useful for correcting hypertriglyceridaemias. Triglycerides as well as LDL-cholesterol are well recognized as an important risk factor for coronary heart disease[1, 2]. HMG-CoA reductase inhibitors such as plavastatin are frequently used for the treatment of hypercholesterolemia, but they are not so effective for hypertriglyceridaemia. Taurine might be used as a drug for correcting hypertriglyceridaemias, also in view of the absence of overt toxicity. The mechanism of triglyceride lowering action of taurine is not known, whereas it is thought that taurine lowers total cholesterol by promoting cholesterol clearance through bile acid conjugation. Unexpectedly, hepatic concentration of cholesterol and triglycerides were not changed by taurine, probably owing to the large amount of cholesterol and fat in the diet. It is possible that taurine promotes the incorporation of VLDL and LDL into liver; however, further studies are needed to define its mechanism of action.

In conclusion, taurine effectively lowers plasma total cholesterol and triglycerides without affecting HDL-cholesterol in hyperlipidemic hamsters. These results outline the possible use of taurine as an effective drug for the treatment of hypercholesterolemia and hypertriglyceridaemia.

REFERENCES

1.   Castelli, W.P., Garrison, R.J., Wilson, P.W.F., Abbott, R.D., Kalousdian, S. and Kannel, W.B., 1986, Incidence of coronary heart disease and lipoprotein cholesterol levels: the Framingham Study. *JAMA* **256**:2835-2838.
2.   Austin, M.A., 1991, Plasma triglyceride and coronary heart disease. *Arterioscler. Thromb.* **11**:2-14.
3.   Gandhi, V.M., Cherian, K.M. and Mulky, M.J., 1992, Hypolipidemic action of taurine in rats. *Ind. J. Exp. Biol.* **30**:413-417.
4.   Yamanaka, Y., Tsuji, K. and Ichikawa, T., 1986, Stimulation of chenodeoxycholic acid excretion in hypercholesterolemic mice by dietary taurine. *J. Nutr. Sci. Vitaminol.* **32**:287-296.
5.   Herrmann, R.G., 1959, Effect of taurine, glycine and _-sitosterols on serum and tissue cholesterol in the rat and rabbit. *Circ. Res.* **7**:224-227.
6.   Goodman, H.O. and Shinabi, Z.K., 1990, Supplemental taurine in diabetic rats: Effect on plasma glucose and triglycerides. *Biochem. Med. Metabol. Biol.* **43**:1-9.
7.   Gaull, G.E., Pasantes-Morales, H. and Charles, E., 1985, Taurine in human nutrition: An overview. *Prog. Clin. Biol. Res.* **179**:3-21.
8.   Nistor, A.N., Bulla, A., Filip, D.A. and Radu, A., 1987, The hyperlipidemic hamster as a model of experimental atherosclerosis. *Atherosclerosis* **68**:159-173.
9.   Sullivan, M.P., Cerda, J.J., Robbins, F.L., Burgin, C.W. and Beatty, R.J., 1993, The gerbil, hamster, and guinea pig as rodent models for hyperlipidemia. *Lab. Anim. Sci.* **43**:575-578.

TAURINE REDUCES ATHEROSCLEROTIC LESION DEVELOPMENT IN APOLIPOPROTEIN E-DEFICIENT MICE

Y. Kondo[1], S. Murakami[1], H. Oda[2], and T. Nagate[1]

[1]Medicinal Research Laboratories, Taisho Pharmaceutical Co. Ltd., Ohmiya, Japan
[2]Dept. of Applied Molecular Biosciences, Nagoya University, Nagoya, Japan

Abstract: The effects of dietary taurine on development of atherosclerotic lesions were investigated using apolipoprotein E (apoE)-deficient mice. Taurine added to regular chow at 2% (w/w), was made freely available to mice for 3 months. Severe hypercholesterolemia and development of atherosclerotic lesions occurred in the apo-E-deficient mice. Taurine treatment decreased the area of Oil red-O positive lipid accumulation in the aortic valve by 31%. In contrast, taurine significantly increased serum atherogenic lipoproteins (LDL + VLDL), without changing HDL cholesterol levels. Although the levels of serum thiobarbituric acid reactive substances (TBARS) in apoE-deficient mice were significantly higher than in wild-type mice, taurine decreased TBARS by 26%. These observations mean that taurine prevents the development of atherosclerosis, independent of serum cholesterol levels. We suggest that anti-oxidative actions may be involved in the anti-atherosclerotic effects of taurine.

INTRODUCTION

Taurine, 2-aminoethanesulfonic acid, is abundant in animal tissues, and has several physiological and pharmacological functions, such as antioxidation, osmoregulation and calcium modulation[1]. The effect of taurine have been studied using rats[2,3], mice[4], rabbits[5], guinea pigs[6] and also in

Taurine 4, edited by Della Corte et al.
Kluwer Academic / Plenum Publishers, New York, 2000.

humans[7]. We have found that 6 month-treatment with taurine decreased serum atherogenic lipoproteins (LDL and VLDL), increased anti-atherogenic lipoprotein (HDL) and prevented the progression of atherosclerosis in C57BL/6J mice fed a high-fat diet. ApoE-deficient mice have both profound hypercholesterolemia and severe atherogenic lesions with many similarities to those found in humans[8,9]. Palinski *et al.* showed that oxidative modification of lipoproteins is involved in atherogenesis of apoE-deficient mice[10]. Several compounds, including anti-oxidants[11-13], angiotensin converting enzyme inhibitor[14] and estradiol[15], have been reported to prevent the formation of aortic lesions in this mouse model. In the present work, we examined the anti-atherosclerotic effects of dietary taurine in apoE-deficient mice using probucol as a standard drug.

MATERIALS AND METHODS

Taurine was synthesized at Taisho Pharmaceutical Co. (Ohmiya, Japan). Probucol and all other chemicals were purchased from Sigma Chemical Co. (St. Louis, MO). Autosera CHO-2 and Cholestetst HDL used were purchased from Daiichi Kagaku Yakuhin (Tokyo, Japan). Lipid peroxide-Test Wako was purchased from Wako Pure Chem. (Osaka, Japan).

Animals and diets

ApoE-deficient male mice and wild-type C57BL/6J male mice were obtained from Jackson Laboratory (Bar Harbor, ME). The standard diet was purchased from Oriental Yeast Co. (Tokyo, Japan). Taurine and probucol were added to the standard diet at 2% (w/w) and 0.1% (w/w), respectively, and made available to mice for 3 months. ApoE-deficient mice (7-10 weeks old) were grouped into 4; Group 1 (apoE-deficient mice), in which the animals were maintained on a standard diet; Group 2 (apoE-deficient mice plus taurine), in which the animals were maintained on a standard diet and 2% taurine (2478 mg/kg/day); Group 3 (apoE-deficient mice plus probucol), in which the animals were maintained on a standard diet and 0.1% probucol (122 mg/kg/day); Group 4 (wild-type C57BL/6J mice), in which the animals were maintained on a standard diet. At the end of the experiment, blood samples from femoral arteries were collected and hearts and aortas were excised, after an overnight fast.

Measurement of serum lipids

Serum was prepared by centrifugation at 3000 rpm for 10 min. Total cholesterol and HDL cholesterol were measured enzymatically, using commercial kits: Autosera CHO-2 and Cholestest HDL, respectively. LDL and VLDL cholesterol was calculated by subtracting HDL cholesterol from the total cholesterol.

Quantitative assessment of atherosclerotic lesions

Quantification of atherosclerotic lesions was carried out as previously described[16,17]. After exsanguination of ether anesthesed mice, the heart was perfused through the left ventricle, first with phosphate-buffered saline and then with 10% buffered formalin, the right atrium were cut to allow the perfusate to flow out. The heart and aorta were immersed in 10% buffered formalin. The formalin-fixed aorta linked to the heart was opened, the adventitia separated and all three aortic valves were removed with fine forceps and scissors, under a stereoscopic microscope. The aortic root thus prepared was stained with Oil red-O. The areas of Oil red-O positive lesions were determined in 2 mm-long portions extending from the bottom of 3-valve arcs. Lesioned areas were analyzed using an image analyzer LEICA Q500 MC (Cambridge, England).

Determination of serum TBARS

Serum levels of thiobarbituric acid reactive substances (TBARS) were measured using commercially available kits: Lipoperoxide-Test Wako. This kit is based essentially on the method of Yagi [18].

Statistical analysis

Results are expressed as mean values ± S.E.M. Student's *t*-test was used to determine statistical differences.

RESULTS

Body weight

During the experiments, there were no significant differences in body weight among each group. Final body weights of apoE-deficient mice, apoE-deficient mice plus taurine, apoE-deficient mice plus probucol, and wild-type C57BL/6J mice were 32.6 ± 0.5, 34.3 ± 0.9, 32.3 ± 0.5, 32.3 ± 0.5 grams, respectively.

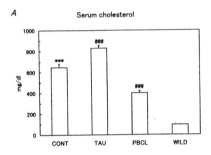

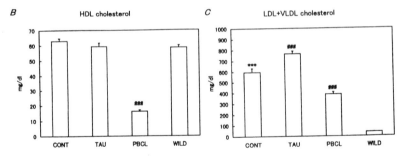

Fig. 1. Effect of taurine on serum cholesterol levels in apoE-deficient mice. ApoE-deficient mice were fed a regular chow (CONT) or regular chow supplemented with 2% taurine (TAU) or 0.1% probucol (PBCL) for 3 months. Wild-type mice (WILD) were also fed a regular chow. The results are expressed as means±S.E.M. for 15 animals. Significant differences; ***p<0.001 (*vs* WILD), ###p<0.001 (*vs* CONT).

Serum cholesterol levels

The effects of dietary taurine on serum cholesterol levels are summarized in Fig. 1. The serum LDL and VLDL cholesterol level of apoE-deficient mice was about 600 mg/dl, that is over 6 times higher than that of wild-type C57BL/6J mice, as described elsewhere[8]. Taurine treatment significantly increased serum LDL and VLDL contents. On the other hand, HDL cholesterol was not influenced by taurine. Probucol significantly decreased both LDL and VLDL, and HDL cholesterol.

Arterial lipid accumulation

As shown in Fig. 2, severe atherosclerotic lesions developed, not only in the aortic valve but also in aortic arch of apoE-deficient mice. Taurine treatment reduced area of Oil red-O positive lesions by 31% (Fig. 2, 3), despite an increase in serum LDL and VLDL cholesterol. In contrast, probucol increased lesioned areas by 24%.

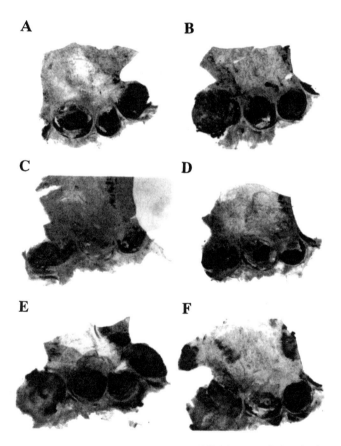

A   B

C   D

E   F

Fig. 2. Photographs of the gross appearances of lipid accumulation in the aortic root. Apo-E-deficient mice were fed a regular chow for 3 months (A, B). Taurine (C, D) or probucol (E, F) was added to regular chow and provided freely to apoE-deficient mice for 3 months. Lipid was stained with Oil red-O. Magnification, X25.

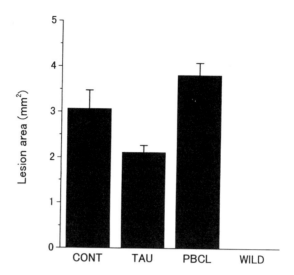

Fig.3. Effect of taurine on arterial lipid accumulation in apo-E-deficient mice. ApoE-deficient mice (CONT) or wild-type C57BL/6J mice (WILD) were fed a regular chow for 3 months. Taurine (TAU) or probucol (PBCL) was added to regular chow and provided freely to apoE-deficient mice. The atherosclerotic surface area stained with Oil red-O was measured in a 2 mm-wide segment from the bottom of arcs of 3 valves. The results are expressed as means ± S.E.M. for 8 animals.

Serum TBARS levels

Serum thiobarbituric acid reactive substances (TBARS) levels in apoE-deficient mice were significantly higher than levels in corresponding wild-type mice (9.07 ± 1.62 *vs* 4.17 ± 0.28 nmol/ml, $p<0.05$). Supplementation of taurine at 0.67 % (w/w) to a regular chow for 3 months decreased the serum TBARS level by 26% in apoE-deficient mice.

DISCUSSION

The present study demonstrates that taurine prevents the development of atherosclerotic lesions in apoE-deficient mice without decreasing serum atherogenic lipoproteins. To elucidate the mechanism by which taurine decreases arterial lipid accumulation, we determined serum thiobarbituric acid reactive substances (TBARS), since anti-oxidative effects of taurine have been reported[1]. The serum TBARS level of apoE-deficient mice was two times higher than that in wild-type mice, suggesting that lipid peroxidation was stimulated in apoE-deficient mice.

Taurine treatment reduced serum TBARS levels by 26%. Recent studies indicated that oxidized LDL participates in the initiation and progression of atherosclerosis[19]. The rapid uptake of oxidized LDL *via* the macrophage scavenger receptor causes transformation of macrophages into the lipid-laden foam cells, characteristic of a fatty streak, an early sign of atherosclerosis. There are numerous reports on the anti-oxidative action of taurine. Taurine reduces malondialdehyde formation[20] and scavenges superoxide radicals in rabbit spermatoza[21], which implies that taurine protects against oxidative damage, under various conditions. Petty *et al.* reported that taurine prevents the development of atherosclerosis without altering serum cholesterol levels, and they suggested the anti-oxidative effects of taurine were involved[5]. Thus, the anti-oxidative action may be associated with the anti-atherosclerotic effects of taurine. Antioxidants such as quercetin[11] and vitamin E[12] have been reported to prevent the progression of atherosclerosis in apoE-deficient mice. The synthetic antioxidant, BO-653 was also shown to be effective in apoE-deficient mice[13]. The hypolipidemic drug, probucol which has anti-oxidative effects, is clinically prescribed for patients with hyperlipidemia and atherosclerosis[22,23]. Although anti-atherosclerotic actions of this drug were seen in Watanabe heritable hyperlipidemic (WHHL) rabbits, an animal model of familial hypercholesterolemia[24,25], probucol accelerates the progression of atherosclerosis in apoE-deficient mice[26]. A new study suggested that pro-atherogenic effects of probucol might be related to increases in plasma fibrinogen and decreased plasma HDL levels[27]. We used probucol as a standard compound and, consistent with previous studies[26,27], probucol markedly decreased serum HDL, LDL and VLDL cholesterol, and it accelerated atherosclerotic lesion formation, while taurine increased atherogenic lipoprotein levels. Hypocholesterolemic effects of taurine have been noted in rats[3,28] and mice fed a high-fat diet[4,16]. Other studies showed that taurine stimulates bile acid synthesis by enhancing cholesterol 7α-hydroxylase activity[3,28], which may be the main mechanism involved in the cholesterol-lowering effects of taurine. We observed a significant decrease in LDL and VLDL, and increase in HDL cholesterol in C57BL/6J mice fed a high-fat diet[16], an event which was accompanied by an increase in hepatic cholesterol 7α-hydroxylase activity. Increased bile acid synthesis induces stimulation of the LDL receptor and cholesterol synthesis in the liver, which leads to increases in LDL uptake from the blood stream. Since apoE-deficient mice lack apolipoprotein E, stimulation of LDL uptake does not occur, and the stimulation of cholesterol synthesis may result in increased hepatic secretion of atherogenic lipoproteins.

In conclusion, our study shows that taurine prevents the development of atherosclerosis without reducing serum atherogenic lipoprotein levels in apoE-deficient mice. Inhibition of lipid peroxidation may be closely linked to the anti-atherosclerotic effects of taurine.

ACKNOWLEDGEMENTS

We thank Drs. S. Haraoka and T. Watanabe, Institute of Basic Medical Sciences, University of Tsukuba, for technical advice regarding the evaluation of arterial lipid accumulation.

REFERENCES

1. Huxtable, R.J., 1992, Physiological actions of taurine. *Physiol. Rev.* **72**:101-163.
2. Sugiyama, K., Kushima, Y., and Muramatsu, K., 1984, Effect of methionine, cysteine, and taurine on plasma cholesterol level in rats fed a high cholesterol diet. *Agric. Biol. Chem.* 48:2897-2899.
3. Murakami, S., Yamagishi, I., Asami, Y., Ohta, Y., Toda, Y., Nara, Y., and Yamori, Y., 1996, Hypolipidemic effect of taurine in stroke-prone spontaneously hypertensive rats. *Pharmacology* **52**:303-313.
4. Nakamura-Yamanaka, Y., Tsuji, K., and Ichikawa, T., 1987, Effect of dietary taurine on cholesterol 7α-hydroxylase activity in the liver of mice fed a lithogenic diet. *J. Nutr. Sci. Vitaminol.* **33**:239-243.
5. Petty, M.A., Kintz, J., and Di Francesco, G.F., 1990, The effects of taurine on atherosclerosis development in cholesterol-fed rabbits. *Eur. J. Pharmacol.* **180**:119-127.
6. Kibe, A., Wake, C., Kuramoto, T., and Hoshita, T., 1980, Effect of dietary taurine on bile acid metabolism in guinea pigs. *Lipids* **15**:224-229.
7. Truswell, A.S., McVeigh, S., Mitchell, W.D., and Bronte-Stewart, B., 1965, Effect in man of feeding taurine on bile acid conjugation and serum cholesterol levels. *J. Atheroscler. Res.* **5**:526-529.
8. Plump, A.S., Smith, J.D., Hayek, T., Aalto-Setala, K., Walsh, A., Verstuyft, J.G., Rubin, E.M., and Breslow, J.L., 1992, Severe hypercholesterolemia and atherosclerosis in apoE-deficient mice created by homologous recombination in ES cells. *Cell* **71**:343-353.
9. Zhang, S.H., Reddick, R.L., Piedrahita, J.A., and Maeda, N., 1992, Spontaneous hypercholesterolemia and arterial lesions in mice lacking apolipoprotein E. *Science* **258**:468-471.
10. Palinski, W., Ord, V.A., Plump, A.S., Breslow, J.L., Steinberg, D., and Witztum, J.L., 1994, ApoE-deficient mice are a model of lipoprotein oxidation in atherogenesis. Demonstration of oxidation-specific epitopes in lesions and high titers of autoantibodies to malondialdehyde-lysine in serum. *Arterioscler. Thromb.* **14**:605-616.
11. Hayek, T., Fuhrman, B., Vaya, J., Rosenblat, M., Belinky, P., Coleman, R., Elis, A., and Ariram, M., 1997, Reduced progression of atherosclerosis in apolipoprotein E-deficient mice following consumption of red wine, or its polyphenols quercetin or

catechin, is associated with reduced susceptibility of LDL to oxidation and aggregation. *Arterioscler. Thromb. Vasc. Biol.* **17**:2744-2752.

12. Pratico, D., Tangirala, R.K., Rader, D.J., Rokach, J., and Fitzgerald, G.A., 1998, Vitamin E suppresses isoprostane generation in vivo and reduces atherosclerosis in apoE-deficient mice. *Nat. Med.* **4**:1189-1192.

13. Cynshi, O., Kawabe, Y., Suzuki, T., Takashima, Y., Kaise, H., Nakamura, M., Ohba, Y., Kato, Y., Tamura, K., Hayasaka, A., Higashida, A., Sakaguchi, H., Takeya, M., Takahashi, K., Inoue, K., Noguchi, N., Niki, E., and Kodama, T., 1998, Antiatherogenic effects of the antioxidant BO-653 in three different animal models. *Proc. Natl. Acad. Sci. U.S.A.* **95**:10123-10128.

14. Hayek, T., Attias, J., Smith, J., Breslow, J.L., and Keidar, S., 1998, Antiatherosclerotic and antioxidative effects of captopril in apolipoprotein E-deficient mice. *J. Cardiovasc. Pharmacol.* **31**:540-544.

15. Bourassa, P.K., Milos, P.M., Gaynor, B.J., Breslow, J.L., and Aiello, R.J., 1996, Estrogen reduces atherosclerotic lesion development in apolipoprotein E-deficient mice. *Proc. Natl. Acad. Sci. U.S.A.* **93**:10022-10027.

16. Sugiyama, F., Haraoka, S., Watanabe, T., Shiota, N., Taniguchi, K., Ueno, Y., Tanimoto, K., Murakami, K., Fukamizu, A., and Yagami, K., 1997, Acceleration of atherosclerotic lesions in transgenic mice with hypertension by the activated renin-angiotensin system. *Lab. Invest.* **76**:835-842.

17. Murakami, S., Kondo-Ohta, Y., and Tomisawa, K., 1999, Improvement in cholesterol metabolism in mice chronic treatment of taurine and fed a high-fat diet. *Life Sci.* **64**:83-91.

18. Yagi, K., 1976, A simple fluorometric assay for lipoperoxide in blood plasma. *Biochem. Med.* **15**:212-216.

19. Ross, R., 1995, Cell biology of atherosclerosis. *Annu. Rev. Physiol.* **57**:791-804.

20. Pasantes-Morales, H. and Cruz, C., 1985, Taurine and hypotaurine inhibit light-induced lipid peroxidation and protect rod outer segment structure. *Brain Res.* **330**:154-157.

21. Alvarez, J.G. and Storey, B.T., 1983, Taurine, hypotaurine, epinephrine and albumin inhibit lipid peroxidation in rabbit spermatozoa and protect against loss of motility. *Biol. Reprod.* **29**:548-555.

22. Yamamoto, A., Matsuzawa, Y., Yokokawa, S., Funahashi, T., Yamamura, T., and Kishino, B., 1986, Effect of probucol on xanthomata regression in familial hypercholesterolemia. *Am. J. Cardiol.* **27**:29H-35H.

23. Davigon, J., Nestruck, A.C., Alaupovic, P., and Bouthillier, D., 1986, Severe hypoalphalipoproteinemia induced by a combination of probucol and clofibrate. *Adv. Exp. Med. Biol.* **201**:111-125.

24. Carew, T.E., Schwenke, D.C., and Steinberg, D., 1987, Antiatherogenic effect of probucol unrelated to its hypocholesterolemic effect: evidence that antioxidants in vivo can selectively inhibit low density lipoprotein degradation in macrophage -rich fatty streaks and slow the progression of atherosclerosis in the Watanabe heritable hyperlipidemic rabbit. *Proc. Natl. Acad. Sci. U.S.A.* **84**:7725-7729.

25. Kita, T., Nagano, Y., Yokode, M., Ishii, K., Kume, N., Narumiya, S., and Kawai, C., 1988, Prevention of atherosclerotic progression in Watanabe rabbits by probucol. *Am. J. Cardiol.* **25**:13-19.

26. Sunny, H.Z., Reddick, R.L., Avdierich, E., Surles, L.K., Jones, R.G., Reynolds, J.B., Quarfordt, S.H., and Maeda, N., 1997, Paradoxical enhancement of atherosclerosis by probucol treatment in apolipoprotein E-deficient mice. *J. Clin. Invest.* **99**:2858-2866.

27. Moghadasian, M.H., McManus, B.M., Godin, D.V., Rodrigues, B., and Frohlich, J.J., 1999, Proatherogenic and antiatherogenic effects of probucol and phytosterols in apolipoprotein E-deficient mice. *Circulation* **99**:1733-1739.
28. Sugiyama, K., Ohishi, A., Ohnuma, Y., and Muramatsu, K., 1989, Comparison between the plasma cholesterol-lowering effects of glycine and taurine in rats fed on high cholesterol diets. *Agric. Biol. Chem.* **53**:1647-1652.

DOES TAURINE PLAY AN OSMOLARITY ROLE DURING ETHANOL INTOXICATION?

F. Lallemand, A. Dahchour, R. J. Ward, and P. De Witte
Université catholique de Louvain, Biologie du Comportement, 1 Place Croix du Sud, 1348 Louvain-la-Neuve, Belgium

INTRODUCTION

The transition of life from an aquatic environment to a terrestrial habitat necessitated the evolution of an inert zwitteronic ion with high solubility and low lipophilicity which could move between intra- and extra-cellular compartments to maintain cellular volume. Taurine was that compound. With higher evolutionary development a wide number of other biological functions have been assigned to taurine, including calcium homeostasis[1], an involvement in the NMDA receptor activity[2], as well as a modulator of hypochlorous acid toxicity within leucocytes[3]. Its role as an osmotic regulator appears to be retained in higher organisms; imposition of an osmotic challenge to animals[4] and certain cell lines[5] show an increase in taurine release to the exterior media in order to maintain cellular volume.

Alcohol administration, either chronic or acute will have a devastating effect upon extracellular and intracellular compartments by virtue of the fact that ethanol rapidly diffuses across all membranes including the blood brain barrier. This could lead to osmotic disturbances in the brain which could perturb the complex relationship between neurotransmitters and their receptors.

The exact role of taurine in the brain remains undefined[6]; it is unlikely to be a neurotransmitter but may modify the action of the excitatory amino acid glutamic acid[7]. Microdialysis experiments of the hippocampus show that taurine is elevated transitorily after an acute injection of ethanol[8] although

Taurine 4, edited by Della Corte *et al.*
Kluwer Academic / Plenum Publishers, New York, 2000.

the exact biological explanation for this increase remains unknown although osmotic alterations may be involved[9,10].

In these present studies we have investigated the effect of acute ethanol injection either 1 g/kg or 2 g/kg upon

a) plasma ethanol concentration

b) brain ethanol concentration

c) plasma taurine concentration

d) hippocampal taurine release

e) hippocampal taurine release after blocking the NMDA receptors with MK801 prior to the ethanol challenge.

MATERIALS AND METHODS

Surgery and HPLC methodology

Male Wistar rats, 200-250g, 6 in each group, were anaesthetised with chloral hydrate (400mg/kg). A guide cannula was inserted into the hippocampus CA2 region using standard stereotaxic techniques (A/P –4.3 mm; M/L 4.0 mm D/V –3.0 mm) of Paxinos and Watson[11]. The dialysis experiments commenced 48h post operation. The probe was connected to a micro-infusion pump (Infusion syringe pump 22, Harvard apparatus) and continuously perfused at 1 µl/min with Ringer solution containing 145 mM NaCl; 4 mM KCl; 1.3 mM $CaCl_2$; buffered to pH 7.2 with phosphate buffer.

Acute ethanol administration

For the microdialysis experiments, rats received either intraperitoneal injection of ethanol, 1 or 2 g/kg, with or without a prior intraperitoneal injection of MK801, 10 mg/kg, 20 minutes before its administration. An appropriate control group was also injected with MK 801, 10 mg/kg, alone. Microdialysate samples were collected every 20 minutes over a 4 hour period.

For the plasma kinetic studies, groups of rats , (6 in each group), were injected with either 1g/kg or 2 g/kg ethanol, blood was removed by cardiac puncture at 20, 40, 80 120 and 240 minutes, placed in tubes either with sodium fluoride for ethanol estimation or heparinised tubes for amino acid analysis. The rats were then killed by cervical dislocation, the brains removed and homogenised in 0.25M sucrose and ethanol content assayed.

The taurine content of either the microdialysates or the plasma samples (after precipitation with 5% sulphosalicylic acid and dilution with phosphate buffer pH 7.0), was assayed by reverse phase HPLC with electrochemical detection using pre column derivatisation with *o*-phthaldialdehyde. The taurine concentration was quantitated by comparison to standards in the range of 5×10^{-6} M with a PC Intergration pack (Kontron Instruments).

Chronic alcohol administration

Male Wistar rats (250 g), individually housed, were pulmonary alcoholised in an isolated plastic chamber (120 x 60 x 60 cm) which contained a mixture of alcohol and air. This mixture was pulsed into the chamber via a mixing system, allowing the quantity of alcohol to be increased every 2 days during the experimental procedure. The animals were kept for 30 days in the alcoholisation chamber. Probes were inserted into the hippocampus region, as described above, and microdialysate samples collected during the first 2h to calculate basal concentrations.

Blood specimens were collected from chronically alcoholised rats at the end of the alcoholisation period. Ethanol concentration in the biological samples was assayed by the production of $NADH_2$ from NAD, when ethanol was oxidised to acetaldehyde via an alcohol dehydrogenase (ADH) -based method (Boerhinger-Mannheim kit).

Statistical Analyses

For the evaluation of statistical differences in the microdialysis experiments, the mean baseline value for taurine was calculated by averaging the concentration of the three perfusate sample values immediately before injection of either ethanol or ethanol + MK801 in the acute studies or during the first 2 hours of withdrawal in the chronic studies. The variation of taurine concentrations in each perfusate was then expressed as a percentage of the baseline value and then analysed by two-way analysis of variance (ANOVA) treatment groups, (i.e. ethanol vs MK801 or MK801 vs Saline) x time, with repeated measures on one factor followed by the least-significant difference test of multiple comparisons (Fischer LSD protected t-test) to determine statistical difference (which was set at $P<0.05$), between each time point and the two drugs (GB-STAT, Dynamic Microsystems, Silver Spring, MD, USA).

RESULTS

Acute ethanol administration

After an acute injection of ethanol, either 1 g/kg or 2 g/kg, the ethanol content in the brain increased in proportion to its plasma concentration, r = 0.94, P <0.01, Fig. 1 and Fig. 2, respectively. Taurine release from the hippocampal brain region was significantly elevated to the baseline value at 20 minutes after 1 g/kg, and between 20 and 80 minutes after 2 g/kg. ethanol concentration.

However such increases of extracellular taurine hippocampal content were not related to the brain ethanol content since, with the lower dose of ethanol, taurine release was evident when the brain ethanol content was 41.8 mg/mg protein; while after the higher brain ethanol levels, 2 g/kg, a significant increase in extracellular hippocampal taurine content was assayed when the brain ethanol content was between 121.1 and 94.6 mg/mg protein. Ethanol administration, 1 g/kg did not significantly alter plasma taurine content. With the higher dose, 2 g/kg there was a significant increase in taurine content at 40 and 80 minutes after injection. However, such changes in plasma taurine levels did not relate to brain hippocampal taurine with either dose of ethanol.

Co-administration of ethanol and MK801

When the NMDA inhibitor MK801 alone was administered i.p. to rats, 10mg/kg, there was a gradual decrease in hippocampal taurine content which reached significance at 100 minutes and was sustained for the duration of the experiment, Fig. 3.

However, when MK801 was administered 20 minutes before ethanol injection, 2 g/kg, no alteration in the taurine release into the hippocampal microdialysate was assayed during the initial stages, taurine only declining significantly after 200 minutes, Figure 4. A similar result was also obtained when MK801 was administered 20 minutes prior to an acute ethanol injection, 1 g/kg (data not shown).

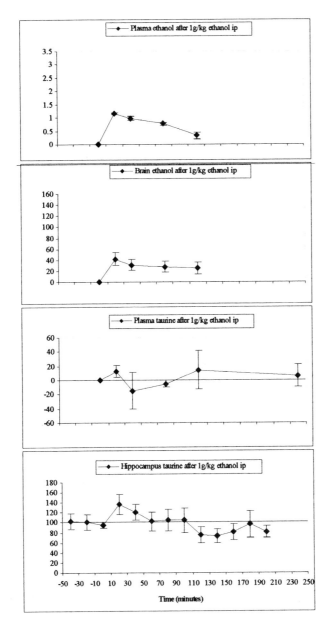

Fig. 1. Alterations in ethanol and taurine in the plasma and brain after administration of 1 g/kg ethanol.

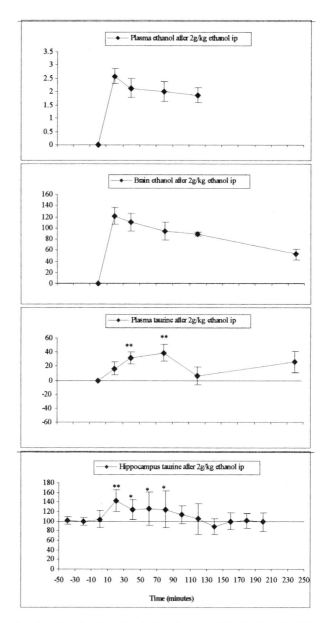

Fig. 2. Alterations in ethanol and taurine in the plasma and brain after administration of 2 g/kg ethanol.

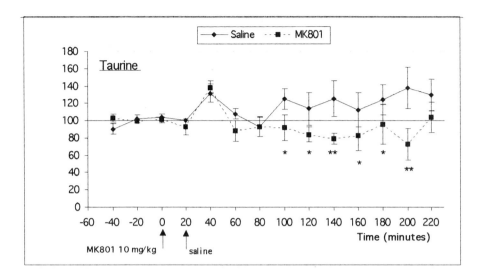

Fig. 3. Percentage change in the baseline TAURINE value after intraperitoneal injection of MK801, 10 mg/kg or saline alone. Significance is calculated by ANOVA between the MK801 and saline injected rats and time for the duration of the microdialysis experiments. * P<0.05; ** P<0.01.

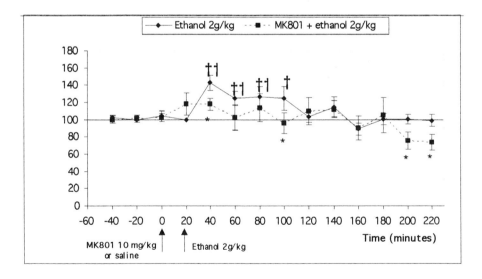

Fig. 4. Percentage change in the baseline TAURINE value after intraperitoneal injection of ethanol, 2 g/kg, or MK801, 10 mg/kg with ethanol, 2 g/kg. Significance is calculated by ANOVA between the two treatments, ethanol and MK801 and time for the duration of the microdialysis experiments. *P<0.05 by comparison to control, † P<0.05, ††P<0.01 by comparison to baseline.

Chronic ethanol administration

When the rats were exposed to chronic pulmonary alcoholisation for 4 weeks, at the end of the alcoholisation period the blood ethanol concentration was 2.79 g/l $\pm$ 0.73 (n = 9). The basal concentration of taurine increased significantly in the hippocampal microdialysate by approximately 2 fold, i.e. from 33.92 $\pm$ 3.46 to 58.67 $\pm$ 9.42 μmol/20 μl.

DISCUSSION

In these present studies we have investigated the effects of acute and chronic ethanol administration upon taurine homeostasis in both the brain and the blood. Our earlier studies had indicated that ethanol induced changes in cellular osmolarity might be, in part, responsible for the transitory rise in taurine after acute ethanol administration. Such increases in extracellular taurine content could be caused by an ethanol-induced hyposmolarity state in the brain cell resulting in the Na^+- dependent transport of taurine across cellular membranes to maintain osmolarity.

What was apparent from these present studies was the differing brain ethanol concentration which elicited an increase in taurine in the extracellular hippocampus media, i.e. 41.8 mg ethanol/mg protein, with the lower ethanol dose (1 g/kg) while the higher ethanol dose (2 g/kg) induced an increase in taurine release at brain ethanol concentrations between 121.1 and 94.6 mg ethanol/mg protein. No change in taurine release was assayed when the brain ethanol concentration reached the lower value of 41.8 mg ethanol/mg protein. Such results indicate that osmolarity might not be the unique cause of the changes in taurine homeostasis observed after acute ethanol administration.

Furthermore, in the plasma, injection of a dose of ethanol, 1 g/kg, did not alter the taurine content for the duration of the experiment. In contrast, with the higher ethanol dose, 2 g/kg, plasma taurine concentration increased significantly at both 40 and 80 minutes. Although some studies have reported a decrease in plasma taurine concentration after 2 g/kg[12], when each animal was individually cannulated via the jugular vein, other studies[13], similarly have reported an increase in taurine in administration of 2 g/kg ethanol. The availability of certain brain amino acids will depend upon their circulating levels as well as their competition with other plasma amino acids for transport carriers across the blood brain barrier. Taurine is able to traverse the blood brain barrier[14], although to a limited extent, such that local synthesis of the sulphonated amino acid could also occur.

When the NMDA receptor was blocked by MK801, doses of ethanol, which alone had elevated hippocampal taurine microdialysate, now elicited no effect upon the release of taurine into the extracellular compartment. Such results would indicate that osmolarity alone is not responsible for the increase in extracellular taurine after either acute or chronic ethanol administration. We would conclude that the NMDA receptor must surely play a role in controlling taurine homeostasis since its inhibition with MK801 abolishes the taurine hippocampal release after ethanol challenge.

Chronic administration of ethanol for 4 weeks caused a two fold increase in the taurine content of the hippocampus extracellular content. The explanation for such changes in taurine homeostasis are as yet unexplained but could indicate a neuromodulatory role for taurine during the extensive adaptation by the brain to the increasing ethanol loading.

Plasma osmolarity is certainly maintained within very narrow limits despite wide fluctuations of plasma osmolyte concentrations, such that for any significant change in plasma osmolarity, alterations in numerous homeostatic controls e.g. kidney function, would be necessary. Further studies of the relationship between changes in tissue osmolarity and taurine homeostasis are clearly warranted to ascertain whether during evolution cellular taurine has retained its ability to act as a regulator of cellular osmolarity in man.

ACKNOWLEDGEMENTS

This work was supported by the Fonds de la Recherche Scientifique et Médicale (1997-2000), IREB and sponsored by LIPHA.

REFERENCES

1. Schaffer, S.W., Azuna, J. and Madura, J.D., 1995, *Amino Acids*. **8** : 231-246.
2. Menendez, N., Herreras, O., Solis, J.M., Herranz, A.S. and Martin del Rio, R.M., 1989, *Neurosci. Lett.*. **102** : 64-69.
3. Cunningham,C., Tipton, K.F. and Dixon, H.B.F., 1998, *Biochem J*.**330:** 933-937.
4. Wade J.V., Olson J.P., Samson F.E., Nelson S.R. and Pazdernik T.L., 1988, *J Neurochem.* **51** : 740-745.
5. Pasantes-Morales, H., Alavez, S., Sanchez-Olea, R. and Moran, J., 1993, *Neurochem. Res.* **18** : 445.
6. Huxtable, R.J., 1992, *Physiol. Rev.* **72** : 101-163.
7. Ward, R.J., Marshall, E.J., Ball, D., Martinez, J. and De Witte, Ph., 1999, *Neurosci. Res. Com.*, **24** : 41-49.
8. Lallemand F., Ward R.J. & De Witte, Ph., 1998, *Alcohol. Clin. Exp. Res.***22**: 175A
9. Dahchour, A., Quertemont, E. and De Witte, Ph., 1994, *Alcohol. Alcoholism.* **29** : 485-487.

10. Dahchour, A., Quertemont, E. and De Witte, Ph., 1996, *Brain. Res.* **735** : 9-19.
11. Paxinos, G.and Watson, C., 1982, *The Rat Brain in Stereotaxic Co-ordinates*, Academic Press, New York.
12. Milakofsky, L., Miller, J.M., and Vogel, W.H., 1986, *Biochem. Pharmacol.* **35** : 3885-3888.
13. Bekairi, A.M., Abulaban, F.S., Tariq, M., Parmar, N.S. and Ageel, A.M., 1987, *Alcohol. Drug. Res.* **7** : 471-479.
14. Benrabh H., Bourre J-M. and Lefauconnier J-M , 1995, *Brain Res* **692**: 57-65.

OSMOTIC SENSITIVITY OF TAURINE RELEASE FROM HIPPOCAMPAL NEURONAL AND GLIAL CELLS

[1,2] James E. Olson and [1] Guang-ze Li

[1] *Department of Emergency Medicine and* [2] *Physiology and Biophysics, Wright State University School of Medicine, Dayton, Ohio*

Abstract: Taurine transport is important for volume regulation of cultured neurons and astroglial cells. Both cell types utilize similar mechanisms for taurine accumulation and efflux. However, taurine lost from cerebellar Purkinje cells *in vivo* is accumulated by adjacent astrocytes during hypoosmotic hyponatremia. To examine mechanisms for transfer of taurine between cell types, we measured relative sensitivities of taurine loss from cultured neurons and astrocytes. Primary cultures of hippocampal neurons and astrocytes were grown from embryonic and neonatal rat brain, respectively. Neurons were used after 10-14 days in culture. Astrocytes were used after 14 days in culture and were grown in the same culture medium used for neurons for 3 days prior to experimentation. Cells were incubated at 37°C for 30 min in isoosmotic (290 mOsm) phosphate-buffered saline (PBS). The PBS was then changed to fresh isoosmotic or to hypoosmotic PBS (270 mOsm or 250 mOsm), made by reducing the NaCl concentration. Cell volume and taurine content were determined immediately before hypoosmotic exposure or 3, 15, or 30 min later. In isoosmotic PBS, astrocytes contained 162 ± 18 nmol taurine/mg protein (mean ± SEM), equivalent to an intracellular concentration of 30.2 ± 2.1 mM. No taurine loss was detectable after 3 or 15 min in either hypoosmotic PBS, but after 30 min in 270 or 250 mOsm PBS, astrocyte taurine was reduced by 8.0% or 22.2%, respectively. Neurons initially contained 114 ± 13 nmol taurine/mg protein, equivalent to an intracellular taurine concentration of 22.2 ± 2.5 mM. After 3 min of exposure to 270 or 250 mOsm PBS, the cells had lost $17 \pm 5\%$ or $25 \pm 4\%$ of their taurine content, respectively. Cell volumes of each cell type were similarly affected by hypoosmotic exposure. We conclude that taurine loss from cultured hippocampal neurons is more sensitive to osmotic swelling than taurine loss from cultured hippocampal astrocytes. This characteristic, if present in cells of the hippocampus *in vivo*, could lead to net transfer of taurine from neurons to glial cells during pathological conditions which cause cell swelling.

Taurine 4, edited by Della Corte *et al.*
Kluwer Academic / Plenum Publishers, New York, 2000.

213

INTRODUCTION

Figure 1 describes a simplified model of spaces in the brain parenchyma. Taurine moves slowly across an intact blood-brain barrier. Taurine is exchanged more rapidly between extracellular and intracellular spaces by transporters and efflux pathways on neurons and glia. The steady state contents of taurine within cellular elements and the concentration of taurine in the extracellular space is determined by activities of the neuronal and glial taurine uptake transporters and rates of release from the cell interiors. An alteration in the rate of one of these transporter pathways will alter the steady state contents and concentration of all three spaces.

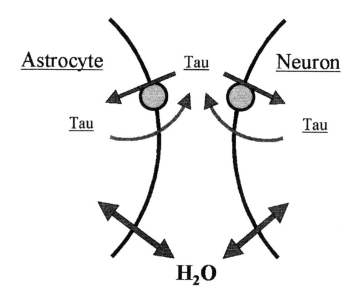

Figure 1. Schematic representation of brain volume spaces and taurine (Tau) transport pathways relevant during hypoosmotic brain edema. Intracellular spaces of astrocytes and neurons are shown on the left and right sides of the drawing, respectively. The extracellular space is represented by the center region.

To explore potential mechanisms which effect redistribution of taurine between neurons and glia of the swollen brain, we examined the sensitivity of changes in taurine content in cultured hippocampal neurons and glial cells to altered extracellular osmolality.

MATERIALS AND METHODS

Cell cultures

All procedures using animals were approved by the Laboratory Animal Care and Use Committee of Wright State University. Neuron cultures were prepared from rat hippocampus using a modification of methods described by Banker and Cowan[1]. Fetuses were removed from anesthetized animals by laparotomy and hippocampi dissected from the brains under sterile conditions. The tissue was softened with 15 min exposure to 0.125% trypsin, triturated with a fire-polished Pasteur pipette and then plated at 60,000 cells/cm^2 onto 35 mm plastic petri dishes coated with polyornithine. The initial plating medium consisted of minimum essential medium containing 50 units/ml penicillin, 50 µg/ml streptomycin and 10% horse serum. After 4-6 hours this was replaced with growth medium, consisting of Neurobasal® plus B27 additives (Life Science Technology, Grand Island, New York) and the same antibiotics. Cultures were maintained at 37°C in an atmosphere of 5% CO_2. Every other day, one-half of the medium was removed and replaced with fresh growth medium. To avoid evaporation of medium and resulting changes in osmolality, petri dishes were placed within a 100 mm petri dish which also contained an open dish of water as described previously[2]. The measured osmolality of the culture medium at the time of experimentation was approximately 290 mOsm.

Cultures of hippocampal astrocytes were prepared according to methods described previously by Pappas and Ransom[3]. The brains of 2-4 day-old rat pups were removed following decapitation and the hippocampi isolated by sterile dissection. Tissue was incubated for a total of 15 min in 0.125% trypsin and triturated. The resulting cell suspension was filtered through an 80 µm nylon mesh. Cells were plated onto 35 mm plastic petri dish at a density of 60,000 cells/cm^2 and grown in culture for two weeks at 37°C in 5% CO_2. At this time, the cells were nearly confluent. The initial medium used for plating contained 20% newborn calf serum. This medium was subsequently changed twice each week to similar growth medium with only 10% newborn calf serum[2]. Three days prior to experimentation, cells were fed with neuron growth medium as described above.

Cell volume and taurine contents

All studies were performed at 37 °C. Cells in culture were rinsed free of growth medium and incubated for 30 min in an isosmotic (290 mOsm) phosphate buffered saline (PBS) containing 137 mM NaCl, 2.7 mM KCl, 1.0

mM CaCl$_2$, 0.5 mM MgCl$_2$, 2.7 mM Na$_2$HPO$_4$, 0.5 KH$_2$PO$_4$, and 5.5 mM glucose (pH=7.3). The PBS then was changed to fresh isoosmotic PBS or to PBS made hypoosmotic (270 mOsm or 250 mOsm) by reducing the concentration of NaCl. Dishes were sampled at the end of the initial incubation in isoosmotic PBS or 3, 15, or 30 min after changing to fresh PBS solutions.

Five minutes prior to sampling, cells were exposed to osmotically matched PBS without glucose but containing 1.0 mM 3-O-methylglucose (3OMG) plus 0.5 μCi/ml [^{3}H]-3OMG and 0.5 μCi/ml [^{14}C]-sucrose. For cells sampled after 3 min in hypoosmotic conditions, this exposure to radioactive tracers began while the cells were in isoosmotic PBS and continued during the hypoosmotic exposure. At the time of sampling, cells were rinsed three times within 15 sec using an osmotically matched sucrose solution containing 1.0 mM phloretin, 0.5 mM Ca(NO$_3$)$_2$ and 10 mM Tris (pH=7.3). Cells then were scraped into 0.6 M HClO$_4$ and the resulting suspension centrifuged at 10,000 x g for 1 min. Radioactivity and taurine contents were determined in the supernatant by liquid scintillation spectroscopy and HPLC, respectively, as previously described[4]. Protein was determined in the cell pellet by the method of Lowry et al.[5]. Cell volume was expressed as the 3OMG space in units of μl/(mg protein) and was corrected for adherent PBS using radioactive sucrose as a marker for the extracellular space.

Data analysis

Data were analyzed by one way ANOVA or repeated measures ANOVA as appropriate. Post hoc analyses were performed using Dunnett's test with measurements in isoosmotic PBS representing the control condition. Significance was indicated for $p < 0.05$.

RESULTS

Cultured astrocytes and neurons had cell volumes of 5.36 ± 0.38 μl/(mg protein) and 4.64 ± 0.15 μl/(mg protein) in isoosmotic PBS, respectively. Both cell types swelled similarly in hypoosmotic solutions and demonstrated subsequent cell volume recovery during the 30 min period of observation.

Taurine contents of glia and neurons are given in Table 1. Because intracellular volume and taurine content were measured for each culture dish, we could calculate initial mean intracellular taurine concentrations of 30.2 ± 2.1 mM and 22.2 ± 2.5 mM in astrocytes and neurons, respectively.

The taurine content of glial cells was not appreciably changed after 3 min of hypoosmotic exposure. After 30 min in 270 mOsm or 250 mOsm PBS, approximately 8% or 18% of the initial taurine content was lost, respectively. In contrast, neuron cultures demonstrated a significant decrease of taurine within 3 min at each hypoosmotic exposure. A further decrease was observed at the 15 min time point for cells in 250 mOsm PBS.

Table 1. Taurine Content of Hippocampal Astrocytes and Neurons during Hypoosmotic Swelling

Osmolality (mOsm)	Hypoosmotic Exposure Time	Astrocyte Taurine Contents (nmol/mg protein)	Neuron Taurine Contents (nmol/mg protein)
290	0 min	162 ± 18	114 ± 13
270	3 min	145 ± 13	94 ± 10*
	15 min	156 ± 15	102 ± 10*
	30 min	149 ± 7*	ND
250	3 min	149 ± 11	85 ± 7*
	15 min	125 ± 25	64 ± 3*
	30 min	126 ± 5*	ND

Values are the mean $\pm$ SEM of 4-6 independent observations.
ND – not determined
* indicates data which are significantly different from the value measured from cells incubated in isoosmotic (290 mOsm) PBS for the same time period ($p<0.05$).

DISCUSSION

These data indicate cultured hippocampal neurons lose taurine more readily than cultured hippocampal astrocytes in response to moderate hypoosmotic exposure. Mechanisms involved in activating the taurine efflux pathways in each cell type have yet to be established; however, in cultured cerebral cortical astrocytes we have previously demonstrated moderate hypoosmotic exposure activates chloride and taurine membrane conductances[6]. Thus, we hypothesize similar mechanisms exist in hippocampal astrocytes and neurons and these pathways are involved in taurine efflux during hypoosmotic exposure. Our data suggest the taurine efflux pathway is activated by smaller osmotic challenges in cultured hippocampal neurons than in cultured hippocampal astrocytes. Alterations in rates of taurine transport into cells also may play a role in redistribution of brain taurine. Further studies addressing this question may be performed using these cell culture systems.

In astrocytes derived from rat cerebral cortex, we previously have shown anion conductances are activated by levels of hypoosmotic exposure which do not activate cation conductances[6]. A complete analysis of ion conductivities activated by hypoosmotic exposure has not been performed for cells cultured from the hippocampus. Therefore, other conductances, including those which cause potassium loss, may be involved in the cell volume responses we have observed. We and others have shown[7,8] the extracellular concentration of potassium does not increase in hippocampal slices exposed to hypoosmotic conditions. This suggests mobilization of intracellular potassium and resulting loss from cells does not occur in vivo under these conditions.

We observed taurine efflux associated with cell swelling in both astrocytes and neurons during these moderate hypoosmotic exposures. In these in vitro conditions, loss of cellular taurine did not alter the taurine concentration in the essentially infinite extracellular space. Thus, taurine efflux contributed to volume regulation of each cell type. However, if these rates of efflux were to occur into the small extracellular space of the osmotically swollen brain in vivo, the extracellular taurine concentration would increase. With maintained rates of taurine accumulation, this would lead to quantitative redistribution of taurine out of neurons and into astrocytes.

ACKNOWLEDGEMENTS

Supported by NIH (NS 37485) and Kettering Medical Center.

REFERENCES

1. Banker, G.A., and Cowan W.M., 1977, Rat hippocampal neurons in dispersed cell culture. *Brain Res.* **126**:397-425.
2. Beetsch, J.W., and Olson, J.E., 1998, Taurine synthesis and cysteine metabolism in cultured rat astrocytes: effects of hyperosmotic exposure. *Am. J. Physiol.* **274**:C866-C874.
3. Pappas, C.A., and Ransom, B.R., 1994, Depolarization-induces alkalinization (DIA) in rat hippocampal astrocytes. *J. Neurophysiol.* **72**:2816-2826.
4. Olson, J.E., 1999, Osmolyte contents of cultured astrocytes grown in hypoosmotic medium. *Biochim. Biophys. Acta* **1453**:175-179.
5. Lowry, O.H., Rosebrough, N.J., Farr, A.L., and Randall, R.J., 1951, Protein measurement with the Folin phenol reagent. *J. Biol. Chem.* **193**:265-275.
6. Olson, J.E., and Li, G., 1997, Increased potassium, chloride, and taurine conductances in astrocytes during hypoosmotic swelling. *Glia* **20**:254-261.
7. Chebabo, S.R., Hester, M.A., Jing, J., Aitken, P.G., and Somjen, G.G., 1995, Interstitial space, electrical resistance and ion concentrations during hypotonia of rat hippocampal slices. *J. Physiol.* **487**:685-697.
8. Kreisman, N.R., and Olson, J.E., 1995, Taurine enhances volume regulation in osmotically swollen hippocampal slices. *Neurosci. Abstr.* **21**:156.

ISOVOLUMIC REGULATION IN NERVOUS TISSUE

A Novel Mechanism of Cell Volume Regulation

O. Quesada, R. Franco, K. Hernández-Fonseca, and K. Tuz

Department of Biophysics, Institute of Cell Physiology, National University of Mexico, Mexico City, 04510, D.F., Mexico

INTRODUCTION

Cell volume regulation is a property present in most animal cell lineages that allows them to recover their original volume after events of swelling or shrinkage. Such events can be caused by changes in external osmolarity or to osmotic gradients generated during normal cell functioning[4,6]. The mechanism of cell volume regulation involves transmembrane fluxes of osmotically active solutes in the necessary direction to counteract the net gain or lose of intracellular water[9]. The process through which cells recover their normal volume after swelling is named Regulatory Volume Decrease (RVD). This consists of the efflux of inorganic osmolytes, such as K^+ and Cl^-, as well as organic compounds such as free amino acids, methyl amines and polyalcohols. These movements create a new osmotic gradient that leads to water efflux and volume recovery.

The simplest and more often experimental paradigm used in RVD studies consists of sudden exposures of cells/tissues to mild-to-acute hyposmotic media (~20 to 50% hyposmotic) and recording different cell parameters (cell volume, osmolyte release, membrane potential, conductances, etc.) These conditions, however, do not occur under normal, physiological circumstances, where changes in external or internal osmolarity are gradual, as the different homeostatic systems are challenged and activated[12]. Even in pathological situations, such as water intoxication or hyponatremia, the activation and eventual surpassing of the encephalic mechanisms of water and electrolyte control lead to *progressive* osmotic changes in the extracellular brain milieu[3].

Taurine 4, edited by Della Corte *et al.*
Kluwer Academic / Plenum Publishers, New York, 2000.

Gradual and continuous changes in external osmolarity (in contrast to the sudden changes usually used) were introduced by Lohr and Grantham[8] to examine volume regulation properties in the S_2 proximal tubules. It was observed that regardless of the progressive reduction in the osmolarity medium *cells do not swell*, provided that the rate of change does not exceed – 3.0 mOsmol/min. The adaptive cell response occurred, however, as cells swell immediately after reintroduction of isosmotic medium[8]. The response was described as *Isovolumic Regulation* due to the lack of change in cell

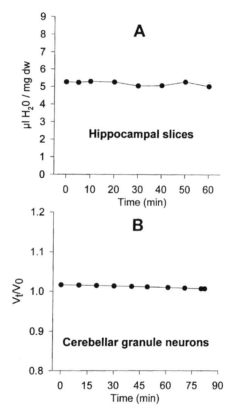

Figure 1. Cell volume changes of hippocampal slices and cerebellar granule neurones exposed to a continuous hyposmotic gradient (300 mOsmol/l $\Rightarrow$ 150 mOsmol/l). A gradient-generating system was constructed as described by Van Driessche *et al.*[13] The rate of change in osmolarity was adjusted at -2.5 or –1.8 mOsmol/min for the experiments with hippocampal slices or cell cultures, respectively. A and B: At time 0 in figures, slices or cerebellar granule neurones were superfused with the osmotic gradient and cell volume was determined at different times. A. Volume changes in hippocampal slices were indirectly estimated by quantification of tissue water content corrected by the interstitial space measured by ^{14}C-inulin distribution. B. Relative cell volume in cerebellar granule neurones was quantified fluorometrically, using calcein-AM as fluorescent dye[1]. Data are means ± SE of 6 (A) or 3 (B) individual experiments.

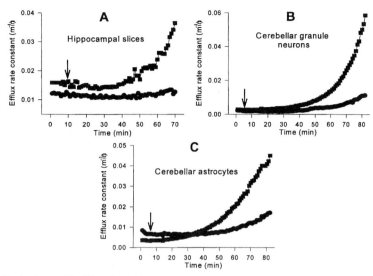

Figure 2. Amino acid efflux from hippocampal slices, cerebellar granule neurones and cerebellar astrocytes elicited by a continuous hyposmotic gradient. Preloaded tissues were initially superfused with isosmotic medium (6-10 min) and then (at arrow) exposed to the osmotic gradient (300 ⇒ 150 mOsmol/l). (■): ^{3}H-taurine; (●): ^{3}H-D-aspartate in all figures. Data are expressed as efflux rate constants and are means ± SE of 6 (A), 8 (B) or 2 (C) individual experiments.

Table 1. Kinetic constants of ^{3}H-taurine and ^{3}H-D-aspartate fluxes from cerebellar granule neurones and hippocampal slices elicited by a continuous hyposmotic gradient

Cerebellar Granule Neurones	k_1 (x10^{-5})	k_2 (x10^{-5})	k_3 (x10^{-5})
External osmolarity range (mOsmol/l)	300 → 249	248 → 203	202 → 156
^{3}H-D-Aspartate	N.A.	10.37 ± 0.53	37.17 ± 1.26
^{3}H-Taurine	5.29 ± 0.33	34.97 ± 1.59	213.38 ± 10.12
Hippocampal Slices	k_1 (x10^{-5})	k_2 (x10^{-5})	k_3 (x10^{-5})
External osmolarity range (mOsmol/l)	300 → 250	249 → 178	175 → 150
^{3}H-D-Aspartate	-2.322 ± 0.111	2.070 ± 0.219	7.138 ± 1.070
^{3}H-Taurine	-4.066 ± 0.542	8.968 ± 0.502	34.829 ± 3.733

The efflux of ^{3}H-taurine and ^{3}H-aspartate under IVR conditions were kinetically analysed adjusting the experimental data of Fig. 1 to lineal regressions in different segments of the curves, as indicated (fractions). Values are the slopes of the adjusted averaged points ± S.E. (n= 4-8) N.A.: Not adjusted

volume. In the present study, we described the occurrence of Isovolumic Regulation in different preparations of nervous tissue, where it is observed, the early activation of taurine and glutamic acid efflux, as well as the relatively late (or absence of) mobilisation of K^+.

RESULTS AND DISCUSSION

Although the experimental model of large and sudden decreases in osmolarity had rendered valuable information to elucidate some basic mechanisms of cell volume control, such changes probably never occur in brain under physiological conditions. This is also true during pathological situations such as chronic hyponatremia, water intoxication or the inappropriate secretion of vasopressin, where the osmolarity changes in the brain interstitial space occurs most likely in a gradual manner, as the osmotic challenge from plasma progressively surpasses the brain homeostatic resistance[3,12]. Thus, the experimental approach of the present work, decreasing gradual and slowly the external osmolarity, could reflect more accurately physiological variations. Under these conditions, Figure 1 shows the lack of change in cell volume when hippocampal slices (A) or cultivated neurones (B) are exposed to the osmotic gradient, regardless of the low, final external osmolarity (~150 mOsmol/l or 50% hyposmotic). This constancy in cell volume appears to result from an active process of volume control accomplished by the adjustment of osmolyte intracellular content and its named Isovolumic Regulation (IVR). This is supported by the swelling observed in cells previously exposed to gradual hyposmotic changes and suddenly returned to isosmotic medium[8].

Figure 2 shows the efflux of ^{3}H-taurine and ^{3}H-D-aspartate from hippocampal slices (A), cerebellar granule neurones (B) or cerebellar astrocytes (C) elicited by a continuous hyposmotic gradient ($300 \Rightarrow 150$ mOsmol/l). In these preparations it is observed first, that efflux of both amino acids is activated early during IVR; second, the lower the external osmolarity is, the faster the amino acid efflux, showing no inactivation phase; third, the release of taurine is larger than that of D-aspartate. A similar higher efflux rate for taurine as compared with other osmolytes has been observed in rat brain *in vivo* upon microdialysis perfusion with hyposmotic solutions[2,11].

In the present study, the release of taurine and D-aspartate during IVR were resolved into three first-order velocity components by fitting linear regressions to different segments of the efflux curves. Table 1 shows the derived kinetic constants for hippocampal slices and cerebellar granule neurones, along with the corresponding external osmolarity ranges. The osmotic intervals for each of the release components of taurine and D- aspartate are remarkably similar, suggesting a common efflux pathway.

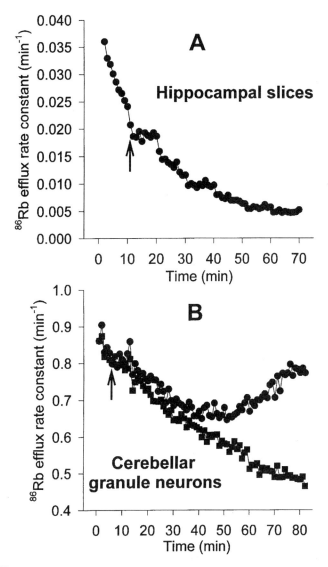

Figure 3. ^{86}Rb efflux from hippocampal slices and cerebellar granule neurones elicited by a continuous hyposmotic gradient. Preloaded tissues were initially superfused with isosmotic medium (6-10 min) and then (at arrow) exposed to the osmotic gradient (300 mOsmol/l ⇒ 150 mOsmol/l). A and B (●): Hyposmotic gradient; B (■): Isosmotic medium. Data are expressed as efflux rate constants and are means ± SE of 6 (A) or 8 (B) individual experiments.

Moreover, such intervals are similar also between preparations, i.e. k_1 describe taurine and D-aspartate fluxes during the same osmotic range in hippocampal slices and granule neurones. This suggests that the efflux mechanisms present in both preparations have similar osmotic sensitivities. The magnitude of the rate release for each amino acid however, is notably different. In the hippocampal slices, k_1 for taurine is about half of that of D-aspartate (Table 1). When the amino acids efflux is described by k_2 and k_3, velocity of taurine efflux is 4 times higher than that of D-aspartate. Similar differences are observed in cerebellar granule neurones: k_2 and k_3 for taurine efflux are about 3.5 and 5.7 times higher than those of D-aspartate. These differences could be due to distinct permeability coefficients through the suggested common pathway, or/and to different availabilities of the intracellular pools. Brain glutamate is extremely active and deeply involved in synaptic transmission and thus, is sequestered into vesicles and other metabolic compartments. Taurine in contrast, is essentially an inert compound, not contributing to protein synthesis nor involved in any metabolic reaction, and is found essentially free in the cytosol[5]. All these results suggest a predominant role for taurine in volume regulation in brain.

Potassium is an important osmolyte due to its high intracellular content. An osmosensitive release of K^+ has been consistently described in cells showing regulatory volume decrease (RVD)[6,9]. In cerebellar astrocytes, it has been shown that K^+ efflux is the rate-limiting factor in an on-going RVD process[10]. In isovolumic conditions, an osmosensitive K^+ outflow has been clearly shown in renal cells. In the distal nephron cell line A6, K^+ efflux is activated with a threshold[13] of 210 mOsmol/l, while in renal proximal tubules K^+ content decreased after superfusion with an osmotic gradient[7]. Figure 3 shows the release of K^+ (traced with ^{86}Rb) during IVR, from hippocampal slices (A) and cerebellar granule neurones (B). In this last preparation, K^+ outflow initially follows a k_1 equal to $4.98 \pm 0.36 \times 10^{-5}$ until the external osmolarity has decreased 90 mOsmol ($\approx 30\%$ hyposmotic). This value is in the same range than that measured in isosmotic conditions ($3.48 \pm 0.15 \times 10^{-5}$). When the external osmolarity has decreased ≈ 65 mOsmol, K^+ efflux is activated and its movement follows a second k, with a value of $39.97 \pm 1.04 \times 10^{-5}$ for the rest of the experiment. In contrast, ^{86}Rb outflow from hippocampal slices *does not change* during IVR, and its efflux runs parallel with that observed under isosmotic conditions. This is an unexpected result, since as previously mentioned, K^+ is a key osmolyte in essentially all cell types. The difference in K^+ efflux during RVD and IVR may be due to the involvement in each case, of different mechanisms of release. In renal proximal tubules, the Na^+-K^+ ATPase seems implicated in IVR[7] but not in RVD. Also, these two processes differ in A6 cells[13]. In addition, unlike in cells in culture, in the hippocampal slices which have an intact cytoarchitecture, buffering of extracellular K^+ by the efficient mechanisms known to exist in brain tissue, could mask an osmosensitive release occurring gradually as during IVR. Due to the key role played by K^+ in nervous excitability, its extracellular levels in brain have to be kept under strict control. Clearly, studies on the occurrence and features of IVR in different cell types are essential for a better understanding of the physiological significance of this mechanism of volume regulation.

ACKNOWLEDGMENTS

This research was supported in part by grants IN-201297 from DGAPA-UNAM and 2262-P from CONACYT, Mexico.

REFERENCES

1. Altamirano, J., Brodwick, M.S. and Alvarez-Leefmans, F.J., 1998. Regulatory volume decrease and intracellular Ca^{2+} in murine neuroblastoma cells studied with fluorescent probes. *J. Gen Physiol.* 112: 145-160.
2. Estévez, A.Y., O'Regan, M.H., Song, D., Phillis, J.W., 1999. Effects of anion channel blockers on hyposmotically induced amino acid release from the in vivo rat cerebral cortex. *Neurochem Res* 24: 447-452.
3. Fraser, C.L., Arieff, A.I., 1997. Epidemiology, pathophysiology, and management of hyponatremic encephalopathy. *Am J Med* 102: 67-77.
4. Häussinger, D., 1996. The role of cellular hydration for the regulation of cell function. *Biochem J* 313: 697-710.
5. Huxtable, R.J., 1992. Physiological actions of taurine. *Physiol Rev* 72 : 101-163.
6. Lang, F., Busch, G.L., Ritter, M., Völki, H., Waldegger, S., Gulbins, E., Häussinger, D., 1998. Functional significance of cell volume regulatory mechanisms. *Physiol. Rev.* 78: 247-306.
7. Lohr, J.W., 1990. Isovolumetric regulation of renal proximal tubules in hypotonic medium. *Ren Physiol Biochem* 13: 233-240.
8. Lohr, J.W., Grantham, J.J., 1986. Isovolumetric regulation of isolated S_2 proximal tubules in anisotonic media. *J Clin Invest* 78: 1165-1172.
9. Pasantes-Morales, H., 1996. Volume regulation in brain cells: cellular and molecular mechanisms. *Metab Brain Dis* 11: 187-204.
10. Pasantes-Morales, H., Murray, R.A., Lilja, L., Morán, J., 1994. Contribution of organic and inorganic osmolytes to volume regulation in rat brain cells in culture. *Neurochem Res* 18: 445-452.
11. Solis, J.M., Herranz, A.S., Herras, O., Lerma, J., Del Río, R.M., 1988. Does taurine act as an osmoregulatory substance in the rat brain? *Neurosci Lett* 91: 53-58.
12. Trachtman, H., 1991. Cell volume regulation: a review of cerebral adaptive mechanisms and implications for clinical treatment of osmolal disturbances II. *Pediatric Nephrology* 5: 743-750.
13. Van Driessche, W., de Smet, P., Li, J., Allen, S., Zizi, M., Mountian, I., 1997. Isovolumetric regulation in a distal nephron cell line (A6). *Am J Physiol* 272: C1890-C1898.

NEW ROLE OF TAURINE AS AN OSMOMEDIATOR BETWEEN GLIAL CELLS AND NEURONS IN THE RAT SUPRAOPTIC NUCLEUS

Nicolas Hussy, Charlotte Deleuze, Vanessa Brès and Françoise C. Moos

Biologie des Neurones Endocrines, CNRS-UMR5101, CCIPE 141 rue de la Cardonille, 34094 Montpellier cedex 5, France

INTRODUCTION

Taurine has an established function as an osmolyte in the nervous system. Present inside cells at high concentration, it is released upon cell swelling induced by a decrease in extracellular tonicity. The loss of taurine is believed to contribute in a major way to the subsequent regulatory volume decrease (RVD) undergone by both neurons and astrocytes[25]. However, some observations indicate that the osmoregulatory role of taurine in the nervous system may be more complicated than sometimes thought. First, the distribution of taurine is not homogenous, as it is concentrated in selected sets of cells in various brain areas. For instance, it is found almost exclusively in Purkinje neurons in the cerebellum[24], primarily in neurons in hippocampal and cortical formations, but prominently in astrocytes in the hypothalamus and brain stem[18,32]. While volume regulation of all brain cells should be critical for proper neuronal function, the uneven distribution of taurine suggests a differential use of this amino acid by the various cell populations. Second, taurine is not an inactive compound as it would be expected for an "ideal" osmolyte. Taurine is an agonist of the inhibitory ligand-gated channels glycine and GABA$_A$ receptors[4,9,11,36], binds to the metabotropic GABA$_B$ receptor[14,17], and alters the electrical activity of neurons and other excitable cells to various extents[9,31]. Therefore, release of taurine from brain cells under hypotonic stress may well serve other function than allowing the efflux of water to counteract cell swelling, a function that is likely to depend on the brain region, on the cell population, as well as on the physiological processes those cells are involved in.

The hypothalamo-neurohypophysial system is a particularly interesting area to study the osmoregulatory role of taurine because it constitutes an osmosensitive structure involved in the regulation of the whole body fluid balance, through the control of the plasma concentration of the neurohormones vasopressin (AVP) and oxytocin (OT). This system is composed of two populations of neurons that synthesize either AVP or OT. AVP is the major antidiuretic hormone, and OT, although best known for its implication in events linked to parturition and lactation, has also a natriuretic action[2]. The soma of these neuroendocrine cells are localized in the hypothalamic supraoptic and paraventricular nuclei (SON and PVN) and send their axons to the neurohypophysis where they release the neurohormones. The modality of the release is tightly correlated to the firing pattern of the neurons, and the activity of both types of neurons is under strong control of the osmolarity of the plasma and cerebrospinal fluid[2,26]. Thus, peripheral or central hypertonic stimuli reinforce, and hypoosmotic stimuli suppress the specific phasic activity of AVP neuron and tonic activity of OT neurons. This regulation results from the complex integration by SON and PVN neurons of the osmosensory information coming from osmoreceptors located both peripherally and centrally, as well as from their own osmosensitivity[1,2]. So far, this intrinsic osmosensitivity was believed to be mainly due to the presence of stretch-inactivated cationic channels on the plasma membrane of the neurons. Hypertonicity-induced cell shrinkage activates these channels leading to a depolarization, and thus an excitation. Cell swelling in hypotonic solution inhibits the neurons via a hyper-polarization resulting from the closing of all spontaneously active channels[1].

In the SON, taurine is prominently accumulated in astrocytes, both in their soma in the ventral glial lamina and in the processes surrounding the dendritic arborizations and the cell bodies of the neurons[6,10]. Only a few neuronal structures display weak taurine-like immunoreactivity. Similarly, in the neurohypophysis, taurine is found primarily in the pituicytes, the specialized astrocyte-like glial cells that engulf the nerve terminals[20,27]. To understand the role of taurine in the osmotic regulation of the hypothalamo-neurohypophysial complex, we studied the properties of release of taurine in the SON, as well as its action on the electrical activity of SON neurons.

RELEASE OF TAURINE FROM SON ASTROCYTES

Application of hypotonic stimuli increases the basal release of both endogenous taurine[12] and pre-loaded [3H]-taurine from acutely isolated supraoptic nuclei from adult male rats (Fig 1A)[7]. Both uptake and release of [3H]-taurine are inhibited by the selective glial metabolic blocker fluorocitrate, whereas they are unaffected by the elimination of all SON neurons by an excitotoxic treatment with NMDA[7], in agreement with the specific astrocytic localization of endogenous taurine in this structure[6]. Because only glial cells take up and release [3H]-taurine in acutely isolated SON, this preparation appears ideally suited to study the *in situ* properties of the osmodependent taurine efflux from astrocytes, keeping the tissue specialization that cultured cells necessarily lose.

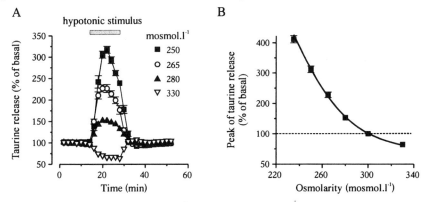

Figure 1. Osmodependent release of [3H]-taurine from acutely isolated rat SON. Methods were as described previously[7]. A, The basal release in isoosmotic conditions (300 mosmol.l⁻¹) is decreased by hypertonic and increased by hypotonic stimuli in a dose-dependent manner. Data are expressed as percent of the basal level of release, and are the mean ± s.e.m. of 8-10 experiments. Error bars are shown when exceeding the size of the symbol. B, Relationship between the peak amplitude of taurine release and the osmolarity of the medium.

The osmodependence of the release indicates a very high sensitivity since an enhancement of release is already observed for osmotic changes around 5 % (Fig 1). Moreover, weak hypotonic stimuli induce sustained increase in release over at least an hour[7], suggesting that regulation of release by physiological variations of osmotic pressure is independent of cell volume regulation. Interestingly, basal release can be decreased by a hypertonic stimulus, indicative of a sustained level of osmodependent release in isotonic conditions (Fig 1). Release is independent of extracellular Ca^{2+} and Na^{+} (Fig 2A), and blocked by the anion channel blockers DIDS, NPPB and DPC (Fig 2B). This discards a vesicular type of release or the implication of the Na^{+}-dependent taurine transporter, and indicates that taurine efflux occurs through volume-activated anion channels, as shown in many other cell types[23,34]. However, taurine release in the SON is insensitive to blockade by the antiestrogen tamoxifen, unlike most taurine permeable volume-activated anion channels described so far (Fig 2B)[23].

Another evidence for the expression of a peculiar anion channel carrying taurine efflux in SON astrocytes is provided by its unique regulation by

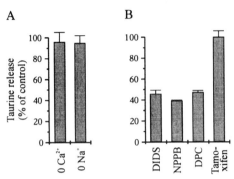

Figure 2. Properties of taurine release. A, Taurine release evoked by a hypoosmotic stimulus of 250 mosmol.l^{-1} is not altered by removal of extracellular Ca^{2+} (and addition of 1 mM EGTA), or of extracellular Na^+ (replaced with NMDG$^+$). B, The response to the same stimulus is inhibited by the chloride channel blockers DIDS (1 mM, n = 4), NPPB (50 μM, n = 6) and DPC (300 μM, n = 6), but not by tamoxifen (30 μM, n = 4) at a concentration known to largely inhibit most taurine-permeable volume-sensitive anion channels[23].

tyrosine kinase. In cultured astrocytes and other cell preparations, activation of volume-dependent Cl$^-$ channels and taurine release has been shown to critically involve protein tyrosine kinases[5,35,38]. We have also recently investigated the effects of agents interfering with tyrosine phosphorylation on taurine release in the SON. We showed that inhibition of tyrosine kinase reduces and inhibition of tyrosine phosphatase increases taurine efflux both in iso- and hypoosmotic conditions[8] (data not shown). Measuring the effects of the two types of inhibitors upon stimuli of various tonicity revealed that these effects resulted from shifts in opposite direction of the relationship between the amplitude of the release and the osmolarity of the medium. This indicates that the level of tyrosine phosphorylation actually regulates the osmosensitivity of taurine efflux[8]. This modulation of the osmotic set point of the release is inconsistent with a direct implication of protein tyrosine kinase in the activation of taurine-permeable Cl$^-$ channels in SON astrocytes, conversely to what has been reported in other cell preparations. Activation of protein tyrosine kinase is known to occur in response to a hypotonic shock, and this should lead to a progressive enhancement of the osmosensitivity of taurine release, and thus to an increase in the efflux at any given osmolarity. This could be a way to maintain a high level of taurine release despite an ongoing RVD. Given the proposed role of taurine as a messenger of the osmotic status from glial cells to neurons in this structure (see below), it would make sense to keep its extracellular concentration high as long as the osmotic pressure has not been corrected.

Cultured astrocytes also release taurine in response to cell swelling induced by high concentrations of external K[+] in isotonic solutions[16,25]. This release differs from that induced by the anisotonic swelling in that it typically displays slower activation kinetics. This time course reflects the slow cell swelling due to the incorporation of water accompanying the progressive uptake of KCl[16,33]. The role of taurine release in these conditions

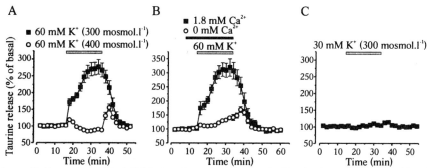

Figure 3. Properties of high K[+]-induced taurine release. A, Basal release is increased by the application of an isotonic solution containing 60 mM K[+] (substituted for Na[+], n = 8), but not by a hypertonic high K[+] medium (obtained by adding 60 mM KCl to the control isotonic solution, n = 4). B, High K[+]-induced release is largely prevented by removal of external Ca[2+] (and addition of 1 mM EGTA, n = 4). C, Application of an isotonic 30 mM K[+] solution has no significant effect on taurine release (n = 4).

is still obscure though, since high K[+]-induced swelling is not followed by RVD[16]. In the SON, application of an isotonic solution containing 60 mM KCl triggers an increase in taurine release, with slower kinetics than that induced by hypoosmolarity (Fig 3A). This response is due to cell swelling because it is largely absent if KCl is applied as a hyperosmotic solution. But conversely to anisotonic swelling-induced release, the efflux activated by high K[+] critically depends on the presence of extracellular Ca[2+] (Fig 3B), indicative of a different mode of activation by the high K[+] solution. Also, in contrast to cultured astrocytes where it shows a high sensitivity to even small increases in external K[+] concentration[19], taurine release in the SON is not stimulated by raising external K[+] up to 30 mM (Fig 3C). Therefore, the increase in extracellular K[+] that accompanies the normal electrical activity of the neurons is unlikely to stimulate release of taurine in the SON. Whether concentrations of external K[+] high enough to stimulate taurine release can be reached in this structure in pathological situations has still to be determined.

REGULATION OF NEURONAL EXCITABILITY BY TAURINE IN THE SON

If taurine is not strictly linked to cell volume regulation in the SON, then it must serve some other function. Since the condition that favors its release, namely extracellular hypotonicity, is known to decrease the firing of SON neurons[3], taurine could participate to the mediation of this inhibition. Indeed, taurine, applied onto acutely dissociated SON neurons, activates strychnine-sensitive glycine receptors in a dose-dependent manner, with a half-maximal concentration of about 400 µM (Fig 4)[12]. Such concentration in the extracellular space could be reasonably reached, especially when the latter is reduced by cell swelling. Activation of glycine receptors opens Cl⁻ channels in the neuronal membrane[12], which results in an hyperpolarization and inhibition of firing[30]. Higher concentrations of taurine (above 1 mM) also weakly activate GABA$_A$ receptors on these neurons[12].

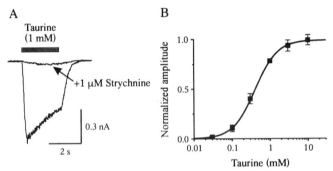

Figure 4. Taurine activates glycine receptors on SON neurons. A, Taurine applied onto a neuron acutely dissociated from rat SON[12] activates a Cl⁻ current that is blocked by pre-incubation with the glycine receptor antagonist strychnine. Recording under voltage-clamp in the whole-cell configuration of the patch-clamp technique. Holding potential is –66 mV. B, Dose-response curve (mean of 5 cells) for the activation of glycine receptors by taurine, giving an EC$_{50}$ of 406 µM. Recordings were obtained in the presence of 3 µM gabazine to prevent activation of GABA$_A$ receptors by high concentrations of taurine (>1 mM)[12].

Such an action of taurine on glycine receptors implies that these receptors should be mobilized in an osmodependent manner to control the excitability of neurons. Indeed we showed, by *in vivo* extracellular recordings of SON neurons in anaesthetized rats under normal hydration conditions, that application of a low concentration of strychnine in the vicinity of the recorded neuron increases the typical phasic activity of AVP neurons (Fig 5A). This effect is more pronounced when strychnine is applied onto a neuron primarily inhibited by a peripheral hypotonic stimulus (Fig 5B). This relief of inhibitory influence by strychnine indicates that glycine receptors participate to the control of the excitability of AVP neurons, and that they are in part responsible for the inhibition induced by the hypotonic stimulus[12].

Whether activation of glycine receptors in these conditions is actually mediated by taurine released by glial cells is not known for sure, but a number of observations strongly argue in favor of it. First, immunohistochemical studies have failed to reveal consistent glycinergic afferent fibers to the SON that could account for the level of glycine receptors expressed in these neurons[28,37]. This is corroborated by electrophysiological recordings of SON neurons in slices, which revealed that all spontaneous and evoked inhibitory postsynaptic potentials (ipsp) and currents (ipsc) are blocked by GABA$_A$ receptor antagonists[13,29,39]. We also recently showed that application of strychnine on hypothalamic slices affects neither the amplitude nor the frequency of miniature ipsc in SON neurons, these ipsc being all GABAergic (unpublished observations). These data are strong evidence for the absence of functional glycinergic synapses within the SON. Second, among the potential agonists of glycine receptors, only taurine is released in the SON in response to hypotonic stimuli. Release of glycine and β-alanine is low under control conditions and is not influenced by external osmolarity[12]. Third, the high osmosensitivity of the release, as well as its sustained kinetics in response to mild hypotonic stimuli are compatible with the sustained and osmodependent recruitment of glycine receptors *in vivo* (Fig 5)[7,12].

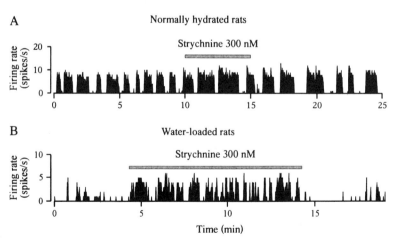

Figure 5. Implication of glycine receptors in the osmoregulation of AVP neurons *in vivo*. A, Extracellular recordings of SON neurons displaying the typical phasic activity of AVP neurons from normally hydrated rat. Pressure injection of 300 nM strychnine onto the recorded neuron reinforces its phasic activity, by increasing the duration of active phases. B, When applied onto a neuron rendered almost silent by i.p. injection of distilled water, strychnine has a greater excitatory effect, often inducing a phasic firing pattern.

CONCLUSIONS

Taurine appears ideally placed to play a role as a glia-to-neuron osmotransmitter in the SON, released from glial cells via volume-dependent anion channels and regulating neuronal activity through activation of apparently extrasynaptic glycine receptors. Because there exists a basal release of taurine in isotonic conditions that can be both potentiated by cell swelling and decreased by cell shrinkage, this inhibitory system lies in a highly dynamic window, and can participate to both inhibition by hypotonicity and excitation by hypertonicity. This is strikingly reminiscent of the osmotic control of the activity of the mechanoreceptor channels present on SON neurons, albeit with a strict opposite osmodependence of the two systems (see above). This confers them a cooperative action, which should increase considerably the sensitivity to minute osmotic changes. The osmotransmitter role of taurine may not be limited to the SON, as a similar mechanism is likely to be found in the PVN, and it may also apply in the neurohypophysis as part of the osmotic control of the release of AVP and OT from the nerve terminals. There, taurine concentrated in the pituicytes is also released upon hypotonic stimulation[20]. We are presently studying whether glycine receptors are present on the axonal terminals. Similarly, the other osmosensitive structures involved in the osmotic regulation of SON and PVN neurons, i.e. the subfornical organ, the *organum vasculosum lamina terminalis* and the median preoptic nucleus, may use the same mechanisms to regulate their activity according to the osmolarity of the extracellular medium. Indeed, neurons in these nuclei express similar type of mechanoreceptors[2], taurine is likely to be localized there in glial cells[32], and glycine receptors are present on the neurons[15].

Is such a role of taurine restricted to the SON and possibly a few osmosensitive structures or could it be extended to other brain areas as well? This is hard to tell today, but some observations may argue for the specificity of the mechanism of action of taurine in the SON. Taurine is variably expressed by glial cells and neurons in the brain and a unified view of taurine function is difficult to imagine. For instance, *in vivo* hypoosmotic stimulus induces taurine to leave cerebellar Purkinje neurons and to be taken up by neighboring glial cells, only the latter showing an increase in cell volume. In contrast, no such transfer of taurine upon similar conditions is observed in the hippocampus, even though taurine is found predominantly in hippocampal neurons[21]. Moreover, our results point to the expression of a particular type of taurine-permeable channel expressed by SON astrocytes, based on its pharmacological properties and its modes of activation and regulation. This may relate to other peculiarities displayed by SON glial cells, such as the unique morphological arrangement of the cell bodies in the ventral glial lamina of the nucleus with processes enwrapping the neurons[10]. This situation makes glial cells ideally located at the interface between extracellular fluids and neurons to sense and transmit changes in osmotic pressure. SON astrocytes also show a uniquely wide distribution of the aquaporin AQP4 over their entire membrane area[22], and the resulting ease of water movement may also contribute to the perception and transmission of the fluid osmotic status. From these particular characteristics, it could be argued that the function of taurine in the SON is likely to be specifically adapted to the functional specialization of the structure.

In conclusion, we provide evidence for a new role of taurine in osmoregulation, as an osmotransmitter implicated in the regulation of the whole body fluid balance through the control of the excitability of hypothalamic neuroendocrine cells.

REFERENCES

1. Bourque, C.W., and Oliet, S.H.R., 1997, Osmoreceptors in the central nervous system, *Annu. Rev. Physiol.,* **59**:601-619.
2. Bourque, C.W., Oliet, S.H.R., and Richard, D., 1994, Osmoreceptors, osmoreception, and osmoregulation, *Front. Neuroendoc.,* **15**:231-274.
3. Brimble, M.J., and Dyball, R.E.J., 1977, Characterization of the responses of oxytocin- and vasopressin-secreting neurones in the supraoptic nucleus to osmotic stimulation, *J. Physiol.,* **271**:253-271.
4. Bureau, M.H., and Olsen, R.W., 1991, Taurine acts on a subclass of $GABA_A$ receptors in mammalian brain in vitro, *Eur. J. Pharmacol.,* **207**:9-16.
5. Crepel, V., Panenka, W., Kelly, M.E.M., and MacVicar, B.A., 1998, Mitogen-activated protein and tyrosine kinases in the activation of astrocyte volume-activated chloride current, *J. Neurosci.,* **18**:1196-1206.
6. Decavel, C., and Hatton, G.I., 1995, Taurine immunoreactivity in the rat supraoptic nucleus: prominent localization in glial cells, *J. Comp. Neurol.,* **354**:13-26.
7. Deleuze, C., Duvoid, A., and Hussy, N., 1998, Properties and glial origin of osmotic-dependent release of taurine from the rat supraoptic nucleus. *J. Physiol.,* **507**:463-471.
8. Deleuze, C., Duvoid, A., Moos, F.C., and Hussy, N., 2000, Tyrosine phosphorylation modulates the osmosensitivity of volume-dependent taurine efflux from glial cells in the rat supraoptic nucleus, *J. Physiol.,* **523**:291-299.
9. Galaretta, M., Bustamante, J., Martín del Río, R., and Solís, J.M., 1996, Taurine induces a long-lasting increase of synaptic efficacy and axon excitability in the hippocampus, *J. Neurosci.,* **16**:92-102.
10. Hatton, G.I., 1999, Astroglial modulation of neurotransmitter/peptide release from the neurohypophysis: present status, *J. Chem. Neuroanat.,* **16**:203-222.
11. Horikoshi, T., Asanuma, A., Yanagisawa, K., Anzai, K., and Goto, S., 1988, Taurine and β-alanine act on both GABA and glycine receptors in *Xenopus* oocyte injected with mouse brain messenger RNA, *Molec. Brain Res.,* **4**:97-105.
12. Hussy, N., Deleuze, C., Pantaloni, A., Desarménien, M.G., and Moos, F., 1997, Agonist action of taurine on glycine receptors in rat supraoptic magnocellular neurones: possible role in osmoregulation, *J. Physiol.,* **502**:609-621.

13. Kabashima, N., Shibuya, I., Ibrahim, N., Ueta, Y., and Yamashita, H., 1997, Inhibition of spontaneous EPSCs and IPSCs by presynaptic GABA$_B$ receptors on rat supraoptic magnocellular neurons, *J. Physiol.,* **504**:113-126.

14. Kamisaki, Y., Wada, K., Nakamoto, K., and Itoh, T., 1996, Effects of taurine on GABA release from synaptosomes of rat olfactory bulb, *Amino Acids,* **10**:49-57.

15. Karlsson, U., Haage, D., and Johansson, S., 1997, Currents evoked by GABA and glycine in acutely dissociated neurons from the rat medial preoptic nucleus, *Brain Res.,* **770**:256-260.

16. Kimelberg, H.K., and Mongin, A.A., 1998, Swelling-activated release of excitatory amino acids in the brain: relevance for pathophysiology, In *Cell Volume Regulation,* Ed F.Lang. 123, pp 240-257. Contrib. Nephrol., Basel, Karger.

17. Kontro, P., and Oja, S.S., 1990, Interactions of taurine with GABA$_B$ binding sites in mouse brain. *Neuropharmacology,* **29**:243-247.

18. Madsen, S., Ottersen, O.P., and Storm-Mathisen, J., 1987, Immunocytochemical demonstration of taurine, In: *The Biology of Taurine: Methods and Mechanisms,* pp 275-284. Eds. R.J. Huxtable, F. Franconi and A. Giotti. Plemum Press: New York.

19. Martin, D.L., Madelian, V., Seligmann, B., and Shain, W., 1990, The role of osmotic pressure and membrane potential in K$^+$-stimulated taurine release from cultured astrocytes and LRM55 cells, *J. Neurosci.,* **10**:571-577.

20. Miyata, S. Matsushima, O., and Hatton, G.I., 1997, Taurine in rat posterior pituitary: localization in astrocytes and selective release by hypoosmotic stimulation, *J. Comp. Neurol.,* **381**:513-523.

21. Nagelhus, E.A., Lehmann, A., and Ottersen, O.P., 1993, Neuronal-glial exchange of taurine during hypo-osmotic stress: a combined immunocytochemical and biochemical analysis in rat cerebellar cortex, *Neuroscience,* **54**:615-631.

22. Nielsen, S., Nagelhus, E.A., Amiry-Moghaddam, M., Bourque, C., Agre, P., and Ottersen, O.P., 1997, Specialized membrane domains for water transport in glial cells: high-resolution immunogold cytochemistry of aquaporin-4 in rat brain, *J. Neurosci.,* **17**:171-180.

23. Nilius, B., Eggermont, J., Voets, T., Buyse, G., Manolopoulos, V., and Droogmans, G., 1997, Properties of volume-regulated anion channels in mammalian cells, *Prog. Biophys. Molec. Biol.,* **68**:69-119.

24. Ottersen, O.P., 1988, Quantitative assessment of taurine-like immunoreactivity in different cell types and processes in rat cerebellum: an electronmicroscopic study based on a postembedding immunogold labelling procedure, *Anat. Embryol.,* **178**:407-421.

25. Pasantes-Morales, H., and Schousboe, A., 1997, Role of taurine in osmoregulation in brain cells: Mechanisms and functional implications, *Amino Acids,* **12**:281-292.

26. Poulain, D., and Wakerley, J.B., 1982, Electrophysiology of hypothalamic magnocellular neurones secreting oxytocin and vasopressin, *Neuroscience,* **7**:773-808.

27. Pow, D.V., 1993, Immunocytochemistry of amino-acids in the rodent pituitary using extremely specific, very high titre antisera, *J. Neuroendoc.,* **5**:349-356.

28. Rampon, C., Luppi, P.H., Fort, P., Peyron, C., and Jouvet, M., 1996, Distribution of glycine immunoreactive cell bodies and fibers in the rat brain, *Neuroscience,* **75**:737-755.

29. Randle, J.C.R., Bourque, C.W., and Renaud, L.P., 1986, Characterization of spontaneous and evoked inhibitory postsynaptic potentials in rat supraoptic neurosecretory neurons *in vitro, J. Neurophysiol.,* **56**:1703-1717.

30. Randle, J.C.R., and Renaud, L.P., 1987, Actions of γ-aminobutyric acid on rat supraoptic nucleus neurosecretory neurones *in vitro, J. Physiol.,* **387**:629-647.

31. Satoh, H., and Sperelakis, N., 1998, Review of some actions of taurine on ion channels of cardiac muscle cells and others, *Gen. Pharmacol.,* **30**:451-463.

32. Storm-Mathisen, J., and Ottersen, O.P., 1986, Antibodies against amino acid neurotransmitters, In: *Neurohistochemistry: Modern Methods and Applications*, pp 107-136. Eds. P. Panula, H. Paivarinta and S. Soinila. Alan R. Liss, Inc.: New York.
33. Strange, K., 1994, Are all cell volume changes the same? *News Physiol. Sci.,* 9:223-228.
34. Strange, K., Emma, F., and Jackson, P.S., 1996, Cellular and molecular physiology of volume-sensitive anion channels, *Am. J. Physiol.,* **270**:C711-C730.
35. Tilly, B.C., van den Berghe, N., Tertoolen, L.G.J., Edixhoven, M.J., and de Jonge, H.R., 1993, Protein tyrosine phosphoprylation is involved in osmoregulation of ionic conductances, *J. Biol. Chem.,* **268**:19919-19922.
36. Tokumoti, N., Kaneda, M., and Akaike, N., 1989, What confers specificity on glycine for its receptor site, *Br. J. Pharmacol.,* **97**:353-360.
37. van den Pol, A.N., and Gorcs, T., 1988, Glycine and glycine receptor immunoreactivity in brain and spinal cord, *J. Neurosci.,* **8**:472-492.
38. Voets, T., Manolopoulos, V., Eggermont, J., Ellory, C., Droogmans, G., and Nilius, B., 1998, Regulation of a swelling-activated chloride current in bovine endothelium by protein tyrosine phosphorylation and G proteins, *J. Physiol.,* **506**:341-352.
39. Wuarin, J.-P., and Dudek, F.E., 1993, Patch-clamp analysis of spontaneous synaptic currents in supraoptic neuroendocrine cells of the rat hypothalamus, *J. Neurosci.,* **13**:2323-2331.

GENE EXPRESSION OF TAURINE TRANSPORTER AND TAURINE BIOSYNTHETIC ENZYMES IN HYPEROSMOTIC STATES

A comparative study with the expression of the genes involved in the accumulation of other osmolytes

Marc Bitoun and Marcel Tappaz
INSERM U 433, Faculté de Médecine RTH Laennec, Lyon, France

INTRODUCTION

When cells are submitted to an hyperosmotic environment they accumulate small organic solutes called osmolytes and thus achieve osmotic equilibrium while maintaining their normal volume. Taurine is held as an osmolyte in mammalian cells[1]. Thus, taurine content increased in astrocytes[2] or MDCK renal cells[3] maintained in an hyperosmotic medium, in the brain of hyperosmotic rats[4,5] or in the renal medulla following antidiuresis[6,7]. Taurine may be accumulated by cells through two possible mechanisms: a) it may be taken up from the extracellular space through a sodium dependent transport mediated by a specific taurine transporter (TauT), b) it may be synthetized from cysteine within the cells through the cysteine sulfinate pathway which involves cysteine dioxygenase (CDO) and cysteine sulfinate decarboxylase (CSD). Astrocyte primary cultures[8,9] as well as MDCK renal cells[3] adapted to an hyperosmotic medium showed an increased taurine transport. In MDCK renal cells the TauT-mRNA was found also elevated[10]. In astrocyte primary cultures hyperosmotic exposure induced a net increase in culture content of taurine which suggested an enhanced rate of taurine synthesis[9].

Extensive investigations on sorbitol, myo-inositol and betaine as osmolytes in renal cell lines showed that their accumulation in hypertonic

Taurine 4, edited by Della Corte *et al.*
Kluwer Academic / Plenum Publishers, New York, 2000.

conditions resulted from an upregulation of either the biosynthetic enzyme or the transporter of the osmolyte. Thus, genomic expression of aldose reductase (AR), the biosynthetic enzyme of sorbitol, sodium-dependent myo-inositol transporter (SMIT) and betaine transporter (called BGT1 for betaine-GABA transporter) were reported to be upregulated by hyperosmolarity[11]. Further investigations led to the concept of an osmotic responsive element (ORE) i.e. a short consensus sequence present in the promoter of the gene[12] on which osmo-induced nuclear proteins bind and act as transcription factors to stimulate gene expression[13].

In the present study we have investigated the gene expression of TauT., CDO and CSD as well as the expression of genes concerning the other osmolytes, namely AR, SMIT and BGT1 in different experimental paradigms of hyperosmotic states in vitro and in vivo. We have thus determined the mRNA levels of TauT, CDO, CSD, SMIT, BGT1 and AR in a) astrocyte primary cultures exposed to an hyperosmotic medium, b) whole brain of rats exposed to acute and chronic salt loading and c) kidney regions of normal rats and rats exposed to diuresis and antidiuresis. mRNA levels were determined through RT-PCR assays that we have developped according to the rat corresponding cDNA sequences available in Genebank. The RT-PCR products were hybridized with a phospholabeled internal probe and then quantified with a phosphorimager. mRNA of cyclophilin, an housekeeping gene, was used as internal standard. The following ratio, radioactivity associated with a given mRNA/radioactivity associated with cyclophilin-mRNA, was determined in each sample.

RESULTS

Gene expression in astrocyte primary cultures exposed to an hyperosmotic chemically defined medium

Primary astrocyte cultures were grown from the cerebral cortex of one-day Sprague-Dawley rats until confluence in DMEM medium containing 10% fetal calf serum for three weeks. The cultures were then adapted over three days to a defined medium without fetal calf serum to eliminate exogenous taurine provided by the serum and then submitted to an hyperosmotic medium.

In rat cortical astrocyte primary cultures exposed to an hyperosmotic medium, 30% above the isosmotic value, a transient between 1.5 to 2.5 fold increase of the mRNA levels of TauT, AR, SMIT was observed (Fig 1). Levels of these mRNAs were back to normal values at 24 hours. Evolution of BGT1-mRNA was similar but the maximum was more pronounced about 6-fold the control value. No significant changes of CSD-mRNA levels were measured. For CDO- mRNA a significant increase was recorded after 16 hours exposure.

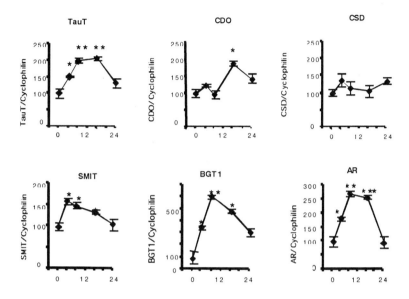

Figure 1. Evolution of TauT-, CDO-, CSD-, SMIT-, BGT1-, and AR-mRNA levels in astrocyte primary cultures exposed to an hyperosmotic medium.
The cultures in a chemically defined medium (300 mosm/l) were exposed at t = 0 to the same medium made hyperosmotic (400 mosm/l) by adding sodium chloride. The results are expressed in percentage of the mRNA levels in isosmotic medium at t = 0 as a function of time (h). * $p < 0.05$ ** $p < 0.001$ *** $p < 0.0001$

Gene expression in brain of salt-loaded rats

The animals were deprived of water for 24 h. They received then one i.p. injection of an hyperosmotic NaCl solution (2 ml of a 1.5 M NaCl solution/100 g body weight) which was repeated once every 24 h. These rats were given tap water containing 5% NaCl *ad libitum*. Control animal received injections of isosmotic NaCl solution and were given tap water *ad libitum*. The animals were scarified 5h (acute salt loading) and 120h (chronic salt loading) after the first NaCl injection. Plasma osmolarity was measured at the time of sacrifice.

In the rat brain mRNA levels of TauT, SMIT and BGT1 increased significantly following an acute salt-loading which led to an about 20% increase of plasma osmolarity. The increase in mRNA levels was more pronounced for SMIT and BGT1 than for TauT. Following chronic salt loading the SMIT-mRNA only remained elevated. CDO-, CSD- and AR-mRNA levels were not significantly modified at any time (Fig 2).

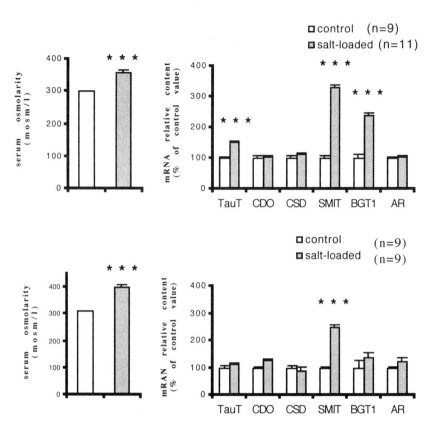

Figure 2. TauT-, CDO-, CSD-, SMIT-, BGT1-, and AR-mRNA levels in whole brain of rats following acute (5 hours) or chronic (5 days) salt loading.
The results are expressed in percentage of the values in control animals. *** p < 0.0001

Gene expression in the kidney of normal rats

We have first comparatively established the regional distribution of the different mRNAs in the kidney of normal rats i.e. maintained in a controlled environment and having free access to tap water. The kidneys were dissected out, quickly frozen on dry ice and cut on a cryostat according to the sagittal plane. Five regions of the kidney were punched out from the frozen tissue using stainless tubes of appropriate diameter.

TauT-, CSD- and to a lesser extend CDO-mRNA appeared all enriched in the OS. This distribution pattern differed from that of SMIT- and BGT1-mRNAs that were found similarly enriched in the IS, IM and papilla and from that of AR-mRNA which appeared strongly expressed in the papilla but undetectable in the OS and cortex. For all genes the lowest expression levels were found in renal cortex (Fig 3).

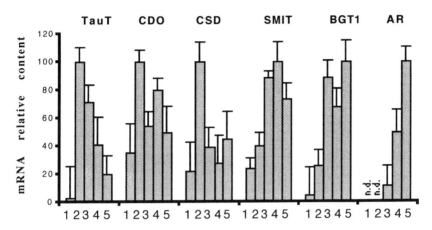

Figure 3. Regional distribution of TauT-, CDO-, CSD-, SMIT-, BGT1- and AR-mRNA in rat kidney. The different mRNAs levels were determined relatively to that of cyclophilin. For each of them, the value 100 was arbitrarily given in the renal region where it was found most enriched. The numbers refer to the different renal regions that were taken out. 1 : cortex. 2 : outer stripe of the outer medulla (OS). 3 : inner stripe of the inner medulla (IS). 4 : inner medulla (IM). 5 : papilla.

Gene expression in the renal papilla of rats undergoing diuresis and antidiuresis

The osmolarity of excreted urine was experimentally modified by submitting the rats to diuresis followed by antidiuresis. The rats were given water containing 10% sucrose for 10 days. This treatment brought about chronic diuresis. The animals excreted large volumes of diluted urine. Half of the rats were maintained under a similar diuresis while the other half were submitted to water deprivation for 2 days. This treatment brought about antidiuresis. The animals excreted small volumes of concentrated urine. In the renal papilla of the anitidiuretic rats a similar between 1.5 to 3 fold increase in mRNA levels of TauT, CDO, CSD, SMIT, BGT1 and AR was observed when compared to the corresponding levels in diuretic rats (Fig 4).

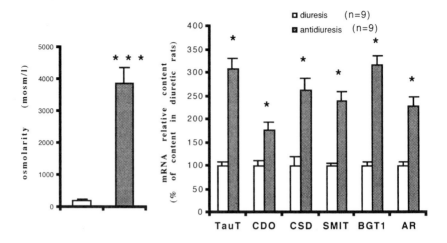

Figure 4. TauT-, CDO-, CSD-, SMIT-, BGT1-, and AR-mRNA levels in the renal papilla
of rats undergoing diuresis followed by antidiuresis.
The results are expressed as percentage of the mRNA levels in the diuretic rats. * p < 0.05
*** p < 0.0001

DISCUSSION

Upregulation by hyperosmolarity of the genes regulating osmolyte cell
content in vitro and in vivo was shown in several reports (see[11] for
references). Usually each report deals with only one gene. As far as taurine
is concerned it was only reported an increased TauT-mRNA level in MDCK
renal cells adapted to hyperosmolarity[10]. Gene expression of CDO and CSD
have not been investigated so far. Our study is the first one in which the
regulation of TauT, CDO, CSD, SMIT, BGT1 and AR genes was
comparatively investigated in various model of hyperosmotic states.
Assaying the mRNA levels of several genes in small amounts of cell culture
or tissue such as the renal papilla prevented the use of conventional
quantification through Northern blot which usually requires around 10-20
μg total RNA per analysis. Through the PCR amplification the assays of all
mRNAs could be carried out in as little as 0.2 μg total RNA extract. The
amplified RT-PCR products were identified by several controls (size,
restriction sites, hybridization of internal probe and finally sequence). The
quantification was performed using internal radioactive probes relatively to
the mRNA level of an housekeeping gene (cyclophilin). The parameters of
the PCR amplification were carefully determined so that the amount of

amplified products were proportional to the amount of initial mRNA over at least a five-fold range.

The main finding of our investigations is that the mRNA-level of both, taurine transporter and taurine biosynthetic enzymes CDO and CSD, can be elevated in cells exposed to an hyperosmotic environment. This result which was clearly observed in the renal papilla of antidiuretic rats suggests that the rat genes of TauT, CDO, and CSD are upregulated by hypertonicity like the genes regulating the cell content of other osmolytes, namely AR, SMIT and BGT1[11]. Accordingly, it is likely that the genes of TauT, CDO and CSD which have not been cloned yet possess in their 5' flanking region an osmotic responsive element[12].

However, the upregulations of TauT gene on the one hand and of CDO and CSD genes on the other hand do not coincide necessarily. Thus the upregulation of TauT gene occurred in astrocytes exposed to a moderately (+ 30%) hyperosmotic medium while no clear and consistent increase of CDO- and CSD-mRNA levels could be shown. Indeed, in our investigations so far the upregulation of CDO and CSD genes has been demonstrated only in the renal papilla of antidiuretic rats i.e. in cells which were exposed to an extremely high osmolarity. It appears thus clearly that the upregulation of CDO and CSD genes is much less sensitive to hyperosmolarity than that of the TauT gene.

In astrocyte primary cultures a transient upregulation of TauT was observed. This results is consistent with previous data on taurine transport by astrocyte primary cultures exposed to an hyperosmotic medium that was found increased after a short exposure[8,9], but unchanged after long term adaptation[14]. The transient upregulations of TauT, SMIT and AR gene were similar. Upregulation of BGT1 was stronger. These results are in striking contrast with those observed in brain of salt-loaded rats. Thus, no elevated AR-mRNA was detected in brain following salt-loading in spite of the presence of a well identified ORE in the promoter of the AR gene[15]. This result suggests that there exist in brain cells in vivo some specific gene repression mechanisms which selectively prevent the upregulation of the AR gene by hyperosmolarity. Apparently those repression mechanisms are not operative in astrocyte primary cultures. The amplitudes of SMIT, BGT1 and TauT upregulation in vivo do not parallel those found in vitro in astrocyte primary cultures. These discrepancies may possibly reflect regional heterogeneity in the gene upregulations in brain as shown previously for SMIT upregulation which appeared widespread but uneven[16]. The low average value for elevated TauT-mRNA in the whole brain may thus indicate a less widespread upregulation. Further regional studies are needed to clarify this point. The increased expression of TauT, SMIT and BGT1 may be at the origin of the increased taurine[4,5], myo-inositol[5,17] and betaine[5] content in brain of salt loaded rats. The lack of upregulation of AR gene expression is consistent with the unchanged brain sorbitol content following chronic hypernatremia[5,17].

Another striking feature of the results obtained in brain of salt loaded rats is the time course of the upregulation. Following chronic salt loading the upregulation of SMIT was maintained after five days while TauT and BGT1 were no more upregulated. This result shows that exposure to plasma hyperosmolarity is only one factor among others to regulate the gene expression in vivo which may or may not be dominant.

In the kidney, as a result of the urine concentrating mechanisms there exists an osmotic gradient between the renal cortex where plasma is filtered and the papilla where urine is excreted. The renal distribution of TauT-, CSD- and CDO-mRNA of normally hydrated rats does not follow the cortico-papillary osmotic gradient. Indeed Tau-T-mRNA, CSD-mRNA and to a lesser extend CDO-mRNA appeared all enriched in the outer stripe of the outer medulla (OS) i.e. in a renal region where interstitial fluid is expected to be isosmotic to plasma. Thus, the enrichment at this level is unlikely to be related to osmoregulation in normal conditions. The high expression of TauT in the OS is primarily related to the reabsorption of taurine by the proximal straight tubules which regulates the taurine body pool [18]. The possible physiological significance of the enriched expression of CDO and CSD in the OS is at the moment unclear. It must be noted that for the genes concerning the other osmolytes, the distribution of AR-mRNA only followed the osmotic gradient.

The amplitude of the cortico-papillary osmotic gradient can be manipulated by submitting the animals experimentally to diuresis followed by antidiuresis. A large difference in urine osmolarity to which the renal papilla is exposed could be thus achieved. The active osmolarity i.e. tonicity is somewhat lower than the measured osmolarity because of the presence of a large amount of urea which contributes largely to the measured osmolarity but is osmotically inactive since it readily crosses the cell plasma membrane. Still the difference in urine tonicity between diuretic and antidiuretic rats which could be estimated around 5-8 fold remains much larger than in any other experimental paradigm. In the papilla of the antidiuretic rats all genes encoding proteins handling osmolyte cell content i.e. biosynthetic enzymes (AR, CDO, CSD) or transporters (SMIT, BGT1, TauT) were similarly upregulated as shown by the similar 1.5 to 3 fold increase in the tissue content of their respective mRNAs. There appeared no selective upregulation of a given gene. The even upregulation of the various genes is consistent with the even about 3-4 fold increase of sorbitol, myo-inositol, betaine and taurine content in the renal papilla of rats submitted to chronic antidiuresis following chronic diuresis[7]. As far as taurine is concerned it is striking that the genes controlling both mechanisms of taurine accumulation, biosynthesis and transport, are concomitantly upregulated. However, the actual contribution of each mechanism to the about 3-fold increased taurine content previously reported in the papilla[7] remains unknown.

CONCLUSION

Our results show that the gene expression of taurine transporter and taurine biosynthetic enzymes cysteine dioxygenase and cysteine sulfinate decarboxylase can be upregulated by hyperosmolarity. This finding strongly suggests that these three genes possess in their 5' flanking region the consensus osmotic responsive element previously identified as necessary for the hyperosmolarity-induced upregulation of the genes, aldose reductase, myo-inositol transporter and betaine transporter involved in the accumulation of the osmolytes, sorbitol, myo-inositol and betaine. Although all genes involved in the accumulation of the various osmolytes can possibly be upregulated by hyperosmolarity, the actual pattern and time-course of their respective upregulations differ markedly in cells or tissues exposed to the same hyperosmotic conditions. This observation clearly indicates that there exist other factors which can selectively prevent the upregulation of these genes by hypertonicity.

ACKNOWLEDGEMENTS

This work was supported by INSERM and DRET (Grant 95/067). Marc Bitoun is the recipient of a DRET fellowship DGA.

REFERENCES

1. Pasantes-Morales H., Quesada O. and Moran J., 1998, Taurine: an osmolyte in mammalian tissues. *Adv. Exp. Med. Biol.* **442**: 209-217.
2. Olson J.E. and Goldfinger M.D., 1990, Amino-acid content of rat cerebral astrocytes adapted to hyperosmotic medium in vitro. *J. Neurosci. Res.* **27**: 241-246.
3. Uchida S., Kwon H.M., Preston A.S. and Handler J.S., 1991, Taurine behaves as an osmolyte in MDCK cells : protection by polarized, regulated transport of taurine. *J. Clin. Invest.* **76**: 656-662.
4. Thurston J.H., Hauhart R.E. and Dirgo J.A., 1980, Taurine: a role in osmotic regulation of mammalian brain and possible clinical significance. *Life Sci.* **26**: 1561-1568.
5. Lien Y.H.H., Shapiro J.I. and Chan L., 1990, Effects of hypernatremia on organic brain osmoles. *J. Clin. Invest.* **85**: 1427-1435.
6. Nakanishi T., Uyama O. and Sugita M., 1991, Osmotically regulated taurine content in rat renal inner medulla. *Am. J. Physiol.* **261**: F957-F962.
7. Sone M., Ohno A., Albrecht G.J., Thurau K. and Beck F.X., 1995, Restoration of urine concentrating ability and accumulation of medullary osmolytes after chronic diuresis. *Am. J. Physiol.* **269**: F480-F490.

8.   Sanchez-Olea R., Moran J. and Pasantes-Morales H., 1992, Changes in taurine transport evoked by hyperosmolarity in cultured astrocytes. *J. Neurosci. Res.* **32**: 86-92.
9.   Beetsch J.W. and Olson J.E., 1996, Hyperosmotic exposure alters total taurine quantity and cellular transport in rat astrocyte cultures. *Biochim. Biophys. Acta* **1290**: 141-148.
10. Uchida S., Kwon H.M., Yamauchi A., Preston A.S., Marumo F. and Handler J.S., 1992, Molecular cloning of the cDNA for an MDCK cell Na$^+$-dependent and Cl$^-$-dependent taurine transporter that is regulated by hypertonicity. *Proc. Natl. Acad. Sci. U.S.A* **89**: 8230-8234.
11. Burg M.B., Kwon E.D. and Kultz D., 1997, Regulation of gene expression by hypertonicity. *Annu. Rev. Physiol.* **59**: 437-455.
12. Ferraris J.D., Williams C.K., Ohtaka A. and Garcia-Perez A., 1999, Functional consensus for mammalian osmotic response elements. *Am. J. Physiol.* **45**: C667-C673.
13. Miyakawa H., Woo S.K., Dahl S.C., Handler J.S. and Kwon H.M., 1999, Tonicity-responsive enhancer binding protein, a Rel-like protein that stimulates transcription in response to hypertonicity. *Proc. Natl. Acad. Sci. U.S.A* **96**: 2538-2542.
14. Beetsch J.W. and Olson J.E., 1993, Taurine transport in rat astrocytes adapted to hyperosmotic conditions. *Brain Res.* **613**: 10-15.
15. Ferraris J.D., Williams C.K., Jung K.Y., Bedford J.J., Burg M.B. and Garcia-Perez A., 1996, ORE, a eukaryotic minimal essential osmotic response element - the aldose reductase gene in hyperosmotic stress. *J. Biol. Chem.* **271**: 18318-18321.
16. Ibsen L. and Strange K., 1996, In situ localization and osmotic regulation of the Na+-myo-inositol cotransporter in rat brain. *Am. J. Physiol.* **40**: F877-F885.
17. Heilig C.W., Stromski M.E., Blumenfeld J.D., Lee J.P. and Gullans S.R., 1989, Characterization of the major brain osmolytes that accumulate in salt-loaded rats. *Am. J. Physiol.* **257**: F1108-F1116.
18. Chesney R.W., 1985, Taurine: its biological role and clinical implications. *Adv. Pediatr.* **32**: 1-42.

INVOLVEMENT OF ION CHANNELS IN ISCHEMIA-INDUCED TAURINE RELEASE IN THE MOUSE HIPPOCAMPUS

Pirjo Saransaari[1] and Simo S. Oja[1,2]

[1]*Tampere Brain Research Center, University of Tampere Medical School, Tampere, Finland;*
[2]*Department of Clinical Physiology, Tampere University Hospital, Tampere, Finland*

INTRODUCTION

Brain ischemia causes a massive release of excitatory amino acids, which then activate glutamate receptors. For instance, the N-methyl-D-aspartate (NMDA) receptor-associated ion channels are opened, elevating intracellular Ca^{2+} and triggering a long-lasting potentiation of the receptor-gated currents[30]. We have recently shown that various cell-damaging conditions, including ischemia, induce a substantial release of taurine in the hippocampus[23,25]. This potentiation is enhanced by glutamate receptor activation[24]. Amino acids are released by means of several mechanisms, including Ca^{2+}-dependent exocytosis from nerve endings and reversal of the functions of Na^+-dependent membrane transporters. It has also been suggested[12] that the enhanced release of excitatory amino acids in ischemia may stem from changes in membrane permeability due to activation of phospholipases by elevated intracellular Ca^{2+}. Amino acid efflux may also be a consequence of the regulatory volume decrease possibly mediated by ion channels and triggered as a response to ischemia-induced cell swelling. The involvement of ion channels in the ischemia-induced release of preloaded [³H]taurine in hippocampal slices from adult (3-month-old) and developing (7-day-old) mice were now studied using a superfusion system.

MATERIALS AND METHODS

NMRI mice of both sexes, aged 3 months (adult) and 7 days, were used throughout. [1,2-³H]Taurine (specific radioactivity 1.07 PBq/mol) was

Taurine 4, edited by Della Corte *et al.*
Kluwer Academic / Plenum Publishers, New York, 2000.

obtained from Amersham International (Bristol, UK). The ion channel inhibitors were purchased from Tocris Cookson (Bristol, UK), except for the Cl⁻ channel blocker diisothiocyano-stilbene-2,2'-disulphonate (DIDS), which was obtained from Sigma (St.Louis, MO). The drug concentrations used were chosen to equal those commonly applied in other neuropharmacological experiments. At these concentrations the effects of the drugs on the ion channels have been reported to be selective for their targets[1,2,4]. All drugs were soluble in the superfusion medium, except for 4-aminopyridine, which was solubilized with 1.5 % dimethylsulfoxide. This solvent had no significant effects on taurine release.

Slices 0.4 mm thick weighing 15-20 mg prepared from the hippocampi with a Stadie-Riggs tissue slicer were first preloaded for 30 min with 10 μM (50 MBq/l) [³H]taurine in preoxygenated Krebs-Ringer-Hepes-glucose medium under O_2 and then superfused for 50 min as described in Kontro and Oja[7]. At 30 min the medium was in many experiments changed to another modified medium. Neural cell damage, designated 'ischemia', was induced by modified experimental conditions: glucose-free medium was bubbled with N_2 gas for 1 h before the experiments and then throughout preloading and superfusion.

The desaturation curves of labeled taurine from the slices were plotted as a function of time on the basis of the radioactivities remaining in the slices after superfusion and recovered in the collected superfusate fractions[7]. The efflux rate constants of taurine for the time intervals of 20 to 30 min (k_1) and 34 to 50 min (k_2) were computed as negative slopes for the regression lines of the logarithm of radioactivity remaining in the slices vs. superfusion time.

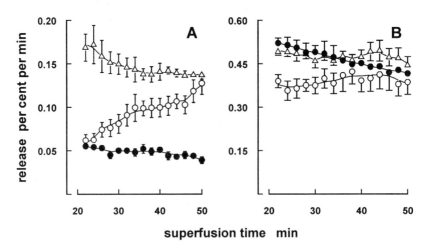

Figure 1. Time-course of taurine release (-●-) from hippocampal slices from 7-day-old (A) and 3-month-old (B) mice in normoxia and in the presence of 0.5 mM DIDS (-○-) and 0.5 mM aminopyridine (-Δ-), applied at the beginning of superfusion. The results are mean values ± SEM of 4-8 independent experiments.

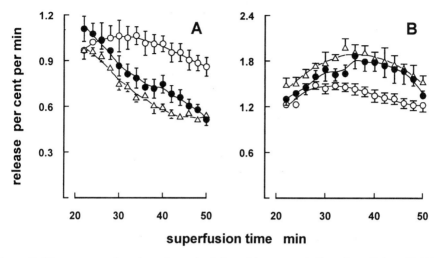

Figure 2. Time-course of taurine release (-•-) from hippocampal slices from 7-day-old (A) and 3-month-old (B) mice in ischemia and in the presence of 0.5 mM DIDS (-○-) and 0.5 mM aminopyridine (-Δ-), applied at the beginning of superfusion. The results are mean values ± SEM of 4-8 independent experiments.

RESULTS

The basal release of taurine was greatly enhanced in ischemic conditions (Figs. 1 and 2). In the adults there was then no K^+ stimulation, whereas in the developing hippocampus K^+ (50 mM) significantly enhanced the release (Table 1). The effects of various ion channel blockers were investigated by adding them to the superfusion medium at the beginning of superfusion, with subsequent K^+ stimulation at 30 min. In normoxia and ischemia the Na^+ channel inhibitor riluzole (2-amino-6-trifluoromethoxybenzothiazole) (0.1 mM) had no effect on the basal release in either age group (not shown). In normoxia the K^+-stimulated release was reduced in the developing hippocampus, whereas in the adults the stimulated release was enhanced by this drug in both normoxia and ischemia (Table 1). Another Na^+ channel blocker, amiloride, had likewise no effect on the basal release, but it potentiated the K^+-stimulated release in the adults in ischemia (Table 1). Aminopyridine (0.5 mM), which inhibits K^+ channels, potentiated the basal release only in normoxia in the immature hippocampus (Fig. 1A), without any significant effects in the adults (Figs. 1B and 2B). The Ca^{2+} channel inhibitor nimodipine had no effects on the basal or K^+-stimulated release in normoxia or ischemia. In the adults the Cl^- channel blocker DIDS (0.5 mM) inhibited the K^+-stimulated release in normoxia (Table 1) and the basal release in ischemia (Fig. 2B). In the developing hippocampus this drug reduced the K^+-stimulated release in

normoxia (Table 1), also potentiating the basal release under both
experimental conditions (Figs. 1A and 2A).

Table 1. Effects of ion channel blockers on the K^+-stimulated release of
taurine in hippocampal slices

Compound (mM)	Efflux rate constants k_2 x 10^{-3} min^{-1}			
	7-Day-old		Adult	
	Normoxia	Ischemia	Normoxia	Ischemia
Control (high K^+)	1.84 ± 0.18	3.68 ± 0.30	2.87 ± 0.10	6.77 ± 0.15
Riluzole 0.1	1.16 ± 0.13*	3.16 ± 0.29	3.66 ± 0.25**	9.69 ± 0.64**
Amiloride 0.1	1.60 ± 0.11	3.18 ± 0.09	2.65 ± 0.06	8.31 ± 0.32**
Aminopyridine 0.5	1.81 ± 0.07	3.16 ± 0.33	2.69 ± 0.15	7.39 ± 0.32
DIDS 0.5	0.80 ± 0.04*	4.53 ± 0.10	2.14 ± 0.30*	5.98 ± 0.35
Nimodipine 0.01	1.79 ± 0.10	3.96 ± 0.11	2.85 ± 0.42	7.34 ± 0.66

The ion channel inhibitors were added at the beginning of superfusion and 50 mM K^+ at
30 min. The results show the mean efflux rate constants k_2 (34-50 min) with SEM. The
number of independent experiments varied from 4 to 14. Significance of differences from the
corresponding controls: *p<0.05, ** p<0.01.

DISCUSSION

The properties of ischemia-induced release of neurotransmitters have
been studied mainly with excitatory amino acids. The increased release of
the inhibitory amino acid taurine is apparently mediated by the same systems
as those of aspartate and glutamate, multiple mechanisms being involved
(see Introduction). In normoxia the Ca^{2+}-dependency of taurine release has
been found to be greater in the immature than the mature hippocampus[23,25,26].
In ischemia, Ca^{2+}-dependent processes are also involved in both basal and
K^+-stimulated release[23,25,26]. On the other hand, the Ca^{2+} channel blocker
nimodipine had now no effect on taurine release under either condition. The
L-type of voltage-dependent Ca^{2+} channels thus do not participate in the
release. Even though Ca^{2+}-dependent processes may nevertheless be
involved, the release could also result from excitotoxicity-induced cell
swelling under cell-damaging conditions. Both K^+ depolarization[21] and
exposure to glutamate receptor agonists[20] induce a swelling-associated
release of taurine from brain slices. The Ca^{2+}-dependent exocytosis of
synaptic vesicles may thus play a minor role in the ischemia-evoked release,
even though glutamate has been claimed to be released by a Ca^{2+}-dependent
process in anoxia *in vitro*[3,5]. The depolarization-induced release probably

contributes to the initial release but is limited by the rapid inactivation and desensitization of both voltage- and glutamate-receptor-gated Ca^{2+} channels[9] and the dependency of exocytotic release on adequate levels of ATP^{19}. Furthermore, ischemia first initiates exocytotic release and then nonexocytotic release of glutamate from cultured cerebellar granule cells[15]. Accordingly, only a part of the ischemia-induced release of taurine in the adult and aged hippocampus has been Ca^{2+}-dependent[23], the major part being Ca^{2+}-independent.

A regulatory volume decrease has also been held to contribute to glutamate and aspartate release during ischemia[13]. In this process swollen cells attempt to regain their normal volumes by releasing osmolytes, including taurine. The swelling-induced increase in taurine release has been demonstrated to be a diffusional process without any involvement of carriers[16,17]. The Cl^- channel antagonist DIDS now reduced the ischemia-induced taurine release in the adults, indicating that this release may occur through anion channels, similarly to the volume-sensitive taurine release in astrocytes and neurons[6,28]. In line with such a conception, diffusion of other amino acids, including aspartate and glutamate, through an anion channel is thought to be partially responsible for the elevated levels of excitotoxic and other amino acids during ischemia[14]. At variance with this situation, in the adult hippocampus DIDS reduced only the K^+-stimulated release under normal conditions. The anion channels may thus be involved in the release only in normoxia, but not in ischemia.

The participation of K^+ channels in ischemic taurine release is unlikely, as the K^+ channel blocker aminopyridine had no effects. Inferences on the involvement of Na^+ channels in the release are complicated by the participation of Na^+ also in carrier-mediated transport operating in a reverse direction in the case of release. In ischemia, both Na^+ channel inhibitors, riluzole and amiloride, influenced the K^+-stimulated release in the adults, indicating a direct involvement of Na^+ channels. On the other hand, both basal and K^+-stimulated release of taurine is markedly affected by Na^+ deficiency in normoxia in mouse cerebral cortical and hippocampal slices[7,11,25,27]. This bespeaks the involvement of Na^+-dependent taurine transporters operating outwards. Indeed, brain tissue possesses a saturable, Na^+-requiring transport system for taurine at neuronal and glial cell membranes in both mature and immature brain tissue, comprising both high- and low-affinity components[10], which could exhibit this kind of behavior. When the Na^+-gradient is dissipated, the preferred direction of transport changes from inward (uptake) to outward (release). Such a Ca^{2+}-independent release for glutamate has been assumed to be activated in certain pathological conditions, such as anoxia[18], but it appeared now to be operating also for taurine in normoxia. In ischemia, neurons are suddenly depolarized. This event is accompanied by a massive increase in the extracellular K^+ concentration and a decrease in the extracellular Na^+ level[29].

The reduction in ischemia-induced taurine release in Na$^+$-free media in hippocampal slices further indicates that this release is mediated by the Na$^+$-requiring carriers[26,27]. In addition, the involvement of carriers in taurine release has been confirmed with structural analogues of taurine in Na$^+$-free media[26,27]. The results clearly demonstrate that a considerable part of the Ca^{2+}-independent taurine release in ischemia is mediated by Na$^+$-dependent transport at both ages. In keeping with this, taurine transporters in the mouse cerebral cortex are found to be still operative in ischemia, though nonsaturable diffusion is also concomitantly greatly increased[22].

The Na$^+$ channel blocker and neuroprotective compound riluzole affected taurine release in both normoxia and ischemia. In addition to inhibiting Na$^+$ channels, riluzole has a variety of other effects in nervous tissue, including inhibition of glutamate release in glutamatergic nerve endings, blocking the postsynaptic effects of glutamate by noncompetitive inhibition of the NMDA receptor[2], and promoting synaptic accumulation of GABA[8]. The enhancement of ischemic taurine release by riluzole may thus be due to a number of factors.

In conclusion, the ischemia-induced release of taurine in both adult and developing hippocampus represents mostly Ca^{2+}-independent efflux mediated by the reversal of Na$^+$-dependent transporters, although Na$^+$ channels may also be involved. In addition to this, the enhanced release in the adults may comprise a swelling-induced component through Cl$^-$ channels. The increase in the extracellular levels of taurine may provide an important protective mechanism against excitotoxicity.

ACKNOWLEDGEMENTS

The skillful technical assistance of Mrs. Irma Rantamaa, Mrs. Oili Pääkkönen and Mrs. Sari Luokkala and the financial support of the Medical Research Fund of Tampere University Hospital and the Academy of Finland are gratefully acknowledged.

REFERENCES

1. Bouchard, R., and Fedida, D., 1995, Closed- and open-state binding of 4-aminopyridine to the cloned human potassium channel Kv1.5. *J. Pharmacol. Exp. Ther.* 275: 864-876.
2. Doble, A., 1996, The pharmacology and mechanism of action of riluzole. *Neurology* 47: S233-S241.
3. Gibson, G.E., Manger, T., Toral-Barza, L., and Freeman, G., 1989, Cytosolic-free calcium and neurotransmitter release with decreased availability of glucose and oxygen. *Neurochem. Res.* 14: 437-443.
4. Hamill, O.P., and McBride, D.W. Jr., 1996, The pharmacology of mechanogated membrane ion channels. *Pharmacol. Rev.* 48: 231-252.

5. Katayama, Y., Kawamata, T., Tamura, T., Becker, D.P., and Tsubokawa, T., 1991, Calcium dependent glutamate release concomitant with massive potassium flux during cerebral ischemia in vivo. *Brain Res.* 558: 136-140.

6. Kimelberg, K., Goderie, S.K., Higman, S., Pang, S., and Waniewski, R.A., 1990, Swelling-induced release of glutamate, aspartate, and taurine from astrocyte cultures. *J. Neurosci.* 10: 1583-1591.

7. Kontro, P., and Oja, S.S., 1987, Taurine and GABA release from mouse cerebral cortex slices: potassium stimulation releases more taurine than GABA from developing brain. *Dev. Brain Res.* 37: 277-291.

8. Mantz, J., Laudenbach, V., Lecharny, J.-B., Henzel, D., and Desmonts, J.-M., 1994, Riluzole, a novel antiglutamate, blocks GABA uptake by striatal synaptosomes. *Eur. J. Pharmacol.* 257: R7-R8.

9. Mody, I., and MacDonald, J.F., 1995, NMDA receptor-dependent excitotoxicity: the role of intracellular Ca^{2+} release. *Trends Pharmacol. Sci.* 16: 356-359.

10. Oja, S.S., and Kontro, P., 1983, Taurine. In *Handbook of Neurochemistry* (A. Lajtha, ed.), Vol. 3, 2nd edn, Plenum Press, New York, pp. 501-533.

11. Oja, S.S., and Kontro, P., 1987, Cation effects on taurine release from brain slices: comparison to GABA. *J. Neurosci. Res.* 17: 302-311.

12. O'Regan, M.H., Smith-Barbour, M., Perkins, L.M., and Phillis, J.W., 1995, A possible role for phospholipases in the release of neurotransmitter amino acids from ischemic rat cerebral cortex. *Neurosci. Lett.* 185: 191-194.

13. Pasantes-Morales, H., 1996, Volume regulation in brain cells: cellular and molecular mechanisms. *Metab. Brain Dis.* 11: 187-204.

14. Phillis, J., Song, W.D., O'Regan, M.H., 1997, Inhibition by anion channel blockers of ischemia-evoked release of excitotoxic and other amino acids from rat cerebral cortex. *Brain Res.* 758: 9-16.

15. Pocock, J.M., and Nicholls, D.G., 1998, Exocytotic and nonexocytotic modes of glutamate release from cultured cerebellar granule cells during chemical ischaemia. *J. Neurochem.* 70: 806-813.

16. Sanchez-Olea, R., Morán, J., Schousboe, A., and Pasantes-Morales, H., 1991, Hyposmolarity-activated fluxes of taurine in astrocytes are mediated by diffusion. *Neurosci. Lett.* 130: 233-236.

17. Sanchez-Olea, R., Peña, C., Morán, J., and Pasantes-Morales, H., 1993, Inhibition of volume regulation and efflux of osmoregulatory amino acids by blockers of Cl⁻ transport in cultured astrocytes. *Neurosci. Lett.* 156: 141-144.

18. Sánchez-Prieto, J., and González, P., 1988, Occurrence of a large Ca^{2+}-independent release of glutamate during anoxia in isolated nerve terminals (synaptosomes). *J. Neurochem.* 50: 1322-1324.

19. Sánchez-Prieto, J., Sihra, T.S., and Nicholls, D.G., 1987, Characterization of the exocytotic release of glutamate from guinea-pig cerebral cortical synaptosomes. *J. Neurochem.* 49: 58-64.

20. Saransaari, P., and Oja, S.S., 1991, Excitatory amino acids evoke taurine release from cerebral cortex slices from adult and developing mice. *Neuroscience* 45: 451-459.

21. Saransaari, P., and Oja, S.S., 1992, Release of GABA and taurine from brain slices. *Prog. Neurobiol.* 38: 455-482.

22. Saransaari, P., and Oja, S.S., 1996, Taurine and neural cell damage: transport of taurine in adult and developing mice. *Adv. Med. Exp. Biol.* 403: 481-490.

23. Saransaari, P., and Oja, S.S., 1997a, Enhanced taurine release in cell-damaging conditions in the developing and ageing mouse hippocampus. *Neuroscience* 79: 847-854.

24. Saransaari, P., and Oja, S.S., 1997b, Glutamate-agonist-evoked taurine release from the adult and developing mouse hippocampus in cell-damaging conditions. *Amino Acids* 9: 323-334.

25. Saransaari, P., and Oja, S.S., 1998a, Release of endogenous glutamate, aspartate, GABA and taurine from hippocampal slices from adult and developing mice in cell-damaging conditions. *Neurochem. Res.* 23: 567-574.

26. Saransaari, P., and Oja, S.S., 1998b, Mechanisms of ischemia-induced taurine release in mouse hippocampal slices. *Brain Res.* 807: 118-124.

27. Saransaari, P., and Oja, S.S., 1999, Characteristics of ischemia-induced taurine release in the developing mouse hippocampus. *Neuroscience,* in press.

28. Schousboe, A., Sanchez-Olea, R., Morán, J., and Pasantes-Morales, H., 1991, Hyposmolarity-induced taurine release in cerebellar granule cells is associated with diffusion and not with high-affinity transport. *J. Neurosci. Res.* 30, 661-665.

29. Somjen, G.G., Aitken, P.G., Balestrino, M., Herreras, O., and Kawasaki, K., 1990, Spreading depression-like depolarization and selective vulnerability of neurons: brief review. *Stroke* 21: 179-183.

30. Szatkowski, M., and Attwell, D., 1994, Triggering and execution of neuronal death in brain ischemia: two phases of glutamate release by different mechanisms. *Trends Neurosci.* 17: 359-365.

MODULATION OF TAURINE RELEASE BY METABOTROPIC RECEPTORS IN THE DEVELOPING HIPPOCAMPUS

Pirjo Saransaari[1] and Simo S. Oja[1,2]
[1]*Tampere Brain Research Center, University of Tampere Medical School, Tampere, Finland;*
[2]*Department of Clinical Physiology, Tampere University Hospital, Tampere, Finland*

INTRODUCTION

Taurine abounds in the hippocampus[14], particularly in the immature hippocampus[26], and taurine-like immunoreactivity has been located in hippocampal interneurons, pyramidal neurons and dentate granule cells[15]. Taurine inhibits the firing of hippocampal pyramidal neurons, increases membrane chloride conductance and causes hyperpolarization[20]. In the hippocampus, the taurine-synthesizing enzyme cysteine sulphinate decarboxylase has also been identified in pyramidal basket interneurons[20]. Taurine has been held to have a special role in immature brain tissue[12,21]. Besides being an osmoregulator and neuromodulator it seems to be essential for the development and survival of neural cells[29].

The excitatory glutamatergic innervation in the hippocampus is modulated by inhibitory GABA-releasing interneurons[8]. These mechanisms could also regulate taurine release. Ionotropic glutamate receptors modify taurine release in the hippocampus[16,25], as do GABA$_A$ receptors in the cerebral cortex and cerebellum[24]. The hippocampal innervation also includes a large family of metabotropic glutamate receptors (mGluRs) coupled to second messenger systems via GTP-binding proteins. At least eight subtypes of them have been cloned, divided into three major groups based on pharmacology, second-messenger coupling and sequence homology[6,22]. Metabotropic GABA$_B$ receptors, which initiate slow inhibition via G-protein-coupled action on Ca^{2+} and K^+ channels[18], also exist in the hippocampus[7]. We have now characterized the regulation of taurine release

Taurine 4, edited by Della Corte *et al.*
Kluwer Academic / Plenum Publishers, New York, 2000.

in the immature hippocampus by agonists and antagonists of metabotropic glutamate and GABA receptors.

MATERIALS AND METHODS

NMRI mice of both sexes aged 7 days were used throughout. Slices 0.4 mm thick weighing 15-20 mg were prepared from the hippocampi with a Stadie-Riggs tissue slicer, preloaded with [³H]taurine and superfused as detailed in Kontro and Oja[12]. The efflux rate constants of taurine for the time intervals of 20 to 30 min (k_1) and 34 to 50 min (k_2) were computed as negative slopes for the regression lines of the logarithm of the radioactivity remaining in the slices vs. superfusion time[12].

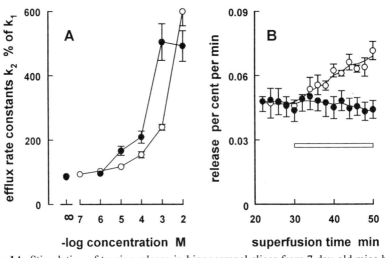

Figure 1A. Stimulation of taurine release in hippocampal slices from 7-day-old mice by glutamate (-l-) and GABA (-m-). **B.** Time-course of taurine release from hippocampal slices from 7-day-old mice in the presence (indicated by the bar) of 0.1 mM DHPG (-m-) and 0.1 mM DHPG + 0.1 mM AIDA (-l-). The results are mean values ± SEM of 4-8 independent experiments.

RESULTS

The release of [³H]taurine from hippocampal slices from 7-day-old mice was concentration-dependently enhanced by glutamate (Fig. 1A) and quisqualate (not shown). The effect of 0.1 mM quisqualate was not altered by the antagonists of group I metabotropic glutamate receptors (S)-2-methyl-4-

carboxyphenylglycine (S-MCGP), (RS)-1-aminoindan-1,5-dicarboxylate (AIDA), (S)-4-carboxyphenylglycine [(S)-4C-PG] or L(+)-2-amino-3-phosphonopropionate (L-AP3) (all 0.1 mM), whereas the ionotropic receptor antagonists (6-cyano-7-nitroquinoxaline-2,3-dione) (CNQX) and 6-nitro-7-sulphamoyl-benzo[f]quinoxaline-2,3-dione (NBQX) reduced the quisqualate-stimulated release (Fig. 2A). Of the group I agonists, (1±)-1-aminocyclopentane-trans-1,3-dicarboxylate (trans-ACPD) was not active, but (S)-3,5-dihydroxyphenylglycine (DHPG) potentiated the release by 28%, an effect significantly reduced by 0.1 mM AIDA (Fig. 1B) and S-MCGP (not shown). The group I antagonists L-AP3, AIDA and (S)-3-carboxy-4-hydroxyphenylglycine [(S)-3C4H-PG] slightly stimulated the release (Fig. 2A).

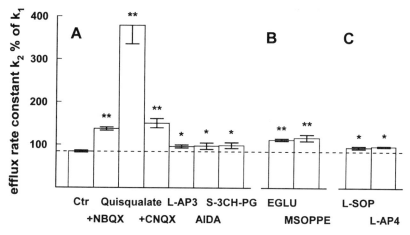

Figure 2. Effects of metabotropic glutamate receptor compounds on taurine release from hippocampal slices from 7-day-old mice. **A.** Group I agonists and antagonists. **B.** Group II antagonists. **C.** Group III agonists. The results are mean values ± SEM of 4-8 independent experiments. Significance of differences from the control (Ctr): *p<0.05, **p<0.01.

Of the group II metabotropic glutamate receptor agonists, (S)-4C-PG (0.1 mM) had no effect on taurine release, but (2S,2'R,3'R)-2-(2',3'-dicarboxycyclopropyl)glycine (DCG IV) markedly potentiated it (1214.3 ± 133.0% of control, mean ± SEM, n = 4). The potentiation by 0.1 mM DCG IV was almost totally abolished by 0.1 mM dizocilpine to 126.3 ± 13.8% of control (n = 4), but the metabotropic antagonists (RS)-2-methyl-4-tetrazolylphenylglycine (MTPG), (2S)-2-ethylglutamate (EGLU), S-MCPG and (RS)-2-methylserine-O-phosphate monophenyl ester (MSOPPE) (all 0.1 mM) had no effect on the DCG IV-stimulated release (not shown). Of this antagonist group, 0.1 mM EGLU and 0.1 mM MSOPPE stimulated basal taurine release (Fig. 2B). The group III agonists L(+)-2-amino-4-phosphonobutyrate (L-AP4) and O-phospho-L-serine (L-SOP) (both 0.1 mM) were slightly stimulatory (Fig, 2C), but their effects were not altered by the antagonists (RS)-2-methyl-4-phosphonophenylglycine (MPPG), (RS)-2-methylserine-O-phosphate (MSOP) or (RS)-2-cyclopropyl-4-phosphonophenylglycine (CPPG) (all 0.1 mM) (not shown).

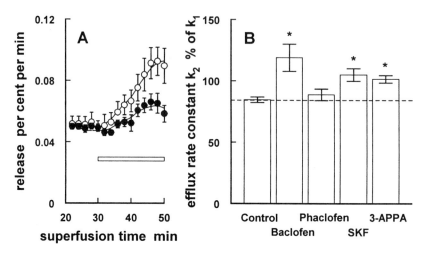

Figure 3A. Time-course of taurine release from hippocampal slices from 7-day-old mice in the presence of 0.1 mM GABA (-■-) and 0.1 mM GABA + 0.1 mM phaclofen (-○-) as indicated by the bar. **B**. The effects of $GABA_B$ compounds on taurine release from hippocampal slices from 7-day-old mice. Results are mean values ± SEM of 4-8 independent experiments. Significance of differences from the control: *p<0.05, **p<0.01.

The release of [^{3}H]taurine from hippocampal slices was concentration-dependently potentiated by GABA (Fig. 1A). The GABA effect was reduced by the $GABA_B$ receptor antagonists phaclofen (Fig. 3A) and saclofen (not shown) with the same efficacy. When Na^+ was omitted from the beginning of superfusion, the efflux rate constant k_2 (34-50 min) increased to 2.71 ± 0.27 x 10^{-3} min^{-1} (n = 4). In this Na^+-deficient medium 0.1 mM GABA was not effective. All tested $GABA_B$ ligands, baclofen, SKF 97541 and 3-aminopropylphosphonate (3-APPA) stimulated taurine release (Fig. 3B). The antagonist phaclofen had no effect on the baclofen stimulation nor did saclofen influence the action of SKF 97541 (not shown).

DISCUSSION

Both excitatory and inhibitory metabotropic receptors modify taurine release in the immature hippocampus. The effects mediated by mGluRs were less pronounced than those of ionotropic glutamatergic agents[16,25]. The variability of the responses to the metabotropic agents may result from both the heterogeneity of mGluRs and the considerable overlap of the selectivity of drugs used[6,22]. Cross-talk between different receptor classes may likewise complicate interpretations. Moreover, a developmental change from inhibition to facilitation has been demonstrated in the control of glutamate release by GluRs[10]. Taurine release may be affected by activation of presynaptic heteroreceptors on nerve terminals or by activation of glutamate autoreceptors, which enhance glutamate release. This subsequently evokes taurine release by means of ionotropic receptors[25]. Multisynaptic mechanisms may also be involved. In this case the agonists would increase the firing rate of local circuit neurons or activate feed-back loops, indirectly stimulating or inhibitoring taurine release.

The enhancement of taurine release by quisqualate is apparently not mediated by metabotropic receptors, since the ionotropic antagonists CNQX and NBQX reduced it. On the other hand, the stimulation by DHPG was reduced by the metabotropic group I antagonists. DHPG has been described[31] as a selective agonist of the mGluR coupled to phospholipase C. mGluR$_1$ expression increases in the brain during early postnatal development with the maturation of neuronal elements[27]. Moreover, the expression of mGlu$_{5a}$-receptor mRNA is higher in early postnatal life than in adults[17]. The DHPG effect in the immature hippocampus could thus be due to this developmental overexpression of group I receptors, which consist of mGlu$_1$ and mGlu$_5$ subtypes. The group I receptors generally increase neuronal excitability and the mGluR I antagonists prevent it[19]. The taurine release enhanced by mGluR I activation may reduce this hyperexcitation, being thus neuroprotective. Although these effects are not very marked they may nevertheless contribute to neuroprotection in the immature hippocampus.

The activation of group II and III mGluRs generally reduce synaptic excitation and they have been thought to function as inhibitory autoreceptors[19,23]. The agonists of both group II and III receptors are neuroprotective[4]. The marked stimulation of taurine release by DCG IV appears not to be mediated by the activation of group II receptors, since their specific antagonists failed to have any effect, but dizocilpine, the potent NMDA receptor antagonist, almost blocked the DCG IV effect. DCG IV also behaves as an NMDA receptor agonist at concentrations higher than 10 μM[32,34]. An inhibitory presynaptic mGluR, sensitive to the group III agonist L-AP4, has been described in synaptosomal preparations[33], consistent with the developmentally regulated depression of synaptic transmission by L-AP4 in the hippocampus[1]. The minor stimulation of taurine release in the immature hippocampus by both L-AP4 and L-SOP may participate in this depression.

GABAergic interneurons generate both GABA$_A$- and GABA$_B$-mediated inhibition in the hippocampus[20]. During development the composition and properties of GABA receptors differ markedly from those expressed in the adult brain. In neonatal hippocampal neurons GABA, acting on $GABA_A$ and $GABA_B$ receptors, depolarizes and hyperpolarizes CA3 pyramidal cells, respectively[5]. During postnatal development the expression of two known $GABA_B$ receptor splice variants is differently regulated[9]. Taurine also interacts in synaptic membranes with baclofen binding to $GABA_B$ sites[13].

The potentiation of taurine release by GABA, antagonized by $GABA_B$ antagonists, suggests that this release could be mediated by presynaptic $GABA_B$ receptors. However, the enhancement by baclofen and SKF 97541 was not reduced by the antagonists phaclofen and saclofen, indicating that other mechanisms are also involved. A major function of $GABA_B$ receptors is to modulate neurotransmitter release, baclofen inhibiting the release processes by different mechanisms[18]. Interpretations are complicated by the existence of $GABA_B$ receptor subtypes having distinct functions and pharmacological properties[2,18]. The receptor heterogeneity is particularly pronounced in the hippocampus[9].

A carrier common to GABA and taurine transport has been inferred to exist at brain plasma membranes[11] and a taurine transporter with such properties has since been cloned from the rat brain[3,28]. The potentiation of taurine release by GABA and GABAergic substances could thus result from a reversal of direction of transporter at cell membranes and by hetero-trans-stimulation. The failure of GABA to enhance taurine release in the absence of Na^+ corroborates such an assumption. However, heteroexchange alone does not explain the inhibition by the $GABA_B$ receptor antagonists.

ACKNOWLEDGEMENTS

The skilful technical assistance of Mrs Irma Rantamaa, Mrs Oili Pääkkönen and Mrs Sari Luokkala and the financial support of the Medical Research Fund of Tampere University Hospital and the Academy of Finland are gratefully acknowledged.

REFERENCES

1. Baskys, A., and Malenka, R.C., 1991, Agonists at metabotropic glutamate receptors presynaptically inhibit EPSCs in neonatal rat hippocampus. _J. Physiol. (Lond.)_ 444: 687-701.
2. Bonanno, G., and Raiteri, M., 1993, Multiple $GABA_B$ receptors. _Trends Pharmacol. Sci._ 14: 259-261.
3. Borden, L.A., 1996, GABA transporter heterogeneity: pharmacology and cellular localization. _Neurochem. Int._ 29: 335-356.

4. Bruno, V., Copani, A., Battaglia, G., Raffaele, R., Shinozaki, H., and Nicoletti, F., 1994, Protective effect of the metabotropic glutamate receptor agonist, DCG-IV, against excito-toxic neuronal death. *Eur. J. Pharmacol.* 256: 109-112.

5. Cherubini, E., Gaiarsa, J.L., and Ben-Ari, Y., 1991, GABA: an excitatory transmitter in early postnatal life. *Trends Neurosci.* 14: 515-519.

6. Conn, P.J., and Pin, J.-P., 1997, Pharmacology and functions of metabotropic glutamate receptors. *Annu. Rev. Pharmacol. Toxicol.* 37: 205-237.

7. Francis, J., Zhang, Y., Ho, W., Wallace, M.C., Zhang, L., and Eubanks, J.H., 1999, Decreased hippocampal expression, but not functionality, of GABA$_B$ receptors after transient cerebral ischemia in rats. *J. Neurochem.* 72: 87-94.

8. Freund, T.F., and Buzsáki, Gy., 1988. Alterations in excitatory and GABAergic inhibitory connections in hippocampal transplants. *Neuroscience* 27: 373-385.

9. Fritschy, J.-M., Meskenaite, V., Weinmann, O., Honer, M., Benke, D., and Mohler, H., 1999, GABA$_B$-receptor splice variants GB1a and GB1b in rat brain: developmental regulation, cellular distribution and extrasynaptic localization. *Eur. J. Neurosci.* 11: 761-768.

10. Herrero, I., Miras-Portugal, M.T., and Sánchez-Prieto, J., 1998, Functional switch from facilitation to inhibition in the control of glutamate release by metabotropic glutamate receptors. *J. Biol. Chem.* 273: 1951-1958.

11. Kontro, P., and Oja, S.S., 1983, Mutual interactions in the transport of taurine, hypo-taurine, and GABA in brain slices. *Neurochem. Res.* 8: 1377-1387.

12. Kontro, P., and Oja, S.S., 1987, Taurine and GABA release from mouse cerebral cortex slices: potassium stimulation releases more taurine than GABA from developing brain. *Dev. Brain Res.* 37: 277-291.

13. Kontro, P., Oja, S.S., 1990, Interactions of taurine with GABA$_B$ binding sites in mouse brain. *Neuropharmacology* 29: 243-247.

14. Kontro, P., Marnela, K.-M., and Oja, S.S., 1980, Free amino acids in the synaptosome and synaptic vesicle fractions of different bovine brain areas. *Brain Res.* 184: 129-141.

15. Magnusson, K.R., Clements, J.R., Wu, J.-Y., and Beitz, A.J., 1989, Colocalization of taurine- and cysteine sulfinic acid decarboxylase-like immunoreactivity in the hippocampus of the rat. *Synapse* 4: 55-69.

16. Magnusson, K.R., Koerner, J.F., Larson, A.A., Smullin, D.H., Skilling, S.R., and Beitz, A.J., 1991, NMDA-, kainate- and quisqualate-stimulated release of taurine from electro-physiologically monitored rat hippocampal slices. *Brain Res.* 549: 1-8.

17. Minakami, R., Iida, K., Hirakawa, N., and Sugiyama, H., 1995, The expression of two splice variants of metabotropic glutamate receptor subtype 5 in the rat brain and neuronal cells during development. *J. Neurochem.* 65, 1536-1542.

18. Misgeld, U., Bijak, M., and Jarolimek, W., 1995, A physiological role for GABA$_B$ receptors and the effects of baclofen in the mammalian central nervous system. *Prog. Neurobiol.* 46: 423-462.

19. Nicoletti, F., Bruno, V., Copani, A., Casabona, G., and Knöpfel, T., 1996, Metabotropic glutamate receptors: a new target for the therapy of neurodegenerative disorders? *Trends Neurosci.* 19: 267-271.

20. Nurse, S., and Lacaille, J.-C., 1997, Do GABA$_A$ and GABA$_B$ inhibitory postsynaptic responses originate from distinct interneurons in the hippocampus? *Can. J. Physiol. Pharmacol.* 75: 520-525.

21. Oja, S.S., and Kontro, P., 1983, Taurine. In *Handbook of Neurochemistry* (A. Lajtha, ed.), Vol. 3, 2nd edn, Plenum Press, New York, pp. 501-533.

22. Pin, J.-P., and Duvoisin, R., 1995, The metabotropic glutamate receptors: structure and functions. *Neuropharmacology* 34: 1-26.

23. Sánchez-Prieto, J., Budd, D.C., Herrero, I., Vázquez, E., and Nicholls, D.G., 1996, Pre-synaptic receptors and the control of glutamate exocytosis. *Trends Neurosci.* 19: 235-239.

24. Saransaari, P., and Oja, S.S., 1992, Release of GABA and taurine from brain slices. *Prog. Neurobiol.* 38: 455-482.
25. Saransaari, P., and Oja, S.S., 1997, Taurine release from the developing and ageing hippocampus: stimulation by agonists of ionotropic glutamate receptors. *Mech. Ageing Dev.* 88: 142-151.
26. Saransaari, P., and Oja, S.S., 1998, Release of endogenous glutamate, aspartate, GABA and taurine from hippocampal slices from adult and developing mice in cell-damaging conditions. *Neurochem. Res.* 23: 567-574.
27. Shigemoto, R., Nakanishi, S., and Mizuno, N., 1992, Distribution of the mRNA for a metabotropic glutamate receptor (mGluR1) in the central nervous system: an in situ hybridization study in adult and developing rat. *J. Comp. Neurol.* 322: 121-135.
28. Smith, K.E., Borden, L.A., Wang, C.-H.D., Hartig, P.R., Branchek, T.A., and Weinshank, R.L., 1992, Cloning and expression of a high affinity taurine transporter from rat brain. *Mol. Pharmacol.* 42: 563-569.
29. Sturman, J.A., 1993, Taurine in development. *Physiol. Rev.* 73: 119-147.
30. Taber, K.H., Lin, C.-T., Liu, J.-W., Thalmann, R., and Wu, J.-Y., 1986, Taurine in hippocampus: localization and postsynaptic action. *Brain Res.* 386: 113-121.
31. Thomsen, C., Boel, E., and Suzdak, P.D., 1994, Actions of phenylglycine analogs at subtypes of the metabotropic glutamate receptor family. *Eur. J. Pharmacol. Molec. Pharm.* 267: 77-84.
32. Uyama, Y., Ishida, M., and Shinozaki, H., 1997, DCG-IV, a potent metabotropic glutamate receptor agonist, as an NMDA receptor agonist in the rat cortical slice. *Brain Res.* 752: 327-330.
33. Vázquez, E., Herrero, I., Miras-Portugal, M.T., and Sánchez-Prieto, J., 1995, Developmental change from inhibition to facilitation in the presynaptic control of glutamate exocytosis by metabotropic glutamate receptors. *Neuroscience* 68: 117-124.
34. Wilsch, V.W., Pidoplichko, V.I., Opitz, T., Shinozaki, H., and Reymann, K.G., 1994, Metabotropic glutamate receptor agonist DCG-IV as NMDA receptor agonist in immature rat hippocampal neurons. *Eur. J. Pharmacol.* 262: 287-291.

EXTRACELLULAR TAURINE AS A PARAMETER TO MONITOR CEREBRAL INSULTS 'ON-LINE'
Time Courses and Mechanisms as Studied In Vivo

Dieter K. Scheller

Janssen-Cilag GmbH, Drug Discovery, Raiffeisenstr. 8, D-41470 Neuss, Germany

Abstract: Taurine increases in the zone surrounding a thrombotic infarct which could be prevented by a neuroprotective drug. Therefore, we aimed at studying the possible release mechanisms since the monitoring of taurine might give valuable information on the progress of cerebral insults and the effect of drugs. A microdialysis membrane was implanted into the cortex of anaesthetised rats: As toxic triggers possibly released by the dying cells in the peri-infarct zone, either a glutamatergic agonist (NMDA) or high potassium were applied via the microdialysis probe. Alternatively, a diluted perfusate was applied to induce cell swelling directly. NMDA antagonists or the NO synthase inhibitor L-NAME were applied locally too. NMDA, NO, high potassium or the hypotonic solution stimulated the release of taurine. The effect of high potassium could be prevented by Ketamine, but not by APV. The effect of NMDA could be inhibited by APV or Ketamine or the NO synthase inhibitor L-NAME. The release of taurine induced by the hypotonic solution could not be be reduced by any of the inhibitors. These data suggest that the release of taurine induced by glutamatergic activity is mediated via the NO cascade. The potassium mediated release seems to be related only in part to glutamatergic activity. Thus, other mechanisms seem to be predominate in potassium mediated swelling. Hypoosmotically induced taurine release is not mediated via the NO cascade and also seems to differ from the aforementioned release mechanisms. In conclusion: Monitoring of extracellular taurine allows to follow pathological events and to differentiate drug effects.

Taurine 4, edited by Della Corte *et al.*
Kluwer Academic / Plenum Publishers, New York, 2000.

INTRODUCTION

The sulfur-containing amino acid taurine has been shown to play an important role during osmoregulatory activity of cells[1]. Within the transition zone surrounding the irreversibly damaged core of an infarct ('penumbra', 'peri-infarct zone'), cell swelling has been described as one of the earliest events preceding cell damage, as detected histologically[2]. Indeed, taurine could be shown to increase within that zone[2]. Furthermore, this increase of taurine could be prevented by the neuroprotective drug Lubeluzole[2]. In order to contribute to the understanding of the drug effect, we tried to isolate the possible mechanisms involved in the pathophysiology of the peri-infarct zone applying various toxins supposed to be released by the dying cells in the peri-infarct zone: Either a glutamatergic agonist (NMDA), a radical precursor (NO) or high potassium were applied via the microdialysis probe. Alternatively, a hypotonic cell swelling was mimicked by diluting the perfusate. To confirm the patho-mechanisms involved we added drugs with a known profile of action.

MATERIALS AND METHODS

The experiments were conducted according to the recommendations of the Declarations of Helsinki and Tokyo and to Guidelines for the Use of Experimental Animals of the European Community. The experimental protocol was approved by the local authorities and the local ethical committee.

Surgical procedure

Wistar male rats (220-280 g) were anaesthetised with urethane (1.80 g/kg) and fixed in a stereotaxic frame. A burr hole of 3 mm diameter was drilled through the skull. A microdialysis (MD) probe (2 mm length, 0.5 mm outer diameter, Carnegie Medicin, Stockholm Sweden) was implanted in the parietal cortex. As perfusate artificial cerebrospinal fluid (aCSF; composition: NaCl 125 mM; KCl 3 mM; $CaCl_2$ 1.1 mM; $MgCl_2$ 0.8 mM; Na_2HPO_4, 0.5 mM; $NaHCO_3$ 25 mM; D-glucose 6 mM) was used at a flow rate of 2 µl/min. Fractions of 20 µl were collected. Within a distance of 50-150 µm a microelectrode at a depth of 1 mm for measurement of direct current (DC) was implanted. Body temperature was maintained at 37°C with a water jacket.

Physiological variables

The following parameters were recorded and monitored: mean arterial blood pressure (using a Statham, Oxnard, CA, USA); heart rate was calculated from the ECG recorded with two electrodes fixed subcutaneously to the left fore paw and the right hind paw; blood samples taken at the beginning and in the course of the experiments were analysed for O_2, CO_2 and acid/base status with an Eschweiler System 2000 (Eschweiler, Kiel, Germany). In order not to withdraw too much blood, blood sampling during the experiment was performed only occasionally. In addition, the EEG was recorded in order to monitor depth of anaesthesia and to confirm the spreading depressions as recorded by the microelectrode.

Experimental protocol

After an equilibration period of 90 min, the microdialysis probe was inserted within the rat cerebral cortex (coordinates: AP – 3 mm, ML 3 mm, DV 2 mm). In a first set of experiments, two short pulses of NMDA (30 sec, 10mM) were applied via the MD probe with an interstimulus interval of 60 min. Ketamine (1 or 20 mM) or APV were added to the perfusate 5 min before the second stimulus was applied. Using a 90 min interstimulus interval, normal CSF was switchted to a CSF containing L-NAME (10 mM) 30 min before the second NMDA stimulus. In a second set of experiments, CSF containing high potassium (128 mM, 3 min) was applied in two subsequent periods with an interval of 90 min with the addition of either APV (0.1 mM), Ketamine (1 or 20 mM) or L-NAME (10 mM) in the perfusate. In a fourth set of experiments, CSF was perfused for 30 minutes after the equilibration period. Afterwards, two short pulses of a hypotonic solution (diluted perfusate CSF/H_2O 7/3; 30 min) were given with an inter-stimulus interval (ISI) of 90 min. L-NAME was added to the perfusate prior to and during the second hypotonic stimulus. All groups consisted of 5 to 7 animals.

Analytical procedure

Amino acids were determined by HPLC with fluorescence detection after automated precolum derivatisation with o-phthtaldialdehyde (HPCL column: 125 x 3 mm Multosphere 100-18-5/FBS, particle size 5μm; mobile phase A: 12 % B in 0.1 mol/l Na-acetate buffer, pH 5.4; mobile phase B: acetonitrile 30 %, methanol 30 % water 40 %; linear step gradient (min/B: 0/10; 17/25; 19/35; 23/50; 24/100; 31/100; 32/10; flowrate 0.6 ml/min).

RESULTS

Local application of NMDA

In a first set of experiments, two short pulses of NMDA (10 mM, for 30 s, via perfusate) were given with an inter-stimulus interval (ISI) of 60 min (Fig. 1). After the first and the second NMDA bolus, a single spreading depression always could be observed. Blood pressure, heart rate and blood gases did not change after application of NMDA.

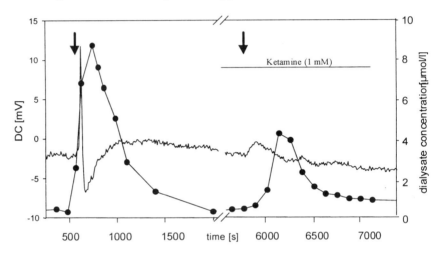

Figure 1. Representative experiment showing the effect of local NMDA application (10 mm, 30 s, indicated by arrow) on DC (line graph) and dialysate concentrations of taurine (filled circles). 60 min after a 1st NMDA stimulus, a 2nd stimulus was applied in the presence of 1 mM ketamine in the perfusate.

Comparing taurine release after the second NMDA-stimulus to the response after the first NMDA stimulus, the ratio was 0.70 ± 0.16 (Fig. 2) for the vehicle treated animals. Application of NMDA did not cause an increase of glutamate in the extracellular space (ECS). Application of the NMDA antagonists APV or Ketamine prevented both, the induction of spreading depressions and the release of taurine after NMDA administration although to a different degree (Fig. 2).

In a second set of experiments the NO-synthase inhibitor l-name (10 mm, via perfusate) was administered 30 min before the second NMDA-stimulus was given (fig. 2). This prolonged the interstimulus interval to 90 min. The ratio of the AUC's (areas under the curve) of the second stimulus in comparison to the first stimulus (0.47 ± 0.20) was significantly different to vehicle treated animals. L-NAME did not affect the induction of spreading depressions.

Local application of high potassium

Local application of high potassium allways induced a single spreading depression. In parallel, glutamate and taurine increased. The recovery of glutamate was very fast whereas the restoration of normal taurine levels required 30 to 40 min. Application of lubeluzole had no effect.

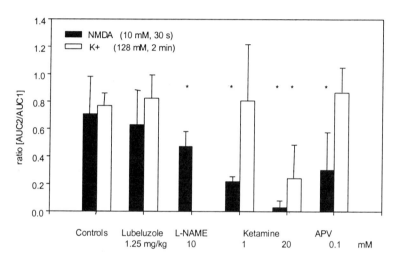

Figure 2. Summary of the effects of Lubeluzole, L-NAME, Ketamine or APV on the extracellular levels of taurine released by the application of either K^+ (128 mM, 3 min, open bars) or NMDA (10 mM, 30 s, black bars). Besides Lubeluzole, which was given i.v., all compounds were added to the perfusate. The ratios of the areas under the curve (AUC) of taurine from the 2nd over the 1st stimulus are given. Significances are indicated by the asterisk (*, $p < 0.05$). The effect of L-NAME on K^+-induced taurine release still has to be determined.

Application of Ketamine inhibited the spreading depression and reduced the amount of taurine released ($p<0.05$, T-test, two-sided). By contrast, APV also inhibited the induction of a spreading depression but had no effect on taurine (or glutamate) release.

Local application of a hypotonic stimulus

Two pulses of a hypotonic solution via the microdialyis probe caused a release of taurine selectively. The release of taurine in response to the second hypotonic stimulus was similar to the response after the first hypotonic stimulus in the control. The ratio of AUC2 over AUC1 was 1,43 ± 0.38 (mean ± SD) for the vehicle treated animals and 1.36 ± 0.35

for animals treated with the R-enantiomer (Fig. 3). The ratio decreased to 0.84 ± 0.14 for lubeluzole treated animals. The effect of lubeluzole was statistically significant (p>0.05, WMWU-test).

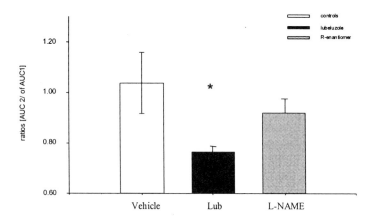

Figure 3. Effect of Lubeluzole or L-NAME on taurine release induced by a hypotonic stimulus. Significant differences of the ratios of the AUCs are indicated by the asterisk (*, p <0.05)

Subsequent to the hypotonic stimuli, a KCl stimulus (120 mM, 3 min) was given which always induced a single spreading depression. The increase of extracellular taurine was smaller than under the hypotonic conditions but reached about the same levels as in other experiments. This verifies that the hypotonic stimulus did not lead to functional or biochemical defects in brain tissue.

DISCUSSION

Application of NMDA caused taurine release as has been shown earlier[3,4]. NMDA receptor antagonists prevented taurine release[3,4] suggesting a mechanism specifically related to activation of that receptor. Co-application of NMDA and the NO synthase inhibitor L-NAME also prevented the release of taurine pointing towards a role of the NO cascade in the taurine release process[3,4]. Thus the NMDA-induced taurine release is (at least to a large extent) mediated via the NO cascade. Consequently, the results as obtained in vitro could be verfied in vivo[3,4,5]. The microdialysis technique, therefore, is useful for a 'proof of principle' in vivo of a supposed mechanisms of action.

Since the neuroprotective drug Lubeluzole has been described to interact with the NO-cascade downstream the NOS[6,7], it was expected that Lubeluzole would interfere with the NMDA-NO-mediated taurine release[2]. However, this could not be observed. Although being highly active in preventing the taurine release in the peri-infarct zone, no effect of the compound besides a small, but significant effect under hypotonic conditions was observed. Since neither L-NAME nor NMDA antagonists were effective preventing the taurine release during the hypotonic stimulus, the underlying mechanism can not be dependent on the NMDA-NO cascade. Thus, the mechanism by which Lubeluzole might have acted in the peri-infarct zone remains unclear. We tend to conclude that the compound exhibited its effect predominantly by limiting ionic changes and/or the accompanying water movement into the cells. This also suggests that the mechanisms leading to cell swelling in the peri-infarct zone remain unclear. It might well be that the spreading depressions and the corresponding metabolic activation and (lactic) acidosis cause the swelling[8]. Alternatively, oxidative stress and/or disarrangments of intracellular Ca^{2+} distribution might induce the cell swelling[9]. These mechanisms deserve further investigations.

The high potassium mediated taurine release could be prevented by Ketamine, but not by APV. This result is somewhat curious because both, APV and Ketamine are NMDA antagonists. However, whereas Ketamine acts at the phencyclidine binding site of the NMDA receptor, APV comptetively interacts at the glutamate binding site. Thus, these data seem to suggest that the K^+ induced taurine release is mediated via the phencyclidine site of the receptor. One would expect that L-NAME should interfere with that K^+ mediated swelling and taurine release. This deserves further investigations. Besides these pathophysiological and pharmacological conclusions, the results also illustrate on how to use the microdialysis technique for the 'proof of a principle in vivo' of previous in vitro investigations.

REFERENCES

1. Pasantes-Morales, H., Moran, J., Schousboe, A., 1990, Taurine release associated to cell swelling in the nervous system. *Progr. Clin.Biol. Res.* **351**:369-376.
2. Scheller, D.K., De Ryck, M., Kolb, J., Szathmary, S., van Reempts, J., Clincke, G., and Tegtmeier, F., 1997, Lubeluzole blocks increases in extracellular glutamate and taurine in the peri-infarct zone in rats. *Eur..J..Pharmacol.* **338**:243-251.
3. Menendez, N., Herreras, O., Solis, J.M., Herranz, A.S., Martine del Rio, R., 1989, Extracellular taurine increase in rat hippocampus evoked by specific glutamate receptor activation is related to the excitatory potency of glutamate agonists. *Neurosci. Lett.* **102**:64-69.
4. Chen, D.Z., Ohkuma, S., Kuriyama, K., 1996, Characteristics of nitric oxide-evoked [3H]taurine release from cerebral cortical neurons. *Neurochem. Int.* **28**:601-607.
5. Shibanoki, S., Kogure, M., Sugahara, M., Ishikawa, K., 1993, Effect of systemic administration of N-methyl-D-aspartic acid on extracellular taurine level measured by microdialysis in the hippocampal CA1 field and striatum of rats. *J. Neurochem.* **61**:1698-1704.
6. Lesage, A.S., Peeters, I., Leysen, J.E., 1996, Lubeluzole, a novel long-term neuroprotectant, inhibits the glutamate-activated nitric oxide synthase pathway. *J..Pharmacol. Exp..Therap.* **279**:759-766.

7. Maiese, K., TenBroeke, M., Kue, I., 1997, Neuroprotection of lubeluzole is mediated through the signal transduction pathways of nitric oxide. *J..Neurochem.* 68:710-714.

8. Kempski, O., Staub, F., Schneider, G.H., Weigt, H., Baethmann, A., 1992, Swelling of C6 glioma cells and astrocytes from glutamate, high K^+ concentrations or acidosis. *Progr. Brain Res.* 94:69-75.

9. Menendez, N., Solis, J.M., Herreras, O., Galarreta, M., Conejero, Martin del Rio, R., 1993, Taurine release evoked by NMDA receptor activation is largely dependent on calcium mobilization from intracellular stores. *Eur. J. Neurosci.* 5:1273-1279.

EFFECTS OF TAURINE AND SOME STRUCTURALLY RELATED ANALOGUES ON THE CENTRAL MECHANISM OF THERMOREGULATION

A structure-activity relationship study

[1]Maria Frosini, [1]Casilde Sesti, [1]Simona Saponara, [2]Alessandro Donati, [1]Mitri Palmi, [1]Massimo Valoti, [3]Fabrizio Machetti and [1]Giampietro Sgaragli

[1]*Istituto di Scienze Farmacologiche, Università di Siena, Italy* [2]*Dipartimento di Scienze e Tecnologie Chimiche e Biosistemi, Università di Siena, Italy* [3]*Dipartimento di Chimica Organica, Università di Firenze, Italy*

Abstract: There is large body of evidences on the role of taurine in the central mechanisms of thermoregulation in mammals, but It is not clear, whether the hypothermic effect of taurine depends on its interaction with GABA receptors or with a specific receptor. In order to answer this question, we have performed a structure-activity relationship study by using both *in vitro* and *in vivo* preparations. μM amounts of taurine or each of 20 analogues were injected intracerebroventricularly in conscious, restrained rabbits while rectal temperature was recorded. Receptor-binding studies, with synaptic membrane preparations from rabbit brain were used to determine the affinities of these compounds for $GABA_A$ and $GABA_B$ receptors. Furthermore, the interaction with presynaptic GABA and taurine uptake systems was studied using crude synaptosomal preparations from rabbit brain. Among the compounds tested, ($\pm$)-*cis*-2-aminocyclohexanesulfonic acid, induced hypothermia, but did not interact with $GABA_A$ and $GABA_B$ receptors neither did it affect GABA and taurine uptake, thus suggesting that its effect on body temperature is not mediated by the central GABA-ergic system. Interestingly, the *trans*-isomer was devoid of effects either *in vivo* or *in vitro*. In order to explain ($\pm$)-*cis*-2-aminocyclohexanesulfonic acid-induced hypothermia, a stereoscopic model was produced showing its possible interactions with a putative taurine brain receptor.

Taurine 4, edited by Della Corte *et al.*
Kluwer Academic / Plenum Publishers, New York, 2000.

INTRODUCTION

Regulation of body temperature involves a delicate balance between the production and loss of heat. Hypothalamus, which contains high concentrations of taurine[1], regulates the set-point at which body temperature is then maintained[7]. Intracerebroventricular (i.c.v.) injection of μM amounts of taurine induces dose-related hypothermia accompanied by reduction of vasomotor tone and peripheral vasodilatation[15], whereas central administration of the purported taurine antagonist 6-aminomethyl-3-methyl-4H-1,2,4-benzothiadiazine-1,1-dioxide (TAG)[21] increases the core temperature of the body[16]. In the rabbit taurine antagonises fever induced by i.c.v. injection of PGE_1 or i.v. *Salmonella typhosa* endotoxin[11]. Furthermore, during heat stress-induced hyperthermia, taurine is released into the cerebrospinal fluid (CSF) probably to counteract the increase in body temperature[4]. In keeping with these findings, taurine has been proposed to be an endogenous cryogen involved in the central mechanisms of thermoregulation. It is not clear, however, which receptor(s) mediates the hypothermic effect of taurine. GABA, muscimol ($GABA_A$ agonist), R(-)baclofen ($GABA_B$ agonist), all cause hypothermia after i.c.v. injection in conscious rabbits. Moreover, since taurine binds both to $GABA_A$ and to $GABA_B$ receptors[9,10], it has been suggested to affect body temperature by interacting with GABA-ergic systems. However, there are striking differences in the overall patterns of effects induced by taurine or GABA. This is particularly true for the electrocorticogram (ECoG) and its relative power spectrum recorded after i.c.v. injection of each compounds. While taurine, in fact, desynchronized the ECoG without affecting its power spectrum, GABA, muscimol and baclofen synchronized ECoG with a huge increase of power in low frequency bands (from 1.1 to 4.6 Hz). These results suggest that the central effects of taurine and GABA could depend on the interaction with different neuronal pathways afferent to brain cortex.

Abbreviations used: **TAU** = 2-aminoethanesulfonic acid, taurine; **AEP** = 2-aminoethylphosphonic acid; **AEA** = 2-aminoethylarsonic acid; **ALA** = β-alanine; **EOS** = ethanolamine-O-sulphate; **MMT** = *N*-methyltaurine; **DMT** = *N,N*-dimethyltaurine; **TMT** = *N,N,N*-trimethyltaurine; **GES** = guanidinoethanesulfonic acid; **ISE** = 2-hydroxyethanesulfonic acid; **ACES** = N-(2-acetamido)-2-aminoethanesulfonic acid; **PIPES** = pipe-razine-N,N'-bis-(2-ethanesulfonic acid); **AMS** = aminomethanesulfonic acid; **OMO** = 3-aminopropanesulphonic acid; **PYR** = pyridine-3-sulfonic acid; **PSA** = piperidine-3-sulfonic acid; **ANSA** = aniline-2-sulfonic acid; **CAHS** = (±)*cis*-2-aminocyclohexane sulfonic acid; **TAHS** = (±)*trans*-2-aminocyclohexane sulfonic acid; **GLY** = glycine; **TAG**=6-aminomethyl-3-methyl-4H-1,2,4-benzothiadiazine-1,1-dioxide.

The present report deals with a structure-activity relationship study (SAR) where μM amounts of taurine or each of 20 analogues were injected i.c.v., while their effects on rectal temperature were monitored. Concurrently, the affinity of these compounds for $GABA_A$ and $GABA_B$ receptors or their interaction with presynaptic GABA and taurine uptake systems were studied *in vitro* by using preparations of rabbit brain.

MATERIALS AND METHODS

In vivo experiments

Adult, male New Zealand albino rabbits bearing a cannula guide chronically implanted into the lateral ventricle, were used[12]. Conscious animals were individually housed in a thermostatted chamber set at neutral temperature (20 °C). Concurrently, rectal temperature (RT) was recorded every 5 min. Compounds dissolved in pyrogen-free water, were administered i.c.v. in μM amounts in a final volume of 10 μl.

In vitro experiments

$GABA_A$ receptors binding assays

The effects of taurine and its analogues on [^{3}H]muscimol binding to $GABA_A$ receptors were studied in highly purified synaptic membranes (WSM) isolated from whole brains of rabbits[20].

$GABA_B$ receptors binding assays

Rabbit brain synaptosomal membranes were prepared and the effects of taurine and its analogues on [^{3}H]GABA binding to $GABA_B$ receptors in the presence of isoguvacine ($GABA_A$ antagonist) were studied according to the method of Hill and Bowery[5].

Uptake assays

To study the effects of taurine analogues on either GABA or taurine uptake, the methods decribed by Watabe et al.[18] and by Hruska et al.[6] were used, respectively.

RESULTS AND DISCUSSION

Whole rabbit brain WSM were able to bind [^{3}H]muscimol with a K_d of 6.5×10^{-9} M; GABA inhibited this binding with an IC_{50} of 1.2×10^{-7} M. The K_d for muscimol and the IC_{50} for GABA are comparable to those already reported for rat[3], cow[10] and mouse[20]. In the presence of 1×10^{-4} M isoguvacine, whole brain synaptic membranes bound [^{3}H]GABA with a K_d of 1.4×10^{-6} M which is quite similar to the value found in the rat[2].

As reported in Table 1, taurine-induced hypothermia might partly depend on its interaction with GABA$_B$ receptors since it inhibited [^{3}H]GABA binding in the presence of isoguvacine at μM concentrations. Among the 20 taurine analogues tested GES, OMO, PIP and CAHS induced a dose-related hypothermia. The effect of GES on RT can be ascribed to its ability to block taurine uptake by rabbit brain crude synaptosomes very efficiently. OMO may act by interacting with GABA$_A$ receptors since it was able to displace [^{3}H]muscimol binding from rabbit brain WSM with an IC_{50} in the nanomolar range, i.e. at concentrations one order of magnitude lower than GABA itself, as already described by other authors[10]. The effects of PIP and CAHS on body temperature do not seem to depend on their interaction with either

GABA$_A$ and GABA$_B$ receptors because of their low affinity for both receptors (>400 μM). Furthermore, they did not affect the uptake systems for either amino acid. This may indicate the existence of a temperature control pathway which does not involve GABA-ergic system in rabbit brain. MMT exhibited a bimodal dose-dependent effect on RT. At low doses it caused a moderate hyperthermia, whereas at the highest dose it induced a mild hypothermia. An in-depth study of the interaction of MMT with GABA receptors is still in progress in this laboratory. AEP, AEA, EOS, PYR, ISE, DMT, TMT and TAG induced a dose-related hyperthermia. As shown in Table 2, only AEP, DMT and TMT exhibited a moderate affinity for GABA$_B$ receptors.

The effect of these compounds may be explained by postulating that they are able to bind to the GABA$_B$ receptor without activating it. In this context, they can be considered as GABA$_B$ antagonist. It has been reported that i.c.v. injection of the GABA$_B$ antagonist phaclofen causes hyperthermia[13]. In terms of their carbon chain length, these compounds are similar to ß-ALA, which may act like a GABA$_B$ agonist/antagonist because of its [^{3}H]-GABA displacing activity at μM concentrations (see Table 3). Unfortunately, this reasoning is hampered by the lack of thermoregulatory activity exhibited by ß-ALA. Thus, it is conceivable that AEP, DMT and TMT act through a mechanism which is independent of the GABA system, as it is the case of AEA, ISE and TAG, which do not interact either with GABA$_A$ or with

GABA$_B$ receptors, and did not affect either the GABA and taurine uptake systems.

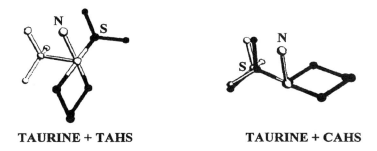

TAURINE + TAHS   **TAURINE + CAHS**

Figure 1. Superimposition of minimum energy structures of taurine + TAHS and Taurine + CAHS. The molecular-mechanic calculations were performed, in vacuum, with the Macromodel 5.0 program using MM2 as force field. Low energy conformers were determined by full energy minimisation, performed on initial conformers generated at random by using a Montecarlo Multiple Minimum (MCMM) algoritm. Black: CAHS or TAHS; white: taurine.

Taking into account the conformation of these compounds, their effects may be explained by postulating the existence of a receptor possessing two recognition sites, one negatively and the other positively charged, interacting with the amino and the sulphonic group, respectively, of taurine (see Figure 1, panel A). The distance between these two docking sites corresponds to that of two bonded carbon atoms. The interaction of taurine with the receptor at both groups is essential, but the receptor binding of the sulphonate group is specific for its tetrahedral shape, since ß-ALA, with its planar carboxylate group, had no effect. When the S atom is replaced with As or P, the resulting arsono and phosphono groups are similar in tetrahedral shape and size to the sulphonic group of taurine. A receptor that has a site for a singly-charged anion may force the arsono- and the phosphono- groups into the singly charged forms $-AsO_3H^-$ and $-PO_3H^-$, so that they can bind. However, the addition of H^+ removes their rotational symmetry, altering the position of the negative charge with respect to the rest of the molecule. This may mean that the rest of the molecule is slightly displaced with respect to the negative receptor site and that the amine group can not interact with it[17]. Similarly, with DMT and TMT, the steric hindrance operated by the methyl groups on the amino N does not allow the positively charged nitrogen to interact with the negatively charged receptor site. Furthermore, replacing of the amine group with very bulky substituents gives rise to compounds that are not active (i.e. ACES, PIPES, Table 3).

The hypothesis of a taurine receptor might help to understand why some of the taurine analogues elicit hyperthermia. If we admit that they are able to

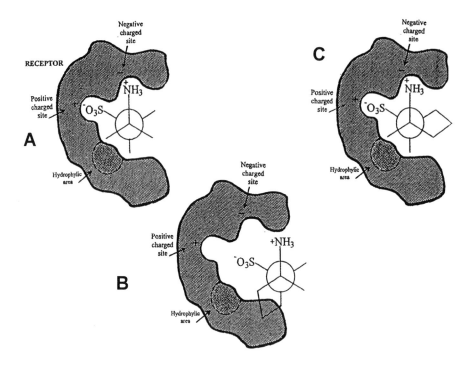

Figure 2. Possible structure of the binding site of the taurine receptor.

block the receptor without activating it, then they will prevent endogenous taurine from exerting a tonic influence on thermoregulatory centres that are directed towards dissipation of body heat. Without this influence the system will be unbalanced and hyperthermia will develop. We can also speculate that the spatial arrangement of the sulfonic and the amine group plays an essential role for the pharmacological activity of taurine derivatives. In fact, ANSA, which is planar, is inactive (Table 3), as it is TAHS, with the amine and sulfonic groups in a di-equatorial position (Table 3), whereas CAHS, with the two functional groups in a axial-equatorial position, induced hypothermia.

To explain the striking difference between CAHS and TAHS activity, we have performed a computer-aided molecular modelling study to calculate their conformation at the minimum energy levels and to compare them to that of taurine. When the spatial conformation of taurine is superimposed on that of CAHS or TAHS along the N-C-C bonds, the position of the cyclohexane ring in both compounds is divergent from that axis by about 90° (Figure 1). Moreover, since the conformation of CAHS is very restricted as

Table 1. Comparative IC_{50} values (μM) for displacement of radiolabelled ligands from $GABA_A$ and $GABA_B$ binding sites and for inhibition of [^{3}H]Taurine and [^{3}H]GABA rabbit brain preparations

Compound and Structure[§]		Effect on R.T.	Binding assay[◆] GABA$_A$	Binding assay[◆] GABA$_B$	Uptake assay[✱] [^{3}H]TAU	Uptake assay[✱] [^{3}H]GABA
TAU	$^+NH_3\text{-}CH_2\text{-}CH_2\text{-}SO_3^-$	↓	300.0	2.3	--	N.A.
GES	$NH{=}C(NH_2)N^+H_2\text{-}CH_2CH_2\text{-}SO_3^-$	↓	132.1	N.A.	3.72	N.A.
OMO	$^+NH_3\text{-}CH_2\text{-}CH_2\text{-}CH_2\text{-}SO_3^-$	↓	0.034	14.2	N.A.	N.A.
PIP	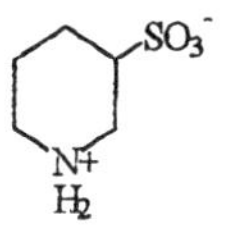	↓	422.3	N.A.	N.A.	N.A.
CAHS	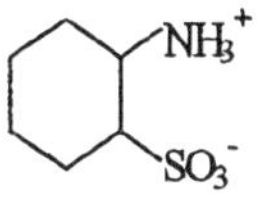	↓	N.A.	N.A.	N.A.	N.A.
MMT	$^+NH_2(CH_3)\text{-}CH_2CH_2\text{-}SO_3^-$	↓↑	⌘	⌘	⌘	⌘

↓= hypothermia, dose-related; ↓↑= bimodal effect (see text); N.A. (Not Active)= IC_{50}>500 μM; ◆ = washed synaptic membranes; ligands used were [^{3}H]muscimol for $GABA_A$ and [^{3}H]GABA + $1x10^{-4}$ M isoguvacine for $GABA_B$ receptors; ✱ = crude synaptosomes; § = at physiological pH; ⌘ = not available.

compared to that of taurine, we can conclude that since CAHS and taurine have a similar effect, the preferred taurine conformation at its biological active site results from a "gauche" position exhibited by both amine and sulfonic groups. Moreover, the hypothetical role played by the position of cyclohexane mojety in the pharmacological activity, suggests that this taurine receptor may posses an hydrophylic area besides the two recognition charged sites (Figure 2, panels B and C). In this model, the lipophylic cyclohexane ring of TAHS is repulsed by the hydrophylic area of the receptor, so that the interaction of its charged functionals groups with the recognition sites on the receptors is hampered (Figure 2, panel B). In contrast, docking of both the amine and sulfonic groups of CAHS to the receptor is not hampered by the cyclohexane moiety (Figure 2, panel C).

Table 2. Comparative IC_{50} values (μM) for displacement of radiolabelled ligands from $GABA_A$ and $GABA_B$ binding sites and for inhibition of [^{3}H]taurine and [^{3}H]GABA uptake rabbit brain preparations

Compound and Structure§	Effect on R.T.	Binding assay♦		Uptake assay✻	
		$GABA_A$	$GABA_B$	[^{3}H]TAU	[^{3}H]GABA
AEP $^+NH_3\text{-}CH_2\text{-}CH_2\text{-}PO_3H^-$	↑	N.A.	9.8	N.A.	N.A.
AEA $^+NH_3\text{-}CH_2\text{-}CH_2\text{-}AsO_3H^-$	↑	N.A.	N.A.	N.A.	N.A.
EOS $^+NH_3\text{-}CH_2\text{-}CH_2\text{-}O\text{-}SO_3^-$	↑	175.6	N.A.	N.A.	N.A.
PYR	↑	62.4	N.A.	N.A.	N.A.
ISE $HO\text{-}CH_2\text{-}CH_2\text{-}SO_3^-$	↑	N.A.	N.A.	N.A.	N.A.
DMT $^+NH(CH_3)_2\text{-}CH_2CH_2\text{-}SO_3^-$	↑	231.2	4.4	N.A.	N.A.
TMT $^+N(CH_3)_3\text{-}CH_2CH_2\text{-}SO_3^-$	↑	120.8	11.3	N.A.	N.A.
TAG	↑	N.A.	N.A.	N.A.	N.A.

↑= hyperthermia, dose-related; N.A.(Not Active)= IC_{50}>500 μM
♦ = washed synaptic membranes; ligand used were [^{3}H]muscimol for $GABA_A$ and [^{3}H] GABA + 1×10^{-4} M isoguvacine for $GABA_B$ receptors; ✻ = crude synaptosomes; § = at physiological pH.

In conclusion, the present SAR study, suggesting the existence of a taurine receptor in mammalian brain, gives support to the hypothesis that taurine plays an important role in the central mechanisms of thermoregulation. The demonstration of the existence of brain receptors for taurine, however, has proved difficult[9,19]. This may have been because of the endogenous taurine which is very difficult to remove from brain preparations used and because selective taurine agonists/antagonists have been so far lacking. The future development of new taurine derivatives may be crucial for the identification and characterization of the brain receptors which are responsible for its pharmacological activity.

Table3. Comparative IC$_{50}$ values (μM) for displacement of radiolabelled ligands from GABA$_A$ and GABA$_B$ binding sites and for inhibition of [^{3}H]taurine and [^{3}H]GABA uptake on rabbit brain preparations

		Effect on R.T.	Binding assay$^\blacklozenge$		Uptake assay*	
	Compound and Structure§		GABA$_A$	GABA$_B$	[^{3}H]TAU	[^{3}H]GABA
AMS	$^+NH_3$-CH$_2$-SO$_3^-$	Inact.	N.A.	N.A.	N.A.	N.A.
β-ALA	$^+NH_3$-CH$_2$-CH$_2$-COO$^-$	Inact.	20.1	4.5	N.A.	N.A.
ACES	CH$_3$COCH$_2$NH$_2^+$-CH$_2$CH$_2$-SO$_3^-$	Inact.	N.A.	N.A.	N.A.	N.A.
PIPES	$^-$O$_3$SCH$_2$CH$_2$HN⟨ ⟩NHCH$_2$CH$_2$SO$_3^-$	Inact.	N.A.	N.A.	N.A.	N.A.
ANSA	NH$_3^+$ / SO$_3^-$	Inact.	N.A.	N.A.	N.A.	N.A.
GLY	$^+NH_3$-CH$_2$-COO$^-$	Inact.	N.A.	N.A.	N.A.	N.A.
TAHS	NH$_3^+$ / SO$_3^-$	Inact.	N.A.	N.A.	N.A.	N.A.

N.A.(Not Active) = IC$_{50}$>500 μM
$\blacklozenge$ = washed synaptic membranes; ligand used were [^{3}H]muscimol for GABA$_A$ and [^{3}H] GABA + 1x10^{-4} M isoguvacine for GABA$_B$ receptors; $*$ = crude synaptosomes; §=at physiological pH.

ACKNOWLEDGEMENTS

This work was supported by contributions of Ministero degli Affari Esteri (Rome, Italy) under law 212/92 and by MURST, Cofin. '98.

REFERENCES

1. Barbeau, A., Inove, N., Tsukada, Y., and Butterworth, R.F., 1975, The neuropharmacology of taurine. *Life Sci.* **17(5)**: 669-677.
2. Bowery, N.G., Hill, D.R., and Hudson, A.L., 1985, [^{3}H](-)Baclofen: an improved ligand for GABA$_B$ sites. *Neuropharmacology* **24(3)**: 207-10.
3. Bureau, M.H., and Olsen, R.W., 1991, Taurine acts on a subclass of GABA$_A$ receptors in mammalian brain in vitro. *Eur. J. Pharmacol.* **207(1)**: 9-16.

4. Frosini, M., Sesti, C., Palmi, M., Valoti, M., Fusi, F., Bianchi, L., Della Corte, L., and Sgaragli, G.P., Changes in CSF composition in heat-stressed rabbits: the possible role of taurine and GABA as cryogens. This book.
5. Hill, D.R., and Bowery, N.G., 1981, 3H-baclofen and 3H-GABA bind to bicuculline-insensitive GABA$_B$ sites in rat brain. *Nature* 12;290(5802):149-52.
6. Hruska, R.E., Padjen, A., Bressler, R., and Yamamura, H.I., 1978, Taurine: sodium-dependent, high-affinity transport into rat brain synaptosomes. *Mol. Pharmacol.* **14(1):** 77-85.
7. Insel, P.A., 1996, Analgesic-antipyretic and antiinflammatory agents and drugs employed in the treatment of gout. In *Goodman and Gilman's - The pharmacological basis of therapeutics*, (Hardman, J.G., Limbird, L.E., Molinoff, P.B., Ruddon, R.W., Goodman Gilman, A., eds) McGraw-Hill, New York, NY, USA, 9th edition, pp. 617-657.
8. Kontro, P., and Oja, S.S., 1987, Co-operativity in sodium-independent taurine binding to brain membranes in the mouse. *Neuroscience* **23(2):** 567-70.
9. Kontro, P., and Oja S.S., 1990, Interactions of taurine with GABA$_B$ binding sites in mouse brain. *Neuropharmacol.* **29(3):** 243-7.
10. Krogsgaard-Larsen, P., Falch, E., Schousboe, A., Curtis, D.R., and Lodge, D., 1980, Piperidine-4-sulphonic acid, a new specific GABA agonist. *J. Neurochem.* **34(3):** 756-9.
11. Lipton, J.M., and Ticknor, C.B., 1978, Central effect of taurine and its analogues on fever caused by intravenous leukocytic pyrogen in the rabbit. *J. Physiol. (London)* **287:** 535-543.
12. Palmi, M., Frosini, M., and Sgaragli, G.P., 1992, Calcium changes in rabbit CSF during endotoxin, IL-1ß and PGE$_2$ fever. *Pharmacol. Biochem. Behav.* **43:** 1253-1262.
13. Sancibrian, M., Serrano, J.S., and Minano, F.J., 1991, Opioid and prostaglandin mechanisms involved in the effects of GABAergic drugs on body temperature. *Gen. Pharmacol.* **22(2):** 259-62.
14. Sesti, C., Frosini, M., and Sgaragli, G.P., 1999, Rabbit rectal temperature and ECoG spectrum changes induced by intracerebroventricular injection of GABA, taurine and analogues. *Pharmacol. Res.* **39:** 61.
15. Sgaragli, G.P., Carlà, V., Magnani, M., and Galli, A., 1981, Hypothermia induced in rabbits by intracerebroventricular taurine: specificity and relationship with central serotonin (5-HT) systems. *J. Pharmacol. Exp. Ther.* **219:** 778-785.
16. Sgaragli, G.P., Frosini, M., Palmi, M., Bianchi, L., and Della Corte, L., 1994, Calcium and taurine interaction in mammalian brain metabolism. *Adv. Exp. Med. Biol.* **359:** 299-308.
17. Sgaragli, G.P., Frosini, M., Palmi, M., Dixon, H.B., Desmond-Smith, N., Bianchi, L., and Della Corte, L., 1996, Role of taurine in thermoregulation and motor control. Behavioural and cellular studies. *Adv. Exp. Med. Biol.* **403:** 527-35.
18. Watabe, S., Yamaguchi, H., and Ashida, S., 1993, DM-9384, a new cognition-enhancing agent, increases the turnover of components of the GABAergic system in the rat cerebral cortex. *Eur. J. Pharmacol.* **238:** 303-9.
19. Wu, J.Y., Liao, C.C., Lin, C.J., Lee, Y.H., Ho, J.Y. and Tsai, W.H., 1987, Taurine receptor in the mammalian brain. *Prog. Clin. Biol. Res.,* **351:**147-56.
20. Yang, J.S. and Olsen, R.W., 1987, gamma-Aminobutyric acid receptor binding in fresh mouse brain membranes at 22° C: ligand-induced changes in affinity. *Mo.l Pharmacol.,* **32(1):**266-77.
21. Yarbrough, G.G., Singh, D.K. and Taylor, D.A., 1981, Neuropharmacological characterization of a taurine antagonist. *J. Pharmacol. Exp. Ther.,* **219:**604-613.

TAURINE-INDUCED SYNAPTIC POTENTIATION
Dependence on extra- and intracellular calcium sources

[1]Nuria Del Olmo, [1]Mario Galarreta, [2]Julian Bustamante, [1]Rafael Martín del Río and [1]José M. Solís

[1]Depto. de Investigación, Hospital Ramón y Cajal, 28034 Madrid, Spain; [2]Depto. de Fisiología, Facultad de Medicina, Universidad Complutense, 28040 Madrid, Spain

INTRODUCTION

Long-lasting changes in synaptic transmission efficacy have been proposed as the functional substrate underlying synaptic plasticity phenomena involved in learning and development[3,5]. We have reported that taurine, an amino acid classically considered to have GABA-like actions in the nervous tissue[13], also induces a long-lasting potentiation of synaptic transmission in the CA1 hippocampal region by a mechanism independent of GABA$_A$ receptors activation[6,7,8]. This synaptic potentiation was evoked by including taurine in the bath perfusion liquid at a concentration of 5 or 10 mM during 20-30 min, and the excitatory postsynaptic potentials (EPSPs) remained potentiated far beyond (at least 3 hours) taurine withdrawal[7,8].

The induction of long-term potentiation (LTP) evoked by high-frequency synaptic stimulation is usually dependent on a rise in the postsynaptic Ca^{2+} concentration following NMDA-type glutamate receptor activation[19]. In contrast, the long-lasting synaptic potentiation induced by taurine (LLP$_{TAU}$) is independent of this type of receptor[7]. In the present study, we have found that Ca^{2+} is also required for the induction of LLP$_{TAU}$ and we have pharmacologically identified the Ca^{2+} sources involved in this taurine-induced potentiation.

Taurine 4, edited by Della Corte *et al.*
Kluwer Academic / Plenum Publishers, New York, 2000.

METHODS

Slice Preparation

Experiments were performed on 400-μm-thick transverse hippocampal slices obtained from adult female Sprague-Dawley rats (200-250 g), following standard methods. Briefly, after the rat was decapitated, its brain was rapidly removed and dropped into ice-cold standard medium (composition in mM: NaCl 119, NaHCO$_3$ 26.2, KCl 2.5, KH$_2$PO$_4$ 1, MgSO$_4$ 1.3, CaCl$_2$ 2.5 and glucose 11) pregassed with 95% O$_2$ and 5% CO$_2$. Once the hippocampi were dissected out, they were sliced by a manual tissue chopper and placed in an interface holding chamber, where they were maintained at room temperature (21-25°C). After at least one hour of recovery, one slice was transferred to a submersion-type recording chamber, where it was continuously perfused (flow rate 1.5-2 ml/min) with the standard medium equilibrated with 95% O$_2$ and 5% CO$_2$. Experiments were performed at 30-32°C.

Solutions

Drugs applied by addition to the standard perfusion solution included, ethylene glycol bis-(b-aminoethyl ether N,N,N′,N′-tetraacetic acid (EGTA), i-isobutoxy-2-pyrrolidino-3-[N-benzyl-anilino] propane (bepridil hydrochloride), kynurenic acid, nickel chloride, nifedipine, picrotoxin, taurine, tetraethyl-ammonium chloride (TEA) and thapsigargin (all from Sigma, St Louis, MO, USA). Taurine was used in these experiments at a concentration of 10 mM. The osmolality increase caused by adding 25 mM TEA chloride to the standard medium was balanced by reducing the NaCl concentration in a equimolar amount. Stock solutions of bepridil (25 mM), nifedipine (50 mM) and thapsigargin (5 mM) were prepared in dimethyl sulfoxide (final concentration 0.02%), stored frozen in the dark, and diluted to its final concentration in the perfusion solution immediately before use. In those experiments in which GABA$_A$ antagonist picrotoxin (100 μM) was present in the bath medium, a cut was made between CA1 and CA3, and the concentrations of Ca^{2+} and Mg^{2+} were increased to 4 mM to prevent epileptiform discharges.

Recording and Analysis of Evoked Synaptic Potentials

Synaptic responses were evoked by stimulating Schaffer collateral-commisural fibers with electrical pulses (20-50 μA, 40 μsec, 0.05-0.066 Hz) applied through bipolar tungsten insulated microelectrodes placed on CA1 stratum radiatum. Electrical pulses were supplied by a pulse generator (A.M.P.I., Mod. Master 8, Jerusalem, Israel). Field EPSP (fEPSP) from the stratum radiatum of the CA1 region, were recorded with a glass micropipette filled with 2 M NaCl (1-2 MΩ), connected to an Axopatch-1D amplifier (Axon Instruments, Foster City, CA, USA).

Evoked responses were digitized at 25-50 kHz (TL-1 interface, Axon Instruments), and stored and analyzed on a 486 IBM compatible computer using pCLAMP software (Axon Instruments). Synaptic strength was calculated using the slope of the initial rising phase of the fEPSP to avoid contamination of the response by the population spike. Data were normalised with respect to the mean values of the responses at the 20 min control perfusion period (in standard medium), before the application of taurine. For these calculations a program developed by J. Bustamante was used. Results are expressed as mean ± SEM. Statistical significance of differences were assessed by one-way or two-way analyses of variance, and two-tailed Student's t-tests.

RESULTS

Taurine (10 mM) perfusion in a standard medium evoked a biphasic depression-potentiation effect on fEPSPs during the 30 min of its application. In addition, taurine perfusion also induced a long-lasting increase of fibre volley amplitude which is a field potential corresponding to the compounded action potential generated in a group of axons. A detailed analysis of fibre volley potentiation induced by taurine is already described[6,7,8]. The fEPSP increase induced by taurine remained potentiated one hour after taurine withdrawal (39.5 ± 1.7% over baseline values), as previously reported[7,8]. However, taurine did not induce fEPSP potentiation when it was added to a medium nominally free of calcium (0 mM Ca^{2+}, 6 mM Mg^{2+} and 100 µM EGTA) (Fig.1). This result indicates that extracellular Ca^{2+} is required for the induction of LLP_{TAU}. In addition, the maintenance of the taurine-induced potentiation also needs the Ca^{2+} released from intracellular stores, because the fEPSP residual potentiation level one hour after taurine withdrawal was greatly reduced when 1 µM thapsigargin, an inhibitor of the endoplasmic reticulum Ca^{2+}-ATPase that depletes intracellular Ca^{2+} stores[22] was present in the perfusion solution (Fig. 1).

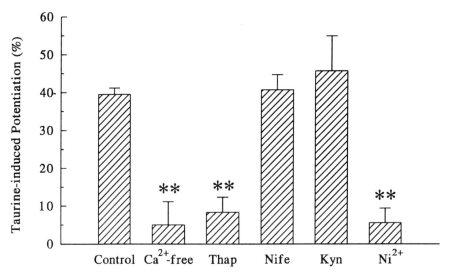

Figure 1. Pharmacological identification of the Ca^{2+} sources involved in the taurine-induced potentiation. Bars represent the fEPSP potentiation level induced by taurine (10 mM, 30 min) measured one hour after taurine withdrawal. Control: taurine was applied in a standard perfusion medium (n=13). Ca^{2+}-free: a medium without Ca^{2+} and containing 6 mM Mg^{2+} and 100 µM EGTA was perfused 20 min before and during taurine application (n=8). Thap: 1 µM thapsigargin was perfused 20 min before, during and 10 min after taurine perfusion (n=10). Nife: 10 µM nifedipine was perfused 20 min before and during taurine (n=5). Kyn: 10 mM kynurenic acid was perfused 20 min before and during taurine (n=6). Ni^{2+}: 50 µM $NiCl_2$ was continuously perfused along the experiment. **P<0.001 compared with control bar. Modified from Del Olmo et al[6].

Having taken into account the three main pathways responsible for the influx of Ca^{2+} in the nervous cells: 1) neurotransmitter-gated Ca^{2+} channels, 2) voltage activated Ca^{2+} channels, and 3) Na^{+}-Ca^{2+} exchanger, and the dependence of LLP_{TAU} on extracellular Ca^{2+}, we have tested, by using selective antagonists, which of these pathways gives rise to Ca^{2+} influx for taurine action. We have previously shown that LLP_{TAU} induction is independent of NMDA receptor activation[7]. In a series of experiments we have further validated this fact by perfusing taurine in the presence of 10 mM kynurenic acid, a low affinity glutamate receptor antagonist. Under these conditions LLP_{TAU} was similar to that evoked in control experiments (Fig.1). LLP_{TAU} was not significantly affected by 10 µM nifedipine (Fig.1), channels. However, LLP_{TAU} was inhibited in the presence of 50 µM Ni^{2+} (Fig. 1), that at this low concentration is a rather selective antagonist of T-type low-voltage activated Ca^{2+} channels[12].

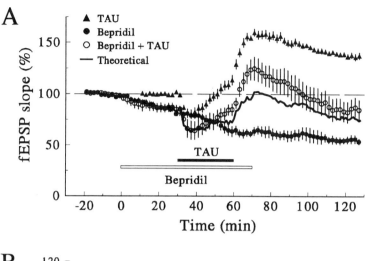

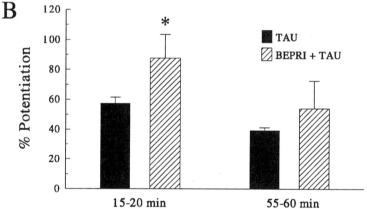

Figure 2. Blockade of Na^+-Ca^{2+} exchanger enhances taurine-induced potentiation. (A) Filled triangles correspond to taurine perfusion experiments in control conditions (n=13). Filled circles correspond to the time course of changes in the fEPSP obtained during exposure to 50 µM bepridil alone (n= 6). Open circles correspond to the experiments (n=8) performed as those indicated by filled circles but in which 10 mM taurine was perfused in the presence of 50 µM bepridil. The continuous line on the graph represents the theoretical time course of fEPSP changes resulting from the perfusion of taurine in the presence of bepridil, assuming no interaction between the effects of both substances. The continuous line was obtained as described in Results. Open and filled horizontal bars indicate bepridil and taurine perfusion periods respectively. (B) Bars represent pooled fEPSP potentiation values measured at 15-20 min and 55-60 min after taurine withdrawal. Black bars correspond to control experiments showed in A as filled triangles. Striped bars correspond to the experiments represented by open circles in A, and were calculated taking as baseline the values obtained during bepridil experiments (filled circles) at those times. Differences between taurine and taurine + bepridil at 15-20 min are statistically significant (P<0.05).

The Na^+-Ca^{2+} exchanger under certain circumstances can operate as a route for calcium entry. Thus, we carried out another series of experiments (n=8) in which the taurine-induced potentiation was studied in the presence of bepridil, an inhibitor of the Na^+-Ca^{2+} exchanger[14]. We firstly tested the effect on the fEPSP of 50 μM bepridil alone. Bepridil caused a progressive decrease of synaptic potentials (64.7 ± 4.9% of basal values at the last five minutes of bepridil perfusion; n=6), that did not recover during the washout period (55.0 ± 3.3% at 55-60 min of bepridil washout) (Fig. 2A). Taurine perfusion in the presence of bepridil elicited a profile of fEPSP slope changes qualitatively similar to that observed in control conditions, giving rise to a depression which superimposed on that evoked by bepridil alone (Fig. 2A). The continuos line in Fig. 2A depicts the theoretical situation in which the effects of bepridil and taurine do not interact, and was calculated every minute considering that if bepridil reduce the fEPSP by DB% and this effect it is potentiated by taurine to PT% of control values, it would give a value of [(100-DB) x PT%]/100. As shown in Fig. 2A, the experimental results obtained when taurine was perfused in the presence of bepridil fell above this line (P<0.001, two-way ANOVA), indicating that bepridil affects positively LLP_{TAU}. In fact, the averaged potentiation measured at 15-20 min after taurine withdrawal was higher in the presence than in the absence of bepridil (Fig. 2B). This result does not support the hypothesis that taurine-induced Ca^{2+} influx is promoted by reversion of Na^+-Ca^{2+} exchanger.

Subsequently, we have examined whether LLP_{TAU} was mechanistically similar to the long-lasting potentiation of the synaptic efficacy induced by perfusing 25 mM TEA, a potassium channel blocker, in the presence of picrotoxin, a $GABA_A$ receptor antagonist. This procedure induces a long-lasting potentiation of synaptic transmission by promoting Ca^{2+} influx through voltage-activated Ca^{2+} channels[2,11]. Figure 3B shows that 10 min TEA application induces a long-lasting increase in fEPSP slope that remains quite stable during the experiment. In another group of slices, 10 mM taurine was perfused after TEA-induced potentiation was generated (Fig. 3A). Under these conditions, taurine potentiated fEPSP more slowly than in control experiments (Fig. 3C). Moreover, this fEPSP increment did not persist during taurine washout (p<0.01; two-way ANOVA), and declined to a potentiation level of 10.4 ± 5.3% at the end of this period, while the fEPSP potentiation, at the same time, induced by taurine in control conditions was of 39.7 ± 7.9% (Fig. 3C).

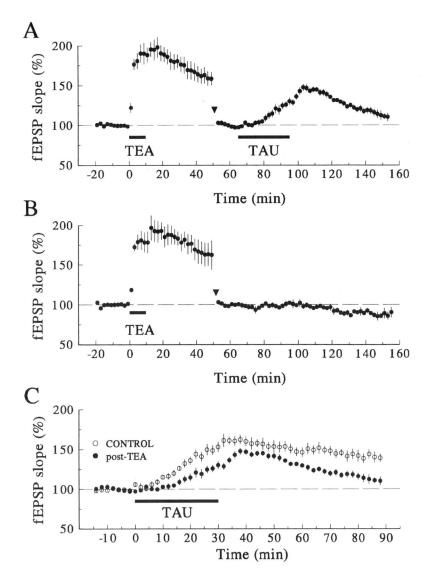

Figure 3. Long-term potentiation induced by TEA partially occludes LLP$_{TAU}$. The experiments were carried out in the presence of 100 μM picrotoxin. (A) Time course of fEPSP changes caused by 25 mM TEA and 10 mM taurine (TAU) in five slices. At the arrowhead, the stimulus strength was reduced to size-match the fEPSP with its baseline value. (B) A group of three experiments was performed as in A but without taurine to verify the stability of TEA-induced LTP. (C) Comparison of LLP$_{TAU}$ induced after TEA application (filled circles) showed in A, with LLP$_{TAU}$ evoked under control conditions (open circles, n=6).

DISCUSSION

This study shows that taurine application persistently increases synaptic transmission in the hippocampal Schaffer-collaterals pathway by a mechanism requiring extracellular Ca^{2+} (see Fig. 1) consistently with a previous finding[6]. Moreover, the induction of also depends on Ca^{2+} released from intracellular stores, as also required for the induction of LTP by high-frequency synaptic stimulation[10]. We have already reported that a postsynaptic rise in the intracellular Ca^{2+} concentration was also necessary for taurine-induced potentiation of the intracellularly recorded EPSP[6]. Thus, the Ca^{2+} requirements for LLP_{TAU} are similar to those involved in other phenomena of long-lasting potentiation of synaptic transmission[5,17]. However, since LLP_{TAU} was not affected by the glutamate receptor antagonists APV[7] or kynurenate, the Ca^{2+} influx activated during LLP_{TAU} appears to be different from that implicated in the LTP evoked by brief pulses of high-frequency stimulation, which induce Ca^{2+} entry through both NMDA receptor activation[19] or the activation of voltage-dependent Ca^{2+} channels[2,9]. Our experiments performed in the presence of nifedipine indicate that L-type high-voltage activated Ca^{2+} channels are not involved in LLP_{TAU}. However, T-type, low-voltage activated Ca^{2+} channels seem to be the Ca^{2+} influx route involved in LLP_{TAU}, as deduced from the inhibitory effect of Ni^{2+}, an antagonist of this type of Ca^{2+} channels. This route of Ca^{2+}, activated by high-frequency synaptic stimulation, appears to be sufficient to induce LTP in the visual cortex[16]. Electrogenic taurine uptake[18] during taurine perfusion could provide the small depolarization necessary to activate the low-voltage activated Ca^{2+} channels. Alternatively, taurine could positively modulate these channels, since it has already been found that taurine induces the appearance of a T-type low-voltage activated Ca^{2+} current in embryonic cardiomyocytes[20].

Another possible route of Ca^{2+} influx is through the Na^{+}-Ca^{2+} exchanger, which in resting conditions of membrane potential and intracellular Na^{+} concentration extrudes Ca^{2+} from the cytoplasm, by using the driving force of the Na^{+} electrochemical gradient[1]. The Na^{+}-Ca^{2+} exchanger could operate in reverse mode, i.e. transporting Ca^{2+} inside the cell, during states of membrane depolarization and high intracellular Na^{+} concentration[21]. Although taurine does not appear to interact directly with the Na^{+}-Ca^{2+} exchanger[15], it might could favour the reverse operation of this exchanger during its Na^{+}-dependent uptake, resulting, therefore, in an intracellular Ca^{2+} increase. This has been proposed to explain the intracellular Ca^{2+} increase induced by taurine in isolated ventricular myocytes[4]. This mechanism, however, can not explain LLP_{TAU}, since in our experiments taurine was able to induce LLP_{TAU} also in the presence of bepridil, a Na^{+}-Ca^{2+} exchanger inhibitor.

There are considerable experimental evidences that the initial intracellular Ca^{2+} concentration rise during LTP generation activates a cascade of Ca^{2+}–dependent enzymes that is critical for the synaptic potentiation maintenance[5]. We have previously demonstrated[6] that when NMDA-mediated LTP was induced in a group of synapses by trains of high-frequency stimulation, the subsequent taurine application evoked a synaptic potentiation that did not last after taurine washout. Similarly, taurine perfusion after induction of LTP by TEA (which activates Ca^{2+} influx through high-voltage activated Ca^{2+} channels) only evoked a rapidly decaying potentiation (see Fig. 3). These occlusion experiments indicate that LLP_{TAU} shares a common feature with other forms of long-term synaptic potentiation, although the Ca^{2+} influx pathway involved in all cases is different.

In conclusion, taurine induces long-lasting increases in synaptic transmission by eliciting Ca^{2+} influx trough low-voltage activated Ca^{2+} channels, and Ca^{2+} release from intracellular Ca^{2+} stores. Some of the Ca^{2+}–sensitive mechanisms triggered by this intracellular Ca^{2+} rise may be similar to those involved in other types of long-lasting synaptic potentiation phenomena.

ACKNOWLEDGEMENTS

This work was supported by the «Fondo de Investigaciones Sanitarias» (Grant 93/0565) and the «Ministerio de Educación y Cultura» (Grant PM95-170). We thank M.J. Asensio and A. Latorre for technical assistance. N. Del Olmo is recipient of a predoctoral fellowship from M.E.C.

REFERENCES

1. Allen, T.J.A., Nobe, D., and Reuter, H., 1989, *Sodium-calcium exchange*. Oxford University Press, New York.
2. Aniksztejn, L., and Ben-Ari, Y., 1991, Novel form of long-term potentiation produced by K^+ channel blocker in the hippocampus. *Nature* **349:** 67-69.
3. Barnes, C.A., 1995, Involvement of LTP in Memory: Are we «searching under the street light»? *Neuron* **15:** 751-754.
4. Bkaily, G., Jaalouk, D., Sader, S., Shbaklo, H., Pothier, P., Jacques, D., D'Orléans-Juste, Cragoe E.J.(Jr.), and Bose, R., 1998, Taurine indirectly increases [Ca]i by inducing Ca $^{2+}$ influx trough the Na^+-Ca^{2+} exchanger. *Mol. Cell. Biochem.* **188:** 187-197.
5. Bliss, T.V.P., and Collingridge, G.L., 1993, A synaptic model of memory: long-term potentiation in the hippocampus. *Nature* **361:** 31-39.
6. Del Olmo, N., Galarreta, M., Bustamante, J., Martín del Río, R., and Solís, J.M., 2000, Taurine-induced synaptic potentiation: role of calcium and interaction with LTP. *Neuropharmacology* **39:** 40-54.

7.　Galarreta, M., Bustamante, J., Martín del Río, R., and Solís, J.M., 1996a, Taurine induces a long-lasting increase of synaptic efficacy and axon excitability in the hippocampus. *J. Neurosci.* **16:** 92-102.

8.　Galarreta, M., Bustamante, J., Martín del Río, R., and Solís, J.M., 1996b, A new neuromodulatory action of taurine: long-lasting increase of synaptic potentials. *Adv. Exp. Med. Biol.* **403:** 463-471.

9.　Grover, L.M., and Teyler, T.J., 1990, Two components of long-term potentiation induced by different patterns of afferent activation. *Nature* **347:** 477-479.

10.　Harvey, J., and Collingridge, G.L., 1992, Thapsigargin blocks the induction of long-term potentiation in rat hippocampal slices. *Neurosci. Lett.* **139:** 197-200.

11.　Huang, Y.-Y, and Malenka, R.C., 1993, Examination of TEA-induced synaptic enhancement in area CA1 of the hippocampus: the role of voltage-dependent Ca^{2+} channels in the induction of LTP. *J. Neurosci.* **13:** 568-576.

12.　Huguenard, J.R., 1996, Low-threshold calcium currents in central nervous system neurons. *Ann. Rev. Physiol.* **58:** 329-348.

13.　Huxtable, R.J., 1989, Taurine in the central nervous system and the mammalian actions of taurine. *Prog. Neurobiol.* **32:** 471-533.

14.　Kaczorowski, G.J., Garcia, M.L., King, V.F., and Slaughter, S., 1989, Development and use of inhibitors to study sodium-calcium exchanger. In *Sodium-Calcium Exchange* (T.J.A. Allen, D. Noble and H. Reuter, eds.), Oxford University Press, New York, pp. 66-101.

15.　Katsube, Y., and Sperelakis, N., 1996, Na^+/Ca^{2+} exchange current: lack of effect of taurine. *Eur. J. Pharmacol.* **316:** 97-103.

16.　Komatsu, Y., and Iwakiri, M., 1992, Low-threshold Ca^{2+} channels mediate induction of long-term potentiation in kitten visual cortex. *J. Neurophysiol.* **67:** 401-410.

17.　Larkman, A.U., and Jack, J.J.B., 1995, Synaptic plasticity: hippocampal LTP. *Curr. Opin. Neurobiol.* **5:** 324-334.

18.　Loo, D.D.F., Hirsch, J.R., Sarkar, H.K., and Wright, E.M., 1996, Regulation of the mouse retinal taurine transporter (TAUT) by protein kinases in Xenopus oocytes. *FEBS Lett.* **392:** 250-254.

19.　Malenka, R.C., and Nicoll, R.A., 1993, NMDA-receptor-dependent synaptic plasticity: multiple forms and mechanisms. *Trends Neurosci.* **16:** 521-527.

20.　Satoh, H., and Sperelakis, N., 1993, Effects of taurine on Ca^{2+} currents in young embryonic chick cardiomyocytes. *Eur. J. Pharmacol.* **231:** 443-449.

21.　Stys, P.K., Waxman, S.G., and Ransom B.R., 1992, Ionic mechanisms of anoxic injury in mammalian CNS white matter: role of Na^+ channels and Na^+-Ca^{2+} exchanger. *J. Neurosci.* **12:** 430-439.

22.　Thastrup, O., Cullen, P.J., Drobak, B.K., Hanley, M.R., and Dawson, A.P., 1990, Thapsigargin, a tumor promoter, discharges intracellular Ca^{2+} stores by specific inhibition of the endoplasmic reticulum Ca^{2+}-ATPase. *Proc. Natl. Acad. Sci. USA* **87:** 2466-2470.

THE EFFECT OF KAINIC ACID AND AMPA ON THE RELEASE OF TAURINE AND GABA FROM THE RAT SUBSTANTIA NIGRA *IN VIVO*

Loria Bianchi, Maria A. Colivicchi, Maria Frosini, Mitri Palmi and Laura Della Corte
Dip. di Farmacologia Preclinica e Clinica "M. Aiazzi Mancini", Università di Firenze, Firenze, Italy, [1]Istituto di Scienze Farmacologiche,Università di Siena, Siena, Italy

INTRODUCTION

One of the main output nuclei of the basal ganglia, a group of sub-cortical nuclei involved in the control of movement and various mnemonic and cognitive functions, is the substantia nigra pars reticulata (SNr). The SNr contains high levels of the inhibitory amino acid transmitter, GABA, which is localised in afferent terminals derived from the striatum and the globus pallidus and it is also the transmitter of the output neurons which project to the superior colliculus, the thalamus and the brain stem and possess local axon collaterals[1,2].

The SNr receives excitatory inputs from the subthalamic nucleus[3], the cortex[4] and the mesopontine tegmentum[5]. Subunits of both the kainate (KA) and AMPA type of non-NMDA excitatory receptors are expressed in the substantia nigra. A high expression of GluR5 and GluR7 subunits of the KA receptor was found in the substantia nigra pars compacta[6], whereas GluR1 and GluR2/3 subunits of the AMPA receptors have been found associated with both dopaminergic and non-dopaminergic neurons[7].

Previous *in vivo* studies, using microdialysis in the freely moving rat, have demonstrated that stimulation of excitatory non-NMDA amino acid receptors, presumably associated with the excitatory inputs, by perfusion with KA, induces a DNQX-sensitive release of GABA[8]. Taurine has also been shown to be present in striatonigral neurons and to be released in the SNr in response

Taurine 4, edited by Della Corte *et al.*
Kluwer Academic / Plenum Publishers, New York, 2000.

to stimulation of striatonigral neurons[9] and local application of KA[10]. These observations suggest that at least some of the released GABA and taurine may be derived from the terminals of striatonigral neurons.. The present experiments were carried out to determine whether the AMPA/KA receptor agonist, KA, and the specific AMPA receptor agonist, AMPA, differentially affect the release of endogenous taurine and GABA.

MATERIALS AND METHODS

The experiments were performed on male Wistar rats (200-250 g body weight). Single cannula microdialysis probes were implanted vertically into the right SNr (1 mm exposed tip), as previously described[8]. Twenty four hours later, the probes, of the now freely moving rats, were perfused with artificial cerebrospinal fluid at a rate of 2 µl/min. After a perfusion period of 2 h the SNr was then perfused with 100 µM kainic acid or AMPA (30 min), alone or in the presence of 100 µM 6,7-dinitroquinoxaline-2,3-dione (DNQX) or 10 µM tetrodotoxin (TTX). Fractions (20 min) were collected for 1 h before, and up to 2 h after the KA perfusion. Following derivatisation and separation (hplc), the amino acids were detected fluorimetrically. The levels of GABA and taurine were expressed as pmol of amino acid/µl of perfusate. Original values were compared by analysis of variance[8].

Table 1. Effect of of DNQX and TTX on GABA and taurine output induced by local application of KA or AMPA in the SNr.

Treatment	n	Stimulated output (Peak value as % of basal value)	
		GABA	Taurine
KA 100 µM	6	720 ± 267*	177 ± 34*
KA 100 µM + DNQX 100µM	5	120 ± 44	109 ± 15
KA 100 µM + TTX 10µM	5	123 ± 41	95 ± 16
AMPA 100 µM	7	371 ± 124*	181 ± 38*
KA 100 µM + DNQX 100µM	7	121 ± 54	111 ± 17
KA 100 µM + TTX 10µM	6	95 ± 16	147 ± 21*

*P<0.05, MANOVA on original values. Figures represent mean ± approx. s.e.m.

RESULTS

Basal output levels of GABA and taurine in the SNr were 0.021 ± 0.003

and 0.406 ± 0.049 (mean ± s.e.m., n=36) pmol\µl perfusate, respectively. Stimulated output levels observed following perfusion of the SNr with 100 µM KA or AMPA, alone or in the presence of 100 µM DNQX or 10 µM TTX, are summarised in Table 1. KA and AMPA induced a statistically significant (p<0.05, MANOVA) increase in the output of GABA and taurine, which were blocked by co-administration of 100 µM DNQX. The presence of 10 µM TTX blocked both the KA- and the AMPA-stimulated release of GABA and the KA-induced release of taurine, whereas the AMPA-stimulated release of taurine was not significantly affected.

DISCUSSION

The main finding of the present study is that endogenous GABA and taurine are released from the SNr in response to perfusion with either KA, as previously observed[10], or AMPA. This release is due mainly to the stimulation of non-NMDA excitatory amino acid receptors as the enhanced release was blocked by DNQX. The present data are consistent with the presence of excitatory amino acids in some of the afferents of the SNr and the presence of non-NMDA, AMPA/kainate, excitatory amino acid receptors[6,7,11,12] in the SNr.

The KA-stimulated release of both taurine and GABA and the AMPA-stimulated release of GABA ware abolished in the presence of TTX, this indicating a dependence on fast sodium channels and thus on axon potential propagation. This observation suggests that the enhanced release in response to KA or AMPA is not a direct effect on GABA- or taurine-containing axon terminals but, more likely, an effect at the level of neuronal cell bodies or dendrites. Thus the released is likely to be derived from the local axon collaterals of GABA neurons in the nigra[1] or due to an indirect action of KA/AMPA on GABA-containing terminals mediated by a class of neuron containing a different transmitter e.g. the dopaminergic neurons. Since there is no evidence for taurine-containing neurons in the SNr, then the latter mechanism is the more likely to apply to the KA-induced release of taurine.

The observation that the AMPA-induced release of taurine was insensitive to TTX suggests that AMPA may directly affect taurine-containing axon terminals. However, since AMPA receptors have been shown to be present also on glial cells[12], the insensitivity to TTX may also indicate a glial origin of the AMPA-stimulated release of taurine.

ACKNOWLEDGMENTS

This work was supported by the European Union (BMH1 CT94-1402 and COST D8), MURST (ex 40%) and Università di Firenze (ex 60%), Italy.

REFERENCES

1. Deniau, J.M., Kitai, S.T., Donoghue, J.P. and Grofova, I., 1982, Neuronal interactions in the substantia nigra pars reticulata through axon collaterals of the projection neurons. *Exp. Brain Res.* **47**:105-113.
2. Bevan, M.D., Bolam, J.P. and Crossman, A.R., 1994, Convergent synaptic input from the neostriatum and the subthalamus onto identified nigrothalamic neurons in the rat. *Eur. J. Neurosci.* **6**:320-334.
3. Rinvik, E. and Ottersen, O., 1993, Terminals of subthalamonigral fibres are enriched with gluatamte-like immunoreactivity: an electron microscopic, immunogold analysis in the cat. *J. Chem. Neuroanat.* **6**:19-30.
4. Naito, A. and Kita, H., 1994 The cortico-nigral projection in the rat: an anterograde tracing study with biotinylated dextran amine. *Brain Res.* **637**:317-322.
5. Lavoie, B. and Parent, A., 1994, Pedunculopontine nucleus in the squirrel monkey: cholinergic and glutamatergic projections to the substantia nigra. *J. Comp. Neurol.* **344**:232-241.
6. Bischoff, S., Barhanin, J., Bettler, B., Mulle, C. and Heinemann, S., 1997, Spatial distribution of kainate receptor subunit mRNA in the mouse basal ganglia and ventral mesenchephalon. *J. Comp. Neurol.* **379**:541-562.
7. Paquet, M. and Smith, Y., 1996 Differential localization of AMPA glutamate receptor subunits in the two segments of the globus pallidus and the substantia nigra pars reticulata in the squirrel monkey. *Eur. J. Neurosci.* **8**:229-233.
8. Bianchi, L., Sharp, T., Bolam, J.P. and Della Corte, L., 1994, The effect of kainic acid on the release of GABA in rat neostriatum and substantia nigra. *NeuroReport* **5**:1233-1236.
9. Della Corte, L., Bolam, J.P., Clarke, D.J., Parry, D., Smith, A.D., 1990, Sites of [^{9}H]taurine uptake in the rat substantia nigra in relation to the release of taurine from the striatonigral pathway. *Eur. J. Neurosci.* **2**:50-61.
10. Colivicchi, M.A., Bianchi, L., Bolam, John P., Galeffi, F., Frosini, M., Palmi, M., Sgaragli, G.P. and Della Corte, L., 1998, The *in vivo* release of taurine in the striatonigral pathway. In *Taurine 3: Cellular and regulatory mechanisms* (Shaffer, S., Lombardini, J.B. and Huxtable R.J.), Plenum Press, New York, pp. 363-370.
11. Bernard, V., Gardiol, A., Faucheux, B., Bloch, B., Agid, Y., Hirsch, E.C., 1996, Expression of glutamate receptors in the human and rat basal ganglia: effect of the dopaminergic denervation on AMPA receptor gene expression in the striatopallidal complex in Parkinson's disease and rat with 6-OHDA lesion. *J. Comp. Neurol.* **368**:553-568.
12. Martin, L.J., Blackstone, C.D., Levey, A.I., Huganir, R.L., Price, D.L., 1993, AMPA glutamate receptor subunits are differentially distributed in rat brain.. *Neuroscience* **53**:327-358.

EFFECTS OF ETHANOL AND GLUTAMATE AGONIST INFUSION ON THE OUTFLOW OF SULPHOACETALDEHYDE

Microdialysis Studies

Selene Capodarca,[1] Colm M. Cunningham,[1] Petra Bartolini,[2]
Keith F. Tipton,[1] Laura Della Corte,[2] Loria Bianchi,[2] Roberta J. Ward,[3]
Abdel Dachour,[3] Etienne Quertemont,[3] Frederic Lallemand,[3] and
Philippe De Witte[3]

[1] *Department of Biochemistry, Trinity College, Dublin 2, Ireland*
[2] *Dipartimento di Farmacologia Preclinica e Clinica, Università degli Studi di Firenze, Italy*
[3] *Biologie du Comportement, Universitè Catolique de Louvain, Belgium*

INTRODUCTION

The reaction of taurine with hypochlorous acid produces *N*-chlorotaurine (taurine chloramine), which is then metabolised to sulphoacetaldehyde (SAA). SSA may then be reduced to isethionic acid. This pathway, which has been shown to occur when neutrophils are activated, is the only known pathway of taurine catabolism in mammalian systems[1]. SAA may be detected by an HPLC method involving pre-column derivatization with 2-diphenylacetyl-1,3-indandione-1-hydrazone (DIH) and fluorescent detection of the resulting adduct[1,2] This procedure resolves SAA from other aldehydes, including succinic semialdehyde (SSA) that arises from the transamination of GABA.

Ethanol ingestion results in an elevation of several aldehydes in the tissues, including those arising from the oxidative deamination of the biogenic amines[3]. This arises from competition of the acetaldehyde formed from ethanol at the levels of the aldehyde dehydrogenases and from the decrease in the $NAD/NADH_2$ ratio that results from ethanol metabolism (see Fig. 1).

Taurine 4, edited by Della Corte *et al.*
Kluwer Academic / Plenum Publishers, New York, 2000.

   Although, sulphoacetaldehyde is not a substrate for the aldehyde
dehydrogenases, the elevated biogenic aldehydes might be expected to
compete for removal by inhibiting the reduction of sulphoacetaldehyde
to isethionic acid, which is catalysed by the aldehyde reductases.

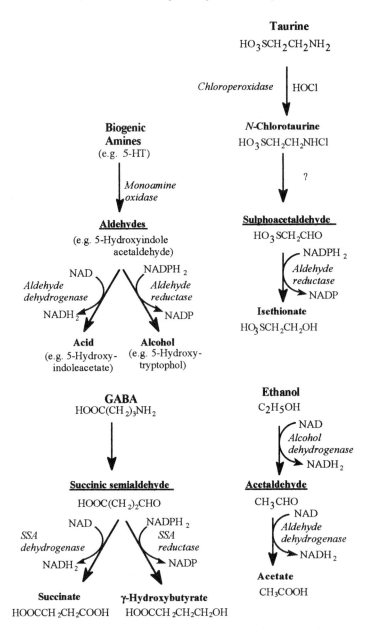

Fig. 1. Metabolism of ethanol and some amines showing potential competition between
aldehydes and through coenzyme redox state.

An increase in taurine levels in hippocampal microdialysate samples has been observed after acute (i.p.) ethanol treatment[4]. This effect may to be related to a tissue protective function of taurine. The aim of the present work was to determine whether ethanol administration affected the levels of SAA.

MATERIALS AND METHODS

Surgery and microdialysis perfusion

In the studies on the effects of NMDA, rats were anaesthetised by an intraperitoneal injection of chloral hydrate (400 mg/kg) and a single microdialysis probe was inserted into the right neostriatum. Stereotaxic coordinates were derived from the atlas of Paxinos and Watson[6]. The animals were allowed to recover and the microdialysis experiments were performed 24 h after the surgery in freely-moving rats. The probes were perfused with artificial cerebrospinal fluid (CSF) containing 140 mM NaCl, 3 mM KCl, 1.2 mM $CaCl_2$, 1 mM $MgCl_2$, 1.2 mM Na_2HPO_4, and 7.2 mM glucose, pH 7.4, through a polyethylene tube (i.d. 0.38 mm) connected to a syringe mounted on a microinfusion pump (CMA/Microdialysis AB, Stockholm, Sweden) set at a flow rate of 2 µl/min. Following a 90 min stabilization period, perfusates were collected every 20 min for a period of 4-5 h. Basal efflux of amino acids was determined in the first three 20 min fractions, after which various doses of NMDA (100 µM, 1mM, 4mM) dissolved in artificial CSF were applied for 20 or 40 minutes, before reversion to normal CSF.

In the ethanol studies microdialysis probes were inserted into the hippocampal region and perfused with Ringer solution (145 mM NaCl, 1.3 mM $CaCl_2$, 4 mM KCl, buffered to pH 7.2 with phosphate buffer, at a flow rate of 1µl/min. Outflow samples were collected over 20 min periods; 5 before the i.p. administration of either ethanol (2 g/Kg) or saline, after which a further 10-15 samples were collected.

HPLC determination

The samples (20 µl) were analysed for both SSA and SAA, following precolumn derivatization with DIH (30mg/100 ml in acetonitrile). This entailed mixing the samples with DIH (1: 4 v/v) and 0.035 equivalents of HCl and incubating the mixture at 50 °C for 90 - 100 min.

For the ethanol studies the HPLC equipment comprised a Beckman 110B delivery system, a Waters 712 WISP automatic sample injector, a

Kontron SFM25 fluorimetric detector and a Spectra Physics SP4290 integrator. The column used was a Waters C18 µBondapack reverse-phase column (3.9 x 300 mm). The column was eluted isocratically with a mobile phase comprising 70 % acetonitrile, 20 % H_2O, 10 % 0.1 M ammonium acetate, pH 6.5.

For the NMDA studies the derivatized perfusates were chromatographed on a Shimadzu C18 column, eluted isocratically with 50 % acetonitrile, 40 % water, 10 % 0.1 M ammonium acetate, pH 5.5, and analysed using a Shimadzu fluorimeter.

RESULTS

Fig. 2 shows the resolution of sulphoacetaldehyde adduct under different elution conditions.

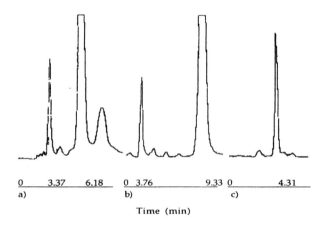

Time (min)

Fig. 2. Resolution of the DIH-derivative of SSA under different elution conditions. The mobile phases comprised acetonitrile / water / ammonium at the following percentages: (a) 50/40/10, pH 6; (b) 60/30/10, pH 6.5; (c) 70/20/10, pH 6.5.

The optimal conditions for the detection of sulphoacetaldehyde were: 50% acetonitrile, 40% H_2O, 10% 0.1 M ammonium acetate, pH 6.0. The excitation and emission wavelengths for maximum fluorescence were 415 and 525 nm, respectively[1].

The optimal conditions for the resolution of SSA and sulphoacetaldehyde were: 70% acetonitrile, 20% H_2O, 10% 0.1 M ammonium acetate, pH 6.5, as shown in Fig. 3.

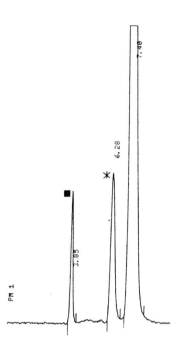

Fig. 3. Resolution of DIH derivatives of SAA and SSA. The separation of the sulphoacetaldehyde-azine peak (■) from the SSA-azine peak (✻) has been reached by using a mobile phase which composition was 70/20/10 (acetonitrile/water/ammonium acetate), pH 6.5.

The presence of glucose in the artificial CSF used in the NMDA studies did not allow the determination of both SAA and SSA, because it gave rise to a broad HPLC peak of the aldose-DIH adduct, that interfered with that of the SSA-DIH adduct. The mobile-phase described above was optimised for the determination of SSA in the presence of glucose.

Fig 4 shows the absorbance and the emission spectra of the DIH adduct with the SSA. The maxima were sufficiently close to those of the adduct of sulphoacetaldehyde to allow excitation and emission wavelength of 415 and 525 nm, respectively, to be used for both their derivatives.

The eluted peak areas (Fig. 4) and heights (not shown) were linear function of aldehyde concentration for both SAA and SSA.

No outflow of either aldehyde could be detected in control samples. There was no detectable outflow of either aldehyde during a period of 200 minutes after ethanol administration (2 g/kg i.p.). Neither did NMDA infusion, *via* the microdialysis probe, result in any detectable liberation of sulphoacetaldehyde.

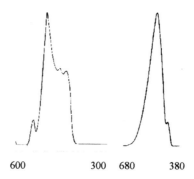

Fig. 4. The fluorescence excitation (left) and emission (right) spectra for the DIH derivative of succinic semialdehyde.

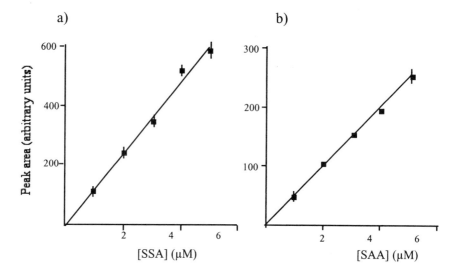

Fig. 5. Peak area versus (a) SSA and (b) SAA concentration for the HPLC- fluorescence assay procedure.

The analytical procedure was found to be capable of quantification of SAA and SSA to concentrations as low as 1.0 μM, with detection limits of 0.4 and 0.5 μM for SAA and SSA, respectively; where a peak height that is consistently 5-10 times greater of the base-line noise is deemed quantifiable, and is 3 times greater than the blank is deemed detectable. Table 1 shows the assay sensitivities for SSA and SAA.

Table 1: Quantification of SAA and SSA by the HPLC-fluorescence assay procedure. Each value is the mean ± s.e.m. of 5-10 separate determinations.

	SUCCINIC SEMIALDEHYDE (SSA)		SULPHOACETALDEHYDE (SAA)	
	[SSA] (μm)	Peak height (mm)	[SAA] (μm)	Peak height (mm)
Base-line	0	2 ± 1	0	2 ± 1
Detection limit	0.5	7 ± 1	0.4	6 ± 2
Quantification limit	1.0	14 ± 1	1.0	15 ± 2

DISCUSSION

It is known that: (a) ethanol ingestion results in elevation of many aldehydes in the tissues[3]; (b) acute ethanol treatment leads to increased taurine levels in brain microdialysate samples[4]; (c) SAA can be detected in activated human neutrophils as a consequence HOCl production[1], and (d) there appears to be a steady-state level of SAA in neutrophils that have not been activated[1], perhaps reflecting a slow rate of H_2O_2, and hence HOCl, production by these cells at rest. The hydrogen peroxide produced by activated neutrophils reacts with chloride ions to form HOCl in a reaction catalysed by myeloperoxidase (chloroperoxidase). HOCl reacts rapidly with taurine to form chlorotaurine and this is then converted to SAA in a reaction that can occur spontaneously but appears to be accelerated by an enzyme[1].

Neutrophil infiltration of brain tissue occurs after traumatic injury[7]. Activated microglia may also be a source of chlorotaurine in the brain. Microglia are the ontogenic and functional equivalents of mononuclear phagocites in somatic tissues[8] and have been shown to migrate and differentiate at sites of inflammation in the CNS and also to participate in phagocytosis[9]. Furthermore, recent studies have shown that brain microglia may contain myeloperoxidase[10].

These considerations suggest that SAA might be formed in the brain as a result of neurotoxic insults, such as those resulting from ethanol or NMDA. Furthermore, the disturbances in aldehyde and coenzyme levels following ethanol ingestion, which results in alterations in the metabolism of the aldehydes derived from the biogenic amines, might also affect the levels of SAA and SSA in the brain.

Neither SAA or SSA was detected in microdialysis samples following acute ethanol treatment (i.p.). Microdialysis should detect only compounds released from the cells and, thus, it is possible that metabolic disturbances resulting from ethanol treatment might occur intracellularly without affecting outflow. NMDA administration also failed to evoke SAA outflow, although this treatment is known to cause taurine release from the neostriatum. This might suggest that administration of this neurotoxin did not result in significant microglial activation or any substantial HOCl formation from any microglial cells that were activated. However, the activation of microglial cells is a relatively slow response and a longer interval of time following toxin treatment may be required for a response to be seen. Experiments in which SAA is determined at longer times after NMDA, and ethanol, administration are in progress to investigate this possibility.

ACKNOWLEDGMENT

We are grateful for support from EU BIOMED (BMH1 CT-1402) and COST (D8/019/98), and from Università di Firenze (Fondo d'Ateneo 60%).

REFERENCES

1. Cunningham C., Tipton K.F. and Dixon H.B.F. ,1998, *Biochem. J.* **330**: 933-937.
2. Bartolini P., Tipton K.F, Bianchi L., Stephenson D., Cunningham C. and Della Corte L., 1999, *Neurobiology* **7**: 159-174.
3. Tipton K.F., Houslay M.D. and Turner A.J., 1977, *Essay in Neurochem Neuropharmacol.* **1**: 103-138.
4. Dachour A., Quertemont E. and De Witte P. ,1994, *Alc. Alcoholism.* **29**: 485-487.
5. Young A.M.J. and Bradford H.F. , 1993, *Biochem. Pharmacol.* **41**: 155-162
6. Paxinos G. and Watson C. , 1982, *The rat brain in stereotaxic coordinates.* Academic Press, New York.
7. Carlos T.M., Clark R.S.B., Franigola-Higgins D., Schinding J.K. and Konachek P.M., 1997, *J. Leukoc. Biol.* **61**: 279-285.
8. Guilian D., 1987, *J. Neurosci. Res.* **18**: 155-171.
9. Perry V.H., Hume D.A. and Gordon S., 1985, *Neurosci.* **15**: 313-326.
10. Reynolds W.F., Rhees J, Maciejewski D., Paladino T., Sieburg H., Maki R.A., and Masliah E., 1999, *Exp. Neurol.* **155**: 31-41.

TAURINE, GLUTAMINE, GLUTAMATE, AND ASPARTATE CONTENT AND EFFLUX, AND CELL VOLUME OF CEREBROCORTICAL MINISLICES OF RATS WITH HEPATIC ENCEPHALOPATHY

Influence of Ammonia

Wojciech Hilgier,[1] Robert O. Law,[2] Magdalena Zielińska,[1] and Jan Albrecht[1]

[1]Dept. of Neurotoxicology, Medical Research Centre, Polish Academy of Sciences, Warsaw, Poland
[2]Dept. of Cell Physiology and Pharmacology, University of Leicester, Leicester, UK

INTRODUCTION

Hepatic encephalopathy (HE) is a complex neurological disorder associated with increased accumulation in the brain of ammonia and, subsequently, with disturbances in the intra- and extracellular content and functioning of the neuroactive amino acids glutamate (Glu), aspartate (Asp), taurine (Tau), and the purportedly neuroinert Glu metabolite - glutamine (Gln) (reviewed in ref. 1). Ammonia at pathophysiologically relevant (low milimolar) concentrations stimulates the release of newly loaded radiolabelled Tau from cultured glia cells[2-4], and of endogenous Tau, Glu, Gln from cerebral cortical minislices, whereby Tau is the most release-prone of the amino acids[5]. Acute HE is often associated with cerebral edema: This occurred in acute HE induced with a hepatotoxin, thioacetamide, which is

Taurine 4, edited by Della Corte et al.
Kluwer Academic / Plenum Publishers, New York, 2000.

305

accompanied by a moderate increase of brain ammonia (to ~0.6 mM concentration), and was detected either by an increased water content of the brain tissue, or decreased specific gravity[6,7].

Our previous studies have shown that an increased water content in the cerebral cortex of rats with acute HE, correlated with an increased efflux of newly loaded radiolabelled Tau in vitro[6] (see Table 1).

Table 1. Efflux of newly loaded [³H] Tau and water content of cerebral cortical slices derived from control rats and rats with acute HE evoked by thioacetamide treatment.

	[³H] Tau efflux (% remaining activity/20 min)	Water content (%)
Control	0.52±0.02	79.15±0.61
HE	0.72±0.02*	82.58±0.46*
% change	+44	+4

*p<0.05 (one-way ANOVA followed by Dunnet's test). Data from Hilgier *et al*[7].

This result was interpreted to reflect the cell volume regulatory transport of Tau out of the cells together with osmolytically obligated water. However, the efflux of endogenous Tau from cerebral cortical minislices triggered by 5 mM ammonium ions was not well correlated with cell volume changes in the slices as assessed by measuring the [¹⁴C] inulin space[5]. Also, the responses of ammonia-induced Tau efflux and cell volume to agents blocking anion channels involved in osmoregulatory Tau efflux were not well correlated with each other, and different from the responses of Tau efflux evoked by hypoosmotic or high potassium media[5]. In this study, therefore, we employed the paradigm used in the studies of the effect of ammonia in vitro, to evaluate the correlation between the efflux of endogenous amino acids and cell volume in cerebrocortical minislices derived from rats with TAA-induced HE. Since HE is accompanied by an increase of total cerebral content of neuroactive amino acids[6] (see also Fig. 1), the question was asked whether and in what degree amino acid efflux is related to amino acid content in the tissue. Since an extra ammonia challenge is often a precipitating factor in HE patients, we compared the effect of HE or/and in vitro treatment of the slices with 5 mM ammonium acetate («HE», «ammonia», «HE + ammonia», respectively) on the amino acid efflux and cell volume.

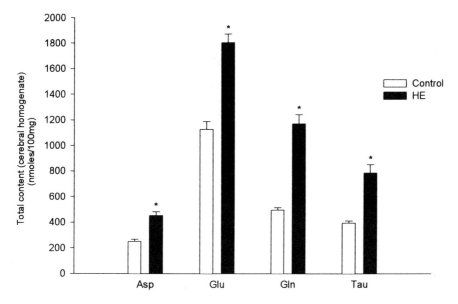

Fig. 1. Cerebral content of amino acids in control and rats with HE induced with thioacetamide. Results are mean ± SD for 3-5 experiments. Data from Hilgier et al.[6] *p<0.05 vs control (one-way ANOVA followed by Dunnet's test).

MATERIALS AND METHODS

HE model

Wistar rats weighing 180-220 g received three i.p. injections of 250 mg/kg body weight of thioacetamide in physiological saline, and were sacrificed 24h after the last injection[6,7]. Control rats received physiological saline at the same time intervals.

Preparation of rat cerebral cortical minislices and efflux assay

A previously described procedure was followed[5]. Cerebral hemispheres of control or HE rats were excised and placed in ice-cold aerated standard medium (SM) containing (mmol/l): NaCl 126, MgSO$_4$ 1.29. NaH$_2$PO$_4$ 1.29, KCl 5, CaCl2 0.8, HEPES 15, D-glucose 10, NaOH 1.7, pH 7.4. Slices (300-400 µm thickness, 4 out of each brain hemisphere) were cut freehand with a chilled razor blade from the frontal part of the cortex. Each slice was weighed to the nearest 50 µg on a torsion balance, and the weight ranged from 5-10 mg. Each slice was placed separately in 350 µl of SM with or without 5 mM ammonium acetate («ammonia»), and incubated at room temperature (18- 21°C) with gentle shaking. At 20 min and again at 40 min, slices were gently transferred to fresh buffer. The extracts obtained after 20, 40 and 60 min of incubation were pooled and freeze-dried. The tissue after incubation was freeze-dried, sonificated in 0.6 M perchloric acid and centrifuged at low speed. The supernatant was neutralized with 1M KOH, centrifuged at low speed, and the supernatant was used for OPA derivatization and HPLC analysis of amino acids.

HPLC analysis of amino acids

Amino acids were analyzed using HPLC with fluorescence detection after derivatization in a timed reaction with o-phtalaldehyde (OPA) plus mercaptoethanol, using the procedure of Kilpatrick[8], as modified previously[5]. Derivatized samples (50 µl) were injected onto 5µm Bio-Sil C18 Hl column (250x4.6 mm, BIO-RAD), with a mobile phase of 0.075 M KH_2PO_4 solution containing 10% v/v methanol, pH 6.2 (solvent A), and methanol (solvent B).

Calculation and expression of the results of efflux tests

Amino acid efflux was expressed as fractional efflux, i.e. the percentage of total remaining amino acid lost during each incubation period. In this way correction was provided for the changes in the total tissue content of the amino acids induced by incubation with ammonia. The initial amino acid content in the slices was calculated from the sum of the effluent and residual tissue contents.

Measurements of cell volume

The steady-state cell volumes in the slices were estimated from the distribution of [^{14}C] inulin (American Radiolabelled Chemicals, Inc.) following 20 min incubation, exactly as previously described[5,9,10].

RESULTS

Initial amino acid content in the slices (Fig. 2)

HE increased the content of Asp (by 36%), Glu (by 77%), Gln (by 29%) and Tau (by 67%). Treatment with 5 mM ammonia in vitro increased the content of Glu, Gln and Tau in a degree comparable to the increase evoked by

HE, but did not affect the Asp content. Ammonia treatment did not significantly potentiate the effects of HE on the amino acid content.

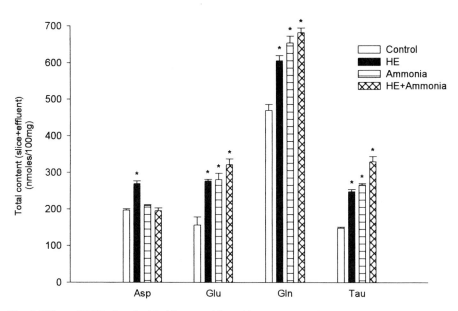

Fig. 2. Effect of HE induced with thioacetamide and/or treatment with 5 mM ammonium acetate ("ammonia") on the amino acid content in the cerebral cortical slices. Results are mean ± SD for 5-6 experiments. *$p<0.05$ vs control (one-way ANOVA followed by Dunnet's test).

Amino acid efflux (Fig. 3)

HE increased the fractional efflux of Asp, Gln, and Tau, without affecting the efflux of Glu. Treatment with 5 mM ammonia more than doubled the efflux of Gln, and produced a four-fold increase of Tau efflux, without affecting Glu or Asp efflux. The effects of ammonia on Asp, Glu, Gln, or Tau efflux were virtually identical in the slices of control and HE rats.

Cell volume (Fig. 4)

No changes in the cell volume were observed in brain slices of HE rats or slices of control rats treated with ammonia. Treatment of slices from HE rats with ammonia led to a significant above control increase of cell volume.

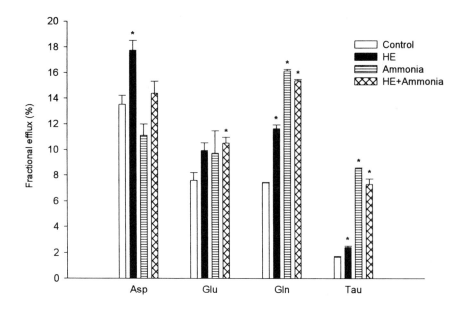

Fig. 3. Effect of HE induced with thioacetamide and/or treatment with 5 mM ammonium acetate ("ammonia") on the fractional amino acid efflux from cerebral cortical slices. Results are mean ± SD for 5-6 experiments. *$p<0.05$ vs control (one-way ANOVA followed by Dunnet's test).

CONCLUSIONS

- HE and ammonia stimulate the efflux of endogenous Tau and other neuroactive amino acids independently of producing changes in the cellular amino acid content, and by mechanisms not coupled to cell volume changes.
- Ammonia does not potentiate the HE-induced changes in amino acid content and transport.
- Cell volume regulation does not appear to be changed in HE rats, but is impaired in HE rats exposed to an extra ammonia challenge.

ACKNOWLEDGEMENTS

This work was supported by a SCSR grant n° 4.P05A.096.14.

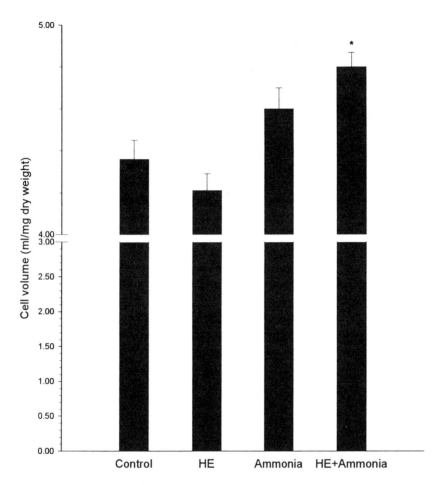

Fig. 4. Effect of HE induced with thioacetamide and/or treatment with 5 mM ammonium acetate ("ammonia") on the cell volume (inulin space) of the cerebral cortical slices. Results are mean ± SD for 7 experiments. *$p < 0.05$ vs control (one-way ANOVA followed by Dunnet's test).

REFERENCES

1. Albrecht, J., 1998, Roles and neuroactive amino acids in ammonia neurotoxicity. *J. Neurosci. Res.* **51**:133-138.
2. Albrecht, J., Bender, A.S., Norenberg, M.D., 1994, Ammonia stimulates the release of taurine from cultured astrocytes. *Brain Res.* **660**:288-292.
3. Faff, L., Reichenbach, A., Albrecht, J., 1996, Ammonia-induced taurine release from cultured rabbit Müller cells is an osmoresistant process mediated by intracellular accumulation of cyclic AMP. *J. Neurosci. Res.* **46**:231-238.
4. Faff-Michalak, L., Reichenbach, A., Dettmer, D., Kellner, K., Albrecht, J., 1994, K$^+$-, hypoosmolarity-, and NH$_4$+-induced taurine release from cultured rabbit Müller cells: The role of Na$^+$and Cl$^-$ions and relation to cell volume changes. *Glia* **10**:114-120.

The role of Na$^+$and Cl$^-$ions and relation to cell volume changes. *Glia* **10**:114-120.

5. Zielińska, M., Hilgier, W., Law, R. O., Gory_ski, P., Albrecht, J., 1999, Effect of ammonia in vitro on endogenous taurine efflux and cell volume in rat cerebrocortical minislices: influence of inhibitors of volume-sensitive amino acid transport. *Neuroscience* **91**:631-638.

6. Hilgier, W., Olson, J. E., 1994, Brain ion and amino acid contents during edema development in hepatic encephalopathy. *J. Neurochem.* **62**:197-204.

7. Hilgier, W., Olson, J. E., Albrecht, J., 1996, Relation of taurine transport and brain edema in rats with simple hyperammonemia or liver failure. *J. Neurosci. Res.* **45**:69-74.

8. Kilpatrick, I. C., 1991, Rapid, automated HPLC analysis of neuroactive and other amino acids in microdissected brain regions and brain slice superfusates using fluorimetric detection. *In Neuroendocrine Research Methods* (Greenstein B. eds), Harwood Academic, London, pp.555-578.

9. Law, R. O., 1994a, Taurine efflux and the regulation of cell volume in incubated slices of rat cerebral cortex. *Biochim. Biophys. Acta* **1221**:21-28.

10. Law, R. O., 1994b, Effects of extracellular bicarbonate ions and pH on volume-regulatory taurine efflux from rat   cerebral cortical slices in vitro: evidence for separate neutral and anionic transport mechanisms. *Biochim. Biophys. Acta* **1224**: 377-383.

THE EFFECT OF TAURINE DERIVATIVE ON LIPID PEROXIDATION AND INTER-HEMISPHERIC ASYMMETRY OF PHOSPHOLIPIDS FROM THE BRAIN SYNAPTOSOMES IN RATS AFTER IMMOBILIZATION STRESS

N. Y. Novoselova, *N. S. Sapronov, *Y. O. Fedotova, *P. A. Torkounov
*Department of Comparative Neurochemistry, Institute of Evolutionary Physiology & Biochemistry, RAS, St-Petersburg, Russia; *Department of Neuropharmacology, Institute of Experimental Medicine RAMS, St-Petersburg, Russia.*

Abstract: Malondialdehyde (MDA) as an indicator of lipid peroxidation (LPO), content and composition of phospholipids (PL) were analyzed in synaptosomes of the rat brain hemispheres using immobilization stress without and with pretreatment with a new synthetic taurine derivative (STD). The stress was accompanied by a decrease MDA content, inversion of the initial asymmetry of total phospholipids (TPL) and modification PL composition in the brain hemispheres of rats. STD administration to the rats after stress antagonized the decrease of the MDA level, attenuates stress induced inversion TPL asymmetry and normalized PL composition in the brain hemispheres. Prevention or diminution of synaptosomal PL changes in the rat brain hemispheres are interpreted as one of the possible mechanisms of the neuroprotective effect of STD.

Taurine 4, edited by Della Corte *et al.*
Kluwer Academic / Plenum Publishers, New York, 2000.

INTRODUCTION

Despite of the wide spectrum of the protective effects of taurine, its pharmacological application is limited by a comparatively low activity and a poor ability to cross the blood-brain barrier. Thus the creation of the basis for understanding the structure–activity relationships of taurine and its analogs is very important.

The morphological, biochemical and functional asymmetry was shown to be characteristic of brain[1-3]. It has been established that the right and the left brain structures differ from their sensitivity to various pathological conditions[4-6]. It is has been suggested that different effects of lesions to the right and the left hemispheres mechanisms of the central nervous system and different compensatory responses of the symmetric brain structures requires an exploitation of side-directed correction procedures involving differential methods of treatment and pharmaco-correction. In this connection, it would be interesting to investigate the protective effect of STD in relation to neurons of the right and left hemispheres. We proposed that one of the mechanisms of the neuroprotective STD effect was its action on the membrane PL. The aim of the present work was to study the content and composition of PL, as well the processes of LPO in synaptosomes from the rat brain hemispheres using immobilization stress without and with pretreatment with STD.

MATERIALS AND METHODS

The experiments were performed on adult albino male rats weighing 300-350 g. The animals were divided into 4 groups: 1 - intact rats , 2 - rats subjected to 3 hours immobilization, 3 - intact rats treated with STD at dose 20 mg/kg, i.p. 3 hours before the decapitation; 4 - rats treated with STD at the same dose 30 min before the immobilization.

All groups of animals (n=4) were decapitated and right and left brain hemispheres were taken for lipid extraction. Extraction of lipids was made according to Folch's procedure[7] using synaptosomal fractions obtained from brain hemispheres homogenates by the method of differential centrifugation on a gradient of sucrose[8]. The main PL classes (PS-phosphatidylserine, PI-phosphatidylinositol, SM-sphingomyelin, PE-phosphatidylethanolamine, and PC- phosphatidylcholine) were separated by two-dimensional micro-TLC on plates (6 cm x 6cm) coated with a suspension of silica gel (5 μm) in 10% Ca_2SO_4 [9]. Chromatography was performed using as a mobile phase chloroform: methanol:28% NH_4OH (65:35:5, v\v) for resolution in the first direction and chloroform:methanol:acetone:CH_3COOH:H_2O (50:10:20:10:5, v\v) for resolution in the second direction[10].   Spots were visualized by iodine vapours and taken for analysis of PL content. The content of total and individual PL was evaluated by the amount of lipid bound inorganic phosphorus[11]. The content of MDA was estimated according to the method of Vladimirov et al.[12]. The protein content was measured by a modification method of Lowry[13]. The result obtained were analyzed using Student's t test.

RESULTS

Analysis of the results obtained failed to show distinctions in the content of MDA between hemispheres of intact rats (Fig.1). However, the left hemispheres was dominant to the content of TPL (Fig.2). The stress was accompanied by a loss MDA content in both hemispheres and a 61.7 % increase of the TPL level in the right hemisphere and by its decrease in the left hemisphere. In terms of the PL composition, a reduction of the PC level by 19.0% and elevation of the PE level by 13.4% in the right hemisphere were observed, whereas in the left hemisphere the reduction of PE of 11.9% and a significant rise of SM, by 140%, were noted (Fig.3). Administration of the STD did not result in significant changes in the content of synaptosomal MDA, in intact rat brain hemispheres or in rat brain hemispheres after stress (Fig.1). Following administration of STD to intact rats decreases of TPL both in the right hemisphere (20.1%) and in the left hemisphere (44.5%) compared to the initial levels were observed (Fig.2).

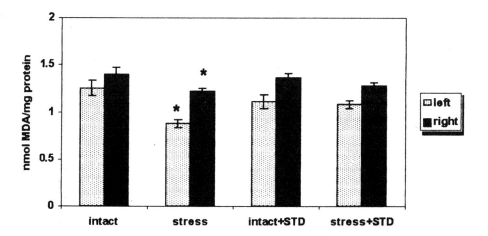

Fig. 1. Effects of STD on MDA content from the brain hemispheres after immobilization stress. cell line. *$p<0.05$ compared to intact rats. Values are means $\pm$ SEM (n = 4).

In terms of the composition of PL, STD administration caused a significant decrease of PE in both hemispheres as well as that of PS in the left hemisphere was. In the rats after immobilization STD resulted in the further elevation of the level of TPL (61.7% - stress; 75.9% - stress + STD) in the right hemisphere but prevented its reduction (43.5% - stress; 25.3% - stress + STD) in the left hemisphere as compared. Administration of STD to the rats after stress completely restored the synaptosomal PL composition (Fig.4).

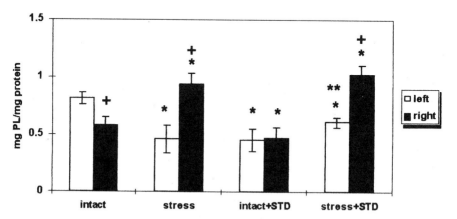

Fig. 2. Effects of STD on PL content from the brain hemispheres after immobilization stress. cell line. *p<0.05 compared to intact rats; **p<0.05 compared to rats after stress; +p<0.05compared to the left hemisphere. Values are means ± SEM (n = 4).

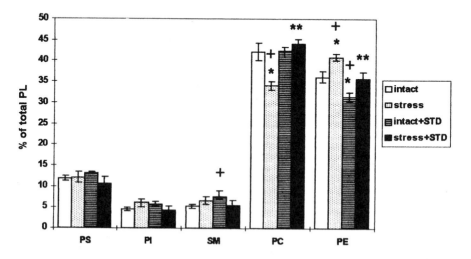

Fig. 3. Effects of STD on PL content from the right brain hemispheres of rats afterimmobilization stress. *p<0.05 compared to intact rats; **p<0.05 compared to rats after stress; +p<0.05 compared to the left hemisphere. Values are means± SEM (n = 4).

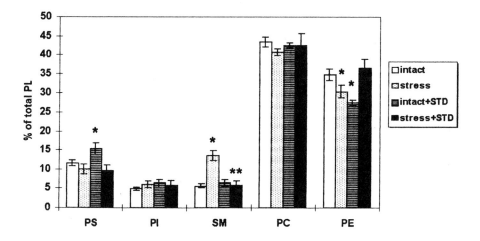

Fig. 4. Effects of STD on PL content from the left brain hemispheres of rats after immobilization stress. *p<0.05 compared to intact rats; **p<0.05 compared to rats after stress. Values are means ± SEM (n = 4).

DISCUSSION

The data obtained demonstrate the presence of an initial asymmetry in the distribution of synaptosomal PL in the rat brain hemispheres, with quantitative predominance in the synaptosomes of the left hemisphere. The asymmetry of PL, which are structural components of cell membranes, appears to reflect the morphological differences between the hemispheres and to give evidence for a greater quantity of neurons in the left hemisphere. Nevertheless, interhemispheric no significant differences in the composition of PL or in the content of MDA were seen in the brains from untreated rats.

The decrease of the MDA level in the synaptosomes from both hemispheres in rats subjected to immobilization may indicate inhibition of LPO. The reason of such inactivation of LPO in response to the stress might be due to large output of cathecholamines and corticosteroids into the blood, since these compounds may also be scavengers of free radicals[14]. The shifts in the content and composition of synaptosomal PL of the brain hemispheres of rats after stress were varied. Stress induced an inversion and enhancement of the initial interhemispheric TPL asymmetry. The changes in TPL changes were greater in the synaptosomes from the in the right hemisphere. In general, this reflects the differences in the biochemical responses of the brain hemispheres at the stress.

Analogous alterations of the synaptosomal PL had been obtained under the conditions of hyperbaric oxygen[15]. It may be assumed that the differences observed correlated with the specific functional characters of the two brain hemispheres. In part, the dominant role of the right hemisphere in the control of negative emotional responses[16] may be related to the more profound shifts of PL in this hemisphere in response to stress. Pretreatment of the rats with the STD antagonized the stress-induced decrease of MDA level in both hemispheres. It can be proposed that STD may normalize the cathecholamine and corticosteroid concentration and to prevent inhibition of LPO in the brain of following stress. STD administration to the rats after stress induced unidirectional shifts of TPL in the brain hemispheres. TPL accumulation occurred in the right hemisphere whereas STD reduced the stress-induced decrease in the left hemisphere. Apparently, the increase of the PL level in synaptosomes results from activation of their synthesis under the action of STD. The between the responses of the brain hemispheres indicates an interhemispheric specificity of STD action. The pretreatment with the STD attenuated the stress-induced inversion of PL asymmetry to some extent. A positive effect of the STD in relation to individual PL classes of the brain hemispheres was found. STD administration reduced the modification of the PL composition as a result of stress, giving rise to normalization of PC and PE in the right hemisphere, as well SM and PE in the left hemisphere.

CONCLUSION

STD administration restored the MDA content and normalized the PL composition in both hemispheres of rats subjected to immobilization stress and it partly antagonized the decrease of PL level in the left hemisphere. These effects may be associated with the membrane-stabilizing action of STD on brain cells. The degree of the protective effect of STD differs between the brain symmetric structures. Prevention or diminution of alterations in brain PL content and composition may be considered as one of the possible mechanisms of the neuroprotective effect of the new taurine derivative. Further investigations may lead to the development exploitation of the effective methods for pharmaco-correction of pathological states of the CNS.

REFERENCES

1. Vartanian G.A., Klement'ev B.I., 1991, The Chemical Symmetry and Asymmetry of the Brain. "Nauka", St.-Petersburg.
2. Bragina N.N., Dobrokhotova T.A., 1981, Functional asymmetry of the human brain, Moscow, "Medizine".
3. Springer S.P., Deutsch G. 1983, Left Brain, Right Brain. University of New York.

4. Opitz B., Sarkisova K. Yu. 1996, Interhemisphere asymmetry of cerebral lipidperoxidation in rats with various types of behavior as a prognostic indicator of their resistance to cerebral ischemia and effectiveness of the anti-ischemic action of substance *P. DAN.* **346**: 275-277.

5. Levshina I.P., Guliava N.V. 1991, Effects of acute stress on lateralitation of lipid peroxidation in the brain depend on the behavioral typology of rats. *Biul-Eksp-Biol-Med.* **111**: 568-570.

6. Ginobili de Martinez M.S., Rodriguez de Turco E.B. and Barrantes F.J. 1986, Asymmetry of diacylglycerol metabolism in rat cerebral hemispheres. *J. Neurochem.* **46**: 1382-1386

7. Folch J., Lees M., Sloane-Stanly G.H. 1957, A simple method for the isolation and purification of total lipids from animal tissues. *J. Biol.Chem.* **226**: 497-509.

8. Lapetina E.G., Soto E.F., De Robertis E. 1967, Gangliosides and acetylcholinesterase iisolated membranes of the rat brain cortex. *Biochem. Biophys. Acta.* **135**:33-43.

9. Vaskovsky V.E., Kostetsky E.V., Vasendin I.M. 1972, A simplified technique for thin-layer microchromatography of lipids. *J. Chromatograph.* **2**, **67**:376-378.

10. Rouser G., Siakotos A.N., Fleisher S. 1966, Quantitative analysis of phospholipids by thin-layer chromatography and phophorus analysis of spots. *Lipids.*, **1**: 85-86

11. Vaskovsky V.E., Kostetsky E.V., Vasendin I.M. 1975, Universal reagent for phospholipid analysis. *J. Chromatograph. l*, **114**: 29-141.

12. Vladimirov Y.A. Archakov A.I. 1972, Lipid peroxidation in biological membranes. Moscow: 241-43.

13. Markwell M.A., Suzanne M.H., Bieber L.L. et al. 1978, A modification of the Lowry procedure to simplify protein determination in membrane and lipoprotein samples. *J. Analyt. Biochem.* **87**: 206-211.

14. Guliava N.V., Luzina N.L., Levshina I.P. and Kryzhanovskii G. N. 1988, The inhibition stage of lipid peroxidation during stress. *Biul-Eksp-Biol-Med.* **106**, 12:660-3

15. Novoselova N. Yu., Moskvin A.N., Torkounov P.A., Sapronov N.S. and Demchenko I.T. 1999, Effect of hyperbaric oxygen on lipid peroxidation and phospholipid content in rat brain. *Biul-Eksp-Biol-Med.* **128**, **9**: 261-3.

16. Bruyer R. 1980, Differential implication of cerebral hemispheres in emotional behavior. *Acta Phychiatr. Belg.*, **3**: 266-84.

TAURINE TRANSPORT MECHANISM THROUGH THE BLOOD-BRAIN BARRIER IN SPONTANEOUSLY HYPERTENSIVE RATS

Young-Sook Kang
College of Pharmacy, Sookmyung Women's University, Seoul, Korea

Abstract: Taurine levels in the brain decrease when an animal is subjected to pathological conditions, such as ischemia-anoxia and seizure, but they tend to increase in hypertension. The present study investigated the blood-brain barrier (BBB) transport of [^{3}H]-labelled taurine in spontaneously hypertensive rats (SHR) using internal artery carotid perfusion (ICAP) at a rate of 4 ml/min for 10, 15 and 30 seconds. The volume of distribution in brain (V_D) and the permeability surface area product (PS) of [^{3}H]-taurine through the BBB in SHR were calculated. The PS value for taurine at 15 s was higher than at the longer perfusion times. This could result from taurine efflux back into blood occurring after 15 s. As in the case of normotensive rats, taurine was shown to enter the brain *via* the sodium and chloride ion dependent carrier system.

INTRODUCTION

Dawson *et al.*[1] reported brain taurine levels of hypertensive rat to increase because of developing hypernaturemia[2] as a result of deficiency in arginine vasopressin. Studies[3-4] using the carotid artery injection technique in animals have shown a saturable transport system for taurine. Sodium and chloride ion dependent taurine transport systems at the blood-brain barrier (BBB) have been identified in normal rat by Tsuji and Tamai[5-6]. The purpose of the present study

Taurine 4, edited by Della Corte *et al.*
Kluwer Academic / Plenum Publishers, New York, 2000.

was to clarify the effect of hypertension on BBB taurine transport by using spontaneously hypertensive and normotensive rats.

MATERIALS AND METHODS

Male Sprague-Dawly rats weighing 230-270g (mean blood pressure; 118 ± 3 mm Hg) were used as controls and the hypertensive rats were 8-12 week old male spontaneously hypertensive rats (SHR, mean blood pressure; 190 ± 3 mm Hg), obtained from Hanlym University Animal Center (Chuncheon, Korea). [^{3}H]-Taurine (NET-541; specific radioactivity 24.1 Ci/mmol) and [^{14}C]-sucrose (NEC-100X; specific radioactivity 442 mCi/mmol) were purchased from New England Nuclear (Bukyung Co., Korea). Rats were anesthetized with an intramuscular injection of ketamine (100 mg/kg body weight) with xylazine (2 mg/kg). The uptake of taurine into the brain was then measured by the brain perfusion technique[7], at a perfusion rate of 4 ml/min.

RESULTS AND DISCUSSION

The [^{3}H]-taurine uptake through the BBB in SHR at different perfusion times is shown in Table 1. The brain volume for the distribution of [^{3}H]-taurine (V_D) values, corrected for the brain volume using the sucrose space, in the SHR were 4.98, 11.7, 13.2 μl/g at 10, 15, 30 s perfusion, respectively. There was a time-dependent increase. But the BBB permeability-surface area product (PS) value was 45% decreased at 30 s perfusion, to 26.5 μl/min/g.

Table 1. Brain uptake of [^{3}H]-taurine and [^{14}C]-sucrose in the SHR group

	V_D (μl/g)		PS (μl/min/g)
Time (s)	[^{3}H]-taurine	[^{14}C]-sucrose	[^{3}H]-taurine
10	9.44 ± 1.78	4.46 ± 0.56	40.4 ± 8.9
15	18.2 ± 0.32	6.47 ± 1.82	47.7 ± 5.9
30	19.4 ± 0.91	6.20 ± 0.53	26.5 ± 1.1

Data are mean values $\pm$ S.E. (n = 3)
V_D, Volume distribution in brain; PS, Permeability surface-area products.
PS values are corrected with the V_D values of [^{14}C]-sucrose, a plasma-volume marker

Figure 1 shows a comparison of PS value of [^{3}H]-taurine in SHR and normal rats; the value in SHR (47.7 ± 5.9 μl/min/g) was 22% decreased compared with that of normal rats (60.9 ± 3.7 μl/min/g) at 15 s perfusion.

The effects of replacements of cations and anions on taurine uptake were examined. As shown in Fig. 2, substitution of sodium ion with choline completely inhibited taurine uptake in both rat groups; specifically the value of V_D in SHR was 46% decreased, to 8.36 ± 1.61 μl/g. Substitution of chloride ion with nitrate also significantly decreased taurine uptake in both rat groups, as shown in Fig. 3. These results suggest that sodium and chloride ions are involved in transport of taurine in both normotensive rat and hypertensive rats.

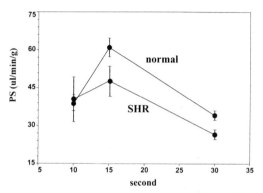

Fig. 1. Blood-brain barrier permeability surface area product of [^{3}H]-taurine, corrected with sucrose space, after ICAP at 4 ml/min for 15 s in SHR and normal rats. Values are means $\pm$ S.E. (n = 3).

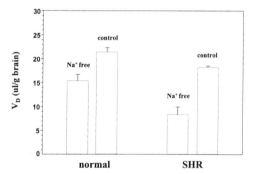

Fig. 2. Inhibition of taurine uptake by brain in SHR and normal rats after internal carotid artery perfusion (4 ml/min) for 15 s in sodium-ion containing and sodium-ion free buffer. Data are mean values $\pm$ S.E. (n = 3).

CONCLUSIONS

The brain volume of distribution (V_D) of [^{3}H]-taurine in SHR was increased time-dependently during 10 to 30 seconds of perfusion, but the values were smaller than those of normal rats. Furthermore, the BBB permeability-surface area product (PS) of [^{3}H]-taurine in SHR showed a maximum value at 15 s of perfusion. This may be explained by taurine starting to efflux back into blood from brain after that time.

[³H]-Taurine uptake to the brain in SHR occurred via inwardly-directed sodium and chloride ion dependent carrier-mediated systems, like those in the normal rat. However, it showed a much greater ion dependency than the taurine uptake in the normal rat.

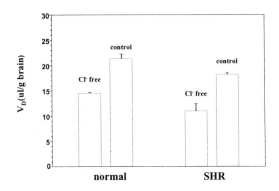

Fig. 3 . Inhibition of taurine uptake by brain in SHR and normal rats after internal carotid artery perfusion (4 ml/min) for 15 s in chloride-ion containing and chloride-ion free buffer. Data are mean values ± S.E. (n = 3).

ACKNOWLEDGEMENTS

This work was supported by the Sookmyung Women's University Research Grants.

REFERENCES

1.   Dawson, R., Wallace, D.R. and King, M.J. 1990, Monoamine and amino acid content in brain regions of Brattleboro rats. *Neurochem. Res.* **15**: 755-761.
2.   Lombardini, J.B. 1992, Recent studies on taurine in the central nervous system, J.B. Lombardini (Eds.) *Taurine*, Plenum Press, New York.
3.   Lefauconnier, J.M. and Trouve, R., 1983, Development changes in the pattern of an amino acid transport at the blood-brain barrier in rat. *Dev. Brain Res.* **6**: 175-182.
4.   Benrabh, H., Bourre, J.M. and Lefauconnier, J.M. 1995, Taurine transport at the blood-brain barrier: an in vivo brain perfusion study. *Brain Res.* **692**:57-65.
5.   Tsuji, A. and Tamai, I. 1996, Sodium- and chloride- dependent transport of taurine at the blood-brain barrier, R. Huxtable et al. (eds.) *Taurine 2*, Plenum Press, New York.
6.   Tamai, I., Senmaru, M., Terasaki, T., and Tsuji, A. 1995, Na+ and Cl- dependent transport of taurine at the blood-brain barrier. *Biochem. Pharmacol.* **50**: 1783-1793.
7.   Takasato, Y., Rapoport, S.I. and Smith, Q.R. 1984, An in situ brain perfusion technique to study cerebrovascular transport in rat. *Am. J. Physiol.* **247**, H484-H493.

PROTECTION OF INTRINSIC NERVES OF GUINEA-PIG DETRUSOR STRIPS AGAINST ANOXIA/GLUCOPENIA AND REPERFUSION INJURY BY TAURINE

Federica Pessina, Giacomo Matteucci, Lucia Esposito, Beatrice Gorelli, Massimo Valoti and Giampietro Sgaragli

Istituto di Scienze Farmacologiche, Via Piccolomini 170, 53100 Siena, Italia

Abstract: There is ample evidence that ischaemia is associated with partial denervation of the detrusor muscle and that this is responsible for much of its abnormal contractile behaviour, resulting in bladder dysfunction (instability). In guinea-pig nerves are very susceptible to the ischaemic damage as compared to the muscle cells. The purpose of this study was to assess the neuroprotection afforded by taurine on guinea-pig detrusor under ischaemic-like conditions. Guinea-pig detrusor strips were subjected for 60 min to ischaemic-like conditions, followed by 150 min reperfusion. Intrinsic nerves underwent every 30 min electrical field stimulation (EFS) by 5-s trains of square voltage pulses of 0.05 ms duration (15 Hz, 50 V). Detrusor strips were perfused with 0.1, 1, 3 or 10 mM taurine during the ischaemia-like exposure and the first 30 min of reperfusion. Taurine (1 and 3 mM) significantly improved the response of the strips to EFS both at the end of ischaemia and reperfusion. On the contrary, neither 0.1 nor 10 mM taurine had significant effects. It is concluded that taurine can partially counteract the ischaemia-reperfusion injury in the guinea-pig urinary bladder.

INTRODUCTION

It has been proposed that in non-human, mammalian urinary bladder at least three types of excitatory neurotransmitter elicit detrusor muscle contraction, acetylcholine, ATP and a component which is insensitive to

Taurine 4, edited by Della Corte *et al.*
Kluwer Academic / Plenum Publishers, New York, 2000.

both atropine and suramin. It has been hypothesised that, in small mammals, purinergic transmission is responsible for initiation of voiding, whereas cholinergic transmission affects the maintenance of voiding, its duration and extent and the period during which most urine is expelled[7,9-10]. In view of predominance of the purinergic component in animals that mark their territory by urination, and which therefore need to urinate frequently in short bursts, such as cat, rat, guinea-pig and dog, and the predominance of cholinergic component in animals which urinate in order to empty their bladder (monkey and human) it has been suggested that the relative proportions of non-cholinergic to cholinergic excitation have arisen during evolution in relation to these needs[2].

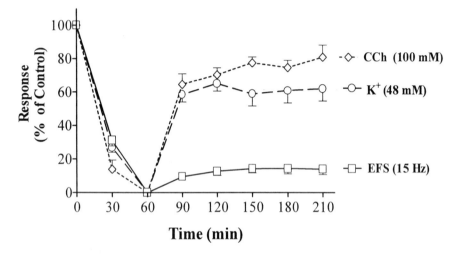

Figure 1. Effects of exposure of tissue to oxygen- and glucose- free solution for 60 min followed by reperfusion on the contractile response of guinea-pig detrusor muscle. Results are expressed as percentage of original control values.
EFS: electric field stimulation; CCh: carbachol.

There is ample evidence that ischaemia is associated with partial denervation of the detrusor muscle and that this is responsible for much of its abnormal contractile behaviour, resulting in bladder dysfunction (instability). It has also been demonstrated that guinea-pig bladder nerves are very susceptible to ischaemic damage, as compared to bladder smooth muscle cells[14]. When guinea-pig bladder strips subjected to 60 min of ischaemia followed by 150 min of following reperfusion, the muscle response evoked by stimulating the tissue with 100 µM carbachol or 48 mM K^+ recovered completely, but the nerve response evoked by electric-field-stimulation (EFS) reached no more than 12 % of the initial response (Fig. 1).

Clinical and experimental data have suggested that neuronal damage resulting from ischaemia-reperfusion is at least partly induced by free radicals and/or lipid peroxidation[1] and that uncontrolled Ca^{2+} influx into the cell leads to the genesis of severe dysfunction during reperfusion of ischaemic tissue[4-11]. Taurine (2-aminoethanesulfonic acid) is an endogenous amino acid which has been proposed to function as a modulator of transmembrane Ca^{2+} transport, an osmoregolator and putatively a free radical scavenger[11-14].

The purpose of this study was, therefore, to examine the effects of taurine and of its analogue 2-aminoethylphosphonic acid (AEP) on the reduction in neurally evoked contractions caused by exposure of guinea-pig bladder smooth muscle strips to ischaemia-reperfusion-like conditions *in vitro*.

METHODS

Strips of guinea-pig bladder smooth muscle (approximately 0.5 x 0.5 x 7 mm) were mounted in small organ baths (0.2 ml) between two platinum ring electrodes and continuously perfused with physiological saline solution (37°C, pH 7.4) at a constant rate of 1.5 ml/min. The set-up is shown schematically in Fig. 2.

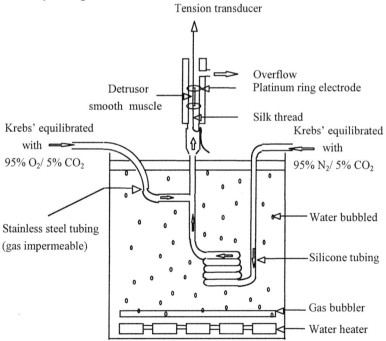

Figure 2. Schematic representation of the experimental apparatus studying the behaviour of isolated detrusor strips subjected to anoxia/glucopenia and reperfusion conditions.

Ischaemia-like conditions were obtained by removing both glucose and oxygen from the bathing solution[14]. To obtain an O_2 tension close to zero, the perfusing solution was continuously supplied to the organ baths using gas-permeable silicone tubes which ran through an enclosed heated water bath equilibrated with 95 % N_2, 5 % CO_2. For normal conditions, the Krebs solution was bubbled with 95 % O_2, 5 % CO_2 and supplied to the organ baths via stainless steel tubes, thus preventing gaseous exchange with nitrogen when passing through the water bath. Under ischaemic conditions, glucose was replaced isosmotically with NaCl and the solution was gassed with 95 % N_2, 5 % CO_2. This arrangement proved necessary, because oxygen-free solution tended to equilibrate with the surrounding atmosphere when passing through the gas permeable plastic tubing in the peristaltic pump. Passing the solution through the silicone tubes in the nitrogen-equilibrated water bath allowed the oxygen tension in the organ baths to drop close to zero.

Guinea-pig detrusor strips were subjected to ischaemia-like conditions for 60 min, followed by 150 min reperfusion. Detrusor strips were perfused with 0.1, 1, 3 or 10 mM taurine and with 3 mM AEP during the ischaemia-like exposure and the first 30 min of reperfusion. Intrinsic nerves underwent electrical field stimulation (EFS) every 30 min by 5-s trains of square wave voltage pulses of 0.05 ms duration (15 Hz, 50 V). The response of strips to EFS was expressed as a percentage of the initial response in normal solution, taken to be 100%.

To assess the relative proportion of cholinergic and NANC components in the response to EFS in guinea-pig bladder, control responses were obtained and strips were incubated for 20 min in medium containing 1 μM atropine, 0.1 mM suramin, 10 μM α,β-Methylene-ATP (α,β-MeATP) or 0.3 μM tetrodotoxin before EFS was repeated. Similar experiments were carried out in strips which had undergone ischaemia in order to assess which type of excitatory innervation was affected by taurine treatment.

Data analysis

Results were expressed as mean ± standard error of the mean (S.E.M.). Statistical analysis of the data was performed by one way analysis of variance (ANOVA) followed by Dunnett's test for multiple comparisons. P values < 0.05 were considered significant.

RESULTS

The response of detrusor muscle strips to EFS in the control, taurine and AEP-treated groups is shown in Figure 3 (a, b and c).

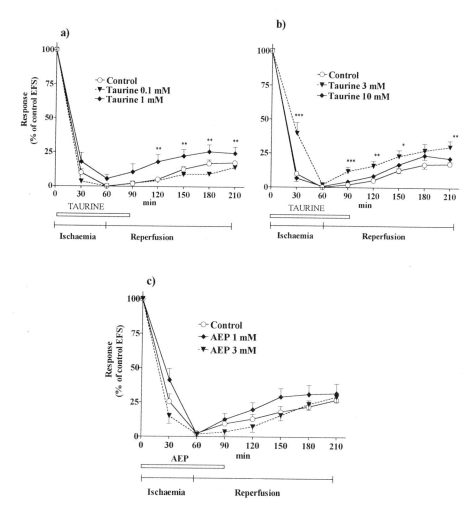

Figure 3. Electric-field stimulation-induced response of guinea-pig detrusor strips, subjected to 60 min of ischemia and 150 min of subsequent reperfusion, in the absence or presence of 0.1 and 1 mM taurine (a), 3 and 10 mM taurine (b) AEP (c). Results are expressed as mean ± S.E.M. of 6 to 10 strips from four bladders in each group. Significant differences from the control group are indicated; *$P < 0.05$; **$P < 0.01$; ***$P < 0.001$. ($n = 4$-8).

During ischaemia the response of the control strips decreased rapidly, being abolished within one hour, whereas in the 3 mM taurine-treated group the response decreased more slowly. The 3 mM taurine-treated group also showed increased recovery during reperfusion, which was significantly different from the control.

The responses of the group treated with 1 mM taurine were also significantly greater during recovery (after 150 min of reperfusion the response was 33.2 ± 4.7 % of the initial response as compared to only 17.3 ± 1.8 % in controls ($P < 0.05$)), but no significant differences in response were seen during the ischaemic period. Strips exposed to 10 mM taurine showed a significantly greater response only at 120 min of reperfusion (30.5 ± 4.3%) as compared to control (15.9 ± 2.2%) ($P < 0.05$); neither the 0.1 mM taurine nor the 1 and 3 mM AEP treated-groups showed any significant differences when compared to controls. Neither taurine nor AEP, at the different concentrations tested (data not shown), significantly affected the electrically-evoked control response under normal conditions.

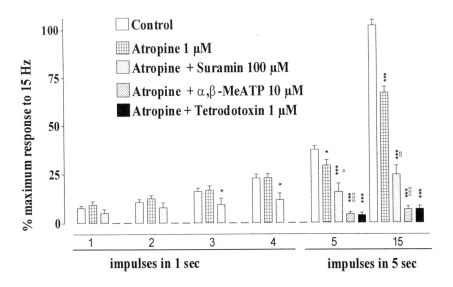

Figure 4. Response of guinea-pig detrusor strips to different parameters of electric-field stimulation under control conditions, in the presence of atropine, atropine and suramin, atropine and α,β-MeATP, atropine and tetrodotoxin. *P < 0.05; **P < 0.01; ***P < 0.001 vs control. °P < 0.05; °°P < 0.01 *vs* atropine. (*n* = 6-8).

Figure 4 shows the proportion of cholinergic and NANC components in the response of guinea-pig detrusor strips to EFS at different frequencies and lengths of stimulation. At 15 Hz the NANC component appears to dominate, in comparison to cholinergic excitation, whereas at low frequencies of stimulation the response appeared to be completely purinergic. In addition after application of suramin a neurally evoked response was still apparent (to the extent of about 20% of the initial response at 15 Hz of stimulation), but this was completely abolished in the presence of α,β-methylene-ATP, the residual response being purely myogenic).

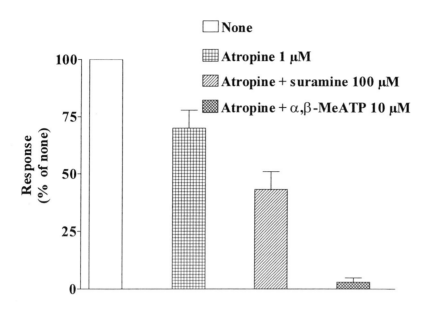

Figure 5. Effects of atropine, atropine plus suramin and atropine plus α,β-MeATP on the response of guinea-pig detrusor strips to EFS (15 Hz) after prior exposure to 60 min of ischaemia conditions followed by 150 min of reperfusion in the absence or in the presence of taurine (3 mM). Responses are expressed as a percentage of the response to EFS after 150 min of reperfusion (6-10 strips of 4 animals). In those strips treated with taurine the drug was applied during the ischaemia and the first 30 min of reperfusion. The untreated and the taurine treated groups are statistically not significantly different. ($n = 6$).

Figure 5 shows the proportions of the cholinergic and NANC components in the response of guinea-pig detrusor strips to EFS (15 Hz) after exposure of the strips to 60 min of ischaemia conditions followed by 150 min of reperfusion both in the absence or in the presence of taurine. Again the NANC component predominated over the cholinergic response as seen in control conditions. Neither taurine nor AEP, at the different concentrations tested, significantly altered the electrically-evoked control response under normal conditions (data not shown).

CONCLUSION

Binding studies have demonstrated that the concentration of P_{2X}-purinoreceptors in the detrusor of small mammals is approximately 1000 times greater than that of either cholinoreceptors or adrenoreceptors[5].

In the present study the anti-trypanosomal polysulphonate naphthylurea, suramin, was found to decrease but not abolish the NANC response in the guinea-pig detrusor strips. Suramin, has been demonstrated to antagonise the P_{2X}-purinoreceptor in the vas deferens[3]. Moreover it displays antagonistic properties on most subtypes of P_2-purinoreceptors[8], but it has poor selectively and low potency. For this reason the effect of α,β-methylene ATP, a stable analogue of ATP which causes a rapid desensitisation of purinoreceptors in the bladder blocking the response to exogenous ATP and non-cholinergic nerve stimulation simultaneously[6], was tested on the atropine-resistant response to EFS in guinea-pig strips. In combination with atropine, desensitisation of P_2-purinoreceptors by α,β-methylene ATP abolished the responses to postganglionic parasympathetic nerve stimulation in the guinea-pig, thus suggesting the involvement of ATP as purinregic neurotransmitter in the guinea-pig bladder and that acetylcholine and ATP are the major, if not the sole, transmitters evoking normal bladder contraction. Moreover, it has been demonstrated that taurine protects the guinea-pig detrusor nerves from ischaemic insult without any significant preference for the cholinergic or the purinergic pathway.

This study has demonstrated the beneficial effects of taurine, but not of AEF, against the neuronal damage to the detrusor *in vitro* that results from ischaemia and reperfusion. Very recently it has been shown that taurine activates Ca^{2+} uptake by rat liver mitochondria through a mechanism which involves stimulation of the uniport system, rather than inhibition of Ca^{2+} release *via* the ion exchangers or modulation of the permeability transition pore of the mitochondrial inner membrane[13]. Moreover, replacing its sulphonic group with a phosphonate, abolished the effect on Ca^{2+} uptake. Such a role of taurine in modifying mitochondrial Ca^{2+} homeostasis might be of particular importance in pathological conditions, such as ischaemia and oxidative stress, that are characterised by cell Ca^{2+} overload.

In the future, the effects of taurine will be studied using the whole bladder organ bath technique, in order to confirm its neuronally protective effect both physiologically, by measuring responses to nerve stimulation and histologically, by examining the structure and protein expression of intramural ganglia after ischaemia and reperfusion.

ACKNOWLEDGEMENTS

Financed by MURST (Cofin '98) and Ministero degli Affari Esteri (L.212), Roma.

REFERENCES

1. Chan, P. H., Schmidley, J. W., Fishman, R. A., and Longar, S. M., 1984, Brain injury, edema, and vascular permeability changes induced by oxygen-derived free radicals. *Neurology* **34**: 315-320.
2. Craggs, M.D., and Stephenson, J.D., 1986, A non-cholinergic urinary bladder mechanism in New World primates. *J. Physiol.* (London) **377**: 341-345.
3. Dunn, P.M., and Blakeley, A.G., 1988, Suramin: a reversible P2- purinoceptor antagonist in the mouse vas deferens. *Br. J. Pharmacol.* **93**: 243-245.
4. Ferrari, R., Agnoletti, L., Comini, L., Gaia, G., Bachetti, T., Cargnoni, A., Ceconi, C., Curello, S., and Visioli O., 1998, Oxidative stress during myocardial ischaemia and heart failure. *Eur. Heart J.* **19** (B): B 2-11.
5. Hoyle, C.H.V., and Burnstock, G., 1993, Postganglionic efferent transmission in the bladder and urethra. In *Nervous control of the urogenital system*. Part of the book series: the autonomic nervous system (C.A. Maggi, ed.), Harwood Academic Publishers, London, pp. 349-381.
6. Kasakov, L., and Burnstock, G., 1982, The use of a slowly degradable analog, alpha, beta-methylene ATP, to produce desensitisation of the P2- purinoceptor: effect on non-adrenergic, non-cholinergic responses of the guinea-pig urinary bladder. *Eur. J. Pharmacol.* **86**: 291-294.
7. Krell, R.D., Mccoy, J.L., and Ridley, P.T., 1981; Pharmacological characterization of the excitatory innervation to the guinea-pig urinary bladder *in vitro*: evidence for both cholinergic and non-adrenergic- non- cholinergic neurotransmission. *Br. J. Pharmacol.* **74**: 15-22.
8. Leff, P., Wood, B.E., and O' Connor, S.E., 1990, Suramin is a slowly-equilibrating but competitive antagonist at P2-receptors in the rabbit isolated ear artery. *Br. J. Pharmacol.* **101**: 645-649.
9. Levin, R.M., Ruggieri, M.R., and Wein, A.J., 1986; Functional effects of the purinergic innervation of the rabbit urinary bladder. *J. Pharmacol. Exp. Ther.* **236**: 452-457.
10. Maggi, C.A., Meli, A., and Santicioli, P., 1987, Neuroeffector mechanisms in the voiding cycle of the guinea-pig urinary bladder. *J. Auton. Pharmacol.* **7**: 295-308.
11. Maxwell, S.R., and Lip, G.Y., 1997, Reperfusion injury: a review of the pathophysiology, clinical manifestations and therapeutic options. *Int. J. Cardiol.* **58**: 95-117.
12. Michalk, D.V., Wingenfeld, P., Licht, C., Ugur, T., and Siar, L.F., 1996, The mechanisms of taurine mediated protection against cell damage induced by hypoxia and reoxygenation. In *Taurine 2. Advances in Experimental Medicine and Biology* (R.J. Huxtable and D.V. Michalk, eds.), Plenum Press, New York and London, pp. 223-232.
13. Palmi, M., Youmbi, G., Fusi, F., Sgaragli, G.P., Dixon, H.B.F., Frosini, M., and Tipton, K.F., 1999, Potentiation of mitochondrial Ca^{2+} sequestration by taurine. *Biochem. Pharmacol.* **58**: 1123-1131.
14. Pessina, F., McMurray, G., Wiggin, A., and Brading, A.F., 1997, The effect of anoxia and glucose-free solutions on the contractile response of guinea-pig detrusor strips to intrinsic nerve stimulation and the application of excitatory agonist. *J. Urol.* **157**: 2375-2380.
15. Schaffer, S.W., Ballard, C., and Azuma, L., 1994, Mechanisms underlying physiological and pharmacological actions of taurine on myocardial calcium transport. In *Taurine in health and Disease. Advances in Experimental Medicine and Biology* (R.J. Huxtable and D.V. Michalk., eds.), Plenum Press, New York and London, pp. 171-196.

MATERIALS AND METHODS

Adult male New Zealand albino rabbits (Charles River, Calco, Como, Italy) weighing 2.0-2.5 Kg were anaesthetised with a mixture of xylazine chloride (10 mg kg^{-1}, i.m.) and ketamine hydrochloride (35 mg kg^{-1}, i.m.) and implanted with cannulas in the *cisterna magna* for the withdrawal of cerebrospinal fluid (CSF) according to the method described by Palmi et al.[22]. 125 μl-CSF fractions were collected at intervals of 25 min from conscious rabbits by a procedure described elsewhere[22]. Plasma samples were obtained from an incision in the ear[9]. Rabbit rectal temperature (RT) was recorded every 5 min by a thermocouple thermometer connected to a personal computer with an Isothermex program (Columbus Instruments, Columbus, Ohio, USA). RT was monitored for at least 1 h before the experimental session.

CSF samples were randomly analysed for amino acid concentrations (taurine, GABA, aspartate and glutamate) by reversed-phase HPLC with o-phthalaldehyde pre-column derivatization[3]. CSF sodium, potassium, magnesium and calcium concentrations were determined by an HPLC-conductimetric detection method[9]. CSF and plasma osmolality was determined with a vapour pressure osmometer (Wescor, Logan, Utah, USA). Protein in CSF was determined by the coomassie blue binding method[6].

Values were expressed as means ± S.E.M.. The statistical significance of differences between values relative to heat stress (125-250 min) and post heat stress (275-375 min) versus data observed during the same period in control rabbits was analysed by analysis of variance (ANOVA). $P < 0.05$ was considered significant.

Experimental protocol

Rabbits (n=10) restrained in a stainless steel cage were individually housed in a chamber thermostatted at 20 °C for 100 min. Subsequently a group of five rabbits was exposed to high ambient temperature (Ta). The temperature of the chamber was raised up to 40 °C in 50 min (150 min after the beginning of the experimental session), was kept at that value for 50 min, and then decreased to 20 °C over 50 min and that temperature was maintained up to the end of the experimental session (375 min). A second group of five rabbits (controls) was kept at room temperature (20 °C) for the same observation period (375 min).

It was assumed that changes in amino acid and cation concentrations in CSF from the *cisterna magna* reflected changes in the brain extracellular milieu.

RESULTS AND DISCUSSION

Heat stress induced a significant increase in RT (Fig. 1), which was accompanied by an increase in CSF and plasma osmolality (Fig. 2). It is well known that acute exposure to high ambient temperatures causes water loss from the cellular and extracellular fluid compartments, resulting in increased plasma osmolality, which in turn stimulates arginine vasopressin (AVP) secretion by the posterior pituitary, *via* activation of cardiopulmonary osmo- and volume-receptors[16,11]. It has been suggested that AVP regulates CSF composition during adaptation of the brain to physiologically acute increases in plasma osmolality[29].

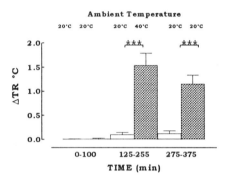

Figure 1. Effect of heat stress on rabbit Rectal Temperature (RT). White bars: five control rabbits kept at an ambient temperature of 20 °C. Grey bars: five rabbits exposed to 40 °C (125-250 min). Values are reported as mean ± s.e.m. Group data of RT were compared statistically by ANOVA. P<0.05 was considered significant. ***P<0.001.

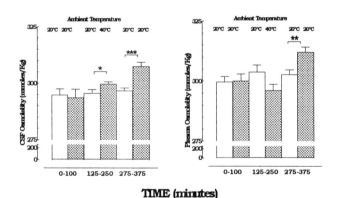

Figure 2. Effect of heat stress on cerebrospinal fluid (left) and plasma (right) osmolarity. White bars: five control rabbits kept at an ambient temperature of 20 °C. Grey bars: five rabbits exposed to 40 °C (125-250 min). Values are reported as mean ± s.e.m. Group data of CSF or plasma osmolarity were compared statistically by ANOVA. P<0.05 was considered significant. *P<0.05, **P<0.01, ***P<0.001.

In the present study, the heat stress-induced increase in plasma osmolality should have stimulated central AVP secretion which in turn might have modulated the composition and osmolality of the CSF.

Withdrawal of CSF from the *cisterna magna* is an unphysiological procedure which causes a progressive increase in CSF protein (Fig. 3), accompanied by a progressive decay in CSF taurine and GABA levels (Fig. 5), as shown in the control rabbits of the present study. The response of cells or organisms to a variety of physiological stresses involves the synthesis of heat stress protein in nervous cells[1].

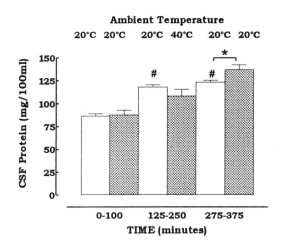

Figure 3. Effect of heat stress on cerebrospinal fluid protein content. White bars: five control rabbits kept at an ambient temperature of 20 °C. Grey bars: five rabbits exposed to 40 °C (125-250 min). Values are reported as mean ± s.e.m. Group data of CSF proteins were compared statistically by ANOVA. P<0.05 was considered significant. *P<0.05; # P<0.05 *vs* their basal (0-100 min) values.

In the rabbits used in this study, the stress arising from handling and restraint may have stimulated the synthesis of such proteins and their release into the extracellular space.

Heat-stress induced increase in CSF calcium levels (Fig. 4), while the levels in plasma were constant during the entire observation period (data not shown). Sodium, potassium and magnesium contents either in CSF and in plasma were unaffected (data not shown). It has been proposed that the set-point of body temperature in mammals is regulated by extracellular changes in calcium concentrations within the hypothalamus[18].

Previous studies from this laboratory suggest the involvement of calcium in the cascade of events which drives thermoregulation towards fever. It has been observed that i.c.v. injection of some calcium antagonists such as verapamil, nifedipine and cinnarizine, induces a dose-related hyperthermia, whereas the calcium agonist Bay-K-8644 causes hypothermia and antagonises the fever induced by *E. coli* endotoxin[21]. Furthermore, i.c.v. injection of human recombinant IL-1ß into conscious rabbits causes a significant and sustained increase in CSF calcium concentration which is accompanied by fever[22,23]. It is conceivable that when thermoregulation is set towards the promotion of heat dissipation, as it happens during the exposure to high ambient temperatures, changes in brain calcium metabolism may follow, giving rise to an increase of CSF calcium concentrations, as observed in the present experiment.

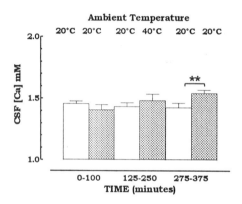

Figure 4. Effect of heat stress on cerebrospinal fluid calcium content. White bars: five control rabbits kept at an ambient temperature of 20 °C. Grey bars: five rabbits exposed to 40 °C (125-250 min). Values are reported as mean ± s.e.m. Group data of CSF calcium contents were compared statistically by ANOVA. $P < 0.05$ was considered significant. **$P < 0.01$.

Heat stress induced changes in brain taurine and GABA release which were reflected by the changes in their concentrations of these components in CSF (Fig. 5). In contrast, CSF aspartate and glutamate levels were unchanged (data not shown).

When one considers the distance separating the possible site of taurine and GABA action (brain extracellular space in the hypothalamus) from the site of taurine and GABA monitoring (*cisterna magna*) and/or the overall diluting effect of the CSF, the effective CSF taurine and GABA changes may be higher than those detected. The origin of the increase in CSF concentrations of taurine cannot be ascribed to blood contamination resulting from a transient opening of the blood-CSF barrier, as CSF glutamate remained constant throughout the experiment, despite the high blood-CSF gradient which approximates a value of 75^{13}.

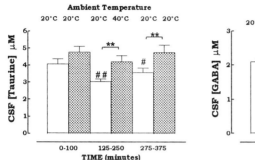 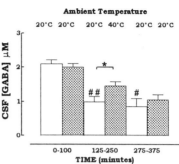

Figure 5. Effect of heat stress on cerebrospinal fluid taurine (left) and GABA (right) contents. White bars: five control rabbits kept at an ambient temperature of 20 °C. Grey bars: five rabbits exposed to 40 °C (125-250 min). Values are reported as mean ± s.e.m. Group data of CSF taurine or GABA contents were compared statistically by ANOVA. P<0.05 was considered significant. *P<0.05, **P<0.01; # P<0.05, ## P<0.01 vs their basal (0-100 min) values.

The results of the present study suggest that the increased output of taurine and GABA from the brain into the CSF is aimed at counteracting the hyperthermia promoted by exposure to heat. The possible role of taurine as an hypothermic regulator during heat stroke in man has been investigated in some laboratories. Bouchama, et al.[5] have shown that heat stroke patients with a RT of 42 °C associated with neurological disorders, exhibit a significant increase in plasma and urinary concentrations of taurine which here were found to return to basal values 24 h after a procedure which induced hypothermia. In dorsal horn slices of rat spinal cord, taurine release has been found to depend on temperature. In particular, basal taurine release is depressed at 8 °C and enhanced at 37 or 40 °C[7]. Tiggers et al.[31] demonstrated that astrocytes and neurones alter taurine efflux in a temperature-dependent fashion. Specifically, taurine efflux from rat hypothalamic astrocytes in culture decreased when the temperature was lowered from 37 to 33 °C, and increased at temperatures above 37 °C (37 °-41 °C). This effect was also shown to occur in cerebellar neuronal cultures.

Furthermore, taurine may also play an important role in fever: i.c.v. injection of taurine in rabbits was shown to reduce the febrile response caused by i.v. injection of *Salmonella typhosa* or leukocytic pyrogen or by i.c.v. injection of endotoxin or prostaglandin E_2[14]. Intracisternal injection of *Escherichia coli* into rabbits, an experimental model of meningitis, caused a significant increase in taurine and GABA concentrations in the extracellular fluid of the posterior frontal cerebral cortex[25]. Rats treated with PGE_2 through a microdialysis probe in the lumbar subarachnoid space, showed a marked increase in CSF concentrations of taurine and GABA[15]. Interleukin-1, a polypeptide cytokine which modulates several functions including fever, increases taurine and GABA release from rat preoptic/mediobasal hypothalamic tissue[8]. Experiments in progress in our laboratory have shown that CSF concentrations of taurine and GABA increase significantly in IL-1ß-induced fever in conscious rabbits.

In conclusion, the results of the present study demonstrate that taurine and GABA are released in the extracellular space of nervous cells during heat stress, probably to counteract the hyperthermia, suggesting these endogenous compounds have a cryogenic role. Taurine, therefore, is not only an osmoregulator of nervous cells, but it may also appear to play an important role in the central mechanisms of thermoregulation.

ACKNOWLEDGEMENTS

This work was supported by contributions of Ministero degli Affari Esteri (Rome, Italy) under law 212/92 and by MURST, (Cofin. '98).

REFERENCES

1. Barbe, M.F., Tytell, M., Gower, D.J., and Welch , W.J., 1988, Hyperthermia protects against light damage in the rat retina. *Science* **241**:1817-1820.
2. Barbeau, A., Inove, N., Tsukada, Y., and Butterworth, R.F., 1975, The neuropharmacology of taurine. *Life Sci.* **17**: 669-677.
3. Bianchi, L., Della Corte, L., and Tipton, K.F., 1999, Simultaneous determination of basal and evoked output levels of aspartate, glutamate, taurine and 4-aminobutyric acid during microdialysis and from superfused brain slices. *J. Chromatogr. B. Biomed. Sci. App.* **723**: 47-59.
4. Blake, M.J., Nowak, T.S., and Holbrook, N.J.,1990, *In vivo* hyperthermia induces expression of HSP70 mRNA in brain regions controlling the neuroendocrine response to stress. *Brain Res. Mol. Brain Res.* **8**: 89-92.
5. Bouchama, A., El-Yazigi, A., Yusuf, A., and Al-Sedairy, S., 1993, Alteration of taurine homeostasis in acute heatstroke. *Crit. Care Med.* **21**: 551-554.
6. Bradford, M., 1976, A rapid and sensitive method for the quantitation of microgram quantities of protein utilizing the principle of protein dye binding. *Anal. Biochem.* **72**: 248-254.
7. Dirig, D.M., Hua, X.Y., and Yaksh, T.L., 1997, Temperature dependency of basal and evoked release of aminoacids and calcitonin gene-related peptide from rat dorsal spinal cord. *J. Neurosci.* **17**: 4406-4414.
8. Felder, C., Refojo, D., Nacht, S., and Moguilevky, J.A., 1998, Interleukin-1 stimulates hypothalamic inhibitory amino acid neurotransmitter release. *Neuroimmunomodulation* **5**: 1-4.

9. Frosini, M., Gorelli, B., Matteini, M., Palmi, M., Valoti, M., and Sgaragli, G.P., 1993, HPLC determination of cation levels in cerebrospinal fluid and plasma of conscious rabbits. *J. Pharmacol. Toxicol. Meth.* **29**: 99-104.
10. Insel, P.A., 1996, Analgesic-antipyretic and antiinflammatory agents and drugs employed in the treatment of gout. In *Goodman and Gilman's - The pharmacological basis of therapeutics*, (Hardman, J.G., Limbird, L.E., Molinoff, P.B., Ruddon, R.W., Goodman Gilman, A., eds) McGraw-Hill, New York, NY, USA, 9th edition, pp. 617-657.
11. Kregel, K.C., Stauss, H., and Unger, T. (1994). Modulation of autonomic nervous system adjustments to heat stress by central ANG II receptor antagonist. *Am. J. Physiol.* **266**: R1985-R1991.
12. Lehmann, A., 1989, Effects of microdialysis-perfusion with anisoosmotic media on extracellular aminoacids in the rat hyppocampus and skeletal muscle. *J. Neurochem.* **53**: 525-535.
13. Lehmann, A., Carlstrom, C., Nagelhus, E.A., and Ottersen, O.P., 1991, Elevation of taurine hippocampal extracellular fluid and cerebrospinal fluid of acutely hypoosmotic rats: contribution by influx from blood? *J. Neurochem.* **56**: 690-697.
14. Lipton, J.M., and Ticknor, C.B., 1978, Central effect of taurine and its analogues on fever caused by intravenous leukocytic pyrogen in the rabbit. *J. Physiol.(London)* **287**: 535-543.
15. Malmberg, A.B., Hamberg, A., and Hedner, T., 1995, Effects of prostaglandin E_2 and capsaicin on behaviour and cerebrospinal fluid aminoacids concentrations of unanesthetized rats: a microdialysis study. *J. Neurochem.* **65**: 2185-2193.
16. Marder, J., Eylath, U., Moskovitz, E., and Sharir, R., 1990, The effect of heat exposure on blood chemistry of the hyperthermic rabbit. *Comp. Biochem. Physiol.* **97A**: 245-247.
17. Murphy, M.T., Lipton, J.M., Loughran, M.B., and Giesecke, A.H., 1985, Postanesthetic shivering in primates: inhibition by peripheral heating and by taurine. *Anestesiol.* **63**: 161-165.
18. Myers, R.D., and Veale, W.L., 1970, Body temperature: possible ionic mechanism in the hypothalamus controlling the set-point. *Science* 170: 95-97.
19. Nagelhus, E.A., Lehmann, A., and Ottersen, O.P., 1993, Neuronal-glial exchange of taurine during hypo-osmotic stress: a combined immunocytochemical and biochemical analysis in rat cerebellar cortex. *Neuroscience* **54**: 615-631.
20. Oja, S.S., and Saransari, P., 1992,Taurine relase and swelling of cerebral cortex slices from adult and developing mice in media of different ionic compositions. *J. Neurosci. Res.* **32**: 551-561.
21. Palmi, M., and Sgaragli, G.P., 1989, Hyperthermia induced in rabbits by organic calcium antagonists. *Pharmacol., Biochem., Behav.* **34**: 325-330.
22. Palmi, M., Frosini, M., and Sgaragli, G.P., 1992, Calcium changes in rabbit CSF during endotoxin, IL-1ß and PGE_2 fever. *Pharmacol. Biochem. Behav.* **43**: 1253-1262.
23. Palmi, M., Frosini, M., Becherucci, C., Sgaragli, G.P., and Parente, L., 1994, Increase of extracellular brain calcium involved in interleukin-1ß-induced pyresis in the rabbit: antagonism by dexamethasone. *Br. J. Pharmacol.* **112**: 449-452.
24. Pasantes-Morales, H., Quesada, O., and Moran, J., 1998, Taurine: an osmolyte in mammalian tissues. *Adv. Exp. Med. Biol.* **442**: 209-217.
25. Perry, V.L., Young, R.S.K., Aquila, W.J., and During, M.J., 1993, Effect of experimental escherichia coli meningitis on concentrations of exitatory and inhibitory amino acids in the rabbit brain: in vivo microdialysis study. *Pediatr. Res.* **34**: 187-191.
26. Schousboe, A., Moran, J., and Pasantes-Morales, H., 1990, Potassium-stimulated release of taurine from cultured cerebellar granule neurons is associated with cell swelling. *J. Neurosci. Res.* **27**: 71-77.

27. Sgaragli, G.P., Carlà, V., Magnani, M., and Galli, A., 1981, Hypothermia induced in rabbits by intracerebroventricular taurine: specificity and relationship with central serotonin (5-HT) systems. *J. Pharmacol. Exp. Ther.* **219**: 778-785.

28. Sgaragli, G.P., Frosini, M., Palmi, M., Bianchi, L., and Della Corte, L., 1994, Calcium and taurine interaction in mammalian brain metabolism. *Adv. Exp. Med. Biol.* **359**: 299-308.

29. Szmydynger-Chodobska, J., Szczepanska-Sadowska, E., and Chodobski, A., 1990, Effect of arginine-vasopressin on CSF composition and bulk flow in hyperosmolar state. *Am. J. Physiol.* **259**: R1250-R1258.

30. Thurston, J.H., Hauhart, R.E., and Dirgo, J.A., 1980, Taurine: a role in osmotic regulation of mammalian brain and possible clinical significance. *Life Sci.* **26**: 1561-1568.

31. Tigges, G.A., Philibert, R.A., and Dutton, G.R., 1990, K^+- and temperature-evoked taurine efflux from hypothalamic astrocytes. *Neurosci. Lett.* **119**, 23-26.

32. Yakimova, K., Sann, H., Schmid, H.A., and Pierau, F.K., 1996, Effects of GABA agonists and antagonists on temperature-sensitive neurones in the rat hypothalamus. *J. Physiol.(London)* **494**: 217-230.

33. Yarbrough, G.G., Singh, D.K., and Taylor, D.A., 1981, Neuropharmacological characterisation of a taurine antagonist. *J. Pharmacol. Exp. Ther.* **219**: 604-613.

INTERACTION OF TAURINE WITH METAL IONS

[1]E. C. O' Brien, [2]E. Farkas and [1]K. B. Nolan
[1]*Department of Chemistry, Royal College of Surgeons in Ireland, St. Stephens Green, Dublin 2;*
[2]*Department of Inorganic and Analytical Chemistry, Lajos Kossuth University, H-4010 Debrecen, Hungary*

INTRODUCTION

Taurine, $^{+}NH_3CH_2CH_2SO_3^{-}$ the sulphonic acid analogue of ß-alanine, is one of the most abundant low molecular weight organic compounds in the animal kingdom[1]. It plays a role in cardiovascular, central and membrane stabilisation and osmoregulation and also has antioxidant and anti-inflammatory effects[2]. It has been reported that taurine modulates many calcium(II)-dependent physiological processes and that the formation of a zinc(II)-taurine complex may account for its effects on several zinc(II)-dependent processes[1]. Despite this, and the fact that taurine is a potential ligand, reports on its complexing behaviour are sparse. This is surprising because of the vast literature on metal complexes of its aminocarboxylic acid analogue which may act as a monodentate ligand, a bidentate ligand to form a six-membered chelate ring or as a bridging ligand (Fig 1).

The amino group of taurine would be expected to bind to metal ions and although the affinity of sulphonate groups on their own for metal ions is weak, the presence of the amino group may serve as an 'anchor' allowing the formation of stable six-membered chelate rings. Thus taurine may

Taurine 4, edited by Della Corte *et al.*
Kluwer Academic / Plenum Publishers, New York, 2000.

coordinate to metal ions in a monodentate or bidentate manner (Fig 2). The only previously reported metal complex of taurine in the solid state was by Ford and Nolan[3] who isolated the complex cis-[Co(en)$_2${NH$_2$(CH$_2$)$_2$SO$_3^-$}Cl]Cl (en = 1,2-diaminoethane) in which taurine acts as a monodentate ligand through its amino group. Several pH-metric solution studies including complexes of taurine with copper(II) and uranyl(VI) ions[4] and mixed ligand complexes of nickel(II) with taurine, DL-methionine and DL-ethionine[5] have been carried out. However these studies were limited due to hydrolysis of the metal cation which occurred at quite low pH values.

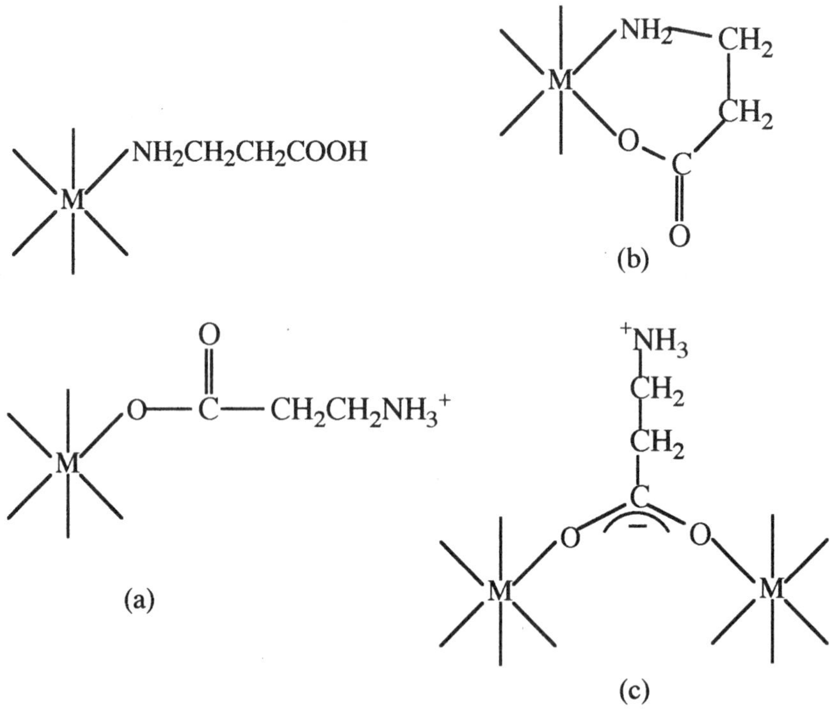

Figure 1. β-alanine as (a) a monodentate ligand, (b) a bidentate chelating ligand and (c) a bridging ligand.

The biological importance of taurine and the fact that metal binding may be important in some of its activity prompted this study of complex formation at physiological pH.

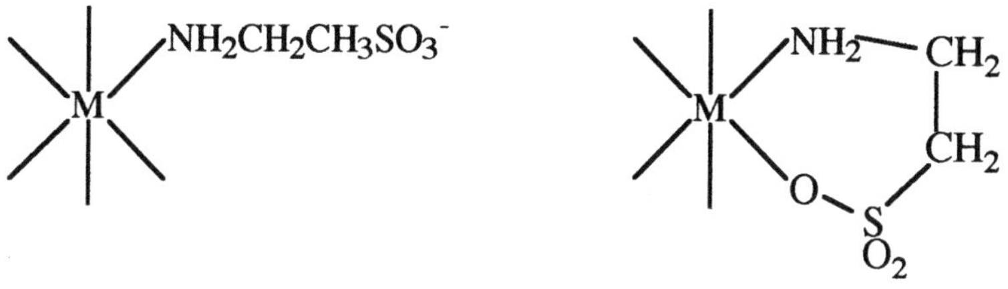

Figure 2. Taurine as (a) a monodentate ligand and (b) a bidentate ligand

REACTION OF TAURINE WITH COPPER(II) - DIPEPTIDE COMPLEXES

It was decided to study ternary complex formation between taurine and the binary complexes of copper(II) with glycylglycine (HGly-Gly) and glycyl-L-aspartic acid (H$_2$Gly-Asp). The binary complexes present at physiological pH are Cu(Gly-GlyH$_{-1}$) and [Cu(Gly-AspH$_{-1}$)]$^-$ (Fig 3). Both of these complexes which are water soluble at relatively high pH values contain aqua ligands in an equatorial coordination site that can be readily replaced by an amine donor such as that present in taurine[6].

The axial coordination sites in the Cu(Gly-GlyH$_{-1}$) complex are occupied by water molecules (not shown in Fig 3) one of which may be replaced by the sulphonate group of taurine. However, in the [Cu(Gly-AspH$_{-1}$)]$^-$ complex the β-COO$^-$ group of the Asp residue occupies one of the axial positions resulting in extensive elongation of the bond length at the other axial position[7]. This elongation eliminates the possibility of any interaction between the sulphonate group of taurine and the second axial position of copper(II) in the ternary copper(II) - Gly-Asp - taurine complex. Thus a comparison of the stability constants of the ternary copper(II) - Gly-Gly - taurine complex in which axial coordination of the sulphonic acid group of taurine is possible with that of the copper(II) - Gly-Asp - taurine complex should indicate if taurine is capable of acting as a bidentate ligand.

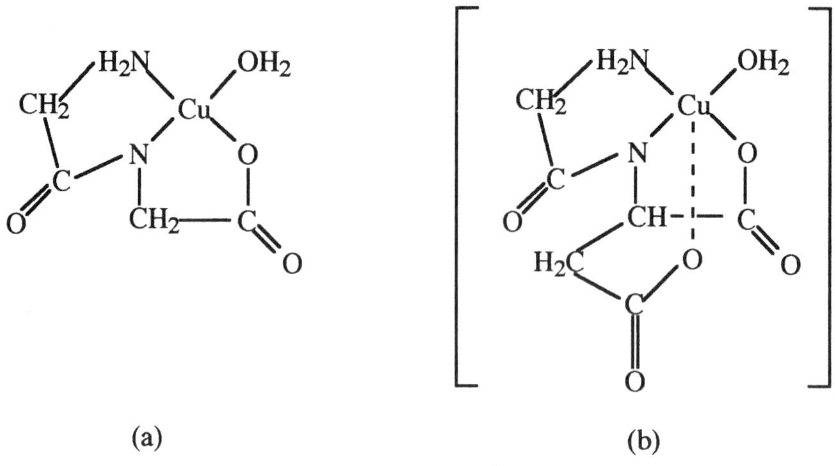

(a)   (b)

Figure 3. Structure of (a) Cu(Gly-GlyH$_{-1}$) and (b) [Cu(Gly-AspH$_{-1}$)]$^-$

PH-METRIC STUDIES

The complex Cu(Gly-GlyH$_{-1}$) may react with taurine to give a ternary species in which taurine acts as a monodentate ligand (Fig 4) or in which the sulphonate group is also axially coordinated and taurine acts as a bidentate ligand. As was previously confirmed, in a study of the interactions of Cu(Gly-GlyH$_{-1}$) and [Cu(Gly-AspH$_{-1}$)]$^-$ with β-alanine, this axial coordination cannot occur in the latter case. The significantly larger stability constant for the formation of the ternary complex [Cu(Gly-GlyH$_{-1}$)β-Ala]$^{2-}$ (log K = 3.63) relative to that of [Cu(Gly-AspH$_{-1}$)β-Ala]$^{2-}$ (log K = 3.02) substantiates this idea[8,9].

pH-metric titrations were carried out over the pH range 2.0 - 11.5 at copper(II) - dipeptide : taurine ratios of 1:1, 1: 2, 1:4 and 1:8. The titration data obtained were evaluated by the PSEQUAD computer program[10] using the known protonation constants of HGly-Gly, H$_2$Gly-Asp and taurine and the previously determined stability constants of the copper(II) - Gly-Gly and copper(II) - Gly-Asp binary systems, Table 1[9,11]. The log K values (Eqns (1) and (2)) were calculated from the log β values for the formation of the respective binary and ternary complexes and the results are summarised in Table 1.

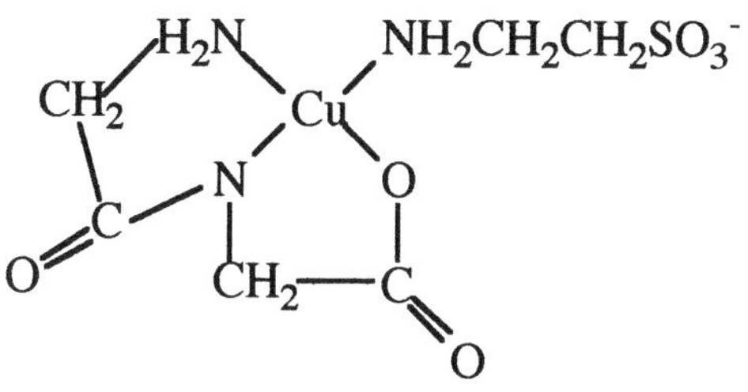

Figure 4. Structure of [Cu(Gly-GlyH$_{-1}$)taurine]$^-$

$$Cu(Gly\text{-}GlyH_{-1}) + L \quad \xrightleftharpoons{K} \quad [Cu(Gly\text{-}GlyH_{-1})L]^- \qquad (1)$$

$$[Cu(Gly\text{-}AspH_{-1})]^- + L \quad \xrightleftharpoons{K} \quad [Cu(Gly\text{-}AspH_{-1})L]^{2-} \qquad (2)$$

The ternary complexes of taurine (log K = 2.95 and 2.68) are less stable than those of β-alanine (log K = 3.6 and 3.02) which is largely due to the decreased basicity of the amino group in taurine relative to β-alanine. The log K value for the formation of the [Cu(Gly-GlyH$_{-1}$)taurine]$^-$ complex (2.95) is only slightly greater than that for the formation of [Cu(Gly-AspH$_{-1}$) taurine]$^{2-}$ complex (2.68) and this may simply reflect the charge difference in the reactant binary complexes i.e. Cu(Gly-GlyH$_{-1}$) is neutral, [Cu(Gly-AspH$_{-1}$)]$^-$ has a negative charge. This suggests that the sulphonate group of taurine is less likely than the carboxylate group in β-alanine to occupy an axial coordination position of the metal ion in [Cu(Gly-GlyH$_{-1}$)taurine]$^-$.

Species distribution curves, calculated from the stability constant data are shown in Figs. 5 and 6. These curves substantiate the finding that formation of the ternary complex begins at pH 6.5 in both systems. Their concentrations reach a maximum at pH 9.0 and decrease after this point as they are replaced by the hydroxo complexes [Cu(Gly-GlyH$_{-1}$)OH]$^-$ and [Cu(Gly-AspH$_{-1}$)OH]$^{2-}$ respectively.

Table 1. Stability constants for the formation of ternary complexes in the copper(II) - dipeptide - taurine systems, I = 0.2 mol dm^{-3} and T = 25.0 °C.

Complex	log β[a]	log K
[Cu(Gly-GlyH$_{-1}$)taurine]$^-$	4.28 (0.03)[b]	2.95 (0.03)
[Cu(Gly-AspH$_{-1}$)taurine]$^{2-}$	4.53 (0.02)	2.68 (0.02)

[a]β = [Cu(LH$_{-1}$)taurine][H$^+$] / [Cu^{2+}][L][taurine] *where L is the peptide*

[b]*Values in parentheses are standard deviations estimated using four titrations per system and 100 data points per titration.*

Literature log K/β values, I = 0.2 mol dm^{-3} KCl, T = 25.0 °C.

Gly-Gly 8.13, 11.30	*Gly-Asp 8.35, 12.66, 15.45*
Cu(Gly-Gly) 5.56	*Cu(Gly-Asp) 6.61, Cu(HGly-Asp) 10.41*
Cu(Gly-GlyH$_{-1}$) 1.33	*Cu(Gly-AspH$_{-1}$) 1.85, Cu(Gly-AspH$_{-2}$) -7.97*
Cu(Gly-GlyH$_{-2}$) -8.04	*Cu(Gly-Asp)$_2$ 11.50, Cu(Gly-Asp)$_2$H$_{-1}$ 4.54*
Cu(Gly-Gly)$_2$H$_{-1}$ 4.46	*Cu$_2$(Gly-Asp)$_2$H$_{-3}$ - 4.20*

VISIBLE SPECTRA

The formation of a ternary copper(II) - Gly-Gly - taurine complex in aqueous solution was confirmed by visible spectroscopy. UV-visible spectra of the copper(II) - Gly-Gly and copper(II) - Gly-Gly - taurine (1:1:1 and 1:1:8) systems were recorded at various pH values. The λ_{max} values and molar extinction coefficients as a function of pH are summarised in Table 2.

The λ_{max} values for the copper(II) - Gly-Gly system do not show a pH dependence in the pH range studied as the two species present in this range are Cu(Gly-GlyH$_{-1}$)H$_2$O and [Cu(Gly-GlyH$_{-1}$)OH]$^-$ both of which are Cu-N$_2$O$_2$ chromophores. In contrast increasing the pH in the copper(II) - Gly-Gly - taurine system results in a blue shift of about 30 nm in the λ_{max} values. This is consistent with the formation of [Cu(Gly-GlyH$_{-1}$)taurine]$^-$ in which three nitrogens and an oxygen are involved in coordination and provides unambiguous evidence for the replacement of an aqua ligand by an amine donor in the equatorial plane[11]. Between pH 8.70 and 10.33 a red shift of 15 nm occurs due to the replacement of [Cu(Gly-GlyH$_{-1}$)taurine]$^-$ by the mixed hydroxo species [Cu(Gly -GlyH$_{-1}$)OH]$^-$.

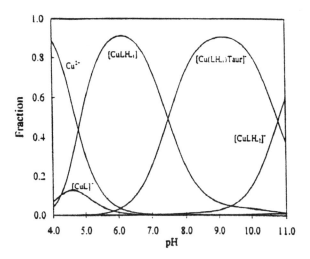

Figure 5. Species distribution curves for the copper(II)-(Gly-Gly)-taurine system in the pH range 4.0 - 11.0, [Cu(II)] = 3.83 mM, [Gly-Gly] = 4.08 mM, [taurine] = 7.95mM (HL = HGly-Gly).

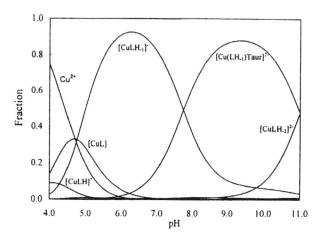

Figure 6. Species distribution curves for the copper(II)-(Gly-Asp)-taurine system in the pH range 4.0 - 11.0, [Cu(II)] = 3.95 mM, [Gly-Asp] = 4.18 mM, [taurine] = 3.42 x 10^{-2} M (HL = HGly-Asp).

Table 2. Electronic absorption spectra (λ_{max} values in nm) for copper(II) - (Gly-Gly) and copper(II) - Gly-Gly - taurine complexes.

pH	Copper(II)-(Gly-Gly) (1 : 1)	Copper(II) - (Gly-Gly) - taurine (1 : 1 : 1)	(1 : 1 : 8)
4.020	—	—	—
5.020	640	640	640
7.250	640	635	625
8.700	640	620	610
10.330	640	630	625

ELECTRON PARAMAGNETIC (EPR) SPECTROSCOPY

EPR spectroscopy allows the assignment of structures to paramagnetic compounds and therefore copper(II) complexes (d^9) which have a single unpaired electron give readily resolved spectra. Spectra of the copper(II) - Gly-Gly (1:1) system in the absence and presence of taurine (5-fold excess) at pH values of 5.0, 7.0 and 9.0 were recorded at liquid nitrogen temperature. The g and A parameters, obtained by computer simulation[12], are shown in Table 3. As reported previously the copper(II) - Gly-Gly system gives complicated EPR spectra due to the presence of a variety of species including isomers[11]. Nevertheless the results of this study provide unambiguous evidence for taurine coordination. At pH 9.0 in the binary system, where the predominant species is Cu(Gly-GlyH_-1), the $g_{||}$ value is 2.270. At pH 9.0 in the ternary system where, according to the species distribution curves maximum ternary complex formation occurs, the $g_{||}$ value is 2.237. The in-plane addition of a stronger donor to copper(II) is characterised in EPR spectroscopy by such a decrease in $g_{||}$[11] thus confirming the replacement of the aqua ligand by the amino group of taurine. In the conversion of Cu(Gly-GlyH_-1)H$_2$O to [Cu(Gly-GlyH_-1)(Gly-Gly)]⁻, which involves a similar change of donor atom set (from N_2O_2 to N_3O), an analogous decrease in was observed i.e. from 2.248 to 2.232 at 140 K[11].

The formation of the ternary complex is accompanied by a decrease in $A_{||}$. Again a similar decrease was observed in the conversion of Cu(Gly-GlyH_-1)H$_2$O to [Cu(Gly-GlyH_-1)(Gly-Gly)]⁻ (reference[11]). This behaviour is the opposite to that expected when a stronger in-plane donor is added. The effect on $A_{||}$ depends on the relative importance of increased covalency of the donor atom which causes $A_{||}$ to decrease or increased hybridisation of the

4s and 3d orbitals having A_1 symmetry which causes $A_{\parallel}$ to increase[13]. The decrease in $A_{\parallel}$ observed indicates that the increase in covalency is the predominant effect.

Table 3. EPR data, for copper(II) - (Gly-Gly) and copper(II) - (Gly-Gly) - taurine solutions at 77 K [a]

Species	pH		$g_{\parallel}$	$A_{\parallel}$
Copper(II)-Gly-Gly	5.0	species 1	2.245	115
		species 2	2.241	176
	7.0		2.254	169.1
	9.0		2.270	169.7
Copper(II)-Gly-Gly-taurine	7.0		2.249	170.9
	9.0		2.237	156.7

[a] Units of A values : 10^{-4} cm^{-1}

CONCLUSION

Taurine is capable of forming moderately stable complexes with copper(II) in which it acts as a monodentate, N-donor ligand. However, it binds to copper(II) less strongly than its carboxylic analogue β-alanine. This is probably due to the decreased basicity of the amino group on taurine. Since zinc(II) and copper(II) have similar coordination behaviour taurine should also bind to zinc(II) under physiological conditions. However taurine would not be expected to bind to class 'a' metal ions such as calcium(II) and magnesium(II).

ACKNOWLEDGEMENTS

This work was supported by the Research Committee of the Royal College of Surgeons in Ireland, Enterprise Ireland (International Collaboration Programme) and EU Cost D8. Professor E. Farkas thanks OTKA T019337 for financial support.

REFERENCES

1. Huxtable, R. J., 1992, *Physiol. Rev.* **72**: 101.
2. Huxtable, R. J., and Sebring, L. A., 1986, *Trends Pharmacol. Sci.* **7**: 481.
3. Ford, P. D., and Nolan, K. B., 1980, *Inorg. Chim. Acta* **43**: 83.

4. Mittal, R. K., Chandra, M., Dey, A. K., 1978, *Monatsh. Chem.* **109**: 953.
5. Maslowska, J., and Chruscinnski, L., 1984, *Polyhedron* **3(12)**: 1329.
6. Sóvágó, I., Kiss, A., Farkas, E., Sanna, D., Marras, P., and Micera, G., 1997, *J. Inorg. Biochem.*, **65**: 103.
7. Cotton, F. A., Wilkinson, G., and Gaus, P. L., In *Basic Inorganic Chemistry.* Wiley, New York, pp. 430-467.
8. Nagypal, I., and Gergely, A., 1977, *J. Chem. Soc., Dalton Trans.*, 1109.
9. Gergely, A., and Farkas, E., 1982, *J. Chem. Soc., Dalton Trans.*, 381.
10. Zekany, L., and Nagypál, I., 1985, in *Computational Methods for the Determination of Formation Constants* (D. J. Leggett, ed.), Plenum, New York, pp. 291-352.
11. Sóvágó, I., Sanna, D., Dess, A., Várnagy, K., and Micera, G., 1996, *J. Inorg. Biochem.* **63**: 99.
12. Rockenbauer, A., and Korecz, L., 1996, *Appl. Magn. Res.* **10**: 29.
13. Rockenbauer, A., 1979, *J. Magn. Reson.* **35**: 429.

ATTENUATION OF OXIDATIVE DAMAGE TO DNA BY TAURINE AND TAURINE ANALOGS

Steve A. Messina and Ralph Dawson, Jr.
Department of Pharmacodynamics, College of Pharmacy, University of Florida, Gainvesville, FL

Abstract: Taurine has been suggested to have cytoprotective actions via a number of different mechanisms. The role of taurine in protecting DNA from oxidative damage has received only limited attention. The aim of the present studies was to test the hypothesis that taurine might act to attenuate oxidative damage to DNA caused by free radicals generated by iron-stimulated catecholamine oxidation in the presence of H_2O_2. Calf thymus DNA (100 µg/tube) was exposed to a reaction mixture containing: ferric chloride (60 µM), H_2O_2 (2.8 mM) and L-dopa (100 µM). Taurine and taurine analogs were added simultaneously to determine their effects to prevent oxidative damage to DNA. The reaction was carried out for 1 hour at 37° C and terminated by rapid freezing in an ethanol/dry ice bath. The DNA was precipitated with ethanol and subsequently hydrolyzed with formic acid under vacuum. The hydroxylated bases were separated by HPLC and detected electrochemically. All experiments were replicated a minimum of 5 times. Taurine (20 mM) was found to reduce (p<0.05) damage to DNA as indexed by reductions in the formation of 5-OH-uracil (49%↓), 8-OH adenine (37%↓), and 8-OH guanine (21%↓). Taurine had minimal effects to reduce the formation of 5-OH cytosine (<7%↓). Taurine (20 mM) also increased total DNA recovery after damage 36-40% and increased total undamaged guanine ~32%. 5-OH Uracil formation could be reduced (p<0.05) by 1 mM taurine and 8-OH-adenine formation was reduced (p<0.05) by 5 mM taurine. Studies were conducted with various amino acid analogs and total base adduct formation was reduced by 20 mM β-alanine (30%↓), lysine (58%↓) and glutathione (88%↓). When tested at 20 mM, both hypotaurine and homotaurine provided greater protection against DNA damage than taurine, whereas isethionic acid provided a similar level of protection as taurine. Using identical conditions as the assays for base hydroxylation, we tested whether inhibition of quinone formation

Taurine 4, edited by Della Corte *et al.*
Kluwer Academic / Plenum Publishers, New York, 2000.

could account for taurine's mechanism of action. Taurine (49%↓), homotaurine (24%↓) and hypotaurine (79%↓) all reduced quinone formation. Thus, inhibition of quinone formation could account for part of taurine's mechanism of action to inhibit oxidative damage, but it could not account for homotaurine's greater efficacy in preventing DNA damage. Overall, these studies show that taurine at concentrations normally found in cells can inhibit oxidative damage to DNA.

INTRODUCTION

Taurine has been shown to have cytoprotective actions against a number of toxic insults[1]. The mechanisms that account for the protective actions of taurine have not always clear, but may involve polyamines[2], osmotic effects[3], antioxidant properties[4] or effects on calcium homeostasis[5]. Taurine has also been shown to possess antimutagenic effects and to reduce various types of damage to DNA[6-8]. A recent study has also suggested that taurine modulates calcium levels within the nucleus[9]. This is consistent with immunoreactive taurine being localised to the nucleus[10] and evidence that taurine may regulate histone H2B phosphoralyation[11]. Taken together, the evidence suggests that taurine may play a previously unappreciated role in regulating nuclear function.

The goal of the present series of experiments was to determine if taurine could attenuate oxidative damage to DNA induced by metal-stimulated oxidation of 3,4-dihydroxyphenylalanine (L-dopa). Previous studies have implicated the oxidation of L-dopa or dopamine as a causative factor in subsequent oxidative damage to DNA[12-14]. Oxidative damage to DNA has been reported in Alzheimer's disease, Parkinson's disease and normal ageing[15]. Taurine levels have been found to decrease in aged rodents[16] and CSF levels of taurine have been reported to be lower in Alzheimer's patients[17]. Furthermore, taurine can decrease metal-stimulated oxidation of catecholamines[18]. If taurine plays a role in normal antioxidant defense systems, an age-related or disease-induced decline in taurine could shift the balance toward a pro-oxidant environment. Therefore, these studies were undertaken to more fully understand the relationship between oxidative DNA damage and taurine's putative cytoprotective role. Studies were performed to determine concentration and structure-activity relationships for taurine and taurine taurine analogs to reduce damage to DNA.

METHODS

Materials

Calf thymus DNA, 8-bromoadenine, isobarbituric acid (5-OH uracil), guanine, FeCl₃-hexahydrate, taurine (2-aminoethane sulfonic acid), β-alanine, isethionic acid (2-hydroxyethanesulfonic acid),) hypotaurine (2-aminoethane sulfinic acid), homotaurine (3-aminopropane sulfonic acid), and glutathione (GSH: Glu-Cys-Gly) were purchased from Sigma Chemical Co. (St. Louis, MO). 2-Amino-6,8,hydroxypurine (8-OH guanine) was purchased from Aldrich (Milwaukee, WI). Lysine and L-dihydroxyphenylalanine (L-dopa) were purchased from Fisher (Fairlawn, NJ). 8-OH-Adenine was synthesized by treatment of 8-bromoadenine with concentrated formic acid (95%) at 150 °C for 45 min and purified by crystallization with water[19]. 2-OH Adenine and 5-OH cytosine were gifts from Dr. Miral Dizdaroglu (National Institute of Standards and Technology, Gaithersburg, MD, U.S.A.

Treatment of DNA

DNA was damaged using a Fenton type reaction with FeCl₃, H₂O₂ and L-dopa, similar to the method described by Spencer *et al.*[12] . Reaction mixtures contained (final concentrations): PBS (phosphate buffered saline) (pH 7.4), calf thymus DNA (100μg/tube), H₂O₂ (2.8mM), FeCl₃ (60μM), and L-dopa (100μM). Taurine or taurine analogs were tested for their effects on oxidative damage to DNA. Control tubes contained only DNA and PBS. The final assay volume was 1 ml, and reaction mixtures were aliquoted into glass borosilicate tubes. Tubes were incubated for 1 hour in a 37 °C shaking water bath. The reaction was stopped by transferring the reaction mixture into siliconized microcentrifuge tubes and then freezing in an ethanol/dry ice bath. Samples were then lyophilized over night using a refrigerated condensation trap and a speedvac condensator (Savant Instruments Inc., Farmingdale, NY). The DNA were resuspended in 100% ethanol and sonicated briefly followed by centrifugation for 30 minutes at 10,000 g at 0 °C. The supernatant was aspirated and pellets were again washed with 100% ethanol and centrifuged for 15 minutes at 10,000 g at 0 °C. The supernatant was again aspirated and pellets were dried under nitrogen and resuspended in 1 ml 60% formic acid. An aliquot was transferred into a cuvette and used to measure total DNA content by reading the absorbance of the samples at 260/280 nm using a Beckman DU700 diode array spectrophotometer (Beckman Instruments, Inc. Fullerton, CA). The sample was hydrolyzed for

45 min at 150 °C using 5 ml vacuum hydrolysis tubes from Pierce (Rockford, IL). Samples were then transferred into borosilicate glass tubes and again lyophilized overnight. The pellets were resuspended in 1 ml mobile phase, sonicated and transferred into siliconized microcentrifuge tubes. The samples were sonicated and then microcentrifuged for 10 minutes. Samples were then diluted with mobile phase if needed, and DNA base adducts were measured using high performance liquid chromatography (HPLC) with electrochemical detection (EC).

The following base adducts were determined in all of the studies: 5-OH uracil, guanine, 8-OH guanine, and 8-OH adenine. For the taurine analog and dose response study, 5-OH cytosine and 2-OH adenine were also measured. Data for the pilot studies were calculated as nmol of adduct/ nmol of guanine per sample. For the taurine analog and dose response studies, DNA content was determined from the sample absorbance at 260 nm (A_{260} of 1.0 $\approx$ 50μg DNA). Data for the taurine analog and dose response studies are presented as nmol of adduct/ mg DNA.

HPLC Analysis of DNA Adducts

The mobile phase consisted of (final concentrations) 50 mM sodium acetate, 1mM EDTA and 2% methanol, pH 5.5[20]. The mobile phase was vacuum filtered using nylon filters (0.45 μm) from MSI (Westboro, MA) and degassed for 20 minutes under helium. The HPLC apparatus consisted of a PM-11 pump, LC-4B amperometric detector, LC22A temperature controller (Bioanalytical Systems, West Lafayette, IN), and Rheodyne model 7125 injector with a 50μl fixed loop. The column used was a Microsorb-MV, C_{18}, 5μm, 4.6mm ID x 25 cm column from Rainin (Woburn, MA). Ag/AgCl reference electrodes and glassy carbon working electrodes with an applied voltage of 0.75 V were used for electrochemical detection. The detector sensitivity was 10 nA. The flow rate was 0.82 ml/min and the column was kept at a constant temperature of 27 °C.

Spectrophotometric Analysis of Quinones

In order to address the mechanism of action of taurine and related analogs in the reduction of oxidative damage to DNA, an experiment was performed to measure quinone formation from the oxidation of L-dopa. Taurine and taurine analogs were all tested at a concentration of 10 mM. Reaction conditions were identical to those described for the DNA damage assays, except at the end of the 15 minute incubation period the samples were immediately read in a spectrophotometer at 490 nm to measure quinones formed from L-dopa oxidation[18]. Quinones derived from L-dopa oxidation were determined under identical reaction conditions in the presence or absence of DNA. These studies were designed to determine any interaction of taurine or taurine analogs with DNA that might influence oxidation rates.

Statistical Analysis

All samples were run in duplicate and each value of N represents an independent replication. An N=5-6 was used in all experiments. In pilot studies, mean concentrations of the adducts (nmol adduct/ nmol guanine per sample) were calculated. The mean adduct concentrations from the damaged samples were compared to the mean adduct concentrations from the damaged with taurine. Statistical comparisons were made to determine the protective effects of taurine in reference to individual adduct concentrations, by using the Mann-Whitney U-test.

In the taurine analog studies and the dose response studies, mean concentrations of the base adducts (nmol/mg DNA) were determined. The mean adduct concentrations from the damaged samples were compared to the mean adduct concentrations from the damaged with taurine or taurine analog samples. One-way analysis of variance (ANOVA) and Dunnett's post-hoc test was used to compare individual treatment or concentration effects.

RESULTS

Preliminary experiments were performed to determine if 20 mM taurine could significantly reduce the formation of hydroxylated DNA base adducts. These experiments showed that our control DNA incubated in PBS and carried through the extraction process did not show significant damage and that taurine added to control incubations had no effect. Furthermore, our protocol to damage DNA produced significant amounts of the hydroxylated bases measured by our HPLC assay and a significant reduction in guanine content. Finally, these pilot studies showed that 20 mM taurine could produce significant reductions in 5-OH uracil ($p<0.05$) and 8-OH guanine ($p<0.01$), but only a nonsignificant decrease in 8-OH adenine (data not shown). While these pilot studies served to validate our general protocol, the quantitation based on total guanine was problematic since it did not accurately reflect DNA recovery. Subsequently, all data were normalized to DNA content measured after the damage protocol since taurine improved DNA recovery.

Studies were performed to determine the concentration dependence of taruine's effect to reduce DNA damage. The first study used concentrations of 5, 10 or 20 mM taurine to assess DNA protection. The results of this experiment are presented in Table 1. Taurine could produce inhibition of damage to 5-OH uracil and 8-OH adenine at a concentration of 5 mM. The recovery of undamaged guanine was also significantly enhanced by taurine (data not shown). Taurine produced no reduction in the formation of 8-OH guanine in this experiment.

Table 1. Concentration dependent reduction in damage to DNA by taurine.

Taurine (mM)	5-OH Uracil	8-OH Adenine	8-OH Guanine
0	59.5 ± 2.7	32.3 ± 1.5	45.9 ± 4.6
5	35.8 ± 1.6*	23.7 ± 0.8*	37.8 ± 4.0
10	34.5 ± 0.9*	23.1 ± 1.0*	42.5 ± 2.9
20	31.1 ± 1.3*	20.7 ± 1.1*	40.4 ± 3.3

Data are expressed as nmoles base adduct/mg DNA ± SEM.
*$p < 0.01$ versus samples incubated with no taurine.

This experiment prompted us to perform a second concentration-response study using taurine concentration of 0.1, 1 and 5 mM to determine the threshold concentration of taurine that would produce protection of DNA. These results are presented Table 2. Taurine produced a significant reduction in 5-OH uracil formation at 1 mM (Table 2). Taurine did not reduce 8-OH guanine or 5-OH cytosine content. Guanine content was not increased by taurine (data not shown). The reduction in 8-OH adenine was not statistically reduced by 5 mM taurine in this experiment, although it was decreased 17%.

Table 2. Concentration dependent reduction in DNA damage by taurine

Taurine (mM)	5-OH Uracil	8-OH Adenine	8-OH Guanine	5-OH Cytosine
0	78.3 ± 5.7	43.8 ± 2.7	72.4 ± 3.2	41.8 ± 3.3
0.1	73.3 ± 3.8	43.1 ± 1.8	78.2 ± 3.8	44.3 ± 1.8
1	56.5 ± 2.2*	40.4 ± 4.2	75.1 ± 4.6	45.5 ± 2.9
5	50.4 ± 0.8**	36.5 ± 2.1	71.1 ± 3.1	41.1 ± 1.3

Data are expressed as nmoles base adduct/mg DNA ± SEM.
*$p < 0.05$, **$p < 0.01$ versus samples incubated with no taurine.

Structure-activity relationships were examined by testing a number of taurine analogs at 20 mM for their ability to reduce damage to DNA. The results of the first experiment are summarized in Figure 1. All analogs signficantly ($p < 0.01$) reduced the the formation of 5-OH uracil and 8-OH guanine (Figure 1). Only lysine and glutathione significantly ($p < 0.01$) reduced damage to guanine (data not shown). All of the compounds tested reduced ($p < 0.05$) 8-OH adenine except β-alanine. When the total base damage was calculated, the order of efficacy to prevent DNA damage was;

glutathione>lysine>β-alanine≥taurine. Taurine reduced 8-OHguanine in this experiment, but the data were calculated based on a ratio of damaged bases to undamaged guanine since glutathione interfered with the UV assay for DNA.

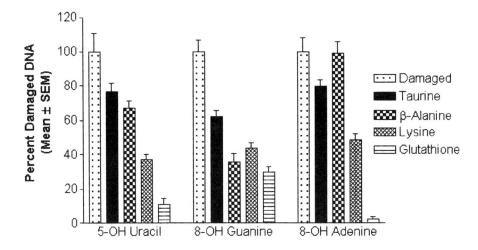

Figure 1. The effect of taurine and related compounds to inhibit oxidative damage to DNA. The data are expressed as a percent of the control samples that were incubated with l-dopa/ferric iron/H_2O_2 (n = 6 independent replications).

A second series of structure-activity experiments were performed and the results are presented in Table 3. Taurine and isethionic acid (ISA) significantly reduced the formation of 5-OH uracil and 8-OH adenine and appeared to produce similar results overall (Table 3). Hypotaurine (hypotau) and homotaurine (homotau) were much better than taurine at reducing oxidative damage to DNA (Table 3). The exception to this was the ability of hypotaurine to increase the amount of 8-OH guanine nearly 3 fold above the level of damage seen in the absence of taurine analogs.

Table 3. Effects of taurine analogs to inhibit oxidative damage to DNA

Analogs	5-OH Uracil	8-OH Adenine	8-OH Guanine	5-OH Cytosine
None	53 ± 3	31 ± 2	45 ± 3	12 ± 1
Taurine	$24 \pm 3*$	$20 \pm 2*$	37 ± 1	12 ± 1
Hypotau	$9 \pm 1*$	8 ± 1	$218 \pm 18*$	$2 \pm 1*$
Homotau	ND	ND	ND	ND
ISA	$24 \pm 2*$	$20 \pm 2*$	44 ± 2	11 ± 1

Data are expressed at nmole base adduct/mg DNA $\pm$ SEM. *$p<0.01$, N=6 replications
ND= no damaged bases detected. All analogs were tested at 20 mM.

A final aspect of the general protective effect of taurine and taurine analogs was the improvement in DNA recovery after the oxidative insult. Based on the studies presented in Tables 1 and 2 taurine showed a concentration dependent increase in DNA recovery after ethanol extraction. Taurine increased the recovery ($77 \pm 3\%$) of DNA with 5 mM being the threshold concentration at which a significant ($p<0.05$) increase was seen over the damaged condition ($62 \pm 2\%$). The effect of taurine analogs on DNA recovery is presented in Figure 2. Consistent with the data in Table 3

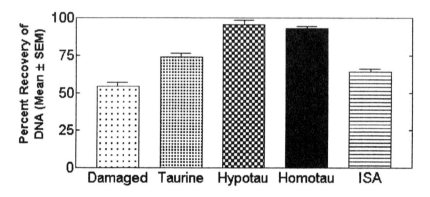

Figure 2. Effect of 20 mM taurine and related analogs on the recovery of total DNA after incubation with l-dopa/ferric iron/H_2O_2. Taurine and the other analogs significantly ($p<0.05$) improved the recovery of DNA.

on oxidative damage to DNA, hypotaurine and homotaurine significantly improved the recovery of DNA. Taurine and isethionic acid also improved the recovery of DNA, but to a lesser extent than homotaurine or hypotaurine. These results suggest taurine and taurine analogs prevented damage to DNA such as strand breaks.

Taurine and taurine analogs (10 mM) were tested for their ability to decrease quinones formed from the oxidation of l-dopa caused by ferric iron and H_2O_2. Quinone formation was measured in the presence and absence of DNA and the results are presented in Figure 3. In the presence of DNA

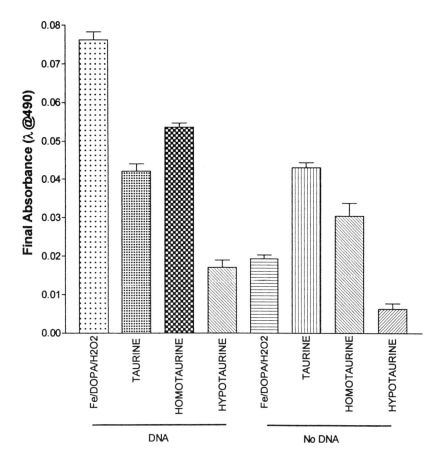

Figure 3. Quinones derived from l-dopa oxidation in the presence and absence of DNA. Taurine and related analogs were tested at 10 mM for their effects on quinone absorbance at 490 nm (n =5).

taurine, hypotaurine and homotaurine all significantly (p<0.05) decreased quinones. Hypotaurine was the most effective and significantly (p<0.01) reduced quinones in the presence or absence of DNA. Previous studies had shown that taurine could significantly reduce detectable quinones generated by incubating l-dopa with ferric iron[18]. The addition of H_2O_2 to l-dopa and ferric iron resulted in Fenton reaction mechanisms which taurine and homotaurine apparently enhanced. Homotaurine alone has been shown to enhance quinone formation from l-dopa in the presence of ferric iron (manuscript submitted). The addition of DNA caused an overall increase in quinone formation, but resulted in taurine and its analogs having inhibitory effects on quinone production.

DISCUSSION

These studies provide the first evidence that taurine can reduce oxidative damage to DNA at concentrations normally found in most mammalian cells. As would be predicted, hypotaurine acted as a better antioxidant and reduced DNA damage and quinone production from ferric iron/H_2O_2-induced l-dopa oxidation. Surprisingly, homotaurine provided even better protection of DNA than hypotaurine even though it does not have the obvious free radical scavenging properties. Both β-alanine and isethionic acid provided similar levels of protection against oxidative damage to DNA as that afforded by taurine. This would appear to suggest that neither the sulfonic acid nor the amino function of taurine is an absolute requirement for DNA protection. Taurine was most effective at reducing the formation of 5-OH uracil and 8-OH adenine. Uracil residues in DNA arise from the spontaneous or chemically induced deamination of cytosine. Interestingly, taurine had no effect on 5-OH cytosine formation. This points out the difficulty in fully appreciating the actions of taurine and taurine analogs on oxidative damage to DNA since our methods only sample a fraction of the plethora of base modifications that occur as the result of DNA oxidation[21].

Taurine and the analogs tested all enhanced DNA recovery after ethanol extraction. In addition to reducing oxidative modification of bases, it is likely that these compounds reduced strand breaks. Taurine has previously been shown to have antimutagenic effects[6-8]. Our results confirm the likelihood that taurine may act to decrease DNA damage mediated by oxidants and alkylating agents.

Taurine has been shown by various methods to produce cytoprotective and antioxidant-like effects[1]. This is despite the fact that taurine is a poor scavenger of oxygen radicals[22]. None the less, at high concentrations taurine can demonstrate some antioxidant properties and also scavenges HOCl via its amino functional group[23-25]. The actions of hypotaurine to inhibit oxidative damage to DNA can easily be attributable to its ability to scavenge hydroxyl radicals[22]. This was also obvious from it ability to inhibit L-dopa oxidation in both the presence and absence of DNA.

The efficacy of homotaurine to protect DNA from oxidative damage would appear to involve mechanisms independent from the direct scavenging of free radicals. Houssier *et al.*[26] studied taurine and other amino acids that act as osmotic agents to protect DNA from cation-induced precipitation. The prevention by organic zwitterions of DNA precipitation by multivalent cations appeared to involve the release of cations from DNA phosphate charges[26]. Metals such as copper and iron are known to bind to DNA and may be involved in the local generation of hydroxyl radical from H_2O_2[27]. The hydroxyl radical is extremely reactive and for oxidative damage to occur the radical must be in close proximity to a base site in DNA. Houssier *et al.*[26] showed that taurine was the most effective organic osmolyte at protecting DNA from precipitation. In fact, increasing the distance between the charged amino group and carboxylic group resulted in a significantly enhanced protection against DNA precipitation due to an increase in the dielectric constant of the solution[26]. It is tempting to speculate that the actions of homotaurine and taurine may be related to decreasing DNA-metal interactions via a mechanism of action similar to that described by Houssier *et al.*[26]. This would explain the ability of homotaurine to be such a good protectant of DNA despite not being a good antioxidant. This could also explain why β-alanine, isethionic acid and taurine provided similar levels of protection due to their similar carbon chain lengths. Clearly, detailed studies are required to evaluate the role that an increase in the dielectric constant may play in taurine actions to protect DNA. It is also likely that the general ability of taurine and related analogs to decrease quinone formation may have contributed to a decrease in DNA damage, but it did not directly predict efficacy.

In general, these studies should be viewed as preliminary since we used calf thymus DNA to model potential *in vivo* interactions of taurine with DNA. We chose iron-stimulated oxidation of L-dopa in the presence of H_2O_2 as our method to induce DNA damage since it was suggested that this type of damage may have relevance for neurodegenerative diseases[12]. Moreover, we had previously shown that taurine could attenuate metal-stimulated quinone formation from catecholamines[18] and predicted that taurine should also prevent this type of oxidative damage to DNA. The fact that taurine is present in the nucleus[10] and has been suggested to modulate nuclear function, further underscores the importance that taurine may serve to stabilize and protect DNA from a number of insults (toxicants, osmotic, oxidative).

Finally, taurine content declines with advanced age in rodents[16]. The free radical theory of aging suggests that increased oxidative stress and a decrease in repair mechanisms and antioxidant defense mechanisms combine to contribute to cellular injury and death. The role of taurine in protecting DNA in tissues such as the brain, heart and skeletal muscle could be especially important given the longevity of nuclear DNA in these tissues. Mitochrondrial DNA is damaged at a much higher rate than nuclear DNA[28]. The role of taurine in the mitochrondria certainly deserves greater attention in regard to any putative functions to protect DNA. A decline in taurine during aging could have adverse effects for many reasons including increased susceptibility to DNA damage.

ACKNOWLEDGMENTS

The authors wish to thank Dr. Miral Dizdaroglu for his kind gift of hydroxylated base standards. The authors also thank Dr. B. Eppler for her technical assistance and L. Mariani for her help with the technical preparation of the manuscript. This work was supported in part with a grant from Taisho Pharmaceutical Co.

REFERENCES

1   Timbrell, J.A., Seabra, V. and Waterfield, C.J., 1995, The in vivo and in vitro protective properties of taurine. Gen. Pharmac. 26: 453-462.
2   Wu, C., Kennedy, D.O., Yano, Y., Otani, S. and Matsui-Yuasa, I., 1999, Thiols and polyamines in the cytoprotective effect of taurine on carbon tetrachloride-induced hepatotoxicity. J. Biochem. Toxicol. 13: 71-76.
3   Wettstein, M. and Haussinger, D., 1997, Cytoprotection by the osmolytes betaine and taurine in ischemia-reoxygenation injury in the perfused rat liver. Hepatology 26: 1560-1566.
4   Redmond, H.P., Wang, J.H., and Bouchier-Hayes, D., 1996, Taurine attenuates nitric oxide- and reactive oxygen intermediate-dependent hepatocyte injury. Arch. Surg. 131 1280-1288.
5   Wang, J.H., Redmond, H.P., Watson, R.W.G., Condron, C and Bouchier-Hayes, D., 1996, The beneficial effect of taurine on the prevention of human endothelial cell death. Shock 6: 331-338.
6   Laidlaw, S.A., Dietrich, M.F., Lamtenzan, M.P., Vargas, H.I., Block, J.B., and Kopple, J.D., 1989, Antimutagenic effects of taurine in bacterial assay system. Cancer Research 49: 6600-6604.
7   Kozumbo, W.J., Agarwal,S., and Koren, H.S., 1992, Breakage and binding of DNA by reaction products of hypchlorous acid with aniline, 1-napthylamine, or 1-naphthol. Toxicol. Appl. Pharmacol. 115: 107-115.
8   Cozzi, R., Ricordy, R., Bartolini, F., Ramadori, L., Perticone, P.,and De Salvia, R., 1995, Taurine and allagic acid: two differently- acting natural antioxidants. Environ. Mol. Mutagens 26: 248-254.
9   Bkaily, G., Jaalouk, D., Hadadd, G., Gros-Louis, N., Simaan, M., Naik, R. and Pothier, P., 1997, Modulation of cytosolic and nuclear Ca2+ and Na+ transport by taurine in heart cells. Mol. Cell Biochem. 170: 1-8.
10 Terauchi, A., Nakazaw,A., Johkura, K., and Usada, N., 1998, Immunohistochemical localization of taurine in various tissues of the mouse. Amino Acids 15: 151-160.
11 Lombardini, J.B.,1998, Increased phosphorylation of specific rat cardiac and retinal proteins in taurine-depleted animals: Isolation and identification of the phospho-proteins. In Taurine 3 (Schaffer, S., Lombardini, J.B. and Huxtable, R.J. eds.) Plenum Press, New York, pp. 441-447.
12 Spencer, J.P.E., Jenner, A., Aruoma, O.I., Evans, P.J., Kaur, H., Dexter, D.T., Jenner, P., Lees, A.J., Mardsen, D.C. and Halliwell, B., 1994, Intense oxidative DNA damage promoted by L-DOPA and its metabolites: Implications for neurodegenerative disease. FEBS Letters 353: 246-250.
13 Morin, B., Davies, M.J. and Dean, R.T., 1998, The protein oxidation product 3,4-dihydroyphenylalanine (DOPA) mediates oxidative DNA damage. Biochem. J. 330: 1059-1067.

14   Snyder, R.D. and Friedman, M.B., 1998, Enhancement of cytotoxicity and clastogenicity of L-DOPA and dopamine by manganese and copper. *Mutation Res.* 405: 1-8.

15   Beckman, K.B. and Ames, B.N., 1998, The free radical theory of aging matures. *Physiol. Rev.* 78: 547-581.

16   Dawson, R., Jr., Liu, S., Eppler, B.and Patterson, T., 1999, Effects of dietary taurine supplementation or deprivation in aged male Fischer 344 rats. *Mech. Age. Dev.* 107: 73-91.

17   Pomara, N., Singh, R., Deptula, D., Chou, J.C.-Y., Schwartz, M.B. and LeWitt, P.A., 1992, Glutamate and other CSF amino acids in Alzheimer's disease. *Am. J. Psychiatry* 149: 251-254.

18   Dawson, R, Jr., Tang, E., Shih, D., Hern, H., Baker, D., and Eppler, B.,1998, Taurine inhibition of iron-stimulated catecholamine oxidation. In *Taurine 3* (Schaffer, S., Lombardini, J.B. and Huxtable, R., eds.), Plenum Press, New York, pp. 155-162.

19   Kaur, H. and Halliwell, B., 1996, Measurement of oxidized and methylated DNA bases by HPLC with electrochemical detection. *Biochem. J.* 318: 21-23.

20   Herbert, K. E., Evans, M.D., Finnegan, M.T.V., Farooq, S., Mistry, N., Podmore, I.D., Farmer, P., and Lunec, J. 1996, A novel HPLC procedure for the analysis of 8-oxoguanine in DNA. *Free Rad. Biol. Med.* 20: 467-473.

21   Marnett, L.J. and Burcham, P.C., 1993, Endogenous DNA adducts: Potential and paradox. *Chem. Res. Toxicol.* 6: 771-785.

22   Shi, X., Flynn, D.C., Porter, D.W., Leonard, S.S., Vallyathan, V., and Castronova, V., 1997, Efficacy of taurine based compounds as hydroxyl radical scavengers in silica induced peroxidation. *Ann. Clin. Lab.* 27: 365-374.

23   Alvarez, J. and Storey, B.T., 1983, Taurine, hypotaurine, epinephrine and albumin inhibit lipid peroxidation in rabbit spermatozoa and protect against loss of motility. *Bio. Reprod.* 29: 548-555.

24   Huxtable, R.J., 1992, Physiologic actions of taurine. *Physiol. Rev.* 72: 101-163.

25   Pasantes-Morales, H., Wright, C.E. and Gaulle, G.E.,1985,   Taurine protection of lymphoblastoid cells from iron-ascorbate induced damage. *Biochem. Pharmacol.* 34: 2205-2207.

26   Houssier, C., Gilles, R. and Flock, S., 1997, Effects of compensatory solutes on DNA and chromatin structural organization in solution. *Comp. Biochem. Physiol.*: 117A, 313-318.

27 Pryor, W.A.,1988, Why is the hydroxyl radical the only that commonly adds to DNA? *Free Rad. Biol. Med.* 4: 219-223.

28 Cortopassi, G.A. and Wong, A., 1999, Mitochondria in organismal aging and degeneration. *Biochim. Biophys. Acta* 1410: 183-193.

TYROSINE-HYDROXYLASE IMMUNO-REACTIVE CELLS IN THE RAT STRIATUM FOLLOWING TREATMENT WITH MPP⁺

M. B. O'Byrne, [1]J. P. Bolam, [1]J. J. Hanley and K. F. Tipton

Department of Biochemistry, Trinity College Dublin 2, Ireland
[1] *MRC Anatomical Neuropharmacology Unit, Mansfield Road, Oxford OX1 3TH, UK*

Abstract: Tyrosine Hydroxylase is the rate-limiting enzyme in the synthesis of dopamine, and as such, it is widely used as a marker of dopaminergic cells. Within the basal ganglia, the dopaminergic cells are located in the substantia nigra pars compacta, and project to the striatum. It is this pathway which degenerates during Parkinson's disease. The data presented here illustrate examples of tyrosine-hydroxylase immunoreactive cells in striatum following intrastriatal injection with the neurotoxin MPP⁺. We further show by electron microscopy that these cells are, in fact, neurons and that they possess ultrastructural features of interneurons.

INTRODUCTION

The basal ganglia comprise a number of sub-cortical nuclei which are involved in a variety of functions, such as cognition, motor programme formation and interfacing between motor and limbic systems. A central function is the control of movement in which the dopaminergic projection from the substantia nigra pars compacta (SNc) to the striatum is intimately involved. The cells of the SNc may be identified by their immunoreactivity

Taurine 4, edited by Della Corte *et al.*
Kluwer Academic / Plenum Publishers, New York, 2000.

for tyrosine hydroxylase (TH), the rate-controlling enzyme in the synthesis of dopamine. The neurotoxin 1-methyl-4-phenylpyridinium (MPP[+]) is known to cause selective degeneration of this nigrostriatal pathway in higher species (see[1]). Recently we have shown taurine to protect dopaminergic neurons against the toxicity of MPP[+] in rat brain coronal slices[2]. As part of a study aimed at determining the whether such protection by taurine could also be demonstrated following stereotaxic administration of MPP[+] into the rat striatum, we observed a small population of striatal cells that were resistant to the toxicity of MPP[+]. These cells were shown to be immunoreactive for tyrosine hydroxylase and as such may contain dopamine. Further studies of some characteristics of this novel population of cells are also presented.

MATERIALS AND METHODS

Stereotaxic administration of MPP[+]

Female Wistar rats (180 g) were anaesthetised with 2.7 ml/kg of a mixture of Hypnoval and Hypnorm in distilled water (1:1:2). Animals received injections of MPP[+] at doses of 0.5 nmol, 5.0 nmol, 30 nmol and 90 nmol to the right striatum, and sham injections of 0.05 M PBS (phosphate-buffered saline) in the left striatum at the co-ordinates AP 9, L 2.6 and DV 5.2 with reference to Bregma, taken from the stereotaxic atlas of Paxinos and Watson[3]. The nose-bar was set at -3.3 mm. Doses were given in a total volume of 1 μl over 2.5 min. Animals were left for a survival period of six days, after which they were anaesthetised with pentobarbitone (Sagatal, Rhone-Poulenc) and perfused with fixative, containing 3 % paraformaldehyde and 0.1 % glutaraldehyde in 0.1 M phosphate buffer. Tissue was then prepared for immunochemical staining. All procedures were carried out according to the Animals (Scientific Procedures) Act, 1986 (UK).

Pre - embedding immunocytochemistry

Sections of the striatum (70 μm) were cut using a vibratome. They were collected and washed several times in 0.1 M PBS (phosphate-buffered saline), pH 7.4. Sections were incubated, in a 1:1000 dilution of anti-TH primary antibody together with 2 % normal serum in 0.05 % Triton X-100, overnight at room temperature. During incubations of sections for electron microscopy the Triton was omitted. Sections were then rinsed 3 x 10 min in PBS at room temperature and incubated in biotinylated secondary antibody for a minimum of one hour. The avidin-biotin-peroxidase complex (ABC) was prepared from a kit (Vectastain, Vector laboratories) according to the manufacturer's instructions and components were allowed to react for 30 min. Sections were then incubated in ABC for one hour. Positive immunostaining was visualised using 3,3'-diaminobenzidine (DAB) as chromogen. A number of control sections were also prepared, in which primary antibody was omitted and the sections then treated as described above; see also[4].

Preparation of tissue for electron microscopy

Sections were treated with 1 % osmium tetroxide, dehydrated and infiltrated with Ducupan resin that was polymerized at 60 °C. Cells and areas of interest were identified and photographed. Sections of 60 nm thickness were taken using the ultra-microtome (Reichert-Jung, Ultracut-E) and collected on Pioloform-coated grids, treated with lead citrate and then examined under the electron microscope.

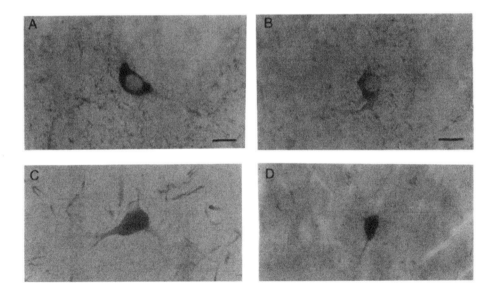

Figure 1. Appearance of TH- immunoreactive MPP^+-resistant neurons in the striatum
The photomicrographs A - D show a sample of the small population of neurons showing immunoreactivity for TH (tyrosine hydroxylase) observed in the striatum after treatment with MPP^+. Perikarya and axons of nigrostriatal neurons can be distinguished in most cases. In D, the section was treated with osmium tetroxide, to enhance contrast. Scale bars = 8 μm

Figure 2. Electron microscopy of TH-immunoreactive cells from MPP⁺-treated striata.

A-C: Light micrographs of two TH-immunoreactive neurons at increasing magnifications and different focal planes. These cells are shown in elctron micrographs D and E.: the arrow indicates the dendritic process originating from the second cell.

D : Electron micrograph of the perikaryon of the upper of the two TH-immunoreactive neurons. Note the indentation of the nuclear membrane (arrow) which is characteristic of striatal interneurons.

E : Electron micrograph of the dendrite of the lower of the two TH-immunoreactive neurons. The denrtite is postsynaptic to a bouton (arrow) that forms an asymmetric synapse.

Scales (μm): A, 20; B, 12.5; C, 5.5; D, 0.83; E, 0.35.

RESULTS

Following MPP$^+$ administration, a small population of surviving cells showing TH-like immunoreactivity (TH-LI) was observed in the striatum. A smaller number of such cells was also observed on t?.e PBS-injected side. The cells were generally bipolar and small in size with perikaryon measuring ~10 μm in diameter, as shown in Figure 1.

Electron microscopy (Fig. 2) revealed these cells had indented nuclei. Synapses were apparent on their dendrites indicating them to be neurons. Fig. 2A shows two TH-labelled neurons. They are surrounded by labelled striatal neuropil and were located in the more ventro-lateral striatum. Both neurons were strongly immunoreactive for TH and the dendritic process originating from one of the cells is clearly visible at the higher magnification of Fig. 2B. The two labelled cells, one overlying the other which is in a different plane of focus, are more clearly discernible at the magnification of Fig. 2C and the nucleus of one of the neurons can be seen, in Fig. 2D, to be indented which is a characteristic of striatal interneurons. The electron micrograph in Fig. 2E shows a dendritic process originating from the lower labelled cell and is postsynaptic to a bouton forming an asymmetric synapse.

DISCUSSION

MPP$^+$ injection resulted in extensive neuronal destruction. The appearance of a small population of cells showing TH-like immunoreactivity in the MPP$^+$-treated striatum may relate to the "switching on" of the gene for TH. Tashiro et al.[5] reported an, apparently similar, resistant population of cells following 6-OHDA (6-hydroxydopamine) or electrothermic lesions to the SNc. A resistant population of cells was observed in another study of 6-OHDA lesioned striatum[6], but these cells were reported to be immunopositive for aromatic-amino-acid decarboxylase (AADC) but not for TH and it was suggested that they might provide a site for exogenous conversion of L-DOPA to dopamine. Recently it has been reported that the striatal TH-immunoreactive cells that appear following 6-OHDA or methamphetamine administration differ in size, morphology and location from those that are immunopositive for AADC or dopamine[7]. We cannot, as yet, say whether the resistant cells identified in the present work are catecholaminergic as we do not know whether they also express other enzymes in the synthetic pathway of the catecholamines. However, electron microscopy revealed these cells to possess charactristics of striatal interneurons. They are thus unlikely to represent the spiny projection neurons of the striatum (see[8]). Dual labelling of cells in the striatum ruled out the co-localisation of TH with those interneurons containing neuropeptide Y (NP-Y)[9]. Dual-labelling studies with TH and other interneuron-localized peptides, such as cholecystokinin, might help further to categorise this novel population of neurons. The demonstration that they also appear, albeit to a lesser extent, following the control injections of PBS in the striatum suggests that, in addition to dopamine lesions, the physical trauma of the injection may also induce their appearance. Investigations into the factors underlying their toxin resistance could be of value in understanding the mechanisms and therapy of neurodegenerative diseases.

ACKNOWLEDGEMENTS

This study was aided by grants from the EU Biomed 1 programme (BMH1-CT94-1402) and the Health Research Board of Ireland.

REFERENCES

1. Tipton, K.F. and Singer, T.P., 1993, Advances in our understanding of the mechanisms of the neurotoxicity of MPTP and related compounds. *J. Neurochem.* **61**, 1191-1206.
2. O'Byrne, M.B. and Tipton, K.F., 2000, Taurine-induced attenuation of MPP⁺ neurotoxicity in vitro. *J. Neurochem.* in press.
3. Paxinos, G and Watson C., 1987 *The rat brain in stereotaxic co-ordinates.* Academic Press, London.
4. Bolam, J.P, 1992, Editor: Experimental Neuroanatomy: A Practical Approach. (Series editors: D. Rickwood and B.D. Hames). Oxford University Press (IRL), Oxford, UK.
5. Tashiro, Y., Kaneko, T., Sugimoto, T., Nagatsu, I., Kikuchi, H. and Mizuno, N., 1989, Striatal neurons with aromatic L-amino acid decarboxylase-like immunoreactivity in the rat. *Neurosci. Lett.* **100**, 29-34.
6. Mura, A., Jackson, D., Manley, S.M., Young, S.J. and Groves, P.M. (1995) Aromatic L-amino acid decarboxylase immunoreactive cells in the rat striatum: a possible site for the conversion of exogenous L-DOPA to dopamine. *Brain Res.* **704**, 51-60.
7. Meredith, G.E., Farrell, T., Kellaghan, P., Tan, Y., Zahm, D.S. and Totterdell, S., 1999, Immunocytochemical characterization of catecholaminergic neurons in the rat striatum following dopamine-depleting lesions. *Eur. J. Neurosci.* **11**, 3585-96.
8. Smith, A.D., and Bolam, J.P., 1990, The neural network of the basal ganglia as revealed by the study of synaptic connections of identified neurones. *TINS,* **13**: 259-285.
9. Aoki, C. and Pickel, V.M., 1988, Neuropeptide-Y-containing neurons in the striatum: ultrastructure and cellular relations with tyrosine-hydroxylase-containing terminals and with astrocytes. *Brain Res.* **459**, 205-225.

INVESTIGATION OF THE THERAPEUTIC EFFICACY OF A TAURINE ANALOGUE DURING THE INITIAL STAGES OF ETHANOL DETOXIFICATION: PRELIMINARY STUDIES IN CHRONIC ALCOHOL ABUSERS

R. J. Ward, J. Martinez[1], D. Ball[1], E. J. Marshall[1] and P. De Witte
Biologie du Comportement, University of Louvain, 1348 Louvain-la-Neuve, Belgium
[1]*Addiction Science Centre, Institute of Psychiatry, London, UK*

INTRODUCTION

Since 1979[1] it has been recognised that taurine, together with some of its precursors, e.g. cysteic acid, and major metabolic products, e.g. taurocholic acid, is able to alter some of ethanol-elicited responses; taurocholic acid decreases ethanol preference, while cysteic acid diminishes circulating ethanol concentrations.

Homotaurine, which is formed by the addition of a methyl group to taurine, Figure 1, has also been identified as an agonist of GABA receptors[2] and will reduce ethanol consumption in spontaneously drinking rats. Further chemical modification of this compound by the addition of an acetyl group, and then calcium bridging of two N-acetyl homotaurine dimers results in the formation of calcium acetyl homotaurine (Fig. 1) which has also been demonstrated to significantly reduces voluntary intake of ethanol, an effect which was inhibited when the $GABA_A$ receptors were blocked by the

Taurine 4, edited by Della Corte *et al.*
Kluwer Academic / Plenum Publishers, New York, 2000.

antagonist bicuculline. Substitution of the cationic ion by sodium totally eliminated the ability of the compound to reduce voluntary alcohol intake[2]. The role of the NMDA receptor in the action of calcium acetyl homotaurinate remains undefined, some studies indicating that acamprosate binds to specific spermidine-senstive sites that modulates the NMDA receptor in a complex manner[3], as well as reducing Ca^{++} fluxes through voltage-operated channels[4]. This drug has now undergone successful clinical trials in Europe and shown to reduce craving in detoxified alcohol abusers such that abstinence is maintained[5,6].

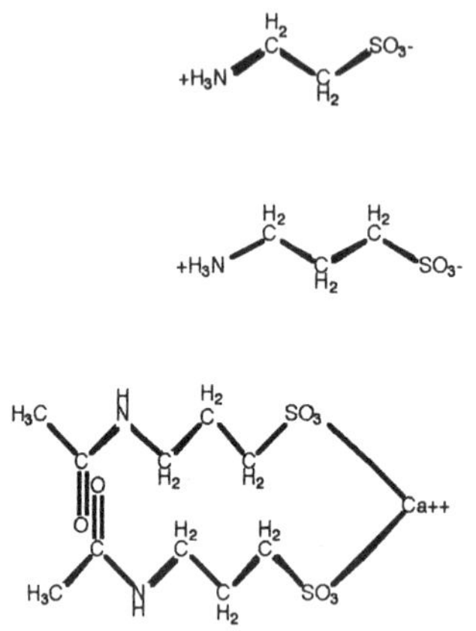

Figure 1. Chemical structure of taurine and some of its analogues

Since calcium acetyl homotaurine may possibly act as an agonist of GABA receptors and an antagonist of NMDA receptors, it was of interest to ascertain whether its administration during the initial stages of detoxification of chronic alcohol abusers, i.e. during the first 7 days of detoxification, might diminish some of the adverse side effects which occur during ethanol withdrawal.

MATERIALS AND METHODS

Studies of chronic alcohol abusers during detoxification

Blood specimens were collected from alcohol abusers n = 32 admitted to the National Alcohol Inpatient Unit at the Royal Bethlum Hospital, Beckenham Kent, UK. Ethical permission for these studies was given by the Bethlum & Maudesley NHS Health Trust Ethical Committee. All patients fulfilled ICD-10 / DSM IV criteria for alcohol dependence. A blood specimen was taken at

the time of the planned admission for the assay of alcohol levels, biochemical parameters as well as haematological indices. Heparinised blood specimens were taken both at the time of admission and 7 days after detoxification for the analysis of taurine and plasma amino acids. Information on both the maximal daily dose and total dose of chlordiazepoxide administered during the alcohol detoxification stage was also collected. In approximately one third of these patients, n = 10, acamprosate, 666 mg x 3 x/day was administered during this detoxification period. In these patients the levels of acamprosate in the plasma after 7 days of administration was assayed by gas chromatography/mass spectrometry[7].
For the analysis of the data, the patients were divided into three groups according to whether 1. Glutamic acid increased after 7 days of detoxification 2. Glutamic acid decreased after 7 days of detoxification and 3. The patients had received acamprosate during the seven days of detoxification.

RESULTS

Patient studies

After division of the patients into the three groups, there were no statistically significant differences between any of the clinical and biochemical parameters assayed even though Group 3 showed higher ethanol intake/day and had greater liver disease as exemplified by the elevated gamma glutamyl transferase, γGT.

	Group 1	*Group 2*	*Group 3*
SADQ	38.4±9.5	38.3±8.9	40±17
Ethanol intake g/day	261±46	284±134	371±190
γGt IU/l	173±325	245±196	443±339

Results are mean ± standard deviation

The plasma amino acids in each of these three groups of chronic abusers of alcohol before detoxification were within their appropriate reference ranges, apart from glutamic acid, which was significantly raised and glutamine which was significantly decreased. There was a strong correlation, between the levels of this excitatory amino acid and the activity of the liver enzyme, γGT, (where the results were available) in each of the groups, r = 0.96; P <0.01; r = 0.81, P <0.01; r = 0.99, P <0.01, respectively, Figure 2.

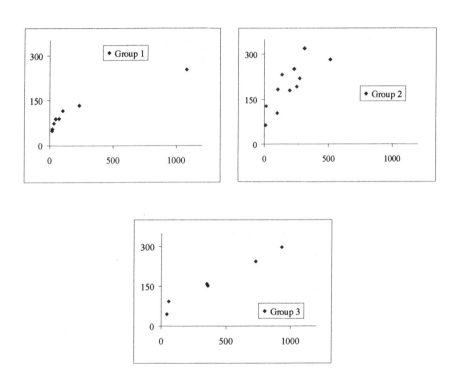

Figure 2. Correlation between glutamic acid (Y axis) and γGT (X axis) in the plasma of the three groups of patients.

After 7 days of detoxification, in group 1, n=10, the concentration of taurine had increased significantly in all patients, many in excess of the quoted reference range, (50-100 μmol/l), Figure 3. In addition, plasma glutamic acid content had increased in parallel. In group 2, n = 12 there was little alteration in the plasma concentrations of either taurine or glutamic acid. In Group 3, n = 10, the patients who had received Acamprosate for 7 days, showed no change in either plasma taurine or glutamic acid concentrations, Figure 3.

Chlordiazepoxide is administered routinely to alcohol abusers during the initial stages of detoxification to lessen many of the unpleasant side effects e.g. tremor, hallucinations, sleeplessness, which occur during these early stages of withdrawal. The total dosage given to the three different groups of alcohol abusers during detoxification were 520 ± 145, 590 ± 189 and 648 ±

213, respectively. The mean level of acamprosate in the plasma of the patients administered this drug for 7 days, i.e. Group 3 was 170 ± 20 ng/ml.

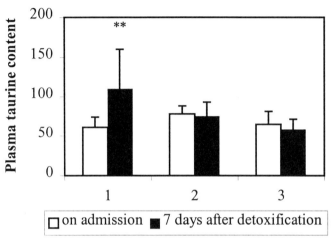

Changes in plasma taurine in three groups of patients after ethanol detoxification.

Figure 3. Changes in mean circulating taurine content, $\pm$ standard deviation in chronic alcohol abusers after 7 days of detoxification ** P<0.01

The administration of acamprosate during these early stages of detoxification had little effect upon the total dosage of chlordiazepoxide administered.

DISCUSSION

In these present studies we have investigated whether the administration of acamprosate during the initial stages of alcohol detoxification in chronic abusers of alcohol might diminish many of the adverse effects, which occur during the initial stages of detoxification in man. During this time, a decreasing dosage of chlordiazepoxide is routinely administered as a sedative to preclude the incidence of a wide number of unpleasant side effects. A diminishment in the dosage of chlordiazepoxide administered during this period when combined with acamprosate might have indicated the beneficial effect of this taurine analogue during the early stages of detoxification. However no change in the chlordiazepoxide dosage of was given, in fact, a marginal increase in the amount of this sedative was noted. Interestingly there was a positive correlation between the maximum dosage of chlordiazepoxide administered and the circulating acamprosate levels at 7 days. The relevance of this finding remains undefined.

Some signs of physical withdrawal such as hyperactivity and hyper-reactivity are reduced by acamprosate[8,9]. Acamprosate when administered to an animal model of chronic alcoholisation is able to reduce the enhanced L-glutamate release within the nucleus accumbens during the initial stages of withdrawal[10], although the drug was given throughout the alcoholisation period which may have influenced NMDA and $GABA_A$ receptor expression. In these present studies acamprosate was only administered to man during the initial stages of withdrawal.

In our previous studies[11] the increased circulating levels of both glutamic acid and taurine between Day 0 and Day 7 were indicative of a significant lower scores on the withdrawal scale after 7 days, CIWA Ar $p<0.01$. The explanation for the increases in these two amino acids in such patients was unclear since with the reduction of alcohol consumption it might have been expected that glutamic acid would be stabilised or even reduced as the alcohol-induced liver disturbances were reduced. The elevation of taurine in detoxifying alcoholics has also been confirmed in later studies[12]; however what biological significance this may have remains to be identified. Acamprosate administration had little effect upon circulating levels of either taurine or glutamic acid. It has been stated that acamprosate is not metabolised, remains inert, and is excreted unchanged. The lack of change in taurine levels, despite the high dose of acamprosate administered, i.e. in excess of 2 g of acamprosate / day, might indirectly confirm the lack of metabolism of this taurine analogue.

The high concentration of taurine in the brain i.e. mM content, would suggest that taurine plays many important biological roles in the brain. Acute ethanol administration will alter taurine homeostasis in various brain regions, identified by a transitory elevation in extracellular taurine content, although its etiology, possibly osmolarity changes[13,14], or alteration in Ca^{++} flux[15] remains unresolved. The exact relevance of circulating taurine content remains unknown. Plasma glutamic acid does not reflect changes, which occur in the brain neurotransmitters. The plasma glutamic acid content clearly correlated with the degree of liver damage as emphasised by the positive correlations between these two parameters in each of the patient groups investigated.

Taurine analogues have shown clinical efficacy in altering many of ethanol-induced behavioural effects in animals and the current use of acamprosate as a drug showing clinical efficacy in the treatment of alcohol abusers after detoxification has been proven. Further studies as to its possible use during the early stages of detoxification are warranted.

ACKNOWLEDGEMENTS

This work was supported by the Fonds de la Recherche Scientifique et Médicale (1997-2000), l'Institut de Recherches Economiques sur les Boissons (IREB) and sponsored by LIPHA.

REFERENCES

1. Messiha F.S.., 1979, *Brain Res Bull* **4**:603-607.
2. Boismare F., Daoust M., Moore N., Saligaut C., Lhuintre J.P., Chretien P. and Durlach J., 1984, *Pharmac Biochem Behav* **21**:787-789 1984.
3. Naassila M., Hammoumi S., Jegrand E., Durbin P. and Daoust M., 1998, *Alcohol Clin Exp Res* **22**:802-809.
4. Spanagel R. and Zieglgansberger W., 1997, *Trends Pharmacol Sci* **18**:54-59 1997
5. Saas H., Soyka M., Mann K., and Zieglgansberger W., 1996, *Arch Gen Psych* **53**:673-680.
6. Whitworth A, 1996, *Lancet* **347**:1438-1442.
7. Girault J., Gobin P. and Fourtillan J.B., 1990, *J Chromatog* **530**:295-305.
8. Gewiss M., Heidebreder C., Opsomer L., Durbin Ph. and De Witte Ph., 1991, *Alcohol Alcoholism* **26**:129-137.
9. Spanagel R., Putzke J., Sterfferl A., Schobitz B. and Zieglgansberger W., 1996, *Eur J Pharmacol* **305**:45-50.
10. Dahchour A. & De Witte Ph., 1999, *Alcohol* **18**:77-81.
11. Ward R.J., Martinez J., Ball D., Marshall E. and De Witte Ph., 1999, *Neurosci Res Commun* **24**:41-49.
12. Badaway A. *et al.*, 1998 *Alcohol Alcohoisml* **34**:616-625.
13. Dahchour A., Quertemont E. and De Witte Ph., 1994, *Alcohol Alcoholism* **29**:485-487.
14. Dahchour A., Quertemont E. and De Witte Ph., 1996, *Brain Res* **735**:9-19.
15. Schaffer S.W., Azuna J. and Madura J.D., *Amino Acids* **8**:231-246.

METABOLISM OF TAURINE TO SULPHOACETALDEHYDE DURING OXIDATIVE STRESS

[1]Colm M. Cunningham and [2]Keith F. Tipton
[1]CNS Inflammation Group, School of Biological Sciences, University of Southampton, Southampton, UK
[2]Department of Biochemistry, Trinity College Dublin, Dublin 2, Republic of Ireland.

INTRODUCTION

Taurine is generally regarded as not being catabolized by mammalian systems. The formation of isethionic acid has been reported in dog heart[1] and in rat brain[2]. The latter study demonstrated the *in vivo* flow of radioactively labelled [^{35}S] from cysteine through cysteine sulphinate, and hypotaurine to taurine, and showed formation of labelled isethionate from [^{35}S]taurine at a very slow rate. Sulphoacetaldehyde was proposed as the most likely intermediate. Sturman *et al.*[3] and Urquhart *et al.*[4] have also demonstrated the presence of isethionate in human urine. Cavallini *et al.*[5] have suggested that this isethionate originates from the same precursor as taurine rather than from taurine itself. (this study demonstrated the formation of isethionate from mercaptoethanol) and Fellman *et al*[6] have suggested that mammalian formation of isethionate from taurine is effected by gut bacteria While these studies provide alternative routes of isethionate formation neither can explain the appearance of radioactively labelled cysteine sulphinate and isethionate in the brain. We have examined the catabolism of

Taurine 4, edited by Della Corte *et al.*
Kluwer Academic / Plenum Publishers, New York, 2000.

taurine by detection of the proposed intermediate sulphoacetaldehyde using the HPLC assay developed for this purpose[7], and by examining the various candidate enzymes for the reduction/oxidation of sulphoacetaldehyde.

METHODS

HPLC assay for sulphoacetaldehyde

The assay depended on the fluorimetric detection of the adduct formed by reaction of sulphoacetaldehyde with 2-diphenylacetyl-1,3-indandione-1-hydrazone (DIH) (30 mg/100 ml) for 90 minutes at 50°C, and seperation of this product from the unreacted derivatizing agent by reversed phase HPLC. This was based on a similar assay used for the detection of acetaldehyde[8]. A Waters C18 μBondapak reverse-phase column (3.9 x 300 mm) was used and this was initially eluted isocratically with 63 % acetonitrile, 27 % H_2O, 10 % 0.1 M ammonium acetate (v/v/v) at a final pH of 6.5. This composition was then varied depending on the seperation required Excitation and emission wavelengths were 415 nm and 525 nm, respectively.

Synthesis of chlorotaurine

Chlorotaurine was synthesized freshly each day by reaction of taurine and NaOCl in a 1:1 molar ratio. Chlorotaurine formation and hypochlorite disappearance were monitored by UV absorbance, absorbance maxima at 252 nm (ε=429 $M^{-1}.cm^{-1}$), and 291 nm (ε=142 $M^{-1}. cm^{-1}$), respectively[9]. The first order decay of chlorotaurine to sulphoacetaldehyde could then be measured by HPLC after incubation of various concentrations of chlorotaurine (0 to 3.65 mM) in 0.1 M phosphate buffer for various times at 37°C, first destroying chlorotaurine by reaction with Nbs⁻ (37 mM, 30 μl).

Isolation and activation of human neutrophils

Neutrophils, isolated using Ficoll (Histopaque 1077), were activated using a 10% latex bead suspension[9]. Reactive oxidant production by the neutrophils (5 x 10^6 cells/ml) was measured by following chemiluminescence at 37°C. A latex bead suspension was used as stimulus to induce phagosome formation[9]. This was done in the presence or absence of taurine to demonstrate the decreased chemiluminescent response upon formation of chlorotaurine. The same reaction conditions were used to investigate sulphoacetaldehyde formation (by HPLC) upon activation in the presence of taurine.

Assays of sulphoacetaldehyde reduction

The HPLC assay was used once again to determine the depletion of sulphoacetaldehyde by rat liver homogenates, in the presence or absence of NADPH/NADH/NADP⁺/NAD⁺ and various aldehyde reductase inhibitors. Aldehyde reducaste activity could also be measured spectrophotometrically by monitoring the oxidation of NADPH/NADH at 340 nm.

RESULTS

Taurine was not deaminated by either monoamine oxidase A or B, by the membrane-bound or soluble forms of semicarbazide sensitive amine oxidase or by any transaminase activity present in rat liver homogenates. Freshly synthesized chlorotaurine was found to form sulphoacetaldehyde by first-order decay, with a rate constant of $9.9 \pm 0.5 \times 10^{-4}.h^{-1}$; half-life approx. 29 days. Incubation with rat liver homogenates (1 mg/ml) resulted in a 30 fold acceleration of sulphoacetaldehyde formation. Catalysis was proportional to protein concentration and hyperbolic with chlorotaurine concentration (Figs. 1 a & b). It was denaturable and was not lost upon dialysis. Furthermore, activation of human neutrophils with latex beads resulted in a marked oxidative burst, as measured by chemiluminescence, which was decreased in the presence of taurine. Similar activation conditions resulted in the time-dependent formation of sulphoacetaldehyde in the presence of exogenous taurine (Fig. 2). Sulphoacetaldehyde was not formed in the absence of taurine or in the absence of latex beads.

Incubation of sulphoacetaldehyde with rat liver homogenates and NADPH resulted in oxidation of the nicotinamide cofactor and depletion of sulphoacetaldehyde (Fig. 3), consistent with aldehyde reductase activity. NADPH could not be substituted by NADH. This depletion could be inhibited by the general aldehyde reductase inhibitor quercetin but not by the high Km aldehyde reductase inhibitor valproate (Fig. 4).

DISCUSSION

Despite claims that mammalian systems cannot metabolize taurine, the results presented here demonstrate at least one potential route of sulphoacetaldehyde and isethionate formation. This suggests a scenario in which chlorotaurine may be formed as a result of the oxidative burst in activated neutrophils. Although chlorotaurine is generally reported to be a long-lived oxidant its stability in vivo has not been studied and thus little is know of its in vivo stability. Our findings suggest the existence of an enzyme which can catalyse the formation of sulphoacetaldehyde from chlorotaurine, and the detection of sulphoacetaldehyde in the hours after in vitro activation of neutrophils supports the existence of such an enzyme. Furthermore we have demonstrated the ability of one or more members of the aldehyde reductase family to accept sulphoacetaldehyde as a substrate, and thus have provided a possible link to previous reports of isethionate generation in mammalian tissues.

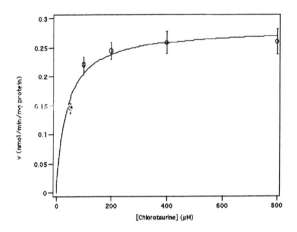

Figure 1. a) Rate of sulphoacetaldehyde formation in rat liver homogenates versus chlorotaurine concentration.

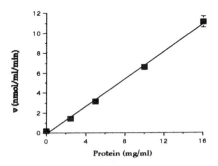

Figure 1. b) Rate of sulphoacetaldehyde formation in rat liver homogenates versus protein concentration

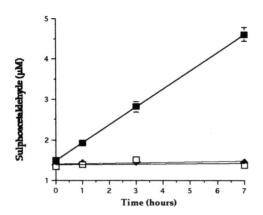

Figure 2. Formation of sulphoacetaldehyde by latex bead-activated human neutrophils *in vitro*, in the presence (■) or absence (□) of taurine.

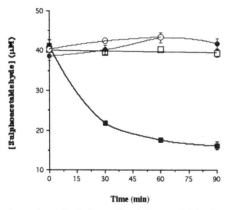

Figure 3. Time-dependent depletion of sulphoacetaldehyde (40μM) in the presence of phosphate buffer (♦), in rat liver homogenate (2mg/ml) containing NADPH (■) or NADH (○) and in the presence of NADPH and denatured homogenate (□).

As the reduction of sulphoacetaldehyde has a high K_m and low V_{max}, the detection of sulphoacetaldehyde may be a good assay for oxidative stress resulting from tissue infiltration by activated neutrophils. As neutrophil infiltration is known to occur in the brain after traumatic brain injury, stroke and infection, some of the neuroprotective functions attributed to taurine may result from sequestration of myeloperoxidase-generated hypochlorite.

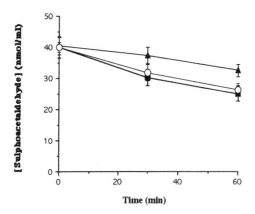

Figure 4. Time-dependent depletion of sulphoacetaldehyde by rat liver homogenates (1 mg/ml) in the presence of 150 µM NADPH alone (■) or in addition to quercetin (20µM, ▲) or valproate (2mM, ○).

ACKNOWLEDGEMENTS

Supported by EU BIOMED (BMH1 CT-1402) and COST (D8/019/98).

REFERENCES

1. Welty J.D., Read, W.O. and Shaw E.H., 1962, *J. Biol .Chem.* **237**: 1160-1161.
2. Peck E.J. and Awapara J., 1967, *Biochim. Biophys. Acta.* **141**: 499-506.
3. Sturman J.A., Hepner G.W., Hofmann A.F. and Thomas P.J., 1975, *J. Nutr.* **105**: 1206-1214.
4. Urquhart N., Perry T.L., Hansen S. and Kennedy J., 1974, *J. Neurochem.* **22**: 871-872.
5. Cavallini D., Dupre S., Antonucci A. and De Marco C., 1978, In: *Taurine and neurological disorders*, pp 29-34. Eds A. Barbeau and R.J. Huxtable. Raven press: New York
6. Fellman J.H., Roth E.S. and Fujita T.S., 1978, In: *Taurine and neurological disorders*, eds Barbeau A. and Huxtable R.J., pp. 19-24, Raven Press, N.Y. .
7. Cunningham C., Tipton K.F. and Dixon H.B.F., 1998, *Biochem. J.* **330**: 933-937.
8. Rideout J.M., Chang K.M. and Peters T.J. , 1986, *Clin. Chim. Acta.* **161**: 29-35.
9. Thomas E.L and Learn D.B., 1988, In: *Peroxidases in Chemistry and Biology,* **2** (Everse, J., Everse, K.E. and Grisham, M.B.eds.), pp. 83-105, CRC press, Boca Raton, Florida
10. Desjardins H., Huber L.A., Parton R.G. and Griffiths G., 1994, *J. Cell. Biol.* **124**: 677-688.
11. Shimamoto, G. and Berk, R.S. (1979) *Biochim. Biophys. Acta.* 569, **287-292**.

TAURINE CHLORAMINE ATTENUATES THE HYDROLYTIC ACTIVITY OF MATRIX METALLOPROTEINASE-9 IN LPS-ACTIVATED MURINE PERITONEAL MACROPHAGES

Eunkyue Park[1], Michael R Quinn[2] and Georgia Schuller-Levis[1]

Departments of [1] Immunology and [2] Developmental Biochemistry, NYS Institute for Basic Research in Developmental Disabilities, Staten Island, NY 10314

INTRODUCTION

Matrix metalloproteinases (MMPs) are a family of zinc- and calcium-dependent endopeptidases capable of proteolytically degrading many of the components of extracellular matrix[6,8]. MMPs are produced by fibroblasts and endothelial cells as well as inflammatory cells, such as macrophages, lymphocytes, neutrophils and eosinophils[2,7,8,9,19]. They are involved in physiological processes such as development, angiogenesis and wound healing. Under pathological conditions, MMPs are involved in tumor invasion and inflammation by transmigration of leukocytes or tumor cells through basement membrane[2,7,10,17]. MMP-9, the focus of this study, is collagenase/gelatinase that cleaves basement membrane collagen type IV and V[3,17,19,20]. MMP-9 helps transmigration of various leukocytes through blood vessels to sites of inflammation [2,7]. Taurine chloramine (Tau-Cl) is produced from taurine by halide-dependent myeloperoxidase when polymorphonuclear leukocytes are activated[12,14]. Previously we

Taurine 4, edited by Della Corte *et al.*
Kluwer Academic / Plenum Publishers, New York, 2000.

demonstrated that Tau-Cl inhibits production of proinflammatory mediators such as nitric oxide (NO), prostaglandin E_2 (PGE_2), and tumor necrosis factor-α (TNF-α) produced in LPS and IFN-γ activated RAW 264.7 cells and murine peritoneal macrophages[5,12,16]. Tau-Cl suppresses the transcription of nitric oxide synthase (iNOS), but not that of PGE_2 and TNF-α in LPS and IFN-γ activated RAW 264.7 cells[13]. We examined the effect of Tau-Cl on the hydrolytic activity of MMP-9 secreted from LPS activated murine peritoneal macrophages in order to further our understanding of the anti-inflammatory properties of Tau-Cl.

MATERIALS AND METHODS

Mice

Specific pathogen-free female C57BL/6 mice (8-12 weeks old) were purchased from Jackson Laboratory (Bar Harbor, ME). Animals were maintained in facilities approved by the American Association for Accreditation of Laboratory Animal Care and in accordance with current regulation and standards of the United States Department of Agriculture, Department of Health and Human Services and the National Institutes of Health.

Murine peritoneal exudate macrophages (PEC)

Mice were injected intraperitoneally with 1 ml of 4% thioglycollate broth (Difco Laboratories, Detroit, MI). PEC were collected four days later by peritoneal lavage with Ca^{++} and Mg^{++} free Hank's Balanced Salt Solution (HBSS, Gibco BRL, Grand Island, NY)[5]. Cell viability was measured by trypan blue dye exclusion (>98%).

Cell culture and activation

One million PEC suspended in Eagle's minimum essential medium (MEM) containing 5 % FCS and penicillin and streptomycin were plated and adhered in 24-well plate for 2 h in 5% CO_2[20]. Cells were washed 2 times with DMEM-F12 without FCS and were treated with FCS-free DMDM-F12 containing 1 µg/ml LPS and either taurine or Tau-Cl for 24 h. After 24-h incubation, culture medium was removed and cells were incubated with medium without FCS and activator for an additional 24 h. Supernatants were collected and analyzed by gelatin zymography.

Gelatin zymography

Gelatin zymography was performed as previously described[1,20]. Briefly, aliquots of supernatants from control and test groups were analyzed by substrate (gelatin) gel electrophoresis. The samples were applied without reduction to a 7.5% polyacrylamide slab gel impregnated with 1 mg/ml gelatin (Sigma Chemical Co., St. Louis, MO). After electrophoresis, the gel was washed at room temperature for 30 min with washing buffer (50mM Tris-Cl, pH 7.5/5 mM $CaCl_2$ /1uM $ZnCl_2$ /2.5% Triton X-100), and then incubated overnight at 37 °C with shaking in the washing buffer adding 1% Triton X-100 instead 2.5 %. The gel was stained with a solution of 0.1 % Coomassie Brilliant blue R-250. In this assay, a clear zone against the blue background indicates the presence of gelatinolytic activity. Molecular weights of gelatinolytic bands were estimated using a prestained m.w. marker (Bio Rad, Hercules, CA). Human active MMP-9 (85 kD) and pro MMP-9 (92 kD) (Oncogene Research Products, Cambridge, MA) were used as a standard.

Preparation of Tau-Cl and Nitrite assay

Tau-Cl was prepared by adding equimolar amounts of NaOCl dropwise into a taurine solution in phenol-free HBSS as previously described[14]. The nitrite assay was performed as previously described[12].

RESULTS AND DISCUSSION

The hydrolytic activity of MMP-9 activated by various amounts of LPS (1ng/ml-1000ng/ml) in murine PEC is shown in Fig 1. EDTA inhibited the hydrolytic activity of MMP-9 by chelating Zn^{++} and Ca^{++} that is required for the reaction. 1,10-phenanthroline, which chelates divalent metals, also inhibited MMP-9 activity. However, PMSF, an inhibitor of serine protease such as trypsin and chymotrypsin, did not inhibit MMP-9 activity, consistent with the characteristics of this enzyme[6,20]. There was no significant difference in production of MMP-9 by various amounts of LPS (10 to 1000ng/ml) except for 1 ng/ml which stimulated less MMP-9 hydrolytic activity. Maximum production of MMP-9 occurred between 18 h and 24 h of activation (Fig 2), with MMP-9 produced as early as 2 h after LPS activation. One µg/ml LPS and 24 h of activation were considered to be optimal and were used throughout these studies.

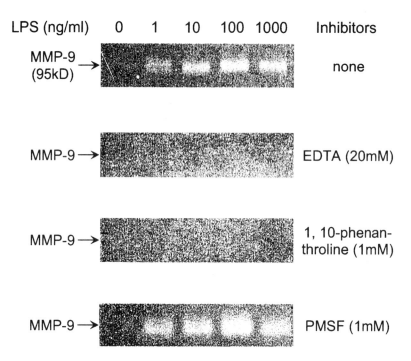

Fig. 1. MMP-9 production and inhibition by various chemicals. Murine peritoneal macrophages were activated with various amount of LPS (1 ng/ml – 1000 ng/ml) in the absence or presence of inhibitors for 24 h. The cells were washed and incubated with FCS-free media for an additional 24 h. Supernatants were collected and applied to a gelatin-impregnated gel. This is a representative experiment of four experiments

The effect of Tau-Cl and taurine on the hydrolytic activity of MMP-9 by LPS activated murine peritoneal macrophages activated with 1μg/ml LPS was determined. The concentration range 0.1-0.8 mM had no effect on the viability of the cells. LPS and either Tau-Cl or taurine were added simultaneously and the MMP-9 activity was analyzed by gelatin zymography. Tau-Cl inhibited the hydrolytic activity of MMP-9 in a dose-dependent manner (Fig 3A). One half mM Tau-Cl inhibited MMP-9 activity greater than 50 % and 0.8 mM Tau-Cl almost abolished MMP-9 activity. (Fig 3A). Taurine was without effect (Fig 3B). These data are similar to our previously reported inhibition of nitrite production by Tau-Cl[5,12].

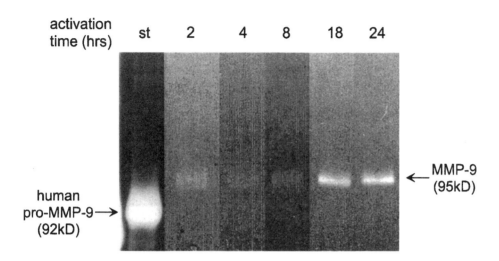

Fig. 2. Kinetics of MMP- 9 production. Murine peritoneal macrophages were activated and incubated at different times with 1 µg/ml of LPS. The cells were washed and incubated for an additional 24 hrs without activator in FCS-free media. Supernatants were collected and applied to a gelatin-impregnated gel. This is one representative experiment of three experiments.

To determine whether the effects of Tau-Cl were mediated by direct inhibition of MMP-9 or inhibition of induction, the cells were pretreated with Tau-Cl before activation with LPS (Fig 4A and 4B) or cells were treated with Tau-Cl after activation with LPS (Fig 5A and 5B). After pretreatment or post-treatment, MMP-9 activity and nitrite production were measured using the same supernatants. As described previously[14], pretreatment of cells with Tau-Cl inhibited nitrite production (Fig 5B). Tau-Cl (0.5 mM) for 2 h and 3 h before LPS activation inhibited nitrite production 50 % and 42 %, respectively. However, MMP-9 was not inhibited significantly by 0.5 mM of Tau-Cl for 2 and 3 h (Fig 4A).

Nitrite production by LPS-activated cells began 8 h after LPS activation (data not shown). No inhibition of nitrite production was observed 8 h after activation (Fig 4B)[12]. These data suggested that Tau-Cl may be inhibiting activation of MMP-9 in LPS activated cells, perhaps by interfering with the signal transduction pathway of MMP-9 instead the direct inhibition of MMP-9 activity.

A

LPS (1µg/ml)   -   +   +   +   +   +
Tau-Cl (mM)   -   -   0.1   0.3   0.5   0.8

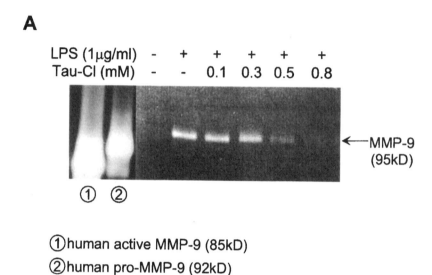

←MMP-9
(95kD)

①   ②

①human active MMP-9 (85kD)
②human pro-MMP-9 (92kD)

B

LPS (1µg/ml)   -   +   +   +   +   +
Taurine (mM)   -   -   0.1   0.3   0.5   0.8

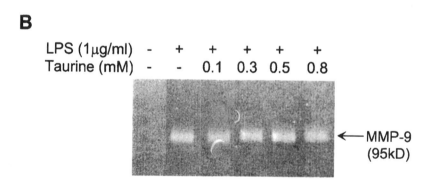

←MMP-9
(95kD)

Fig. 3. The effects of Tau-Cl (**A**) and taurine (**B**) on MMP-9 production in LPS-activated murine peritoneal macrophages. The cells were activated simultaneously with 1 µg/ml of LPS and various amounts of Tau-Cl or taurine (0.1-0.8mM) for 24 h and supernatants were discarded. After an additional 24-h incubation in FCS-free medium without activator, supernatants were collected and applied to a gelatin-impregnated gel. **A** is a representative experiment of ten experiments and **B** is a representative of three experiments.

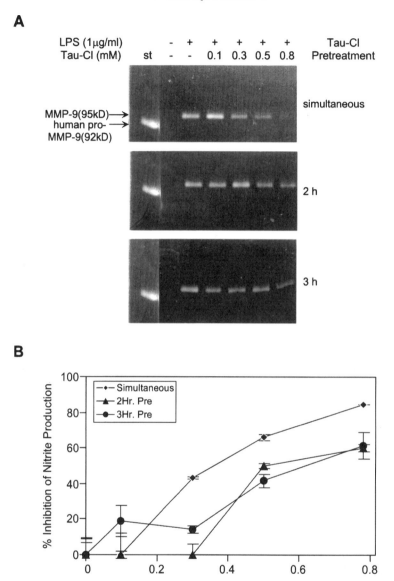

Fig. 4. The effects of MMP-9 production on pretreatment of murine peritoneal macrophages with Tau-Cl. The cells were incubated with various concentrations of Tau-Cl (0.1-0.8mM) for 2 and 3 hrs before adding the activator (1 µg/ml of LPS). After the cells were washed with FCS-free media to remove Tau-Cl, they were activated with LPS for 24 h. The first supernatant was discarded. After an additional 24-h incubation, the supernatants were collected and applied to a gel (**A**) and nitrite production determined using Griess reagent (**B**). This is one representative experiment of two experiments.

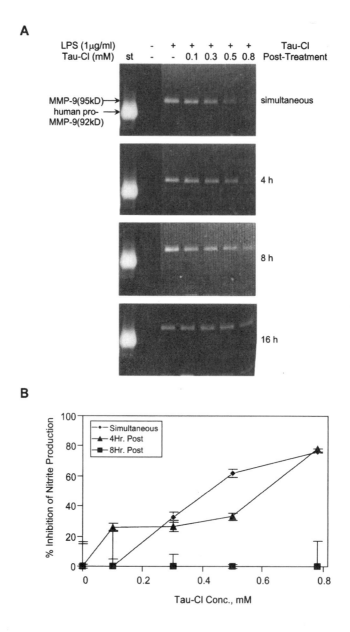

*Fig. 5.*The effects of MMP-9 production on post-treatment of Tau-Cl in LPS-activated murine peritoneal macrophages. The cells were activated with 1μg/ml of LPS and Tau-Cl added at various times (4, 8 and 16 h) after treatment with LPS. Other procedures were the same as Fig. 3. Supernatants were applied to the gel (**A**) and nitrite production determined using Griess reagent (**B**). This is one representative experiment of two experiments.

While nitrite production was inhibited 62 % and 60 % by 0.8 mM Tau-Cl for 2 and 3 h, respectively, MMP-9 was inhibited only marginally by 0.8 mM Tau-Cl. These data also indicated that Tau-Cl might suppress signal transduction for nitrite production but not for MMP-9 prior to LPS activation. Simultaneous treatment of PEC with Tau-Cl and LPS resulted in the greatest inhibition of nitrite production and MMP-9 secretion. MMP-9 production by LPS-activated murine peritoneal macrophages was no longer inhibited 16 h post LPS activation (Fig 5A).

Multiple second messenger system such as protein kinase C (PKC)[15], pertussis toxin sensitive guanine nucleotide kinase (PKG)[4] and protein tyrosine kinase (PTK)[18] appear to cooperate in mediating the pleiotropic effects of LPS on macrophages. Xie *et al.* demonstrated MMP-9 production stimulated by LPS in murine PEC was inhibited by H-7, an inhibitor of PKC, by genistein and tryphostin, specific inhibitors of PTK and was inhibited somewhat by HA1004, an inhibitor of PKG[20]. Inhibition of MMP-9 by Tau-Cl may be mediated by suppressing these kinases. Although Tau-Cl inhibits of nitric oxide synthase at the the transcription level[13], transcriptional regulation of MMP-9 by Tau-Cl remains to be determined. Since Tau-Cl inhibits MMP-9 production, a proinflammatory protease as well as proinflammatory mediators such as TNF-α, nitric oxide, PGE_2 IL-6, IL-8 and superoxide anion[5,11,12,13,16], Tau-Cl produced at the site of inflammation by polymorphonuclear leukocytes may play a pivotal role in down–regulating the inflammatory response.

ACKNOWLEDGEMENT

This work was supported by the Office of Mental Retardation and Developmental Disabilities of the State of New York and by U.S. Public Health Service Research Grant Award HL-49942 from National Institutes of Health. We thank Dr. Bernd C. Kieseier for helpful discussions.

REFERENCES

1. Birkedal-Hansen, H. and Taylor, R.E., 1982, Detergent-activation of latent collagenase and resolution of its component molecules. *Biochem. Biophy. Res. Comm.* **107**:1173-1178.
2. Delcaux, C., Delacourt, C., d'Ortho, M.P., Boyer, V. and Lafuma, C., 1996, Role of gelatinase B and elastase in human polynuclear leukocytes migration across basement membrane. *Am. J. Respi. Cell. Mol. Biol.* **14**:288-295.
3. Hibbs, M. S., Hoidal, J. R. and Kang, A.H., 1987, Expression of a metalloproteinase that degrade native type V collagen and denatured collagens by cultured human alveolar macrophages. *J. Clin. Invest.* **80**:1644-1650.

4.   Jackway, J.P. and DeFranco, A.L., 1986, Pertusis toxin inhibition of B cell and macrophage response to bacterial lipopolysaccharide. *Science* **234**:743-746.

5.   Kim, C., Park, E., Quinn, M.R. and Schuller-Levis, G., 1996, The production of superoxide anion and nitric oxide by cultured murine leukocytes and the accumulation of TNF-α in the conditioned media is inhibited by taurine chloramine. *Immunopharm.* **34**:89-95

6.   Kumagai, K., Ohno, I., Okada, S., Ohkawara, Y., Suzuki, K., Shinya, T., Nagase, H., Iwata, K. and Shirato, K., 1999, Inhibition of matrix metalloproteinase prevents allergen-induced airway inflammation in a murine modle of asthma. *J. Immunol.* **162**:4212-4219.

7.   Leppert, D., Waubant, E., Galardy, R., Bunnett, N. W. and Ha , S. L., 1995, T cell gelatinases mediate basement membrane transmigration in vitro. *J. Immunol.* **154**:4379-4389.

8.   Mautino, G., Oliver, N., Chanez, P., Bousquet, J. and Capony, F., 1997, Increased release of matrix metalloproteinase-9 in bronchoalveolar lavage fluid and by alveolar macrophages of athsmatics. *Am. J. Respi. Cell. Mol. Biol.* **17**:583-591.

9.   Murphy, G. and Docherty, A.J.P., 1992, The matrix metalloproteinase and their inhibitors. *Am. J. Respir. Cell. Mol. Biol.* **7**:120-125.

10. O'Connor, C. M., amd FitzGerald, M. X., 1994, Matrix metalloprotease disease. *Throx* **49**:602-609

11. Park. E., Alberti, J., Quinn, M.R. and Schuller-Levis, G., 1998. Taurine chloramine inhibits the production of superoxide anion, IL-6 and IL-8 in activated human polymorphonuclear leukocytes. Adv. Exp. Med. Biol. 442:177-182

12. Park, E., Quinn, M.R., Wright, C. and Schuller-Levis, G. 1993, Taurine Chloramine inhibits the syntheses of nitric oxide and the release of tumor necrosis factor in activated RAW 264.7 cells. *J. Leukoc.* **54**:119-124.

13. Park, E., Schuller-Levis, G. and Quinn, M.R., 1995, Taurine chloramine inhibits production of nitric oxide and TNF-α in activated RAW 264.7 cells by mechanisms that involve transcriptional and translational events. *J. Immunol.* **154**:4778-4784.

14. Park, E., Schuller-Levis, G., Jia, J. and Quinn, M.R., 1997, Preactivation exposure of RAW 264.7 cells to taurine chloramin attenuates subsequent production of nitric oxide and expression of iNOS mRNA. *J. Leukoc. Biol.* **61**:161-166.

15. Prpic, V., Weiel, J.E., Somers, S.D., DiGuiseppi, J., Gonias, S.L. and Pizzo, S.V., 1987, Effects of baterial lipopolysaccharide on the hydrolysis of phophatidylinositl-4, 5-biphosphate in murine peritoneal macrophages. *J. Immunol.* **139**:526-533.

16. Quinn, M.R., Park, E. and Schuller-Levis, G., 1996, Taurine chloramine inhibits prostaglandin E_2 in activated RAW 264.7 cells by post-transcriptional effects on inducible cyclooxygenase expression. *Immunol. Let.* **50**:185-188.

17. Stahle-Backdahl, M., Inoue, M., Guidice, G.J. and Parks, W. C., 1994, Type IV collagenase mRNA by eosinophils associated with basal cell carcinoma. *J. Invest. Derma.* **99**:497-503.

18. Weistein, S.L., Gold, M.R., and DeFranco, A.L., 1991, Bacterial lipopolysaccharide stimulates protein tyrosin phosphorylation in macrophages. *Proc. Natl. Acad. Sci.* USA **88**:4148-4152.

19. Welgus, H.G., Campbell, E.J. Cury, J.D., Eisen, A. Z., Senior, R. M., Wilhelm, S. M. and Goldberg, G. I., 1990, Neutral metalloproteinase produced by human mononuclear phagocytes: Enzyme profile, regulation and expression during cellular development. *J. Clin. Invest.* **86**:1496-1502.

20. Xie, B., Dong, Z., and Fidler, I. 1994, Regulatory mechanisms for the expression of Type IV collagenases/gelatinases in murine macrophages. *J. Immunol.* **152**:3637-3644.

SYNTHESIS OF TAURINE ANALOGUES FROM ALKENES
Preparation of 2-aminocycloalkanesulfonic acids

Fabrizio Machetti, Martina Cacciarini, Fernando Catrambone, Franca M. Cordero, *Maria Frosini, Francesco De Sarlo

*Dipartimento di Chimica Organica «U. Schiff», and Centro di Studio sulla Chimica e la Struttura dei Composti Eterociclici e loro Applicazioni, C.N.R., Università di Firenze, Via G. Capponi 9, I-50121 Firenze, Italy. * Istituto di Scienze Farmacologiche, Università di Siena, Italy*

Abstract: (±)*trans* 2-Aminocyclohexanesulfonic acid and (±)*trans* 2-aminocyclo-pentanesulfonic acid were prepared from cyclohexene and cyclopentene respectively by sulfur monochloride addition, followed by oxidation to 2-chlorosulfonic acid and substitution of chlorine.

The study of the biological activity of taurine requires the availability of analogues which could provide useful information for understanding the mechanisms underlying the many-sided biological effects of this amino acid. Therefore it is important to seek new and more convenient synthetic methods to obtain 1,2-aminosulfonic acid derivatives.

We considered the addition of S_2Cl_2 to alkenes as the first step of a sequence leading to 2-aminosulfonic acids. Cyclopentene (**1**) or cyclohexene (**2**) were treated with sulfur monochloride (S_2Cl_2), as previously reported[1], and then the crude mixture of sulfides (**3**, respectively **4**) was added to *m*-chloroperbenzoic acid (MCPBA) in methylene chloride and the corresponding 2-chlorosulfonate (**5**, respectively **6**) were obtained (Scheme 1).

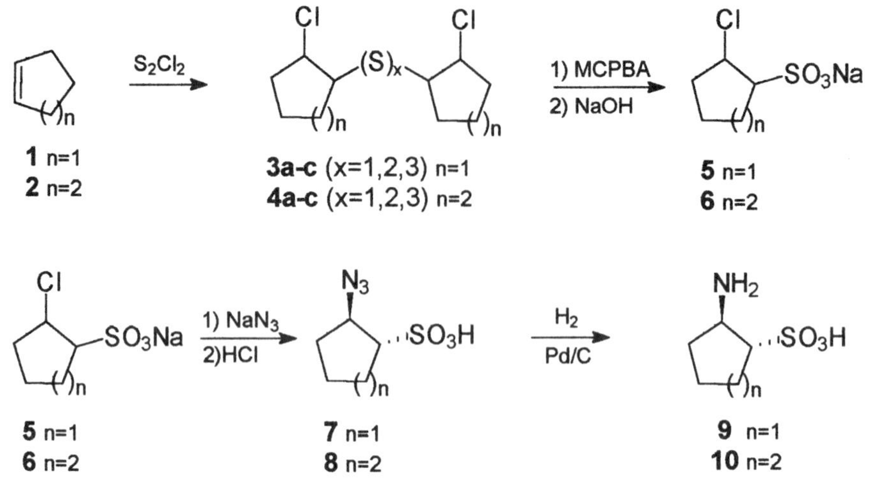

Scheme 1. Conversion of the 2-chlorosulfonates **5** and **6** into the amine derivatives **9** and **10** could not be achieved directly, as a basic reagent (NH_3, K phthalimide) causes mainly elimination. Substitution by an azido group to the intermediates **7** and **8**, followed by catalytic hydrogenation, afforded the products **9** (37% from cyclopentene) and **10** (50% from cyclohexene) respectively. Even this procedure gave some (20%) elimination product. The products **9** and **10** were *trans*, as indicated by the coupling constants in ^{1}H NMR.

Attempted application of the above procedure to norbornene afforded *3-chloronorbornane-2-sulfonic acid sodium salt* (mixture of stereoisomers), but their conversion into the corresponding azido derivatives was difficult.

Since the direct reaction of alkenes with SO_3 and CH_3CN has been reported,[2,3] we examined the practical utility of the DMF-SO_3 complex as a sulfonating agent together with the CH_3CN as a nitrogen source. We performed model experiments to test the potentiality of this three-components one-pot reaction. Low yields of *trans* 2-aminocyclohexanesulfonic and of 2-aminohexanesulfonic acid were obtained from cyclohexene and 1-hexene, respectively.

EXPERIMENTAL

2-Chlorosulfonic acids; 5 and 6, sodium salts. Sulfur monochloride (50 mmol) was added dropwise to an excess (150 mmol) of cyclopentene (**1**) or cyclohexene (**2**) over 2 h. After stirring at room temperature for 3 days, then concentrating *in vacuo*, the chlorosulfides **3** and, respectively, **4** were obtained as clear syrups (ratios 35:25:40 for **3a:3b:3c** and 3:5:2 for **4a:4b:4c**, as determined by GC-MS).

The sulfide mixture (**3** or **4**) was added dropwise to an ice-cold solution containing 5 times its weight of MCPBA (70%) in dichloromethane. The reaction mixture was allowed to warm to room temperature, and refluxed for 5 h, then extracted 4 times with water. The white solid was removed and the clear aqueous phase concentrated and neutralised with 0.5 M NaOH solution. The chlorosulfonic acids **5** or **6** were obtained as sodium salts on concentration to dryness.

2-Aminosulfonic acids; **9** and **10**. Sodium azide (2:1 molar ratio) was added to a 5% solution of 2-chlorosulfonate (**5** or **6**) in dimethylformamide (DMF). After reflux (3h) the solvent was evaporated, the solid residue was dissolved in 2M HCl and concentrated again, to give the *2-azidosulfonic acids* **7** or, respectively, **8**, containing minor amounts of the unsaturated sulfonic acids.

The crude product, **7** or **8**, was hydrogenated in methanol with Pd/C (10%) under hydrogen at atmospheric pressure for 3 days. The crude solid, obtained after removal of the catalyst and of the solvent, was dissolved in water and passed trough a column of Dowex 50 WX4 H^+ form. After evaporation of the solvent the solid was recrystallized: **9**, mp 324 °C dec. (lit.,[3] 330 °C dec); **10**, mp 408 °C dec (lit.,[2b] 410 °C dec).

3-Chloronorbornane-2-sulfonic acid sodium salt. Sulfur monochloride (10 mmol) was added dropwise to an excess (30 mmol) of norbornene in *n*-hexane (2 ml) with ice-cooling. The mixture was set aside for 4 days at room temperature, then concentrated. A mixture of chlorosulfides was obtained, in the ratios 4:2:1 of mono-, di- and tri-sulfide. Oxidation was carried out as described for the sulfides **3** or **4** to give the title compound.
Attempted substitution with sodium azide was difficult, giving low yields of the corresponding azidosulfonic acid.

ACKNOWLEDGMENTS

The authors thank M.U.R.S.T.-Cofin98 (Nuove Metodologie e Strategie di Sintesi di Composti di Interesse Biologico) for financial support. F. M. thanks the Università di Firenze for a two years post-doctor fellowship.

REFERENCES

1. Lautenschlaeger, F. , Schwartz, N.V., 1969, *J. Org. Chem.* **34**:3991.
2. a) Broussalian, G. L., U. S. Patent 3: 337, 457 (1967); b) Arlt, D., Neth. Appl. 6, 611, 091; *Chem. Abs.* **67**, 11213 (1967) c) Broussalian, G. L., U. S. Patent 3, 287, 389 (1966); *Chem.Abs.* **66**, 37459 (1967) d) Broussalian, G. L., U. S. Patent 3, 303, 137 (1967); *Chem. Abs.* **66**, 67063 (1967).
3. Liebowitz, S. M., Lombardini, J. B., Salva, P. S., 1987, *Biochem. Pharmacol.* **36**:2109.

EFFECTS OF GUANDINOETHANE SULFONATE ON CONTRACTION OF SKELETAL MUSCLE

C. Cuisinier, P. Gailly, M. Francaux and J. Lebacq
Faculty of medicine, Université catholique de Louvain, Louvain-la-Neuve, Belgium

Abstract: Guanidinoethane sulfonic acid (GES), a chemical and biological analog of taurine, decreases rat muscle taurine content when added to drinking water. Over the same period, GES appears in muscle[8]. GES supplementation is often used to study the effect of taurine depletion on physiological mechanisms, without taking into account the possible actions of GES. The purpose of the present study was to investigate the specific actions of GES on contraction of skeletal muscle. In mice EDL muscle, the time delay needed to observe a 20% force decrease after the end of a tetanic stimulation was higher in GES-supplemented than in control muscle. This observation in GES-supplemented muscle could be explained by the action of taurine or GES on several targets, beside others the rate of Ca^{2+} uptake by sarcoplasmic reticulum (SR) and the Ca^{2+} sensitivity of myofilaments. SR of rat EDL was isolated by successive centrifugations. The effect of 20 mM taurine or GES on the rate of Ca^{2+} uptake by SR was measured with the fluorescent Ca^{2+} indicator fura-2. The results show that the rate of Ca^{2+} uptake by SR is not modified in the presence of taurine or GES. The Ca^{2+} sensitivity of myofilaments was studied in chemically skinned fibers in the presence of 20 mM taurine or GES. Both taurine and GES increased the myofilament sensitivity to Ca^{2+}. Thus, the prolonged relaxation time of GES-supplemented muscle can be attributed to an increase in myofilament sensitivity to Ca^{2+}. This higher sensitivity is not due to a decrease in muscle taurine content but rather to an increased GES concentration.

INTRODUCTION

Taurine is highly concentrated in organs such as liver, brain, heart and skeletal muscle but relatively scarce in extracellular fluids[9]. The gradient across cell membranes is maintained by a regulated taurine transport system[4]. Several agonists able to compete with taurine transport have been discovered : β-alanine, hypotaurine, guanidinopropionic acid and guanidinoethane sulfonic acid (GES)[8]. The latter is frequently used as a taurine-depleting agent when added 1% (w/v) to animal drinking water. Taurine depletion is accompanied by GES accumulation in tissues. In rat ventricular papillary muscle, GES supplementation induces a prolonged relaxation time after contraction. This observation was attributed to the taurine depletion[11]. To the best of our knowledge, the effects of taurine on contractility of skeletal muscle have never been studied. The results of the present study show that taurine depletion by GES supplementation prolongs the relaxation time of skeletal muscle. This increase cannot be explained by muscle taurine content decrease but is due to the presence of GES in muscle cells that modifies myofilament sensitivity to calcium.

MATERIALS AND METHODS

GES Supplementation of Mouse Skeletal Muscle

Male C57 B6 mice, aged 8 weeks, were separated into 2 groups. The GES group (n = 9) was provided with drinking water containing 1% (w/v) guanidinoethane sulfonic acid (GES) for 6 weeks. The control group (n = 10) had free access to pure water without any supplementation.

Mechanical Measurements

Extensor digitorum longus (EDL) muscles were removed from control and GES-supplemented animals, mounted on a force measuring system and immersed in continuously gassed (95% O_2- 5% water-saturated CO_2) Krebs solution (118 mM NaCl, 25 mM $NaHCO_3$, 5mM KCl, 1 mM $MgSO_4$, 2.5 mM $CaCl_2$ and 5 mM glucose). Optimal muscle length for maximal force development was determined and the muscles were stimulated for 400 ms at a frequency of 100 Hz by capacitor discharges of alternating polarity. The force developed during the tetanus and the relaxation period was recorded. The time delay needed to observe a 20% force decrease after the end of the stimulation ($t_{20\%}$) was calculated.

Sarcoplasmic Reticulum Isolation and Calcium Uptake Measurements

Both EDL muscles of male Wistar rats aged 3 months were quickly removed and homogenized at 0°C in 1 ml of a buffer solution (200 mM sucrose, 10 mM sodium azide, 1 mM EDTA, 40 mM L-histidine, pH 7.8). The light fraction of sarcoplasmic reticulum (SR) was isolated according to the method of O'Brien[13]. The rate of Ca^{2+} uptake by the SR was determined with the Ca^{2+} fluorescent indicator fura-2[5,10]. The assay was initiated by injection of 15 μg SR in 200 μl assay medium which contained 10 μM Ca^{2+}, 5 μM fura-2 free acid, 20 mM Tris. HCl, 5 mM potassium oxalate, 5 mM $MgCl_2$, 5 mM Na_2ATP, 5 μM TPEN, 80 mM KCl, pH 7.0. The extravesicular Ca^{2+} concentration was measured for 200 s following initiation of the assay. Results are presented as the rate of Ca^{2+} uptake by isolated SR vesicles in the presence of different extravesicular free Ca^{2+} concentrations. Two conditions were tested by adding 20 mM taurine or 20 mM GES. The control conditions were conducted in presence of 20 mM sucrose.

Skinned Fiber Preparation and Ca^{2+} Sensitivity Determination

Bundles of rabbit psoas muscle were chemically skinned as described by Brenner[2]. Bundles of about 10 fibers were dissected and mounted on a force transducer. Sarcomere length was measured by laser diffraction[1] and adjusted to 2.2 μm. Fibers were stimulated by immersion in a solution containing 0.54 μM buffered free Ca^{2+}, 5 mM EGTA/CaEGTA, 10 mM imidazole, 5 mM MgATP, 10 mM caffeine, 20 mM taurine or 20 mM GES, pH 7.2 and ionic strength τ/2 adjusted at 138 mM. Fiber bundles were relaxed between force measurements. The effects of 20 mM taurine (n = 6) and 20 mM GES (n = 6) on force developed by myofilaments were tested in a random order. In control conditions, taurine or GES was replaced by 20 mM KCl (n = 6). Results are presented as a percentage of maximal force developed in the presence of 10 μM Ca^{2+}.

Statistics

Data are presented in figures as means ± SD. Changes in variables of interest were assessed by a mean comparison test. Level of significance was set at $p < 0.05$ (*; ** and *** indicate $p < 0.01$ and $p < 0.001$ respectively).

RESULTS

Skeletal muscles of GES treated and control mice were isolated from the animals and stimulated *in vitro*. GES-treated muscles showed a significantly longer $t_{20\%}$ than control muscles ($p < 0.01$) (Figure 1).

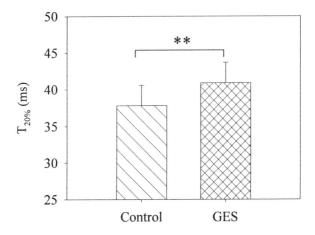

Figure 1. Time delay (ms) needed to observe a 20% force decrease after the end of the stimulation of control and GES supplemented mice EDL muscles (mean $\pm$ SD). ** $p < 0.01$.

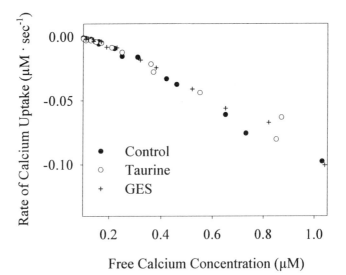

Figure 2. Rate of Ca^{2+} uptake by isolated SR vesicles in the presence of different extravesicular free Ca^{2+} concentration. The measure was conducted with the fluorescent Ca^{2+} indicator fura-2 in the presence of 20 mM taurine (o) or 20 mM GES (+). The control assay was conducted in the presence of 20 mM sucrose (●).

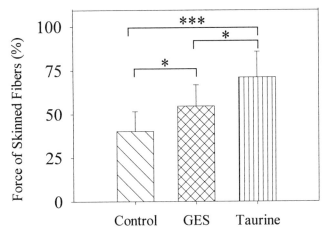

Figure 3. Force developed by bundles of chemically skinned rabbit psoas. Force was measured when fibers were activated in solutions containing 0.54 μM buffered free Ca^{2+} and 20 mM taurine or 20 mM GES. Results are expressed in percent of maximal force developed when fibers were submersed in solution containing 10 μM free Ca^{2+}. *$p < 0.05$, ***$p < 0.001$.

Several muscle compartments were then isolated in order to explain this observation. First, SR vesicles were prepared as described previously. The effect of 20 mM taurine or GES on the rate of Ca^{2+} uptake by the SR was examined with the Ca^{2+} fluorescent indicator fura-2. Figure 2 shows that no difference in the rate of Ca^{2+} uptake was observed in the presence of 20 mM taurine or GES. Furthermore, effects of 20 mM taurine or GES on Ca^{2+} sensitivity of myofilaments were studied. Both taurine ($p < 0.001$) and GES ($p < 0.05$) increased the force developed by chemically skinned fibers activated in buffer solution containing 0.54 μM Ca^{2+}. The increase of myofilament sensitivity due to taurine was higher than the effect observed with GES ($p < 0.05$) (Figure 3).

DISCUSSION

Rats treated with GES show a decreased muscle taurine content (-2.1 μmol·g ww^{-1} in leg) and an increase of 2.7 μmol·g ww^{-1} in GES content[8]. The treatment also increases relaxation time of EDL muscle (Figure 1) which is consistent with a similar observation made on cardiac muscle[11]. To study the effects of GES supplementation on muscle contraction, both the effects of muscle taurine content decrease and those of GES increase have to be considered. Different targets could be responsible for this prolonged relaxation time after tetanic contraction, other than the rate of Ca^{2+} uptake by SR and the Ca^{2+} sensitivity of myofilaments. The former is not responsible for the $t_{20\%}$ increase, as neither GES nor taurine modify the rate of Ca^{2+} uptake by the SR (Figure 2). Huxtable[7] reported that when extravesicular Ca^{2+} concentration was 0.1 mM, taurine increased Ca^{2+} transport activity by rat skeletal muscle SR. Remtulla *et al.*[14] has shown in cardiac muscle that taurine did not affect ATP-dependent calcium uptake. The results of the present investigation suggest that the increase in $t_{20\%}$ is due to an increase in myofilament sensitivity to Ca^{2+} (Figure 3). Physiological taurine concentrations increase myofilament sensitivity to calcium in triton-skinned trabeculae[15], in skinned skeletal muscle of crayfish[3] and in mammalian skeletal muscle (Figure 3). Figure 3 shows that GES also increases myofilament sensitivity to Ca^{2+}, although the effect of taurine is larger than that of GES. Higher values of $t_{20\%}$ observed in GES supplemented muscles cannot be due to a decrease in taurine muscle content but rather to an increase in GES content which itself can explain the increased myofilament sensitivity to Ca^{2+} and the increased relaxation time. Nevertheless, other factors, involving for example energy supply changes could also slow down muscle relaxation after GES supplementation. Decrease in heart[12] and in skeletal muscle (own unpublished observations) creatine phosphate contents have been reported. Although GES is a possible substrate for creatine kinase[6], the energy flux through this reaction is probably reduced. Clearly, these assumptions would require further investigations.

REFERENCES

1. Allen, J.D., and Moss, R.L., 1987, Factors influencing the ascending limb of the sarcomere length-tension relationship in rabbit skinned muscle fibres. *J. Physiol.* 390: 119-136.
2. Brenner, B., 1998, Muscle mechanics II: skinned muscle fibres. In *Current methods in muscle physiology. Advantages, problems and limitations* (H. Sugi, ed.), Oxford University Press, Oxford, pp.22-69.
3. Galler, S., Hutzler, C., and Haller, T., 1990, Effects of taurine on Ca^{2+}-dependent force development of skinned muscle fibre preparations. *J. Exp. Biol.* 152: 255-264.
4. Ganapathy, V., and Leibach, F.H., 1994, Expression and regulation of the taurine transporter of cultured cell lines of human origin. In *Taurine in Health and disease* (R. Huxtable), Plenum Press, New York, pp.51-57.
5. Grynkiewicz, G., Poenie, M., and Tsien, R.Y., 1985, A new generation of Ca^{2+} indicators with greatly improved fluorescent properties. *J. Biol. Chem.* 260: 3440-3450.
6. Huxtable, R.J., 1992, Physiological actions of taurine. *Physiol. rev.* 72: 101-163.
7. Huxtable, R., and Bressler, R., 1973, Effect of taurine on a muscle intracellular membrane. *Biochimi. Biophys. Acta* 323: 573-583.
8. Huxtable, R.J., Laird, H.E., and Lippincott, S.E., 1979, The transport of taurine in the heart and the rapid depletion of tissue taurine content by guanidinoethyl sulfonate. *J. Pharmacol. Exp. Ther.* 211: 465-471.
9. Jacobsen J.G., and Smith L.H., 1968, Biochemistry and physiology of taurine and taurine derivatives. *Physiol. Rev.* 48: 424-511.

10. Kargacin, M.E., Scheid C.R., and Honeyman T.W., 1988, Continuous monitoring of Ca^{2+} uptake in membrane vesicules with fura-2. *Am. J. Physiol.* 245: C694-C698.
11. Lake, N., Splawinski, J.B., Juneai, C., and Rouleau, J.L., 1990, Effects of taurine depletion on intrinsic contractility of rat ventricular papillary muscles, *Can. J. Physiol. Pharmacol.* 68: 800-806.
12. Mozaffari, M.S., Tan, B.H., Lucia M.A., and Schaffer S.W., 1986, Effect of drug-induced taurine depletion on cardiac contractility and metabolism, *Biochem. Pharm.* 35: 985-989.
13. O'Brien, P.J., 1990, Calcium sequestration by isolated sarcoplasmic reticulum: real-time monitoring using radiometric dual-emission spectrofluorometry and the fluorescent calcium-binding dye indo-1. *Mol. Cell. Biochem.* 94: 113-119.
14. Remtulla, M.A., Katz, S., and Applegarth D.A., 1978, Effect of taurine on ATP-dependent calcium transport in guinea-pig cardiac muscle. *Life Sciences.* 23: 383-390.
15. Steele, D.S., Smith, G.L., and Miller, D.J., 1990, The effects of taurine on calcium uptake by the sarcoplasmic reticulum and calcium sensitivity of chemically skinned rat heart. *J. Physiol.* 422: 499-511.

PROTECTION BY TAURINE AND STRUCTURALLY RELATED SULFUR-CONTAINING COMPOUNDS AGAINST ERYTHROCYTE MEMBRANE DAMAGE BY HYDROGEN PEROXIDE

Prabhat K. Pokhrel and Cesar A. Lau-Cam
College of Pharmacy and Allied Health Professions, St. John's University, 8000 Utopia Parkway, Jamaica, New York 11439, USA

INTRODUCTION

Taurine (TAU) is a sulfur-containing nonprotein β-amino acid whose abundance in excitable mammalian cells and tissues susceptible to oxidation appears to underlie a protective role against oxidant-induced cellular damage[7,23,42]. Numerous studies, conducted on a variety of biological samples and under *in vitro* and *in vivo* conditions, have verified that TAU, in spite of a lack of ready oxidizibility, can express antioxidant actions such as serving as a trap for hypochlorous acid $(HOCl)$[51,59], decreasing the formation of malondialdehyde[25,32,53] and of reactive oxygen species[3,28,53,55], inhibiting the depletion of natural anti-oxidants[6,30,55], minimizing oxidant-mediated protein and ATPase activity losses[6] (Banks *et al.*, 1992), and reversing the decrease in antioxidant enzyme activities under oxidative stress[3,16,55]. As a result of these actions, TAU may contribute to the maintenance of cell viability[1,6,33], the modulation of cell necrosis and apoptosis[38,56], the attenuation of cell damage[21,31,42,55], and the curtailment of tissue fibrosis[21,53].

The purpose of the present investigation was to examine the effects of TAU against oxidative cell injury using a simple and reproducible *in vitro* test system and a readily available model oxidant[48]. To this end, fresh rat erythrocytes (RBCs) were exposed to the actions of hydrogen peroxide (H_2O_2) in the presence and absence of TAU, and indices of membrane damage and of lipid peroxidation were measured thereafter. In addition, the results obtained with TAU were compared with those derived from a group of eleven additional compounds (Table 1), representing analogs and homologs of TAU, to verify whether the antioxidative properties of TAU are dependent on specific structural features of its molecule.

MATERIALS AND METHODS

Chemicals

All the amino acids and sulfur-containing compounds used in the study were obtained from local commercial sources: TAU, hypotaurine (HYTAU), isethionic acid (ISA), N-acetylcysteine (NACYS), L-cysteic acid (CA) and aminomethanesulfonic acid (AMSA) were from Sigma Chemical Co., St. Louis, MO; ethanesulfonic acid sodium salt (ESA) and 1,2-ethanedisulfonic acid (EDSA) were from Aldrich Chemical Co., Milwaukee, WI; β-alanine (BALA) and N-(2-acetamido)-2-aminoethanesulfonic acid (ACES) were from TCI America, Portland, OR; N-methyltaurine (MTAU) was from Pfaltz & Bauer, Waterbury, CT; 2-mercaptoethanesulfonic acid (MESA) was from ICN Biomedicals, Aurora, OH; α-sulfo-β-alanine (ASBA) was from USB, Cleveland, OH; and homotaurine (HMTAU) was from Acros Organics, Pittsburgh, PA. The samples of glucose, disodium phosphate, monosodium phosphate, GSH, cromolyn sodium (CROM), H_2O_2 (30%), 2-thiobarbituric acid (TBA), 5,5'-dithiobis-(2-nitrobenzoic acid) (DTNB), methemoglobin (metHb), sodium azide and 1,1,3,3-tetraethoxypropane (TEP) were from Sigma Chemical Co.. All other chemicals were of analytical reagent grade and were purchased from Aldrich Chemical Co. or J.T. Baker, Philipsburgh, NJ.

Animals

Male Sprague-Dawley rats, weighing 200-250 g, were purchased from Taconics Farms, Germantown, NY. The animals were housed in plastic cages, 6 per cage, in a temperature-regulated ($23\pm1°C$), constant humidity, room. During a 7-day acclimation period, the animals had free access to a pelleted diet (Purina Rat Chow, Ralston, Purina, St. Louis, MO) and tap water.

RBC Suspension

Rats were decapitated just before an experiment, and their blood samples collected into heparinized polyethylene test tubes immersed in ice. RBC suspensions were prepared according to the method of Sharma and

Premachandra[43]. After centrifugation of the blood sample at 700 x g for 10 min, and removal of the plasma and buffy coat by aspiration, the pellet was suspended in ice-cold isotonic phosphate-buffered saline (PBSG, 10 mM phosphate buffer pH 7.4-150 mM NaCl) solution, supplemented with 5 mM glucose and 1 mM sodium azide, and the suspension centrifuged at 700 x g for 10 min. After repeating this washing step once, the RBCs were resuspended in PBSG to a hematocrit of 20%, and stored on ice pending an experiment.

Table 1. List of sulfur-containing compounds tested in the present study

Compound	Abbreviation	Chemical formula
Hypotaurine	HYTAU	$H_2N-CH_2-CH_2-SO_2H$
Taurine	TAU	$H_2N-CH_2-CH_2-SO_3H$
Homotaurine	HMTAU	$H_2N-CH_2-CH_2-CH_2-SO_3H$
N-Methyltaurine	NMTAU	$H_3C-NH-CH_2-CH_2-SO_3H$
3-Mercaptoethanesulfonic acid	MESA	$HS-CH_2-CH_2-CH_2-SO_3H$
Aminomethanesulfonic acid	AMSA	$H_2N-CH_2-SO_3H$
Ethanesulfonic acid	ESA	$H_3C-CH_2-SO_3H$
1,2-Ethanedisulfonic acid	EDSA	$HO_3S-CH_2-CH_2-SO_3H$
Isethionic acid	ISA	$HOH_2C-CH_2-SO_3H$
α-Sulfo-β-alanine	ASBA	$H_2N-CH_2-CH(COOH)-SO_3H$
L-Cysteic acid	CA	$H_2N-CH(COOH)-CH_2-SO_3H$
N-Acetyl cysteine	NACYS	$HS-CH_2-CH(NHCOCH_3)-COOH$
ß-Alanine	BALA	$H_2N-CH_2-CH_2-COOH$
N-(2-Acetamido)-2-aminoethane-sulfonic acid	ACES	$H_2N-CO-CH_2-NH-CH_2-CH_2-SO_3H$

Test Solutions

Solutions of TAU, CROM and the other compounds listed in Table 1 were prepared freshly in PBSG to contain 480 mM. A working H_2O_2 solution was prepared by diluting the commercial solution with water to a final concentration of 20 mM. This solution was kept on ice when not in use.

Incubation Procedure

To a test tube marked "*Test*", the RBC suspension (~587 μL) and a solution of the test compound (~313 μL) were added. After gentle mixing, the mixture was incubated at 37°C for 15 min, mixed with the H_2O_2 solution (100 μL), incubated again at 37°C for 30 min, and centrifuged at 700 *x g* for 15 min. The supernatant was used for the assay of free hemoglobin (Hb), met-hemoglobin (metHb), potassium ions (K^+), and lactic dehydrogenase (LDH) activity. The pellet was used for the assay of malondialdehyde (MDA) and reduced glutathione (GSH). Parallel experiments were carried out in test tubes labeled "*Control Compound*" (containing RBC suspension, a test compound and PBSG), "*Control H_2O_2*" (containing RBC suspension, PBSG and H_2O_2) and "*Control Buffer*" (containing RBC suspension and PBSG). In some cases a "*Washed Test*" tube was prepared in identical manner as the "*Test*" tube except that after incubating the RBCs with a sulfur-containing compound, the cells were washed with PBSG prior to the addition of H_2O_2. In experiments where BALA was included, this compound was incubated with the RBC suspension at 37°C for 15 min before adding a test compound and/or H_2O_2.

Assay of Free Hb

The amount of Hb lost into the extracellular medium was measured spectrophotometrically at 540 nm using a commercial assay kit (Procedure No. 525-A, Sigma Chemical Co.) based on the oxidation of Hb and its derivatives (excepting sulfHb) to metHb with Drabkin's reagent. The concentration of Hb in the sample was derived from a calibration curve constructed from serial dilutions of a Hb standard preparation (Part No. 525-18, Sigma Chemical Co.), and was reported as mg/dL.

Assay of MetHb

MetHb was measured spectrophotometrically at 630 nm by the method of Rodkey and O'Neal[39]). In this case, the absorbance was read before and after the addition of alkaline KCN to the sample, to convert metHb to cyanmetHb. The amount of metHb present was derived from the difference between the two readings and a calibration curve for metHb prepared on the day of the analysis. Results were expressed in μg/dL.

Assay of MDA

MDA was determined as TBA reactive substances (TBARS) by the end-point assay method of Buege and Aust[10], following the reaction of MDA with TBA to yield a colored product whose absorbance was measured spectrophotometrically at 535 nm. The amount of TBARS present was calculated by reference to a calibration curve of MDA prepared on the day of the analysis from serial dilutions of a TEP stock solution, and was reported as ng/g of Hb.

Assay of K$^+$

The amount of K$^+$ released into the extracellular fluid was measured potentiometrically using a K$^+$-specific electrode (Cole-Palmer, Vernon Hills, IL). The concentration of electrolyte was derived from a standard curve for K$^+$ prepared with a commercial solution of KCl. The results were expressed as ppm.

Assay of GSH

The intracellular concentration of GSH was measured spectrophotometrically by the method of Ellman[18], which is based on the reaction of a sulfhydryl-containing compound with DTNB to form a color product whose absorption can be read at 412 nm. The concentration of the analyte was derived from a calibration curve for GSH, and was expressed as μmol/g of Hb.

Statistical Analysis of the Data

The experimental results were analyzed for statistical significance by unpaired Student's t-test, one-way analysis of variance (ANOVA), and Tukey's post-hoc test using a commercial software program (StatMost for Windows 2.01™ from Datamost Corporation, Salt Lake City, UT).. Differences were considered as statistically significant at p≤0.05. Values were reported as the means ± SEM for n = 6.

RESULTS

The incubation of RBCs with H_2O_2 (5-40 mM) led to a concentration-dependent loss of Hb, LDH, and K$^+$ and formation of metHb (Figs. 1 and 2). In contrast, pretreating the RBCs with TAU (25-175 mM) resulted in a concentration-dependent attenuation of the effects of H_2O_2 (20 mM), with the changes becoming significant at TAU additions equal or above 75 mM (Figs. 3 and 4). From the results of a study where a suspension of RBCs was incubated with fixed concentrations of TAU (150 mM), and H_2O_2 (20 mM) for different intervals of time, it was determined that the cellular changes brought about by H_2O_2 became maximal after about 15 min (Figs. 5 and 6). Hence, this incubation time was used in all subsequent experiments.

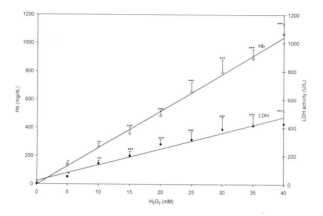

Figure 1. Effects of various concentrations of H_2O_2 (5-40 mM) on the leakage of LDH and Hb from RBCs (20% hematocrit). Values are given as the mean ± SEM (n = 6). Values differed significantly from a control treatment, PBSG alone, at *p<0.05, **p<0.01 and ***p<0.001.

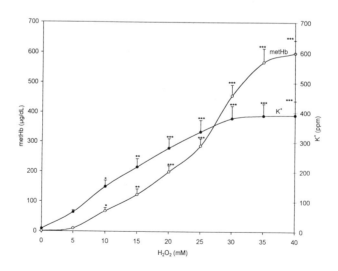

Figure 2. Effect of various concentrations of H_2O_2 (5-40 mM) on the leakage of K^+ and metHb from RBCs (20% hematocrit). Values are given as the mean ± SEM (n = 6). Values differed significantly from a control treatment, PBSG alone, at *p<0.05, **p<0.01 and ***p<0.001

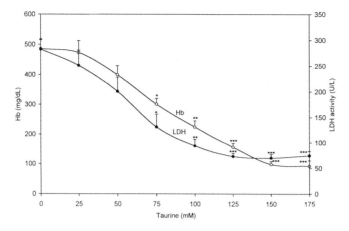

Figure 3. Effects of various doses of TAU (25-175 mM) on H_2O_2 (20 mM)-induced leakage of Hb and K^+ from RBCs (20% hematocrit). Values are given as the mean ± SEM (n = 6). Data were analyzed by unpaired Student's t-test. Values differed significantly from H_2O_2 alone, at *p<0.05, **p<0.01 and ***p<0.001.

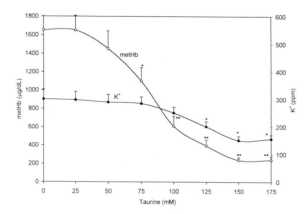

Figure 4. Effects of various doses of TAU (25-175 mM) on H_2O_2 (20 mM)-induced leakage of metHb and K^+ from RBCs (20% hematocrit). Values are given as the mean ± SEM (n = 6). Data were analyzed by unpaired Student's t-test. Values differed significantly from H_2O_2 alone, at *p<0.05 and **p<0.001.

The majority of the sulfur-containing compounds tested reduced the loss of K^+ caused by H_2O_2 by as much as 47%. In this regard, the various protecting compounds could be divided into those exerting a marked attenuation (i.e., >40% reduction, p<0.01: CA, TAU, ASBA, ACES, HMTAU, EDSA) and those exerting a moderate attenuation (i.e., >30% but <40% reduction, p<0.01: AMSA, ESA, ISA, HYTAU). In contrast, neither MESA nor MTAU counteracted the effect of H_2O_2 (Table 2).

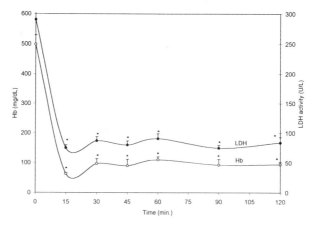

Figure 5. Effects of preincubation time with TAU (150 mM) on H_2O_2 (20 mM)-induced leakage of LDH and Hb from RBCs (20% hematocrit). Values represent the mean ± SEM (n = 6). Data were analyzed by unpaired Student's t-test. All values differed significantly from H_2O_2 alone, at *p<0.01.

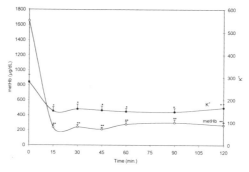

Figure 6. Effects of preincubation time with TAU (150 mM) on H_2O_2 (20 mM)-induced leakage of metHb and K^+ from RBCs (20% hematocrit). Values represent the mean ± SEM (n = 6). Data were analyzed by unpaired Student's t-test. All values differed significantly from H_2O_2 alone, at *p<0.01 and **p<0.001.

The protection offered by the various sulfur-containing compounds against H_2O_2-related losses in LDH and Hb generally followed the trend noted for the loss of K^+ (Table 2). Reductions of LDH losses ranged from 62 to 94% (p<0.001), with the greatest protection (>90% reduction) coming from ASBA and CA. Reductions of Hb release ranged from 88 to 95% (p<0.001). In both instances, MESA and MTAU accentuated the damaging actions of H_2O_2 to a significant extent (p<0.01).

Most of the test compounds protected RBCs against H_2O_2-related formation of metHb, with TAU, ISA, EDSA, ESA and ASBA offering the greatest protection (>75% reduction, p<0.001), and ACES, AMSA, HMTAU and HYTAU offering lesser protection (>40% <70% reduction, p<0.01).

Table 2. Effects of preincubation with TAU (150 mM) on H_2O_2 (20 mM)-induced leakage of Hb, metHb, LDH and K^+ from RBCs (20% hematocrit)[a]

Compound	Hb (mg/dL)	MetHb (μg/dL)	LDH (U/L)	K^f (ppm)
CROM	17 ± 1 [e,l]	35 ± 2 [e,l]	7 ± 1 [e]	5 ± 1 [e,l]
CA	34 ± 2 [e,k]	778 ± 69 [e,h,l]	21 ± 2 [e,f]	149 ± 10 [d,h]
MESA	701 ± 17 [d,h,l]	3589 ± 98 [e,h,l]	220 ± 32 [d,h]	308 ± 24 [h,l]
MTAU	609 ± 18 [d,h,l]	4501 ± 115 [e,h,l]	255 ± 15 [c,h]	332 ± 14 [h,l]
EDSA	56 ± 3 [e,g]	389 ± 46 [e,h,l]	41 ± 2 [e,g]	161 ± 21 [d,h]
AMSA	22 ± 3 [e,k]	782 ± 13 [e,h,l]	106 ± 8 [e,h]	181 ± 19 [d,h]
ACES	42 ± 3 [e,f,j]	618 ± 89 [e,h,l]	34 ± 5 [e,g]	156 ± 17 [d,h]
ASBA	18 ± 3 [e,l]	485 ± 13 [e,h,l]	18 ± 2 [e,f]	156 ± 22 [d,h]
HMTAU	46 ± 4 [e,f,j]	1091 ± 99 [e,h,l]	45 ± 3 [e,g]	161 ± 20 [d,h]
HYTAU	51 ± 4 [e,f]	1136 ± 58 [e,h,l]	50 ± 7 [e,g]	190 ± 20 [d,h]
ESA	62 ± 2 [e,g]	414 ± 84 [e,h,l]	66 ± 2 [e,h]	189 ± 21 [d,h]
NACYS	90 ± 13 [e,h,k]	ND[b]	101 ± 16 [e,h]	201 ± 26 [d,h]
ISA	69 ± 7 [e,g]	312 ± 5 [e,h,j]	55 ± 6 [e,g]	191 ± 22 [d,h]
TAU	59 ± 3 [e,g]	239 ± 29 [e,h]	69 ± 11 [e,h]	151 ± 18 [d,h]
H_2O_2	486 ± 29 [h,l]	2004 ± 276 [h,l]	282 ± 20 [h]	280 ± 30 [h,k]
PBSG	16 ± 2 [e,l]	25 ± 0 [e,l]	8 ± 0.5 [e]	2 ± 0 [e,l]

[a]Values represent mean ± SEM (n = 6). Data were analysed by unpaired Student's t-test and one-way ANOVA followed by Tukey's *post hoc* test. Values differed significantly at [c]p <0.05, [d]p< 0.01, [e]p<0.001 vs. H_2O_2, [f]p<0.05, [g]p<0.01, [h]p<0.001 vs. PBSG and [j]p<0.05, [k]p<0.01, [l]p<0.001 vs. TAU. [b]Not determined due to assay interference.

As inferred from the formation of MDA and the depletion of cellular GSH, a brief exposure of RBCs to H_2O_2 in the presence of sodium azide (to negate the action of catalase) resulted in lipid peroxidation. TAU and its congeners attenuated these effects to different extents, with most test compounds displaying a moderate inhibitory action (22-28%, p<0.01: TAU, HMTAU, CA, ACES, ISA, ESA, AMSA, EDSA) on H_2O_2-mediated MDA formation, or a lesser, but still significant, reduction (ca. 13%, p<0.01: HYTAU, ASBA). Likewise, the various sulfur-containing compounds could be divided according to the magnitude of their protective actions against H_2O_2-related GSH depletion into those exerting a significant inhibitory action (i.e., 43-49%, p<0.01: HYTAU, ISA, TAU) and those with a much weaker action (14-25%, p<0.05: ESA, ACES). MESA and MTAU magnified both indices of lipid peroxidation.

To further gauge the effectiveness of TAU against H_2O_2-induced lipid peroxidation, their effects were compared with those of NACYS and CROM. NACYS is a known biosynthetic precursor of GSH[34] as well as a biological antioxidant[5]. On the other hand, CROM is regarded as a mem-

Table 3. Effects of preincubation with sulfur-containing compounds and cromolyn sodium (150 mM) on H_2O_2 (20 mM)-induced MDA formation and GSH depletion in RBCs (20% hematocrit)[a]

Compound	MDA (nM/g Hb)	GSH (μM/g Hb)
CROM	$1.69 \pm 0.01^{+++,•••}$	$5.01 \pm 0.30^{+++,•}$
CA	$1.21 \pm 0.04^{***,+++}$	$5.84 \pm 0.50^{*,+}$
MESA	$1.87 \pm 0.08^{+++,•••}$	$4.65 \pm 0.08^{+++,•}$
MTAU	$1.98 \pm 0.09^{**,+++,•••}$	$4.54 \pm 0.09^{+++,••}$
EDSA	$1.30 \pm 0.03^{**,+++}$	$5.48 \pm 0.10^{++,•}$
AMSA	$1.27 \pm 0.08^{**,+++}$	$5.83 \pm 0.20^{*,++}$
ACES	$1.25 \pm 0.02^{**,+++}$	$5.77 \pm 0.28^{*,++}$
ASBA	$1.46 \pm 0.09^{+++,•}$	$5.88 \pm 0.51^{*,+}$
HMTAU	$1.19 \pm 0.05^{***,+++}$	$6.01 \pm 0.50^{*,+}$
HYTAU	$1.45 \pm 0.03^{+++,•}$	$7.18 \pm 0.30^{**}$
ESA	$1.28 \pm 0.08^{**,+++}$	$6.01 \pm 0.38^{*,+}$
NACYS	$0.78 \pm 0.01^{***,+++,•••}$	$7.40 \pm 0.50^{***}$
ISA	$1.27 \pm 0.2^{**,+++}$	$6.99 \pm 0.10^{**}$
TAU	$1.20 \pm 0.04^{**,+++}$	$6.89 \pm 0.40^{**}$
H_2O_2	$1.66 \pm 0.05^{+++,•••}$	$4.81 \pm 0.40^{+++,•}$
PBSG	$0.36 \pm 0.02^{***,•••}$	$8.10 \pm 0.80^{***}$

[a]Values represent the mean ± SEM (n = 6). Data were analyzed by unpaired Student's t-test and one-way ANOVA, followed by Tukey's *post hoc* test. Values differed significantly at $*p <0.05$, $**p< 0.01$, $***p<0.001$ vs. H_2O_2, $+p<0.05$, $++p<0.01$, $+++p<0.001$ vs. PBSG and $•p<0.05$, $••p<0.01$, $•••p<0.001$ vs. TAU.

Table 4. Effects of TAU and CROM on H_2O_2-induced leakage of Hb, LDH and K^+ with and without a washing step of the RBCs prior to the addition of H_2O_2[a]

RBC sample	Hb (mg/dL)	LDH (U/L)	K^+ (ppm)
Washed	$15.01 \pm 1.67^{***}$	$5.12 \pm 0.09^{***}$	$3.87 \pm 0.02^{***}$
Unwashed	$16.01 \pm 1.77^{***}$	$4.11 \pm 0.51^{***}$	$2.09 \pm 0.01^{***}$
Washed/H_2O_2	478.15 ± 35.14	294.65 ± 32.18	289.01 ± 35.62
Unwashed/H_2O_2	486.21 ± 28.57	282.02 ± 19.87	280.08 ± 30.03
TAU/Washed/H_2O_2	480.98 ± 27.64	279.65 ± 17.26	284.36 ± 33.65
TAU/Unwashed/H_2O_2	$59.24 \pm 3.23^{***}$	$68.89 \pm 5.01^{***}$	$151.04 \pm 18.05^{**}$
CROM/Washed/H_2O_2	$24.15 \pm 65^{***}$	$9.25 \pm 2.01^{***}$	$8.71 \pm 0.11^{***}$
CROM/Unwashed/H_2O_2	$17.04 \pm 1.91^{***}$	$7.09 \pm 1.87^{***}$	$5.00 \pm 0.01^{***}$

[a]Values are given as the mean ± SEM (n = 6). Values differed significantly from unwashed/H_2O_2 treatment at $**p < 0.01$ and $***p < 0.001$.

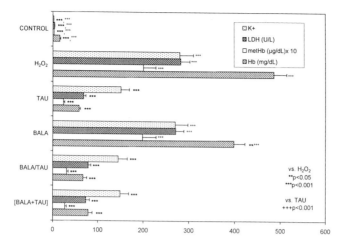

Figure 7. Effects of TAU and/or BALA, (150 mM) on H_2O_2 (20 mM)-induced leakage of K^+, LDH, metHb and Hb from RBCs (20% hematocrit). Values represent the mean ± SEM (n = 6). Data were analyzed by unpaired Student's t-test and one-way ANOVA, followed by Tukey's post hoc test.

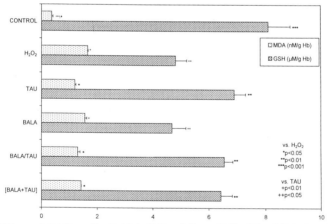

Figure 8. Effects of TAU and/or BALA (150 mM) on H_2O_2 (20 mM)-induced MDA and GSH in RBCs (20% hematocrit). Values represent the mean ± SEM (n = 6). Data were analyzed by unpaired Student's t-test and one-way ANOVA, followed by Tukey's post hoc test.

brane stabilizing agent, especially for mast cells[17,26]. Although TAU appeared better than NACYS in preventing the leakage of K^+, LDH and Hb, it was less effective in suppressing the formation of MDA and in preserving the intracellular loss of GSH (values for metHb were not compared since NACYS interfered with the assay method for this substance). The opposite trend was observed for CROM, namely a greater protective action than TAU against membrane leakage and a lesser effect on indices of lipid peroxidation (Tables 2 and 3).

DISCUSSION

RBCs are suited as an experimental model for studying the generation of active oxygen species and for assessing the effects of oxidative stress on cell integrity. Their membranes are rich in polyunsaturated fatty acids, they are constantly exposed to high concentrations of oxygen, contain a powerful transition metal catalyst[12], and are not amenable to repair or regeneration[52]. On the other hand, H_2O_2 is a long-living, lipophilic prooxidant that is present in RBCs as a result of enzyme catalysis, the dismutation of the superoxide anion, or drug-induced oxidative denaturation of oxyHb[4,40]. Upon reacting with surface thiol groups, H_2O_2 can bring about changes in membrane barrier properties[54]; or after entering cells, it could react with transition metals such as iron to generate the highly reactive hydroxyl radical[36] or with hemoproteins to form lipid peroxidating species[24]. In this manner, this oxidant will damage and degrade Hb[45] and cytoskeletal proteins[46,47], perturb the morphology[9,46] and barrier properties of the membrane[58], and induce hemolyis[13]. In this respect, two major defenses are available to RBCs against H_2O_2, the NADPH-GSH dependent glutathione peroxidase/reductase system and catalase[19]. GSH can spare RBCs from damage by reactive intermediates such as free radicals and peroxides or against electrophiles like quinones and hemin through the reactivity of its sulfhydryl group[44]. Catalase acts as a H_2O_2 scavenger whose activity, like that of the glutathione peroxidase/reductase system, is highly dependent on the availability of NADPH[19]. In the event of an overwhelming concentration of H_2O_2 or of an impaired redox cycling of GSH, both catalase activity[14] and GSH levels[8,14] can become significantly diminished and the RBC membrane susceptible to lipid peroxidation[54].

This study found TAU to attenuate the damaging and peroxidative effects of H_2O_2 on RBCs as manifested by the exit of K^+, LDH and Hb, the formation of MDA and metHb, a loss in catalase activity, and the depletion of GSH (Tables 2 and 3). These protective actions of the TAU molecule could be altered both in direction and in magnitude by certain structural modifications. For instance, greater protection against the losses of LDH, Hb and GSH than TAU was attained with the immediate higher homolog, HMTAU, than with the immediate lower homolog, AMSA. Enhanced protection against the leakage of intracellular components, but not against oxidative changes, was noted upon the introduction of either an α-COOH (like in ASBA) or β-COOH (like in CA) group. ESA, the deaminated analog of TAU, demonstrated a weaker action against K^+ leakage, the form-ation of metHb and MDA, and the loss of GSH due to oxidative stress. Whereas no significant changes in TAU were attained with ISA, the β-hydroxyl analog, or with EDSA, the β-sulfonic analog; N-2-acetamido substitution (like in ACES) had a positive influence on membrane perm-eability but a negative one on the antioxidative properties of the parent molecule. In contrast, MESA, the β-sulfhydryl analog and MTAU, the N-methylated analog of TAU, not only did not protect but also accentuated the altering actions of H_2O_2. When compared with TAU, HYTAU, the corresponding sulfinate analog, was shown to be about equal in suppressing MDA, less effective in preventing K^+ loss and metHb formation, and better in precluding the loss of both LDH and GSH.

At least two earlier studies have examined the actions of TAU on RBC membranes. In one, TAU was found to inhibit the hemolysis of dog RBCs induced by a treatment with HOCl by acting as a trap for oxidized chlorine from HOCl. In the other, TAU was found to prevent the lysis of either dog RBCs or egg yolk lecithin liposomes loaded with glu-cose following an exposure to an oxygen-radical generating system[27]. As TAU did not prevent the oxidative degradation of lecithin it was proposed that its protective actions were the result of a direct, stabilizing interaction with membrane proteins rather than from antioxidation[27]. In the light of these reports, the experiments summarized in Table 4 were performed. In essence, RBCs that had been incubated with TAU were rinsed with isotonic PBSG prior to the addition of H_2O_2 to determine whether or not TAU remained on the RBC surface. Subsequently, indices of membrane leakage were measured in the supernatant fraction. From the data presented in the same Table it is evident that the interaction of TAU with the RBC membrane is a rather weak one as protection by TAU was eliminated by a rinsing with PBSG. This behavior of TAU contrasts markedly with that verified for CROM, whose protective actions on the membrane were not only greater than those conferred by TAU but, more importantly, impervious to the rinsing step (Table 4). Further evidence that TAU is acting on the RBC surface was derived from an experiment in which RBCs were pretreated with BALA, an inhibitor of TAU uptake. From the results shown in Figures 7 and 8, it was evident the actions of TAU on H_2O_2-related cellular alterations were not modified by BALA, a finding that is consistent with an extracellular mechanism.

It has been proposed that many of the biological actions of TAU are membrane-based, involving the participation of Ca^{2+}, and occurring through an interaction with membrane phospholipids[23]. Experiments with buffered suspensions of resealed erythrocyte ghosts found the membrane fluidity to be unaffected by TAU alone, but to decrease it[29] in the presence of Ca^{2+}. TAU possesses two structural features relevant to its biological properties, its ionic nature and its steric and electronic similarity to the zwitterionic membrane phospholipids, PC and PE[22]. At the physiological pH of 7.4, TAU exists as a zwitterion of a strong acid, represented by its sulfonate group (pK$_a$ of 1.5) and the highly ionized (>95%) amino group (pK$_b$ = 8.74). Thus, TAU may weakly ion pair with the positively charged head groups of neutral phospholipids to influence their packing density and the strength of the van der Waals interactions among lipid chains and help to counteract decrease in packing, increase in fluidity and increase in hydrophobicity[35] caused by H$_2$O$_2$. Not surprisingly, TAU exhibits low permeability through bilayer membranes and its cellular uptake is predominantly through the participation of an active and saturable transport system[23,49]. An alternative view regarding the interaction of TAU with the membrane surface has been elegantly put forth in a model by Schaffer and Azuma[41] that assumes a direct ion-pair interaction between TAU and charged membrane proteins or between TAU and a phospholipid-protein complex to cause changes in membrane function. This concept was recently invoked in a study with human RBCs that found TAU to be able to restore the activity of the membrane-bound protein Na-K+ ATPase lost to an exposure to ozone or to a cholesterol enrichment[37].

The results of the present work point to a membrane-related protective effect for TAU against H$_2$O$_2$-induced alterations. Although protection was verified for all compounds bearing a sulfonate group plus or minus an additional functional group with potential for interacting with membrane components through dipolar forces, the actions of the deaminated and disulfonated analogs of TAU were about equal to TAU in their membrane stabilizing abilities. Hence, one can also assume that only one sulfonate group is needed for a positive interaction with the membrane surface. Furthermore, whereas most of the structural modifications of the TAU molecule evaluated here provided a greater and more consistent effect on the leakage of macromolecules like Hb and LDH than on that of K$^+$, the same was not true as it relates to oxidative indices. In the latter case, differences in the magnitude of the protective effects varied within a narrow range or did not always correlate with the potency order observed for indices of membrane leakage.

Certain structural modifications at the amino end, such as N-alkylation and substitution by a sulfhydryl group, led to results that were unexpected and not amenable to a simple explanation. A hemolyzing effect for the sulfhydryl analog of TAU is plausible as other reagents containing this sulfur functionality are known to induce the formation of aqueous leaks in the RBC membrane by promoting the cross linking of membrane proteins[15], an event that could also facilitate the entry of H$_2$O$_2$ into the cell to cause a massive depletion of GSH. Furthermore, protection against cell leakage, but not against oxidative changes, were improved by the addition of either an α- or β-COOH group to the TAU structure, which at physiolgical pH will represent an additional negatively charged center for ion pairing with oppositely charged phospholipid head groups or proteins. The same results were observed with the N-2-acetamido substituted sulfonate analog, ACES, probably because of additional contribution by the C-O bond of the carbonyl group, which is weakly polarized and therefore has a partial ionic character[57].

The antioxidant capacity of TAU has been compared with that of chemically-related compounds, especially to its sulfinate biochemical precursor HYTAU and its structural analog BALA. By assessing the inhibition of spontaneous chemiluminescence in rat brain homogenates[11] and the formation of TBARS in spermatozoa suspensions[1] and rod outer segments[32,33], both TAU and BALA were earlier shown to be either less effective or ineffective[32] as an antioxidant than HYTAU. Likewise, under cell-free experimental conditions TAU did not readily react with superoxide radical, hydrogen peroxide or hydroxyl radical, and its reaction product HOCl, TAU chloramine, was still sufficiently oxidizing to inactivate α_1-antiproteinase[2]. Moreover, in cell-free oxidizing systems HYTAU displayed excellent scavenging properties toward the hydroxyl radical and HOCl while remaining unreactive toward superoxide radical[2] and H_2O_2. The present evidence indicates that in a RBC-H_2O_2 experimental model like the one used here, TAU may be a better membrane stabilizer than HYTAU and, that except for its action on GSH depletion, at least as good a protector against MDA and metHb formation than its sulfinate analog (Tables 2 and 3). In this respect, this laboratory has previously verified that at equimolar concentrations, TAU, HYTAU and BALA can provide about equal protection to the rat liver against MDA formation and GSH depletion induced by an acute dose of ethanol[50]. Hence, some of the differences between the antioxidant actions for TAU and HYTA reported here and those emanating from other laboratories could be due to differences in the test system and oxidant used.

In summary, the results gathered in this investigation suggest that *in vitro* TAU can protect RBCs against membrane and biochemical changes resulting from exposure to a high concentration of H_2O_2 by acting both as an antioxidant and as a membrane stabilizer. The protection by TAU and its congeners is found to center on the sulfonate group and to remain after the introduction of structural modifications such as chain lengthening or shortening, deamination, carboxylation, N-2-acetamido substitution, reduction to a sulfinate group, or replacement of a hydroxyl or sulfonic group for the amino group. Yet a role for the amino group of TAU cannot be dismissed since structural modifications at this end like N-alkylation or replacement by a sulfhydryl group translated into a total loss of all protective action offered by the parent compound.

REFERENCES

1. Alvarez, J.G., and Storey, N.T., 1983, Taurine, hypotaurine, epinephrine and albumin inhibit lipid peroxidation in rabbit spermatozoa and protect against loss of motility. *Biol. Reprod.* 29:548-555.
2. Aruoma, O.I., Halliwell, B., Hoey, B.M., and Butler, J., 1988, The antioxi-dant action of taurine, hypotaurine and their metabolic precursors. *Biochem. J.* 256:251-255.
3. Azuma, J., Hamaguchi, T., Ohta, H., Takihara, K., Awata, N., Sawamura. A., Harada, H., and Kishimoto, S., 1987, Calcium overload-induced myo-cardial damage caused by isoproterenol and by adriamycin: possible role of taurine in its prevention. In *The Biology of Taurine. Methods and Mechan-isms* (R.J. Huxtable, F. Franconi, and A. Giotti, eds.), Plenum Press, New York, NY, pp. 167-179.
4. Babior, B.M., 1981, Oxidizing radicals and red cell destruction. In *The Function of Red Blood Cells: Erythrocyte Pathobiology* (D.H.F. Wallach, ed.), Alan R. Liss, New York, NY, pp. 171-190.
5. Baker, H.W., Brindle, J., Irvine, D.S., and Aitken, R.J., 1996, Protective effect of antioxidants on the impairment of sperm mobility by activated polymorphonuclear leukocytes. *Fertil. Steril.* 65:411-419.
6. Banks, M.A., Porter, D.W., Martin, W.G., and Castranova, V., 1992, Taurine protects against oxidant injury to rat alveolar pneumocytes. In *Taurine* (J.B.Lombardini, S.W. Schaffer, and J. Azuma., eds.), Plenum Press, New York, NY, pp. 341-354.
7. Baskin, S.I., Wakayama, K., Banks, M.A., Porter, D.W., and Salem, H., 1997, Antioxidant effects of hypotaurine and taurine. In *Oxidant, Antioxidants and Free Radicals* (S.I. Baskin and H. Salem, eds.), Taylor & Francis, Bristol, PA, pp. 193-202.
8. Baysal, E., Sullivan, S.G., and Stern, A., 1988, Influence of exogenous iron and ascorbate on H_2O_2-induced glutathione oxidation in red cells. *Biochem. Int.* 17:211-215.
9. Brunauer, L.S., Moxness, M.S., Huestis, W.H., 1994, Hydrogen peroxide oxidation induces the transfer of phospholipids from the membrane into the cytosol of human erythrocytes. *Biochemistry* 33:4527-4532.
10. Buege, J.,A., and Aust, S.D., 1978, Microsomal lipid peroxidation, In *Methods of Enzymology*, vol. 52, (S. Fleischer and L. Packer, eds.), Academic Press, New York, NY, pp. 302-310.
11. Cañas, P., Guerra, R. and Valenzuela, A., 1989, Antioxidant properties of hypotaurine: comparison with taurine, glutathione and β-alanine. *Nutr. Rep. Int.* 39:433-438.
12. Clemens, M.R., and Waller, H.D., 1987, Lipid peroxidation in erythrocytes. *Chem. Phys. Lipids* 45:251-268.
13. Cohen, G., and Hochstein, P., 1963, Glutathione peroxidase: the primary agent for the elimination of hydrogen peroxide in erythrocytes. *Biochemistry* 2:1420-1428.
14. Cohen, G., and Hochstein, P., 1964, Generation of hydrogen peroxide in erythrocytes by hemolytic agents. *Biochemistry* 3:895-900.
15. Deuticke, B., Lütkemeier, P., and Sistemich, M., 1984, Ion selectivity of aqueous leaks induced in the erythrocyte membrane by crosslinking of membrane proteins. *Biochim. Biophys. Acta* 775:150-160.
16. Ebrahim, A.S., and Sakthisekaran, D., 1997, Effect of vitamin E and taurine treatment on lipid peroxidation and antioxidant defense in perchloroethylene-induced cytotoxicity in mice. *Nutr. Biochem.* 8:270-274.
17. Eleno, N., Gajate, E., Macias, J., and Garay, K.P., 1999, Enhancement by reproterol of the ability of disodium cromoglycate to stabilize rat masto-cytes. *Pulm, Pharmacol. Ther.* 12:55-60.
18. Ellman, G.L., 1959, Tissue sulfhydryl groups. *Arch. Biochem. Biophys.* 82: 70-77.
19. Gaetani, G.F., Galiano, S., Canepa, L., Ferraris, A.M., and Kirkman, H.N., 1989, Catalase and glutathione peroxidase are equally active in detoxifi-cation of hydrogen peroxide in human erythrocytes. *Blood* 73:334-339.

20. Giri, S.N. and Wang, Q., 1992, Taurine and niacin offer a novel therapeutic modality in prevention of chemically-induced pulmonary fibrosis in hamsters. In *Taurine* (J.B.Lombardini, S.W. Schaffer, and J. Azuma., eds.), Plenum Press, New York, NY, pp.329-340.
21. Gordon, R.E., Shaked, A.A., and Solano, D.F., 1986, Taurine protects hamster bronchíoles from acute NO_2-induced alterations. A histologic, ultrastructural and freeze-fracture study. *Am. J. Pathol.* 125:585-600.
22. Huxtable, R.J., and Sebring, L.A., 1986, Towards a unifying theory for the actions of taurine. *Trends Pharmacol. Sci.* 7:481-485.
23. Huxtable, R.J., 1992, Physiological actions of taurine. *Pharmacol. Rev.* 72:101-163.
24. Kanner, J., and Harel, S., 1986, Initiation of membrane lipid peroxidation by activated metmyoglobin and methemoglobin. *Arch. Biochem. Biophys.* 237: 314-321.
25. Kaplan, B., Aricioglu, A., Erbas, D., Erbas, S., and Turkozkan, N., 1993, The effects of taurine on perfused heart muscle malondialdehyde levels. *Gen. Pharmacol.* 24:1411-1413.
26. Mancel, E., Drouet, M., and Sabbah, A., 1999, Membrane stabilizers, *Allerg. Immunol.* (Paris). 31:103-105.
27. Nakamura, T., Ogasawara, M., Koyama, I., Nemoto, M., and Yoshida, T., 1993, The protective effect of taurine on the biomembrane against damage produced by oxygen radicals. *Biol. Pharm. Bull.* 16:970-972.
28. Nakashima, T., Seto, Y., Nakajima, T., Shima, T., Sakamoto, Y., Cho, N., Sano, A., Iwai, M., Kagawa, K., Okanoue, T., and Kashima, K., 1990, Calcium-associated cytoprotective effect of taurine on the calcium and oxygen paradoxes in isolated rat hepatocytes. *Liver* 10:167-172.
29. Nakashima, T., Shima, T., Sakai, M., Yama, H., Mitsuyoshi, H., Inaba, K., Matsumoto, Y., Sakamoto, K., Kashima, K., and Nishikawa, H., 1996, Evidence of a direct action of taurine and calcium on biological membranes. *Biochem. Pharmacol.* 52:173-176.
30. Obrosova, I.G., and Stevens, M.J., 1999, Effect of dietary taurine supple-mentation on GSH and NAD(P)-redox status, lipid peroxidation, and energy metabolism in diabetic precataractous lens. *Invest. Ophtalmol. Vis. Sci.* 40: 680-688.
31. Pasantes-Morales, H., Wright, C.E., and Gaull, G.E., 1984, Protective effect of taurine, zinc and tocopherol on retinol-induced damage in human lympho-blastoid cells. *J. Nutr.* 114:2256-2261.
32. Pasantes-Morales, H., and Cruz, C., 1985, Taurine and hypotaurine inhibit light-induced lipid peroxidation and protect rod outer segment structure outer segments. *Brain Res.* 330:154-157.
33. Pasantes-Morales, H., Wright, C.E., and Gaull, G.E., 1985, Taurine protec-tion of lymphoblastoid cells from iron-ascorbate induced damage. *Biochem. Pharmacol.* 34:2205-2207.
34. Penttilä, K.E., 1990, Role of cysteine and taurine in regulating glutathione synthesis by periportal and perivenous hepatocytes. *Biochem. J.* 269:659-664.
35. Pradhan D., Weiser, M, Lumley-Spanski, K., Frazier, D., Kemper, S., Williamson, P., and Schiegel, R.,A., 1990, Peroxidation-induce perturbations of erythrocyte lipid organization. *Biochim. Biophys. Acta* 1023:398-404.
36. Puppo, A., and Halliwell, B., 1988. Formation of hydroxyl radicals from hydrogen peroxide in the presence of iron. *Biochem. J.* 249:185-190.
37. Qi, B.L.F., Yamagami, T., Naruse, Y., Sokejima, S., and Kagamimori, S., 1995, Effects of taurine on depletion of erythrocyte membrane Na-K ATPase activity due to ozone exposure or cholesterol enrichment. *J. Nutr. Sci. Vita-minol.* 41:627-634.
37. Redmond, H.P., Wang, J.H., and Bouchier-Hayes, D., 1996, Taurine attenuates nitric oxide and reactive oxygen intermediate-dependent hepatocyte injury. *Arch. Surg.* 131:1280-1287.

39. Rodkey, F.L., and O'Neal, J.D., 1974, Effects of carboxymethemoglobin on the determination of methemoglobin in blood. *Biochem. Med.* 9:261-270.
40. Saltman, P., 1989, Oxidative stress: a radical view. *Semin. Hematol.* 26:249-256.
41. Schaffer, S.W. and Azuma, J., 1991, Review: Myocardial physiological effects of taurine and their significance. In *Taurine* (J.B.Lombardini, S.W. Schaffer, and J. Azuma., eds.), Plenum Press, New York, NY, pp.105-120.
42. Schuller-Levis, G., Quinn, M.R., Wright. C.E., and Park, E., 1994, Taurine protects against oxidant-induced lung injury: Possible mechanism(s) of action. In *Taurine in Health and Disease* (R. Huxtable and D.V. Michalk, eds.), Plenum Press, New York, NY, pp. 31-39.
43. Sharma, R., and Premachandra, B.R., 1991, Membrane bound hemoglobin as a marker of oxidative injury in adult and neonatal red blood cells. *Biochem. Med. Met. Biol.* 46:33-44.
44. Shviro, Y., and Shaklai, N., 1987, Glutathione as a scavenger of free hemin. A mechanism of preventing red cell membrane damage. *Biochem. Pharmacol.* 36:3801-3807.
45. Snyder, L.M., Sauberman, N., Condara, H., Doilan, J., Jacobs, J., Szyman-ski, I., and Fortier, N.L., 1981, Red cell membrane response to hydrogen peroxide-sensitivity in hereditary xerocytosis and in other abnormal red cells. *Br. J. Haematol.* 48:435-444.
46. Snyder, L.M., Fortier, N.L., Trainor, J., Jacobs, J., Leb, L., Lubin, B., Chiu, D. Shohet, S., and Mohandas, N., 1985, Effect of hydrogen peroxide expo-sure on normal human erythrocyte deformability, morphology, surface characteristics, and spectrin-hemoglobin cross-linking. *J. Clin. Invest.* 76: 1971-1977.
47. Snyder, L.M., Fortier, N.L., Leb, L., McKenney, J., Trainor, J., Sheerin, H., and Mohan-das, N., 1988, The role of membrane protein sulfhydryl groups in hydrogen peroxide-mediated membrane damage in human erythrocytes. *Biochim. Biophys. Acta* 937:229-240.
48. Stocks, J., and Dormandy, T.L., 1971, The autoxidation of human red cell lipids induced by hydrogen peroxide. *J. Haematol.* 20:95-111.
49. Tallan, H.H., Jacobson, E., Wright. C.E., Schneidman, K., and Gaull, G.E., 1983, Taurine uptake by culture human lymphoblastoid cells. *Life Sci.* 33: 1853-1860.
50. Theofanopoulos, V., Pokhrel, P., and Lau-Cam, C.A., 1997, Structure-activity determi-nants of the effectrs of taurine (2-aminoethanesulfonic acid) and related sulfur-containing compounds on ethanol-induced oxidative stress in the rat (Abstract). *Amino Acids* 13:87-88.
51. Thomas, E.L., Grisham, M.B., Melton, D.F., and Jefferson, M.M., 1985, Evidence for a role of taurine in the in vitro oxidative toxicity of neutrophils toward erythrocytes. *J. Biol. Chem.* 260:3321-3329.
52. Tozzi-Ciancarelli, M.C., Di Massimo, C., D'Orazio, M.C., Mascioli, A., Di Giulio, A., and Tozzi, E., 1990, Effect of exogenous hydrogen peroxide on human erythrocytes. *Cell. Mol. Biol.* 36:57-64.
53. Trachtman, H., Futterweit, S., and Bienokowski, R.S., 1993, Taurine prevents glucose-induced lipid peroxidation and increased collagen production in cultured rat mesangial cells. *Biochem. Biophys. Res. Commun.* 191:759-765.
54. van der Zee, J., Dubbelman, T.M.A.R., and van Steveninck, J., 1985, Peroxide-induced damage in human erythrocytes. *Biochim. Biophys. Acta* 818:38-44.
55. Venkatesan, N., and Chandrakasan, G., 1994, In vivo administration of taurine and niacin modulate cyclophosphamide-induced lung injury. *Eur. J. Pharmacol.* 292:75-80.
56. Wang, J.H., Redmond, H.P., Watson, R.W.G., Condron, C., and Bouchier-Hayes, D., 1996, The beneficial effect of taurine on the prevention of human endothelial cell death. *Shock* 6:331-338.
57. Whitfield, R.C., 1966, *A Guide to Understanding Basic Organic Reactions*, Houghton Mifflin Co., Boston, MA, p. 16.
58. Williams, M., Lagerberg, J.W.M., van Steveninck, J, and van der Zee, J., 1995, The effect of protoporphyrin on the susceptibility of human erythro-cytes to oxidative stress: expo-sure to hydrogen peroxide. *Biochim. Biophys. Acta* 1236:81-88.

59. Wright, C.E., Lin, T.T., Lin, Y.Y., Sturman, J.A., and Gaull, G.R., 1985, Taurine scavenges oxidized chloride in biological systems, In *Taurine: Biological Actions and Clinical Perspectives* (S.S. Oja, L. Ahtee, P. Kontro, and M.K. Paasonen, eds.), Alan R. Liss, New York, NY, pp. 137-147.

REGULATION OF HIGH AFFINITY TAURINE TRANSPORT IN GOLDFISH AND RAT RETINAL CELLS

L. Lima, S. Cubillos and A. Guerra

Laboratorio de Neuroquímica, Centro de Biofísica y Bioquímica, Instituto Venezolano de Investigaciones Científicas (IVIC), Apdo. 21827, Caracas 1020-A, Venezuela

ABSTRACT

Adaptive regulation and modulation by phosphorylation are mechanisms by which some cells control taurine transport. Goldfish and rat retinal cells were incubated with the activator of protein kinase C, phorbol 12,13-dibutyrate (PDBu), or the inhibitor of protein phosphatases, okadaic acid (OKA). OKA, 1 nM, inhibited the uptake of taurine at short period of incubation in goldfish retinal cells, and at low concentrations in rat retinal cells incubated with the inhibitor for 1 h. PDBu treatment did not produce significant effects. Isolated Müller cells from the goldfish retina presented a clear adaptive regulation and a decrease of taurine uptake by increasing phosphorylation either by the stimulation of PKC with PDBu or the inhibition of phosphatases with OKA.

INTRODUCTION

Taurine plays a role as a trophic agent in the retina[17,19,23,35]. Among the mechanisms studied for this effect are an increase in calcium fluxes[21] and a modification in phosphorylation[22]. Taurine must stimulate outgrowth by entering cells, as the inhibitor of taurine uptake, guanidinoethane sulfonate (GES), reduces its trophic effect[25].

Some evidence of taurine transport regulation without relation to a mechanism of action has been reported. For instance, illumination modulates

Taurine 4, edited by Della Corte *et al.*
Kluwer Academic / Plenum Publishers, New York, 2000.

fluxes of taurine in the cat retina[33], denervation increases taurine uptake in rat skeletal muscle[12], proliferative and non-proliferative areas of the retina possess differential taurine uptake[20], optic axotomy reduces taurine uptake in goldfish retinal cells[7], aging decreases the high-affinity component of taurine uptake in mouse brain[28], osmolarity of the medium in kidney cells regulates taurine transport[16], and decrease in the activity of placental taurine transporters results in low plasma concentration in the fetuses [27].

Regulation by phosphorylation has been described for other amino acid transport systems. For instance, the high-affinity brain glutamate transporter, GLAST-1, is inhibited by direct phosphorylation of the molecule via a non-protein kinase C (PKC) consensus site[3]. Arginine transporter in Caco-2 intestinal cells is stimulated by PKC, elevating maximal activity, V_{max}, in undifferentiated cells and also increasing the diminished V_{max} in differentiated cells[29]. In embryonic kidney cells expressing the mouse glycine transporter GLYT1B, the activation of PKC diminishes the uptake of glycine[32]. On the other hand, PKC stimulation increases γ-aminobutyric acid (GABA) uptake in *Xenopus* oocytes transfected with the cloned rat brain transporter, GAT1[4].

There are interesting reports concerning taurine transport in various tissues. The stimulation of PKC by phorbol 12-myristate 13-acetate (PMA) inhibits the uptake of taurine in human colon carcinoma cell lines[1]. The authors showed an increase in the V_{max} and also in the Michaelis-Menten constant, K_m, as well as blockade of the effect by the inhibitor of PKC, staurosporine. This study made clear that leucine and lysine transport were unaffected and that inhibition of translation or transcription, as well as lack of integrity of cytoskeletal structure, do not change the decrease observed with PMA. This activator of PKC significantly decreases taurine uptake in a dose- and time-dependent manner in astrocytic but not in neuronal cell lines[36]. This might be due to either differential structure or the differential environment of these cells. The research on taurine transport regulation in Ehrlich ascites tumor cells indicates that PKC inhibits its function, and that protein kinase A (PKA) stimulates it[26].

Adaptive regulation, known to occur for the amino acid transport system A, has also been shown for taurine transport system in various tissues. For example, taurine exposure of human choriocarcinoma cells produces a decrease in V_{max} and a slight decrease in K_m for the substrate, inhibitors of translation and transcription decrease this effect[13]. The renal tubular epithelium presents adaptive regulation, which was demonstrated in man, mouse, rat, dog, and pig. The effect involves decrease or increase in initial rate activity of taurine transport[9]. Direct experiments have been done with cDNA encoding for taurine transporter from LLC-PK1 and MDCK kidney cell lines[16]. These authors reported that activation of PKC has no effect on adaptive regulation in the renal system.

OBJECTIVES AND EXPERIMENTAL APPROACHES

The relevance of taurine in retinal function and the importance of its uptake for maintaining such high levels in this tissue, plus the evidence of specific regulatory mechanisms for taurine uptake described in a variety of cells, made us to explore the possible adaptive regulation and the role of phosphorylation in transport activity. The use of fresh cells or primary cultures is helpful in understanding the physiological mechanisms underlaying the modulation of taurine transport. The following materials and methods were used to accomplish the purposes.

Animals, and Dissociation and Culture of Retinal Cells

Goldfish (*Carassius auratus*), measuring 4-5 cm, from a local commercial breeder (Fauna Roosevelt, Caracas, Venezuela) were kept in an aquarium in the laboratory under 12:12 h light cycle for 1 to 3 weeks before used. The fish were dark-adapted for 30 min and anesthetized with tricaine (0.05%) prior to enucleation of the eye. The eyes were rinsed in Locke's solution free of Ca^{2+} and Mg^{2+} and the retina was dissected, placed in the same solution with 0.25% trypsin, and incubated at 25°C for 30 min. Mechanical dissociation was performed gently with a Pasteur pipette. The material was passed through a mesh. Male adult Sprague-Dawley rats were housed under 12:12 h light cycle for at least 72 h prior to experiments. The eyes were removed after decapitation and the retina was processed for obtaining isolated cells by the same procedure described above. In most of the experiments, fresh retinal cells were utilized for measurement of [^{3}H]taurine uptake. In some experiments with goldfish the cells were washed, resuspended (approximately 300,000 cells/ml) in nutrient Leibovitz medium (L-15, SIGMA, 2 ml per dish), plus 10 mM HEPES, 10% fetal calf serum, and 0.1 mg/ml of gentamicin (Sigma), and then incubated at 25°C for 24 h to 8 days[27]. Prior to plating on poly-L-lysine coated flasks, aliquots of each cell preparation were counted in a hemocytometer after staining with Trypan blue (viability, 90-96%).

Uptake Experiments in Isolated and in Cultured Cells

[^{3}H]Taurine (Amersham, 24.1 Ci/mmol; ~200,000 dpm) was used with non-radioactive taurine in concentrations from 0.003 to 100 μM. Time-dependent and inhibition experiments utilizing β-alanine, hypotaurine, or GABA, 0.1 nM to 3 mM, were performed in the presence of 40 nM [^{3}H]taurine. The cell preparation (250 μl ≅ 300,000 per ml) was preincubated at 25°C for 2 min in Ringer solution. Incubation was started by the addition of the substrate. After 2 min the process was stopped by filtration through glass fiber filters (Whatman GF/B). The measurements were performed in duplicates with values within ±5% of the mean value. For cells in culture, 1, 3, 5 or 8 days, in the absence or in the presence of 4 mM taurine, the medium was removed and the flasks were washed with incubation solution. Refringent cells were counted, [^{3}H]taurine was added to the attached cells, and incubation was performed for 2 min at room temperature. The process ended by removal of the incubation solution. The preparation was washed, and 1 ml of 0.1 M NaOH was added and maintained for 2 h. An aliquot was then placed in a scintillation vial with Aquasol for counting radioactivity. The results are expressed in pmol or fmol of [^{3}H]taurine/10^5 cells.

Culture in the Presence of Protein Kinase Activator and Phosphatase Inhibitor

The cells were exposed for 10, 20, 30, 60, 90 and 120 min to the PKC activator, phorbol 12,13-dibutyrate (PDBu) 0.001-100 nM, for 1 h[10] (Calbiochem, La Jolla, CA, USA) or the protein phosphatase inhibitor, okadaic acid (OKA) 10 nM or 0.01-100 nM for 1 h[2] (Research Biochemicals International, Natick, MA, USA). The drugs were dissolved in dimethyl sulfoxide, final maximal concentration 0.05%.

Experiments on Adaptive Regulation

Some experiments were done by culturing the cells in the presence of 4 mM taurine, an optimal concentration for axonal outgrowth[19]. The cells were kept in culture for 1, 2, 5 and 8 days and uptake experiments were performed as described above. In another set of experiments, fragments of retina were plated on uncoated plastic flasks and kept for 72-96 h in Leibovitz medium with 10% fetal calf serum. After this period of time the fragments were removed by extensive rinsing with medium and the attached cells were cultured for 15 to 20 days. These cells, reacting to glial fibrillary acidic protein antibody and presenting the morphology of Müller cells[6], were cultured in the absence or in the presence of 500 μM taurine, and used for uptake experiments.

Statistical Analysis

Analysis of variance and Student's *t*-test were performed with the program Primer of Biostatistics[5]. Statistical significance was considered if p < 0.05. Non-linear fitting was performed by Prisma 2.0. Kinetic constants,

V_{max} and K_m, were calculated either by Lineweaver-Burk plots or by curvilinear analysis. Each value represents the mean ± SEM.

RESULTS

Goldfish Retinal Cells

Protein kinase C activator and phosphatase inhibitor effects: The incubation of goldfish retinal cells just after dissociation in the presence of 1 μM PDBu or 10 nM OKA for 10 to 120 min did not significantly modify the uptake of [^{3}H]taurine as compared to corresponding controls incubated in the absence or the drugs for the same period of time. Neither PDBu nor OKA present in the incubation of goldfish retinal cells for 1 h in concentrations from 0.001 to 100 nM produce significant changes as compared to controls. In experiments in which the retinal cells were plated for 24 h, incubated for 10 to 120 min in the presence of 1 μM PDBu and used for [^{3}H]taurine uptake assays, no significant alterations were observed. The only significant change was of V_{max} in the presence of 10 nM OKA after 10 min of preincubation (Table 1), without changing in K_m, but not when the incubation was extended for longer periods of time. The V_{max} for the uptake of [^{3}H]taurine into isolated glial cells previously incubated in the presence of 0.1 μM PDBu or 10 nM OKA for 1 h was significantly decreased with respect to control, however, no modifications were observed in the K_m for the substrate (Table 2).

Table 1. High affinity [^{3}H]taurine uptake into goldfish retinal cells cultured for 24 h and preincubated in the presence of 10 nM OKA for different periods

Preincubation time (min)	V_{max} (fmol/10^5 cells)	
	Control	OKA
10	32.10 ± 4.86	17.60 ± 4.80*
20	32.40 ± 5.17	24.20 ± 5.80
30	26.60 ± 2.80	24.60 ± 3.20
60	25.80 ± 3.10	24.40 ± 6.60
90	31.80 ± 4.40	24.60 ± 7.80
120	28.60 ± 4.40	18.40 ± 5.20

n = 6 *P < 0.05 respecting corresponding control

Table 2. High affinity [³H]taurine uptake into goldfish Müller cells cultured and exposed to 1 μM PDBu or 10 nM OKA for 1 h

	K_m (μM)	V_{max} (pmol/10^5cells)
Control	1.0 ± 0.4	3.5 ± 0.1
PDBu	0.8 ± 0.3	$2.1 \pm 0.6*$
OKA	0.9 ± 0.2	$2.7 \pm 0.5*$

n = 3; *P < 0.05 versus control

Culture in the presence of taurine. The total isolated cells of the goldfish retina in the presence of 4 mM taurine and 10% fetal calf serum or only in the presence of the serum decreased the uptake of [³H]taurine after 2 days in culture. No significant changes were observed at longer periods of incubation. The affinity for the substrate, as represented by K_m, was not significantly altered by the presence of taurine in the medium (Table 3). On the other hand, the incubation of isolated glial cells in the presence of taurine produced a decreased in the V_{max} without modifications in K_m for the uptake of [³H]taurine (Table 4).

Table 3. High affinity [³H]taurine uptake into goldfish retinal cells cultured in the absence (Control) or in the presence of 4 mM taurine (Taurine)

Days in Culture	V_{max} (fmol/10^5 cells)	
	Control	Taurine
1	15.01 ± 1.67	20.71 ± 3.88
2	$7.65 \pm 1.53*$	$12.15 \pm 1.02*$
5	10.38 ± 1.86	14.46 ± 2.10
8	14.74 ± 1.19	19.11 ± 2.15

n = 6; *p < 0.05 versus day 1

Table 4. High affinity [³H]taurine uptake into goldfish Müller cells cultured in the absence (Control) or in the presence of 500 μM taurine (Taurine)

Treatment	V_{max} (pmol/10^5 cells)
Control	4.2 ± 0.7
Taurine	$2.3 \pm 0.5*$

n = 3; *p < 0.05 versus control

Table 5. High affinity [³H]taurine uptake into rat retinal cells in the presence of the PKC activator, PDBu, or the phosphatase inhibitor OKA

Concentration	V_{max} (fmol/10^5 cells)	
(nM)	PDBu	OKA
Control	6.38 ± 1.06	6.11 ± 0.43
0.001	7.68 ± 0.90	7.05 ± 0.86
0.003	8.19 ± 0.84	5.44 ± 0.18*
0.01	8.70 ± 0.70	5.78 ± 0.71*
0.03	8.25 ± 0.60	4.79 ± 0.36*
0.1	7.48 ± 1.58	4.92 ± 0.48*
0.3	6.50 ± 1.51	5.17 ± 0.73
1	6.17 ± 0.36	5.13 ± 0.26
3	6.02 ± 0.35	5.14 ± 0.36
10	6.04 ± 0.05	5.59 ± 0.07
30	6.49 ± 1.06	5.35 ± 0.53
100	5.80 ± 0.55	6.29 ± 0.37

n = 6; *$p < 0.05$ versus control

Rat retinal cells

Protein kinase C activator or phosphatase inhibitor effects. Adult retinal cells incubated in the presence of 1 µM PDBu or 10 nM OKA for 10 to 120 min did not present any significant difference in [³H]taurine transport with respect to corresponding control. The treatment of these cells, cultured for 24 h, for 1 h with different concentrations of PDBu did not produce significant modifications. However, OKA produced a dose-dependent decreased in the uptake of the amino acid into rat retinal cells (Table 5).

DISCUSSION

Most of the studies on the regulation by phosphorylation mentioned above have been carried out in cell lines. These being a homogeneous population of cells, the results are quiet clear. The interest of this work has been to study the possible mechanisms by which the high-affinity entrance of taurine into retinal cells is regulated. Although the presence of taurine in the retina has been shown mainly in the photoreceptor layer and in synaptic buttons close to ganglion cells[18], the fact that [³H]taurine is transported by axons of the optic nerve to the brain of the goldfish[7,11], or the rat[30] might be an indication of the uptake, but not the storage, of taurine in ganglion cells, a population of cells also contributing to the results of this work.

High-affinity uptake of [^{3}H]taurine is regulated in several tissues, as indicated in the Introduction, even if the content of taurine in these tissues is considerably high[7,20,24]. In addition, a low-affinity component of [^{3}H]taurine uptake is present in most of the cells in which this transport system has been characterized[20,31,34].

The treatment of retinal goldfish cells just after enzymatic isolation or after 24 h in culture with the PKC activator PDBu or the inhibitor of phosphatase OKA gave mainly negative results, except for OKA at short periods of incubation, which might be the consequence of using a mixed population of cells, neurons and glial cells. However, the fact that clear reduction of the uptake was produced in Müller cells in primary culture indicates that the regulation of transport of taurine in some cells of the goldfish retina might have physiological significance. PKC, however, is known to regulate taurine transport in glioma GL15 cell line[37]. It migh be that phosphatases, rather than PKC, are involved in maintaining certain degree of phosphorylation of taurine transporter, since the inhibitory effect elicited by increasing level of phosphorylation of taurine transporter was observed by the treatment with OKA. Moreover, the concentration-dependent response to OKA could indicate that protein phosphatase 2A is playing a more determined role, since the sensitivity of this phosphatase to OKA is higher than that of the protein phosphatase 1.

Adaptive regulation produced by taurine-starved cells has been shown in renal epithelial cell lines[14], in *Xenopus* oocytes expressing poly(A)+RNA isolated from rat kidney cortex[8], in the human placental choriocarcinoma cells[13], and in human colon carcinoma cell lines[38]. The adaptive response requires the synthesis of proteins and of RNA, as cycloheximide and actinomycin, inhibitors of transduction and transcription, respectively, prevent the adaptation of MDCK cells[15]. In primary tissue cultures from the goldfish retina the reduction observed after 2 days with taurine supplemented medium was also obtained in control cells, these cells were cultured in the presence of fetal calf serum. The content of taurine in fetal calf serum could be between 100 to 500 μM, and that is probably the reason of this observation. However, is an indication of the possibility of adaptive regulation of β-amino acid transport system in the retina and encouraged the study of this process in a more homogeneous population of cells, such as the glial retinal cells. As it has been shown, the presence of taurine in the medium reduces the uptake in Müller cells in culture, and further studies should be done in order to investigate the mechanisms by which this reduction occurs. It was previously reported that the uptake of [^{3}H]taurine into goldfish retinal cells isolated from animals which optic nerve was axotomized for several days, decrease[7] and one of the possibilities of this regulation could be the phosphorylation of the transporter.

ACKNOWLEDGMENTS

This work was supported by the Grant S1-723 from Consejo Venezolano de Investigaciones Científicas y Tecnológicas (CONICIT). We appreciate the secretarial assistance of Mrs. Isabel Otaegui.

REFERENCES

1. Brandsch, M., Miyamoto, Y., Ganapathy, V. and Leibach, F.H., 1993, Regulation of taurine transport in human colon carcinoma cell lines (HT-29 and Caco-2) by protein kinase C. *Am. J. Physiol.* 264:G939-946.
2. Cohen, P., 1989, The structure and regulation of protein phosphatases. *Ann. Rev. Biochem.* 58:453-508.
3. Conradt, M. and Stoffel, W., 1997, Inhibition of the high-affinity brain glutamate transporter GLAST-1 via direct phosphorylation. *J. Neurochem.* 68:1244-1251.
4. Corey, J.L., Davidson, N., Lester, H.A., Brecha, N. and Quick, M.W., 1994, Protein kinase C modulates the activity of a cloned gamma-amino butyric and transporter expressed in *Xenopus* oocytes via regulated subcellular redistribution of the transporter. *J. Biol. Chem.* 269: 14759-14767.
5. Glantz, S.A., 1988, Primer of Biostistics, McGrall-Hill, New York.
6. Goureau, O., Hicks, D., Courtois, Y. and De Kozak, Y., 1994, Induction and regulation of nitric oxide synthase in retinal Müller glial cells. *J. Neurochem.*, 63:310-311
7. Guerra, A., Urbina, M. and Lima, L., 1999, Modulation of taurine uptake in the goldfish retina and axonal transport to the tectum. Effect of optic crush or axotomy. (in press).
8. Han, X., Budreau, A.M. and Chesney, R.W., 1997, Functional expression of rat renal cortex taurine transporter in *Xenopus laevis* oocytes adaptive regulation by dietary manipulation. *Pediatr. Res.* 41: 624-631.
9. Han, X., Budreau, A.M. and Chesney, R.W., 1998, Molecular cloning and functional expression of an LLC-PKI cell taurine transporter that is adaptively regulated by taurine. Taurine 3, *Adv. Exp. Med. Biol.* 442:261-268.
10. Hortelano, S., Genaro, A.M. and Bosca, L., 1992, Phorbol esters induce nitric oxide synthase activity in rat hepatocytes. Antagonism with the induction elicited by lipopolysaccharide. *J. Biol. Chem.* 267:24937-24940.
11. Ingoglia, N.A., Sturman, J.A., Lindquist, T.D. and Gaull, G.E., 1976, Axonal migration of taurine in the goldfish visual system. *Brain Res.* 115:535-539.
12. Iwata, H., Obara, T., Kim, B.K. and Baba, A., 1986, Regulation of taurine transport in rat skeletal muscle. *J. Neurochem.* 47:158-163.
13. Jayanthi, L.D., Ramamoorthy, S., Mahesh, V.B., Leibach, F.H. and Ganapathy, V., 1995, Substrate-specific regulation of the taurine transporter in human placental choriocarcinoma cells (JAR). *Biochem. Biophys. Acta* 1235:351-360.
14. Jones, D.P., Miller, L.A. and Chesney, R.Q., 1990. Adaptive regulation of taurine transport in two continuos renal epithelial cell lines. *Kidney Int.* 38: 219-226.
15. Jones, D.P., Jiang, B. and Chesney, R.W., 1994, Regulation of taurine transport by external taurine concentration and medium osmolality in renal tubular cells in culture, in Taurine inn Health and Disease, Vol. 359 (Huxtable R.J. and Michalk D. eds), 131-138, Plenum Press, New York.
16. Jones, D.P., Miller, L.A. and Chesney, R.W., 1995, The relative roles of external taurine concentration and medium osmolality in the regulation of taurine transport in LLC-PK1i and MDCK cells. *Pediatr. Res.* 37:227-232.

17. Lake, N., 1988, Taurine depletion leads to loss of rat optic nerve axons. *Vision Res.* 28:1071-1076.
18. Lake, N. and Verdone-Smith, C., 1989, Immunocytochemical localization of taurine in the mammalian retina. *Curr. Eye Res.* 8:163-173.
19. Lima, L., Matus, P. and Drujan, B., 1988, Taurine effect on neuritic outgrowth from goldfish retinal explants. *Int. J. Develop. Neurosci.* 6:417-420.
20. Lima, L., Matus, P. and Drujan, B., 1991, Differential taurine uptake in central and peripheral regions of goldfish retina. *J. Neurosci. Res.* 28:422-427.
21. Lima, L., Matus, P. and Drujan B., 1993, Taurine-induced regeneration of goldfish retina in culture may involve a calcium-mediated mechanism. *J. Neurochem.* 60: 2153-2158.
22. Lima, L. and Cubillos, S., 1998, Taurine might be acting as a trophic factor in the retina by modulating phosphorylation of cellular proteins. *J. Neurosci. Res.* 53:377-384.
23. Lima, L., Obregón, F., and Matus, P., 1998, Taurine, glutamate and GABA modulate the outgrowth from goldfish retinal explants and its concentrations are affected by the crush of the optic nerve. *Amino acids* 15:195-209.
24. Malandro, M.S., Beveridge, M.J., Kilberg, M.S. and Novak, D.A., 1996, Effect of low-protein diet-induced intrauterine growth retardation on rat placental amino acid transport. *Am. J. Physiol.* 271:C295-303.
25. Matus, P., Cubillos, S. and Lima, L., 1997, Differential effect of taurine and serotonin on the outgrwoth from explants or isolated cells of the retina. *Int. J. Develop. Neurosci.* 15:785-793.
26. Mollerup, J. and Lambert, I.H., 1996, Phosphorylation is involved in the regulation of the taurine influx via the β-system in Ehrlich ascites tumor cells. *J. Membr. Biol.* 150:73-82.
27. Norberg, S., Powell, T.L. and Jansson, T., 1998, Intrauterine growth restriction is associated with a reduced activity of placental taurine transporters. *Pediatr. Res.* 44:233-238
28. Oja, S.S. and Saransaari, P., 1996, Kinetic analysis of taurine influx into cerebral cortical slices from adult and developing mice in different incubation conditions. *Neurochem. Res.* 21:161-166.
29. Pan, M. and Stevens, B.R., 1995, Protein kinase C-dependent regulation of L-arginine transport activity in Caco-2 intestinal cells. *Biochem. Biophys. Acta* 1239:27-32.
30. Politis, M.J. and Ingloglia, N.A., 1979, Axonal transport of taurine along neonatal and young adult rat optic axons. *Brain Res.* 166:221-231.
31. Saransaari, P. and Oja, S.S., 1994, Taurine in the developing cat: uptake and release in different brain areas. *Neurochem. Res.* 19:77-82.
32. Sato, K., Adams, R., Betz, H. and Schloss, P. 1995, Modulation of a recombinant glycine transporter (GLYT1b) by activation of protein kinase C. *J. Neurochem.* 65:1967-1973.
33. Schmidt, S.Y., 1978, Taurine fluxes in isolated cat and rat retinas: effects of illumination. *Exp. Eye Res.* 26: 529-535.
34. Schmidt, S.Y., and Berson, E.L., 1978, Taurine uptake in isolated retinas of normal rats and rats with hereditary retinal degeneration. *Exp. Eye Res.* 27:191-198.
35. Sturman, J.A., Moretz, R.C., Frech, J.H., and Wisnewski, H.M., 1985. Postnatal taurine deficiency in the kittens results in a persistance of the cerebellar external granule cell layer: correction by taurine feeding. *J. Neurosci. Res.* 13:521-528.
36. Tchomkeu-Nzouessa, G.C. and Rebel, G., 1996, Activation of protein kinase C down-regulates glial but not neuronal taurine uptake. *Neurosci. Lett.* 206:61-64.
37. Tchomkeu-Nzouessa, G.C. and Rebel, G., 1996, Characterization of taurine transport in human glioma GL15 cell line: regulation by protein kinase C. *Neuropharmacology* 35:37-44.
38. Thiruppathi, C., Brandsch, M., Miyamoto, Y., Ganapathy, V. and Leibach, F.H., 1992, Constitutive expression of taurine transporter in a human colon carcinoma cell line. *Am. J. Physiol.* 263:G625-631.

EFFECTS OF OSMOTIC AND LIGHT STIMULATION ON ³H-TAURINE EFFLUX FROM ISOLATED ROD OUTER SEGMENTS AND SYNTHESIS OF TAURET IN THE FROG RETINA

Andranik M. Petrosian,[1] Jasmine E. Haroutounian,[1] Kjell Fugelli,[2] and Hilde Kanli[2]

[1]Buniatian Inst. of Biochemistry of Natl. Acad. Sci. of Armenia, Sevag St. 5/1, Yerevan 14, Armenia; [2]Div. of Gen. Physiology, University Oslo, P.O. Box 1051 Blindern, 0316 Oslo 3, Norway

Abstract: After injection of ³H-taurine into eyeballs of frogs and maintenance for 3 h in darkness by a gentle shaking, an almost homogenous fraction of rod outer segments (ROS) was prepared. About a 22% decrease in tonicity caused by reducing NaCl in isotonic 225 mOsm normal solution caused a rapid increase in the rate coefficient of efflux of ³H-taurine from the ROS fraction. The peak level of increased efflux rate coefficient was 7-times higher than the basal isotonic level. This indicates that taurine could contribute essentially to the volume regulation, either via selective channels or a carrier transporter-mediated pathways. For further clarifying if taurine fluxes in the ROS are sensitive to the light, other experiments were performed. Neither light stimulation of dark-adapted ROSs fractions or dark stimulation of weakly illuminated ROSs revealed any detectable changes in the efflux rate coefficient of ³H-taurine. These results indicate that light-induced taurine efflux, if present in the ROS, must be small, compared with hypoosmotic induced efflux. Thus the question of light-induced release of taurine from ROS still remains to be clarified. In the second part of this study, using TLC (thin layer chromatography) in combination with ³H-taurine measurements we have tried to clarify whether frogs (Rana ridibunda) eye structures can synthesize tauret (retinylidenetaurine). In isolated retinal preparations almost no any noticeable radioactivity was detected compared with background level. The capability of the eye structures to synthesize tauret from ³H-taurine was revealed in the second whole eye injection experiment. About 0.3 % of the total ³H-taurine pool taken up was converted into ³H-tauret in the dark-adapted frog retina. In the retina of frogs adapted to light compared with those which were dark adapted tauret quantities were remarkable lower - on average about half. These results are in agreement with our recent data obtained by HPLC, which indicate tauret levels several times higher in the dark-adapted frog retinae compared with those after long lasting light adaption. Taking into account these results one can conclude that the main structure able to synthesize ^{3}H-tauret is probably pigment epithelium rather than retina.

INTRODUCTION

Despite many investigations[3,5,9,11-14,20,21], studies on volume regulation and light-induced release of taurine in the retinal structures still generate interest. Taurine release in the retina depends on potassium, ouabain, electrical and light stimulus, veratridine glutamate, kainate, and volume changes[3,5,9,11-14,20,21]. Because of its large amounts (10-40 mM[4,7,10,22]) and high water solubility, taurine is an almost ideal osmolyte in photoreceptor cells[9,12,13], which contains 50-80% of the taurine pool of the retina[4,6,10,22]. Maintenance of steady volume of the photoreceptors is based on both specialized microcytoskeletal elements and on dynamic osmolyte constituents such as taurine. In this study we have investigated action of hypoosmotic stress on ^{3}H-taurine efflux from isolated ROSs. We have studied also the effect of illumination on the ^{3}H-taurine efflux from isolated ROSs. Answers to these questions are crucial with respect to the hypothesis[18,19] as to whether tauret - taurine conjugated with retinaldeyde - is involved in the transport of both 11-cis retinal and all-trans retinal between photoreceptors and pigment epithelium. In particular on the basis of this hypothesis one can predict that light stimulation of dark-adapted ROSs must induce ^{3}H-taurine release from ROSs. The only data in the literature on this subject are in agreement with such an assumption and indicate a direct increase of ^{35}S-taurine efflux from isolated ROSs fraction under light stimulation[20]. A related issue is to understand the origin of tauret, an endogenous substance recently revealed in the retina and pigment epithelium of the eye[16,17]. Knowledge of this may clarify the role of taurine in the transport of vitamin A in the eye. Is tauret synthesis in the retina is enzyme-mediated or a spontaneous chemical reaction? The significance of nonenzymatic chemical reactions of the amino group of taurine with 11-*cis*- and *trans*-retinals in this process can not be excluded. Recently, it has been shown that taurine has a high reactivity with aldehydes, which suggests the possibility of an inhibiting effect of taurine against the modification of protein, as well as an antioxidative effect through the reactions of taurine with aldehydes *in vivo*[8]. On the other hand it is of interest also to elucidate the site of tauret synthesis and how it depends on the state of adaptation.

MATERIAL AND METHODS

Animals and Tissue Preparation

Frogs (*Rana ridibunda*) were used throughout this study. Experiments were carried out in October - November. Animals were maintained at 10-20^0C in transparent jars made of glass and subjected to standard light cycles. Frogs were dark-adapted overnight or for at least 3 h before experiments. After dark or light adaptation (about 4 h under white 1000 lx illumination). at 16 - 20^0C all procedures of tissue preparation were performed in dim red light or under white illumination, correspondingly. After decapitation eyes were enucleated, the retinas with the pigment epithelium were immediately placed either in physiological solution or in liquid nitrogen.

Isotope Measurements

The first experiments were to investigate osmotic and light stimulated release of ^{3}H-taurine from isolated rod outer segments after injection into the eyeball of dark-adapted frogs of 3 µl ^{3}H-taurine. Animals were kept in darkness for about 3 h, retina were removed and by a gentle shaking procedure a rod outer segment (ROS) fraction was prepared. After washing in ^{3}H-taurine free buffer, ROS fraction was placed on a filter in the perfusion chamber. Pumping was carried out under dark condition at a velocity of 0.5ml/min, and samples were collected at 1 min intervals. To control osmolarity, an osmometer (KNAUER, Germany) was used. Hypotonicity was achieved by reducing NaCl in normal solution from 105 mM to about 78 mM to achieve an osmolarity of 180 mOsm, which is about 20% lower than the control of 224 mOsm. For light stimulation, steady illumination from a light microscope was used. After collection, to each sample was added scintillator Ultima Gold Packard and radioactivity in cpm was measured by a Packard TRI-CARB Liquid Scintillation Spectrometer. The rate coefficient for efflux of taurine (k_e, min^{-1}) was calculated for each of the 10 - 20 min periods according to the formula[1] $k_e = M_n/t(C_n + 0.5M_n)$, where M_n is the content of ^{3}H-taurine in the medium at the end of n_{th} sampling period (cpm.min^{-1}), C_n is the total cellular ^{3}H-taurine content at the end of the last sampling period (min). During the second part of experiments we have used TLC in combination with ^{3}H-taurine measurements to clarify possible ^{3}H-tauret synthesis in the eye structures under dark and light conditions.

To investigate tauret synthesis in the frog retina, isolated retinae obtained either from dark- or light-adapted frogs were incubated for 3 h in solution (mM): NaCl 105, KCl 2.5, CaCl$_2$ 1.0, MgSO$_4$ 0.1, glucose 5, and HEPES 5 at pH 7.5, to which has been added ^{3}H-taurine 2.5 x 10^{-7} M (Amersham, Great Britain, with specific radioactivity 29.0 Ci/mmol.) and high grade purity taurine (Sigma Chemical Co., St. Louis, USA) to make a final taurine concentration 1.0×10^{-4} M. During this experiment, $3\mu l$ ^{3}H-taurine has been directly injected into the eyeball of dark-adapted frogs. Frogs were then kept in the darkness or under 2000 lx illumination for about 3 h. After this, retinas were removed and quickly placed in liquid nitrogen. Frozen samples were then lyophilized using a Levbold Heraeus liophilizator. Dried samples were grounded in absolute methanol with a glass-glass homogenizer. Methanol homogenate of the retinas and pigment epithelium were extracted on ice for 30 min, centrifuged for 3 min at 5000 g and 2 µl samples were subsequently 5x spotted on Merck or Armsorb sheets of silica gel. On a separate track nearby, samples of synthetic tauret as a standard, dissolved in methanol, also was spotted. For development chloroform:methanol::trifluoroacetic acid in ratio of 20:6:1 or 10:1:0,1 was used. For visual control the blots in some experiments were visualized by treatment with I_2 vapor. After development silica gel sheets were cut into separate longitudinal strips, along which samples of methanol extracts of retina and pigment epithelium were separated. Each such strip has been accurately divided into 7 pieces. Each piece was separately counted for radioactivity.

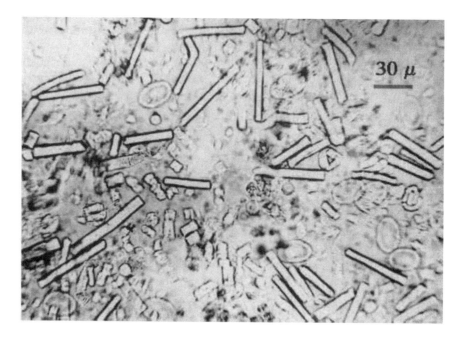

Figure 1. Light micrograph of rod outer segment fraction

RESULTS AND DISCUSSION

The ROS preparation was almost homogenous (Fig. 1). During perfusion of the isolated ROS fraction at 8th min of perfusion, about 22% decrease in tonicity caused by reducing NaCl concentration in isotonic 225 mOsm normal solution caused a rapid increase in the rate coefficient of efflux of ^{3}H-taurine from the ROS fraction (Fig. 2). The peak level of increased efflux rate coefficient is 7 times higher than basal isotonic level. It is evident that taurine may be involved in volume regulation of the ROS structure. This result is in agreement with data from literature concerning the role of taurine as a major osmolyte in photoreceptor cells[9,12,13]. Previously it was established that lowering osmotic pressure to 50 mOsm registers light-induced electrical activity in the Plll component of ERG in the isolated frog retina[15]. This demonstrates unusual high osmoregulation capacity of ROS. It seems likely that taurine outward flux could contribute essentially to the volume regulation, either via selective channels or a carrier-transporters mediated pathways.

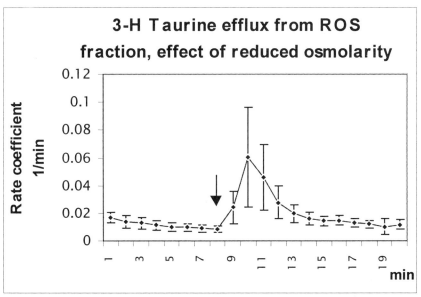

Figure 2. The effect of reduction of osmolarity on the rate coefficient (min^{-1}) for taurine efflux from dark adapted isolated rod outer segments. Values represent the mean ± SEM from six separate experiments.

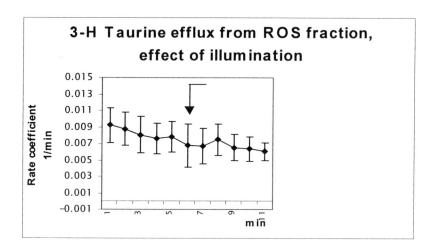

Figure 3. The effect of steady white illumination on the rate coefficient (min⁻¹) for taurine efflux from dark adapted isolated rod outer segments. Values represent the mean ± SEM from four separate experiments.

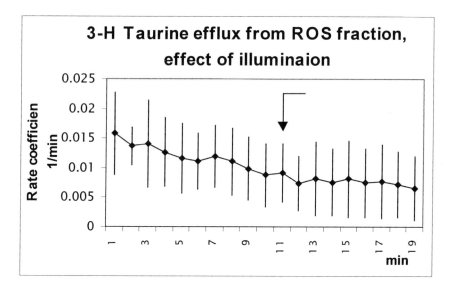

Figure 4. The effect of dark stimulation on the rate coefficient (min⁻¹) for taurine efflux from weakly illuminated isolated rod outer segments. Values represent the mean ± SEM from four separate experiments.

A related question, if taurine fluxes in the ROS are sensitive to the light, is especially crucial in light of the essential role of taurine in retinoid transport between photoreceptor and pigment epithelial cells in the retina[1].

For further clarifying if taurine fluxes in the ROS are sensitive to light other experiments were performed. Either light stimulation of dark-adapted ROS fractions (Fig. 3) or dark stimulation of weakly illuminated ROSs (Fig. 4) did not reveal detectable changes in the efflux rate coefficient of ^{3}H-taurine. However, a hardly noticeable decrease in the bias of the efflux curve in first case is rather an indication that light stimulation with 40000 lx for 90 s induced a small increase in the release of ^{3}H-taurine from the dark adapted ROS fraction (Fig. 3).

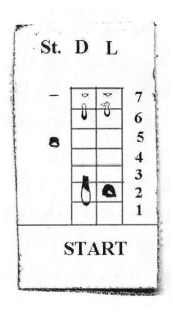

Figure 5. TLC map after separation of methanol extract of retina. Strips # 1, 5, 7 identify taurine, tauret and free retinoids. Tauret has R_f=0.65, strip # 5.

Taking into account ^{3}H-taurine specific activity, concentration of ^{3}H-taurine, and concentration of rhodopsin in the ROS, one can look for an increase in the efflux of ^{3}H-taurine from ROS fraction, as predicted by the tauret-based retinoid transport hypothesis. Compared with hypoosmotically induced efflux, it must be lower than the technical capabilities applied in the current study. This is also in agreement with the data about quantities of endogenous tauret in the retina and estimates of its synthesis represented in this issue. On the other hand there is well-established light induced release of radioactive taurine from whole isolated retina in different species[3,11,12,14,21]. So the question about light induced release of taurine from ROS still remains to be clarified. Probably the type of light stimulation, pumping velocity and other factors must be checked carefully. These results indicate that light induced taurine efflux, if present in the ROS, must be very small, compared with hypoosmotic induced efflux.

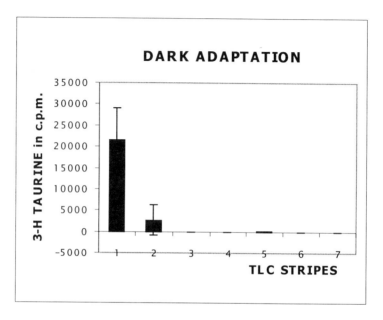

Figure 6. Distribution of ^{3}H-taurine, ^{3}H-taurine derivatives and ^{3}H-tauret on 1 - 7 strips on the silicagel sheets, after separation of methanol extracts of retina and pigment epithelium. Tauret (R_f=0.0.65), strip # 5. Whole eye injection experiment. Dark adaptation. The number of experiments N=8 Confidence interval 95%.

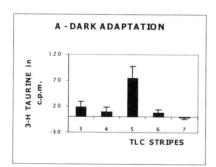
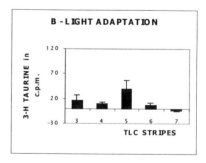

Figure 7. Distribution of ^{3}H-taurine, ^{3}H-taurine derivatives and ^{3}H-tauret on 3 - 7 strips on the silicagel sheets after separation of methanol extracts of retina and pigment epithelium. Tauret (R_f=0.0.65), strip # 5. Whole eye injection experiment. A - Dark adaptation. B - light adaptation. The number of experiments N=8 Confidence interval 95%.

In the second part of this study using TLC in combination with ^{3}H-taurine measurements we have tried to clarify whether frog (*R.. ridibunda*) eye structures can synthesize tauret. According to standard tauret R_f =0.65 corresponds to position of trans-tauret which is laying on strip #5 of silica gel sheet on which separation has been performed. In these samples no noticeable radioactivity has been obtained from either dark- or light-adapted frogs compared with background level in the case of isolated retinal preparations. Either this result indicates that isolated retina is unable to synthesize tauret under either condition or that the quantity of tauret synthesized is too small, and beyond the sensitivity of the method used in this study. Another possibility is that the tauret formed breaks down during the incubation period. Capability of the eye structures to synthesize tauret from ^{3}H-taurine has been revealed during the second whole eye injection experiment, when 3 μl ^{3}H-taurine was directly injected into the eyeball of dark adapted frogs. About 0.3 % of total uptake ^{3}H-taurine pool was converted into ^{3}H-tauret in the dark-adapted frog retina (Fig. 5). In this case radioactivity in the strip #5 (R_f=0.65 corresponding to tauret) was higher than in frog retina which was adapted to light under 2000 lx illumination (Fig. 6). It is important to emphasize that this result is in agreement with our recent data obtained by HPLC[17] indicating several time higher endogenous tauret in the dark-adapted frog retina compared with those of light-adapted. Before comparing results of the first and second experiments obtained during this study correspondingly on the isolated retina and in case of whole eye injection experiments it is reasonable to compare ^{3}H-taurine concentration for both cases. Let us estimate ^{3}H-taurine concentration in the frog eyeball during second experiment. Because frog eyeball diameters average about 5 mm, and have almost round-shaped lens of diameter about 3 mm and total thickness of sclera and retinal layer is no more than 0.6 mm, the free volume in the eyeball is approximately around 30 mm^3. Taking into account this value, specific activity and amount injected ^{3}H-taurine one can conclude that ^{3}H-taurine concentration within the eye during incubation was around 3.5 10^{-6} M during second experiment, which is almost 30 times lower compared with first experiment. Meanwhile specific radioactivity is about 12 times higher. Taking into account all these results one can conclude that the main structure able to synthesize ^{3}H-tauret is probably rather pigment epithelium than retina. These investigations are opening new possibilities for further investigations to understand the mechanism of the origin of tauret in the vertebrate eye and its transport in and between retina and pigment epithelium.

ACKNOWLEDGMENTS

We are grateful to Professor Mohamed Abdullah (Lab Marine Biology of Oslo University) for help throughout this study with lyophilization and Dr.

Ilona Stepanyan (Institute of Zoology of the NAS of Armenia) for assistance with preparing microphotographs. We both from Armenia are very thankful to Professor Laura Della Corte supporting our participation in Taurine Symposium '99. This work was mainly supported by Oslo University and partially by Ministry of Science and High Education of Armenia, Grant 96-681.

REFERENCES

1. Caldwell P.C. and Keynes R.D., 1969, The exchange of ^{22}Na between frog sartorius muscle and the bathing medium. In:*laboratory techniques in Membrane Biophisics.* Ed. by Passow H. and Stampli R., pp-63-68. Berlin, Springer Verlag.
2. Gabrielian, K., Wang, H.M., Ogden, T.E., and Ryan, S.J. 1992, In vitro stimulation of retinal pigment epithelium proliferation by taurine, *Curr.Eye Res.* 11:481-487.
3. Haroutounian, J.E., Petrosian, A.M., 1982. The action of illumination level on the efflux of ^{3}H-taurine from the isolated frog retina. *Armenian Biol. J (Russ).* 35: 259-264.
4. Kennedy A.J. and Voaden M. J., 1974, Free amino acids in the photoreceptor cells of the frog retina. *J. Neurochem.*, 23:1093-1095.
5. Kennedy A.J. and Voaden M. J., 1976, Studies on the uptake and release of radioactive taurine by the frog retina. *J.Neurochem.*, 27:131-137.
6. Lake N, Verdone-Smith C., 1989, Immunocytochemical localization of taurine in the mammalian retina. Current Eye Research 8:163- 173.
7. Lima L, Drujan B, Matus P., 1990, Spatial distribution of taurine in the teleost retina and its role in retinal tissue regeneration, in Prog.in Clin.Biol.Res."Taurine: Functional Neurochemistry, Physiology and Cardiology", Pasantes-Morales, H., Martin, D. L., Shain, W., Martin del Rio, R.,eds Wiley-Liss, Inc. New York, Vol. 351 pp 103-112.
8. Ogasawara, M., Nakamura, T., Koyama, I., Nemoto, M., and Yoshida, T. 1993, Reactivity of taurine with aldehydes and its physiological role, *Chem. Pharm .Bull. (Tokyo),* 41:2172-2175.
9. Oja S.S., and Kontro P., 1983, Taurine. In: Handbook of Neurochemistry (Ed. by Lajtha A.). Plenum Press, New York, 3:501-533.
10. Orr H.T, Cohen A.I, Lowry O.H., 1976, The distribution of taurine in the vertebrate retina. *J .Neurochem.* 26:609-611.
11. Pasantes-Morales, H., Guesada, O., and Carabez, A., 1981, light stimulated release of taurine from retinas of kaininc acid-treated chicks. *J. Neurochem.*, 36:1583-1586.
12. Pasantes-Morales, H., Lopez-Escalera, R., and Macedo, M.D. 1989, Taurine and ionic fluxes in photoreceptors, in:" Extracellular and Intracellular Messengers in the Vertebrate Retina", Redburn, D.A. and Pasantes-Morales, H. eds., Alan R.Liss, New York, pp.87-104.
13. Pasantes-Morales, H., Moran, J., and Schousboe, A. 1990, Taurine release associated to cell swelling in the nervous system, in: "Taurine: Functional Neurochemistry, Physiology, and Cardiology", Pasantes-Morales, H., Martin, D.L., Shain, W. and del Rio, R.M. eds., Wiley-Liss, New York, pp.369-376.
14. Pasantes-Morales, H., Salceda, R. and Lopez-Colome, A.M., 1980, The effect of colchicine and cytochalasin B on the release of taurine from chick retina. *J. Neurochem.*, 34:172-177.
15. Petrosian A.M. and Haroutounian J.E., 1998 In: "Taurine: Basic and Clinical Aspects" The role of taurine in osmotic, mechanical and chemical protection of the retinal rod outer segments.(Ed. by Schaffer S., Lombardinin B., & Huxtable R.) pp.407-413. Plenum Publ. Corp.

16. Petrosian A.M., Haroutounian J.E., Zueva L.V., (1996). Tauret: A taurine-related endogenous substance in the retina and its role in vision. In: " Taurine: Basic and Clinical Aspects" (Ed. by Huxtable R., Azuma J., Kuriyama K., Nakagawa M. & Baba A.) p.333-42. Plenum Publ. Corp.

17. Petrosian A.M,. Haroutounian J.E., Gundersen T.E, Blomhoff R., Fugelli K. and Kanli H., 1999, New HPLC evidences on endogenous tauret in the retina and pigment epithelium. (In press - Taurine 4).

18. Petrosian, A.M. and Haroutounian, J.E. 1988, On the possibility of retinal transportation in rods by tauret through special channels. *13th All-Union conf. on Electron Microscopy*, (Abstract), Zvenigorod , Moscow.

19. Petrosian, A.M. and Haroutounian, J.E. 1990, Tauret: further studies of the role of taurine in retina, in: "Taurine: Functional Neurochemistry, Physiology, and Cardiology", Pasantes-Morales, H., Martin, D.L., Shain, W. and del Rio, R.M. eds., Wiley-Liss, New York,   pp.471-475.

20. Salceda, R., Lopez-Colome, A.M., and Pasantes-Morales, H. 1977, Light-stimulated release of ^{35}S-taurine from frog retinal outer segments, *Brain Res.* 135:186-191.

21. Schmidt, S.Y., 1978, Taurine fluxes in isolated cat and rat retinas: Effects of illumination Exp.Eye Res.,26:529-535.

22. Voaden, M. J., Lake, N., Marshall, J., Morjaria, B. 1977. Studies on the Distribution of Taurine and Other Neuroactive Amino Acids in the Retina. *Exp. Eye Res.* 25:249-257.

NEW HPLC EVIDENCE ON ENDOGENOUS TAURET IN RETINA AND PIGMENT EPITHELIUM

Andranik M. Petrosian, Jasmine E. Haroutounian, Thomas E. Gundersen[1], Rune Blomhoff[1], Kjell Fugelli[2] & Hilde Kanli[2]

Buniatian Inst. of Biochemistry of Natl. Acad. Sci. of Armenia, Sevag St. 5/1, Yerevan 14, Armenia; [1] Inst. for Nutrition Res., University of Oslo P.O.Box 1046, 0316 Oslo, Norway; [2] Div. of Gen. Physiology, University of Oslo P.O.Box 1051 Blindern, 0316 Oslo 3, Norway

Abstract: This investigation was improve the separation for tauret (retinylidene taurine) and to compare its content in the retina under dark and light adaptation. To prevent tauret hydrolysis, retinal samples were quickly frozen and lyophilized. Methanol extracts of dried retina and pigment epithelium from both dark- or light-adapted frogs, *Rana ridibunda,* were injected onto HPLC. Synthetic standard tauret appeared at 4.7 min after the solvent front. At the same time, an endogenous substance was eluted from the mixed retinal and pigment epithelial samples. The UV spectra of this endogenous compound matched with the spectra of synthetic tauret obtained under identical conditions, with $\lambda_{max} = 446$ nm at peak. We conclude that the HPLC system used permitted full separation of tauret from the methanol extracts of the retina and pigment epithelium. TLC and further HPLC analysis have shown that tauret quantities were several times higher in the retina and pigment epithelium of the frogs adapted to dark compared with those light-adapted (about 4 h under 1000 lx illumination). Tauret based vitamin A transport is probably involved in other systems as well, where along with its other known beneficial effects taurine probably is necessary to facilitate vitamin A transport.

INTRODUCTION

Taurine is a heat-stable, water-soluble, amino acid. Taurine plays an important role in osmoregulation[2,20] and metabolism of lipids via conjugation with bile acids. Emulsification of lipids by taurocholic acid is one of the facilitating mechanisms in transport of the lipids in the liver and intestines[17]. The possibility of participation of taurine in retinoid transport would be another example of taurine converting a lipid-soluble substance into a water-soluble conjugate. Since the first description of tauret (retinylidene taurine)

Taurine 4, edited by Della Corte *et al.*
Kluwer Academic / Plenum Publishers, New York, 2000.

as a water-soluble conjugate of the retinaldehyde of vitamin A with taurine, the suggestion has been made that it might be an essential constituent in the transport of retinoids in the eye between pigment epithelium and retina[21-23,25,26]. There may be special canalized structures in rod outer segments which are involved in such transport process[21-23,25,26]. Furthermore, TLC and HPLC analysis has confirmed that tauret is an endogenous substance in the retina and pigment epithelium[21-23,25,26]. To improve the separation process, in order to quantify HPLC tauret measurements and to clarify if there are differences in tauret content in the retina under dark and light adaptation, the current investigation has been undertaken.

MATERIAL AND METHODS

Animals and Tissue Preparation

Frogs (*Rana ridibunda*) were used throughout this study. Experiments were carried out in October - November. Animals were maintained at 10-20°C in transparent jars made of glass and subjected to standard light cycles. Frogs were dark-adapted overnight or at least 3 h before experiments. After dark or light adaptation (about 4 h under white 1000 lx illumination) at 16-20° C, all procedures of tissue preparation were performed in dim red light or under white illumination, correspondingly. After decapitation, eyes of frogs were enucleated, the retinas with the pigment epithelium were carefully detached and immediately placed in liquid nitrogen.

High-Pressure Liquid Chromatography

Frozen tissues were lyophilized and kept O_2-free, in the dark, at -20°C or below for further TLC and HPLC analysis. After lyophilization, frozen samples were ground in absolute methanol either by a glass-glass homogenizer or with an ultra-turrax. Methanol homogenate of the retinas and pigment epithelium were extracted on ice for 30 min, centrifuged for 3 min at 5000 g and 100 µl samples were injected on the HPLC system. The separating system consisted of a 250 x 4.6 mm and a 150 x 4.6 mm Supelcosil C-8 connected in series and a nonaqueous mobile phase of 0.01% acetic acid in methanol delivered at a flow of 1 ml/min. The pump was a Shimadzu LC-10AD. Detection was performed with a Shimadzu SPD-M10A diode array detector operated in 3D mode. For TLC analysis, samples of methanol extracts of the retina and pigment epithelium and samples of synthetic tauret dissolved in methanol were separately spotted on Merck or Armsorb sheets of silica gel. *trans-* and 11-*cis*-Tauret synthesized by a one-step reaction were used throughout this study. The chromatograms were developed with chloroform:methanol::trifluoroacetic acid in a ratio of 20:6:1 or 10:1:0,1. The blots were visualized by spraying with sulfuric acid and heating or by treatment with I_2 vapor.

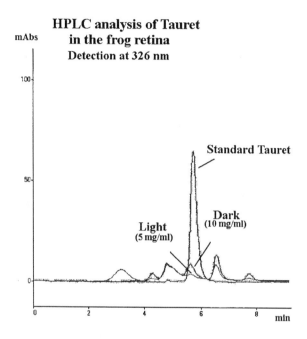

Figure 1. HPLC profiles on Supelcosil C-8 columns. Standard synthetic and endogenous tauret from dark- and light-adapted retina. Chromatograms are shown at 326 nm.

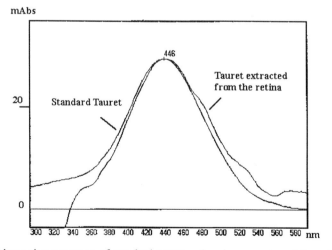

Figure 2. Absorption spectrum of standard tauret and endogenous tauret from dark-adapted retina, separated by HPLC in protonated forms.

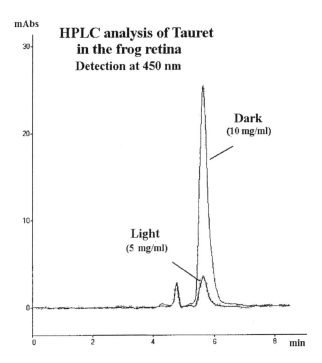

Figure 3. HPLC profiles on Supelcosil C-8 columns. Methanol extracts of retina from dark and light adapted frogs. Chromatograms are shown at 450 nm.

RESULTS AND DISCUSSION

HPLC analysis established that synthetic, standard tauret dissolved in methanol was eluted at 4.7 min, shortly after the solvent front. During analysis of methanol extracts of lyophilized retina mixed with pigment epithelium obtained from dark- and light-adapted frogs, it was established that all samples contained a substance eluting at the same time as standard tauret (Fig 1). The UV spectra of this endogenous compound were recorded on-line and compared with the spectra of synthetic tauret obtained under identical conditions showed a perfect match (Fig. 2). The λ_{max} of this peak was 446 nm. Chromatograms are shown at 326 nm or at 450 nm. HPLC as well as TLC analysis have shown that in the retina of frogs adapted to light under 1000 lx illumination compared with frog retinas which were dark-adapted, tauret quantities were remarkable lower (Fig. 3). So one can conclude that HPLC system which has been used in this study allow a full separation of tauret from methanol extracts of the retina. Levels of tauret in the retina and pigment epithelium are probably around a few hundred μM. Free retinoid content in ocular tissues are normally low. However, in situations, for example, such as strong illumination after prolonged dark adaptation, the level of *all-trans*-retinal rises rapidly. *Trans*-retinal can be quickly removed via formation of *trans*-tauret. By this way ROS damage can be prevented. Following illumination, release of taurine from isolated ROS[28] is more likely a response to prompt formation of the tauret. Its release from ROS supports that tauret is an endogenous substance in the retina and pigment epithelium. These results, with our previous measurements for separation of tauret from retinal extracts using two other HPLC systems[23], strongly support that tauret is an endogenous substance in the retina. The high concentration of taurine found in ROS[35] is probably one determinant promoting this process. Taurine-associated retinoid transport in the opposite direction, from pigment epithelium to photoreceptors, may be essential in processes related to regeneration of rhodopsin. The fact that tauret in the dark-adapted frog retina is several times higher than in the retina of the frogs adapted to light (kept under white illumination 1000 lx for 4 h; Fig. 3) indicates dynamic changes of tauret in the retina. However, whether there is a bursting elevation of tauret as a consequence of elevation of free *all-trans*-retinal in ROS immediately after illumination needs to be investigated.

We were unable to record the expected light-induced release of ^{3}H-taurine from isolated ROS fraction[24]. This needs further study. In any case, it seems now we can understand why about a 50% drop of taurine levels in the retina causes severe retinal damage[9,11,19,30,32], noticeable reduction of rhodopsin level in rat photoreceptors[12] and a drop of amplitude of the a- and b-waves of the electroretinogram (ERG)[12,30].

It is interesting to note that during this study efforts have been made to evaluate tauret content by both TLC and HPLC methods in frog liver and kidney. Investigations by both methods have revealed noticeable tauret in kidney. Probably one can conclude that beside the retina, vitamin A transport based on tauret is involved also in other systems. In this respect, it is quite possible that taurine-based vitamin A transport is important in other tissues and organs. The tauret:taurine ratio in liver stellate cells is interesting from this point of view. Stellate cells, representing 7% of the total cells in the rat liver and about 9% in the human liver, are vitamin A-storing cells[36]. Also, taurine is high[10] in human sperm and seminal fluid which are rich in vitamin A. One can expect very high level of taurine in these cells and systems and probably relatively high tauret levels. Probably taurine-based vitamin A transport is a clue to understand biological phenomena such as the increase of taurine content in the developing brain[33] or for understanding beneficial influence of taurine on preterm infants[3] as well as the presence of taurine in human and mammals milk[33,34]. Taurine may be a conditionally essential nutrient because of its ability to facilitate vitamin A transport by forming tauret. This can provide the child with necessary amount of available vitamin A in the intense phase of growth. In favour of such assumption are data obtained during evaluation of the influence of taurine deficiency in long-term total parenteral nutrition (TPN) on eye function[3]. In children and one-third of adults on TPN, ERG abnormalities and vitamin A increase in the plasma are seen[3]. Elevation of retinol levels in taurine deficiency has been observed[14] also in the kitten liver. In dilated cardiopathy, plasma taurine concentrations are only 38% of the normal level. In contrast, retinol levels in plasma are 15 to 40% higher than the levels in healthy cats[7]. Dietary supplementation with 13-*cis*-retinoic acid of methionine-supplemented rats cause marked elevation in hepatic taurine levels[29]. These associations between taurine and retinoids in blood, liver and retina are in agreement with the hypothesis that taurine may be involved in the transport of retinoids in vertebrate tissue[25,26]. Other data support such an assumption: Taurine augmented proliferative responses of T cells from both young and old mice[18]; At 0.1 mM, taurine significantly increased thymidine incorporation in chick cultured B cells and enhanced their proliferation[27]; Taurine specifically stimulated proliferation of human and rabbit retinal pigment epithelium (RPE)[8] and stimulated phagocytosis in retinal pigment epithelium[4]; Taurine has a positive effect on pre-implantation development of mouse embryos *in vitro*[5]; Taurine induced regeneration of goldfish retina in culture[15,16]; A heat-stable taurine-containing component (< I kD) in the retina is suggested to regulate rod photoreceptor development *in vivo*[1]; By immunohistochemistry, high taurine levels have been found in the larva and in the first half of the pupal stage and gradually decreases towards the end of metamorphosis in honeybee. In photoreceptor cells of the compound eye, taurine immunoreactivity can be detected from the fifth larval instar onwards, prior to differentiation of other components of the ommatidium[6]. Conversely, taurine deficiency induced losses of the major sarcomeric proteins, myosin and actin, in rats[13]. These data strongly support that by conjugation with taurine, water-insoluble retinoids can more readily be delivered to target tissues and then freed by hydrolysis. This type of delivery may be an additional mechanisms in the transport of retinoids, which otherwise are known to be transported by retinoid binding proteins[31].

ACKNOWLEDGMENTS

We are grateful to Professor Mohamed Abdullah (Marine Biology Lab of Oslo University) for help with lyophilization. We both from Armenia are very thankful to Professor Laura della Corte and Professor Ryan Huxtable for supporting our participation in the Taurine Symposium '99. This work was mainly supported by Oslo University and partially by Ministry of Science and High Education of Armenia, Grant 96-681.

REFERENCES

1. Altshuler, D., Lo Turco, J.J., Rush, J., and Cepko, C. 1993, Taurine promotes the differentiation of a vertebrate retinal cell type in vitro, *Development*, 119:1317-1328.
2. Awapara J. 1962 Free amino acids in invertebrates: a comparative study of their distribution and metabolism In Amino Acid Pools pp. 158-175, Elsevier Publ Company Amsterdam, London, New York
3. Chesney R.W., Helms R.A., Christensen M., Budreau A. M., Han X. & Sturman J.A., 1998, The role of taurine in infant nutrition. In: "Taurine: Basic and Clinical Aspects" (Ed. by Schaffer S., Lombardinin B., & Huxtable R.) p.463-476. Plenum Publ. Corp.
4. Dontsov, A. E., Sakina, N. L., Boulton, M. A., 1991, Taurine stimulates phagocytosisin retinal pigment epithelium. *Biol. Membranes,(Russ.)* 8:1197-1198
5. Dumoulin J.C., Evers J.L., Bras M., Pieters M.H., and Geraedts J.P., 1992. Positive effect of taurine on preimplantation development of mouse embryos *in vitro. J. Reprod. Fert*il. 94:373-380
6. Eichmuller S., Schaffer S., 1995. Sensory neuron development revealed by taurine immunocytochemistry in the honeybee. *J. comp. Neurol.* 352:297-307
7. Fox, P.R., Trautwein, E.A., Hays, K.C., Bond, B.R., Sisson, D.D., and Moise, N.S. 1993, Comparision of taurine, α-tocopherol, retinol, selenium and total triglycerides and cholesterol concentrations in cats with cardiac disease and in healthy cats, *Am. J. Vet. Res.* 54:563-569.
8. Gabrielian, K., Wang, H.M., Ogden, T.E., and Ryan, S.J. 1992, In vitro stimulation of retinal pigment epithelium proliferation by taurine, *Curr.Eye Res.* 11:481-487.
9. Hayes, K.C., Garey, R.E., and Schmidt, S.Y. 1975, Retinal degeneration associated with taurine deficiency in the cat, *Science* 188:945- 951.
10. Holmes R.P., Goodman H.O., Shihabi Z.K., JarowJ.P., 1992 The taurine and hypotaurine content of human semen. *J. Androl.* 13: 289-292
11. Imaki, H., Jacobson, S.G., Kemp, C.M., Knighton, R.W., Neuringer, M., and Sturman, J. 1993, Retinal morphology and visual pigment levels in 6- and 12- month-old rhesus monkeys fed a taurine-free human infant formula, *J. Neurosci. Res.* 36:290-304.
12. Lake, N. 1989, The role of taurine in retinal function, in::" Extracellular and Intracellular Messengers in the Vertebrate Retina", Redburn, D..A. and Pasantes-Morales, H. eds., Alan R.Liss, New York. pp.61-86.
13. Lake. N 1993 Loss of cardiac myofibrils: mechanism of contractile deficits induced by taurine deficiency. *Am. J. Physiol.* 264: I323-1326.
14. Lehmann, A., Knutsson, L., and Bosaeus, I. 1990, Elevation of retinal levels and suppression of alanine aminotransferase activity in the liver of taurine-deficient kittens, *J. Nutr.* 120:1163-1167.
15. Lima L, Drujan B, Matus P., 1990, Spatial distribution of taurine in the teleost retina and its role in retinal tissue regeneration, in "Taurine: Functional Neurochemistry, Physiology and Cardiology", Pasantes-Morales, H., Martin, D. L., Shain, W., Martin del Rio, R.,eds Wiley-Liss, Inc. New York, Vol. 351 pp 103-112.
16. Lima L, Matus P., Drujan B., 1993, Taurine-induced regeneration of goldfish retina in culture may involve a calcium-mediated mechanism. *J. Neurochem.* 60:2153-2158.
17. Nakashima T., Shima T., Mitsuyoshi H., Inaba K., Matsumoto N., Sakamoto Y. and Kashima K., 1996 Taurine in the Liver: The function of taurine conjugated with bile acids In: "Taurine 2: Basic and Clinical Aspects", pp. 85-92 (Ed. by Huxtable R., Azuma J., Kuriyama K., Nakagawa M. & Baba A.). Plenum Press, New York & London.
18. Negoro S. and Hara H., 1992. The effect of taurine on the age-related decline of the immune response in mice: the restorative effect on the T cell proliferative response to costimulation with ionomycin and phorbol myristate acetate. *Adv. Exp. Med. Biol.* 315: 229-239.

19. Pasantes-Morales, H., Quesada, O., Carabez, A., and Huxtable, R.J. 1983, Effects of the taurine transport antagonist, guanidinoethane sulfonate and betta-alanine, on the morphology of the rat retina, *J. Neurosci.Res.* 9:135-143.
20. Pasantes-Morales, H., Queseda O., Moran, J., 1998. Taurine: An osmolyte in mammalian tissues In: "Taurine 3: Cellular and regulatory mechanisms", pp. 209-217 (Ed. by Schaffer S., Lombardinin B., & Huxtable R.J.). Plenum Press, NY & London.
21. Petrosian A.M. and Haroutounian J.E., 1998. The role of taurine in osmotic, mechanical and chemical protection of the retinal rod outer segments. In: "Taurine: Basic and Clinical Aspects", pp. 407-413 (Schaffer S., Lombardinin B., and Huxtable R.J. Eds). Plenum Publ.Corp., New York and London.
22. Petrosian A.M. and Haroutounian J.E., 1998. Whether connections between disks and cellular membrane in the retinal rod outer segments are channelized. In: Proceedings of the 14th Internat. Congress on Electron Mycroscopy, Cancun, Mexico, Eds. H.A. Calderon Benavides and M. Jose Yacaman. Inst. Phys Publ. *Biol Sci.,* 7:249-250.
23. Petrosian A.M., Haroutounian J.E., Zueva L.V., 1996. Tauret: A taurine-related endogenous substance in the retina and its role in vision. In: " Taurine: Basic and Clinical Aspects" (Ed. Huxtable R.J., Azuma J., Kuriyama K., Nakagawa M. & Baba A.) p.333-42. Plenum Publ. Corp, New York.
24. Petrosian A.M,. Haroutounian J.E.,. Gundersen T.E, Blomhoff R., Fugelli K. and Kanli H., 1999. Effects of osmotic and light stimulation on ^{3}H-taurine efflux from isolated rod outer segments and synthesis of tauret in the frog retina. *This volume.*
25. Petrosian, A.M. and Haroutounian, J.E. 1988, On the possibility of retinal transportation in rods by tauret through special channels. *13th All-Union conf. on Electron Microscopy*, (Abstract), Zvenigorod, Moscow.
26. Petrosian, A.M. and Haroutounian, J.E. 1990, Tauret: further studies of the role of taurine in retina, in: "Taurine: Functional Neurochemistry, Physiology, and Cardiology", Pasantes-Morales, H., Martin, D.L., Shain, W. and del Rio, R.M. eds., Wiley-Liss, New York, pp.471-475.
27. Porter D.W., Kaczmarczyk W., Martin W.G., 1993. The effect of taurine on the incorporation of thymidin by chick B cells. *Comp. Biochem. Physiol. B.* 106:251-254.
28. Salceda, R., Lopez-Colome, A.M., and Pasantes-Morales, H. 1977, Light-stimulated release of ^{35}S-taurine from frog retinal outer segments, *Brain Res.* 135:186-191.
29. Schalinske, K.L., Steele, R.D. 1991, !3-cis-Retinoic acid alters methionine metabolism in rats, *J. Nutr.* 121:1714-1719.
30. Schmidt, S.Y., Berson, E.L., Watson, G., and Huang, C. 1977, Retinal degeneration in cats fed casein. III.taurine deficiency and ERG amplitudes, *Inves.Ophtal.Vis.Sci.* 16:673-678.
31. Sivaprasadarao A. and Findlay J.B.C., 1994, The retinol-binding protein superfamily. In: Vitamin A in health and disease. Rune Blomhoff ed. Marcel Dekker, Inc, New York, Basel, Hong Kong pp 87-117.
32. Sturman, J..A. 1990, Taurine deficiency, in: "Taurine: Functional Neurochemistry, Physiology , and Cardiology", Pasantes-Morales, H., Martin, D.L., Shain, W. and del Rio, R.M. eds., Wiley-Liss. New York, pp.385-395.
33. Sturman, J.A., Hayes K.C., 1980, The biology of taurine in nutrition and development. *Adv. Nutr. Res.*, 3:231-299
34. Takanami M. and Miura K.L., 1963, *Biochem. Biophys. Acta*, 72:237-
35. Voaden, M. J., Lake, N., Marshall, J., Morjaria, B. 1977. Studies on the distribution of taurine and other neuroactive amino acids in the retina. *Exp. Eye Res.* 25:249-257.
36. Wake K., 1994, Role of perisinusoidal stellate cells in vitamin A storage. In: Vitamin A in health and disease. Rune Blomhoff ed. Marcel Dekker, Inc, New York, Basel, Hong Kong pp 73-86.

CHARACTERIZATION OF TAURINE UPTAKE IN THE RAT RETINA

[1]Julius D. Militante and [1,2]John B. Lombardini

Departments of Pharmacology[1] and Ophthalmology & Visual Sciences[2],
Texas Tech University Health Sciences Center, Lubbock, Texas, 79430, *USA*

INTRODUCTION

Taurine is a free amino acid found in mammalian tissues and its pharmacological depletion in various animal models has been demonstrated to cause visual deficits and morphological degeneration in the retina[1]. Taurine is an important modulator of cellular processes whose most unique characteristic is the unusually high mM concentrations in which it is found intracellularly[2]. In the retina, concentrations as high as 79 mM have been measured[3] with most of the taurine being exogenous in nature[1]. The steep gradient across the cell membrane requires a very efficient transport system that is carefully regulated.

Taurine uptake in the retina has been demonstrated, in various species, to have 2 saturable components of differing affinity to taurine[2]. The transport systems exhibit a 10- to 100-fold difference in their affinites for taurine, as measured by the taurine concentration at which a half-maximal velocity is attained (K_m). In the retina, visual transduction is initiated in the rod outer segments (ROS) of the photoreceptor cells[4] to which the high-affinity taurine uptake component has been specifically linked[5,6]. However, the kinetics of taurine uptake in the ROS have not been studied in detail. In view of the

Taurine 4, edited by Della Corte *et al.*
Kluwer Academic / Plenum Publishers, New York, 2000.

visual deficits associated with pharmacological inhibition of taurine uptake, particularly with the taurine analogue guanidino-ethanesulfonate (GES), the study of uptake kinetics in the ROS is crucial in the understanding of the role of taurine in the visual transduction process.

In rat astrocytes and human GL15 glioma cells, activation of protein kinase C (PKC) by phorbol myristate acetate (PMA) results in the inhibition of taurine uptake[7-9], suggesting that PKC modulation of taurine transport may be important in the central nervous system. Chelerythrine (CHT) is a potent PKC inhibitor that has been found to antagonize the effects of taurine in the retina[10,11]. This study aims to characterize taurine uptake in the retina, particularly in the ROS, in terms of transport kinetics and PKC regulation, using pharmacological agents such as GES and CHT.

MATERIALS AND METHODS

Materials

Radiolabelled [³H]taurine was purchased from New England Nuclear. Chelerythrine chloride (CHT) was obtained from LC Laboratories. Phorbol myristate acetate (PMA) and staurosporine (STAU) were purchased from Sigma.

Preparation of tissue samples

Whole retinal homogenate and isolated ROS preparations were obtained according to previously established protocols[12]. Briefly, retinae were dissected from frozen adult rat eyes, washed and pooled in Krebs-bicarbonate-Ringer (KBR) solution, maintained on ice. The retinae were homogenized with a handheld glass mortar and pestle in KBR solution.

To prepare isolated ROS, the retinae were vortex-mixed gently for 10 seconds to detach the ROS. The tissue and cell debris were allowed to settle and the supernatant containing the suspended ROS was collected. The procedure was repeated with the pelleted retinae to maximize the recovery of ROS. The pooled supernatant was centrifuged and the pelleted ROS was resuspended in KBR solution.

Taurine uptake assay

The assay used in measuring taurine uptake was performed as previously described[13]. Briefly, the reaction was performed in KBR solution and for uptake kinetic experiments, unlabelled taurine was added in varying concentrations (10 μM - 10 mM). For other experiments, unlabelled taurine was added to a final concentration of either 50 μM or 1.5 mM. Equal amounts of [³H]taurine were added to each test tube (1 μCi) and incubated in a 37 °C water bath for 2 minutes. The reaction was initiated by the addition of an equal amount of tissue in each test tube to make a final volume of 250 μl per tube and the reaction was allowed to proceed for 7 minutes before the reaction was terminated with the addition of 3 ml ice-cold buffer and filtration through a Millipore apparatus. The radioactivity on the filter paper was measured in a scintillation counter.

Data analysis

Non-linear regression analysis was performed on uptake data following the method described by Neal & White[14] and using GraphPad PRISM^{TM} software. Linear regression analyses were also performed on transformed data (Eadie Hofstee plots) using the same software.

RESULTS

Michaelis-Menten constants (K_m and V_{max}) were estimated using non-linear analysis of the uptake data (Table 1). Two kinds of saturable uptake systems were identified in whole retinal preparations: a high-affinity transporter and a low-affinity transporter. In contrast, only one uptake system was identified in the ROS which seemed to correspond to the high-affinity component (Table 1).

Table 1. Taurine uptake kinetics (Mean ± SEM)

Sample	K_m	V_{max} (mol x 10^{-6}/g)
Whole retinal preparation		
	133 ± 47 μM	0.73 ± 0.27
	2.74 ± 1.3 mM	3.15 ± 0.42
ROS	140 ± 8 μM	2.46 ± 0.08

Taurine uptake was measured in the whole retinal preparation in the presence of both 50 μM and 1.5 mM taurine to distinguish between the function of the high- and of the low-affinity uptake, respectively, though it is assumed that both components contribute to total taurine uptake to some degree at all taurine concentrations. GES inhibited taurine uptake at both taurine concentrations (Table 2). Half-maximal inhibition (IC_{50}) at 50 μM taurine was 262 ± 49 μM GES, while at 1.5 mM taurine it was 0.89 ± 0.13 mM. With the isolated ROS, the high-affinity uptake system was measured in the presence of 50 μM taurine, and this system was inhibited with GES ($IC_{50} = 352 \pm 89$ μM) (Table 2) with the same potency as with high-affinity uptake in the whole retinal preparation.

Table 2. Taurine uptake (pmol/μg) in the presence of either 50 μM or 1.5 mM taurine: effect of GES treatment

	50 μM Taurine			1.5 mM Taurine		
Whole retinal preparation						
μM GES	Mean ± SEM		N	mM GES	Mean ± SEM	N
0	0.28 ± 0.02		4	0	2.34 ± 0.21	4
0.5	0.23 ± 0.03		4	0.8	1.28 ± 0.14*	3
1	0.20 ± 0.02		4	1.5	0.85 ± 0.17*	4
5	0.10 ± 0.06*		4	5.0	1.02 ± 0.15*	4
7.5	0.06 ± 0.01*		3	20.0	0.89 ± 0.23*	4
ROS						
μM GES	Mean ± SEM		N			
0	0.54 ± 0.08		3			
0.5	0.54 ± 0.05		3			
1	0.65 ± 0.04		3			
5	0.31 ± 0.07		3			
7.5	0.15 ± 0.05*		3			

* $P< 0.05$, compared to 0 GES (One-way ANOVA and Duncan's multiple range test)

The regulation of high-affinity taurine uptake in the ROS (in the presence of 50 μM taurine) by modulators of PKC activity was studied by measuring taurine uptake in the presence of 50 μM taurine. PMA and STAU were used as a PKC activator and inhibitor, respectively. PMA at 0.8, 8, 16 μM concentrations produced no significant effect on taurine uptake. Similarly, STAU at 2, 20, 40 μM concentrations produced no significant effects. Taurine uptake was measured both in the presence and in the absence of 1.0

mM ATP to ensure that ATP depletion during membrane preparation is not a confounding factor.

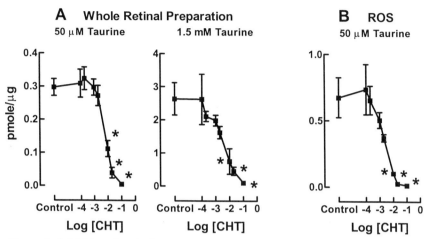

Figure 1. Inhibition of retinal taurine uptake by chelerythrine (CHT). (*P < 0.05 compared to their respective controls, Student's t-test; N = 3 for all comparisons).

Curiously, chelerythrine (CHT), a PKC inhibitor[15], inhibited taurine uptake when measured in both whole retinal preparations and in ROS. Inhibition of taurine uptake was observed in the presence of 50 μM taurine (Fig 1A; IC_{50} = 7.2 ± 2.2 μM) and in the presence of 1.5 mM taurine (Fig 1A; IC_{50} = 7.3 ± 3.4 μM) in whole retinal preparations. CHT also inhibited taurine uptake in isolated ROS (Fig 1B; IC_{50} = 2.4 ± 0.4 μM). CHT inhibition of taurine uptake in the ROS was found to be non-competitive in nature when the effects of 5 μM CHT were measured through a narrow range of taurine concentrations (25-750 μM)(Fig 2) with CHT producing a change in V_{max} and no change in K_m.

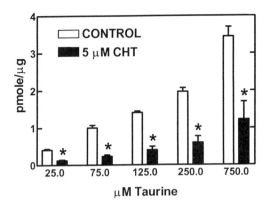

Figure 2. Concentration-dependent taurine uptake in ROS in the presence and absence of 5 μM CHT. Final concentrations of taurine ranged from 25 μM to 750 μM. (*$P < 0.05$ compared to their respective controls, Student's t-test; $N = 3$ for all comparisons).

DISCUSSION

The demonstration of 2 saturable uptake systems with at least a 10-fold disparity in affinity for taurine in the retina is similar to data presented in other reports[2]. Though the high-affinity uptake system has been linked to the photoreceptor cell layer previously[5,6], in this report we demonstrate that only one uptake system is functional in the ROS. It is unclear if high-affinity uptake is exclusive to the ROS or not, but most definitely it is the transport system of significance in the ROS. Taurine uptake was inhibited by GES under all conditions studied.

The high-affinity uptake system is not susceptible to modulation by PKC as measured in ROS, similar to taurine uptake in rat and human neuronal cells[7-9]. It appears that PKC modulation is effective in glial cells but not in neuronal cells, as photoreceptor cells can be considered cells of neuronal origin. The inhibition of taurine uptake by CHT must then be considered independent of PKC inhibition. This inhibition may be the mechanism of action behind the PKC-independent antagonism of the effects of taurine on the phosphorylation of a ~20 kDa retinal protein[10,16] and on ATP-dependent Ca^{2+} uptake in the retina[11]. It is unclear what is the mechanism of action behind this effect of CHT.

ACKNOWLEDGMENTS

This research was supported in part by grants from the RGK Foundation, Austin, Texas, and the Taisho Pharmaceutical Co., Ltd. Tokyo, Japan. Grateful appreciation is extended to Dr. James C. Hutson, Dr. Sandor Gyorke, Dr. John C. Fowler, Dr. Howard K. Strahlendorf, Miss Janet Koss, Miss Anne Carpenter and Mrs. Yevgeniya Lukyanenko for their help in obtaining rat eye samples.

REFERENCES

1. Lombardini, J.B., 1991, Taurine: retinal function, *Brain Res. Rev.* **16**:151-169.
2. Huxtable, R.J., 1989, Taurine in the central nervous system and the mammalian action of taurine, *Prog. Neurobiol.* **32**:471-533.
3. Voaden, M.J., Lake, N., Marshall, J., Morjaria, B., 1977, Studies on the distribution of taurine and other neuroactive amino acids in the retina, *Exp. Eye Res.* **25**:219-257.
4. Baylor, D., 1996, How photons start vision, *Proc Natl Acad Sci* 93:560-565.
5. Lake, N., Marshall, J., Voaden, M.J., 1978, High affinity uptake sites for taurine in the retina, *Exp Eye Res* 27:713-718.
6. Schmidt, S.Y., Berson, E.L., 1978, Taurine uptake in isolated retinas of normal rats and rats with hereditary retinal degeneration, *Exp. Eye Res.* **27**: 191-198.
7. Tchoumkeu-Nzouessa, G. C., Rebel, G., 1996 Regulation of taurine transport in rat astrocytes by protein kinase C: role of calcium and calmodulin, *Am. J. Physiol.* 270:C1022-C1028.
8. Tchoumkeu-Nzouessa, G. C., Rebel, G., 1996, Activation of protein kinase C down-regulates glial but not neuronal taurine uptake, *Neurosci. Lett.* **206**: 61-64.
9. Tchoumkeu-Nzouessa, G. C., Rebel, G., 1996, Characterization of taurine transport in human glioma GL15 cell line: regulation by protein kinase C, *Neuropharm.* **35**:37-44.
10. Lombardini, J.B., 1993, The inhibitory effects of taurine on protein phosphorylation: comparison of various characteristics of the taurine-affected phosphoproteins present in the rat retina, brain and heart, In Huxtable, R.J., Michalk, D. (eds): "Taurine in Health and Disease," New York: Plenum Press, pp. 9-17.
11. Militante, J.D., Lombardini, J.B., 1998, Effect of taurine on chelerythrine inhibition of calcium uptake and ATPase activity in the rat retina, *Biochem. Pharmacol.* **55**: 557-565.
12. Militante, J.D., Lombardini, J.B., 1998, Pharmacological characterization of the effects of taurine on calcium uptake in the rat retina, *Amino Acids* **15**:99-108.
13. Militante, J.D., Lombardini, 1999, Taurine uptake activity in the rat retina: protein kinase C-independent inhibition by chelerythrine, *Brain Res* 818:368-374.
14. Neal, M.J., White, R.D., 1978, Discrimination between descriptive models of L-glutamate uptake by the retina using non-linear regression analysis, *J. Physiol.* 277:387-394.
15. Herbert, J.M., Augereau, J.M., Gleye, J., Maffrand, J.P., 1996, Chelerythrine is a potent and specific inhibitor of protein kinase C, *Biochem. Biophys. Res. Comm.* **172**: 993-999.
16. Lombardini, J.B., 1992, Effects of taurine on the phosphorylation of specific proteins in subcellular fractions of the rat retina, *Neurochem. Res.* **17**:821-824.

CALCIUM UPTAKE IN THE RAT RETINA IS DEPENDENT ON THE FUNCTION OF THE CYCLIC NUCLEOTIDE-GATED CHANNEL: PHARMACOLOGIC EVIDENCE

[1]Julius D. Militante and [1,2]John B. Lombardini
Departments of [1]Pharmacology and [2]Ophthalmology & Visual Sciences,
Texas Tech University Health Sciences Center, Lubbock, Texas 79430, USA

INTRODUCTION

The modulation of calcium (Ca^{2+}) uptake into the retina by taurine, a free amino acid found in high concentrations in mammalian tissues, has been extensively studied through the years[1]. Ca^{2+} uptake in the retina or in retinal subfractions is usually measured by incubating retinal samples with radiolabelled Ca^{2+} under various conditions, filtering the reaction mixture through filter paper and measuring the remaining radioactivity on the paper. Through this method, ATP is demonstrated to stimulate Ca^{2+} uptake into whole retina preparations, synaptosomal fractions, mitochondrial fractions and in isolated rod outer segments (ROS) under conditions of low Ca^{2+} concentrations (10-100 μM). Taurine is known to potentiate this stimulation but has no significant effects in the absence of ATP. It is, thus, assumed that total Ca^{2+} uptake is a combination of different Ca^{2+} components, each one possessing a distinct Ca^{2+} binding and/or transport system, and that the effects of taurine are likewise varied and differ accordingly.

Taurine 4, edited by Della Corte *et al.*
Kluwer Academic / Plenum Publishers, New York, 2000.

The Ca^{2+} component specific to the ROS that is modulated by taurine is of particular interest because of the potential role taurine may play in the visual transduction process. The ROS is the section of the photoreceptor cell which is involved in the transduction of light signals, a process which involves the inward flux of Ca^{2+} through cyclic nucleotide-gated (CNG) channels[2]. These channels allow the flux of cations through the membrane and are found primarily in the ROS[3]. Taurine uptake and efflux in the ROS are known to be modulated by light in the frog, chick and rat retina[4-6], suggesting that taurine may be involved in the same light signalling process.

It is clear that taurine increases ATP-dependent Ca^{2+} uptake in the ROS[7,8]. The involvement of cation channel activation in Ca^{2+} uptake in the retina has not been addressed, particularly that of the CNG channel in the ROS. Understanding the mechanism of action of taurine requires understanding of Ca^{2+} uptake in the ROS. This study aims to identify the mechanism behind Ca^{2+} uptake in the retina, specifically the involvement of CNG channels in the ROS, and the site at which taurine acts to stimulate Ca^{2+} uptake.

MATERIALS AND METHODS

Chemicals

LY83583 (6-anilino-5,8-quinolinedione) and zaprinast (M&B 22,948) was purchased from RBI, Natick, MA. Rp-8-Br-PET-cGMPS was purchased from Biolog Life Science Institute, La Jolla, CA. BCA protein assay reagent was obtained from Pierce Chemicals, Rockford, IL. $^{45}CaCl_2$ was obtained from New England Nuclear, Boston, MA.

Preparation of the retinal membrane homogenate and isolated rod outer segments

Tissue samples were prepared as described previously[9]. Briefly, retinal tissue was teased out of the eye cup in 0.32 M sucrose while on ice. The tissue was centrifuged for 15 minutes at 16,000 x g, washed in 20 mM bicarbonate, recentrifuged as before and then washed in sodium-bicarbonate buffer [NaHCO$_3$, 50 mM, NaCl, 50 mM; KCl, 50 mM; KH$_2$PO$_4$, 1.2 mM; MgCl$_2$, 2 mM[10]] with CaCl$_2$ added to a final concentration of 10 μM. The tissue was recentrifuged, resuspended in sodium bicarbonate-CaCl$_2$ buffer and gently homogenized.

For the isolation of rod outer segments (ROS), 0.3 M mannitol was used instead of 0.32 M sucrose. Retinal tissue was dissected out as before and the ROS were removed by vortex-mixing for 6s, allowing the tissue to settle, and then decanting the supernatant which contained the ROS. The supernatant was centrifuged at 16,000 x g for 15 minutes and the pellet was then suspended in sodium-bicarbonate-CaCl$_2$ buffer. The remaining tissue components were discarded.

Ca^{2+} uptake assay

The assay was performed as described previously[9]. Briefly, ATP and taurine were added to the reaction mixture in the appropriate concentrations, including identical amounts of ^{45}CaCl$_2$. Retinal homogenate (100-300 μg) or ROS (30-100 μg) was added to start the reaction and the reaction was terminated by filtering on a Millipore glass fiber filter. The filter was then counted for radioactivity with a scintillation counter.

Statistical Analysis

Statistical analyses were performed using the GraphPad Prism and InStat software. Data were analyzed using the one-way analysis of variance (ANOVA) or linear regression analysis. Post-hoc analysis was accomplished using the Duncan's multiple range test.

RESULTS

Cadmium (Cd^{2+}) is an non-selective blocker of Ca^{2+} currents in neurons at low micromolar concentrations (2-20 μM)[11], and was used in this study as a blocker of Ca^{2+} channels in the retina. Cd^{2+} was effective in inhibiting Ca^{2+} uptake in the whole retinal preparation at 5 μM, indicating that a significant portion of the Ca^{2+} uptake measured was dependent on the activation of Ca^{2+} channels. The stimulatory effect of taurine on retinal Ca^{2+} is known to be wholly dependent on the presence of ATP. At this low concentration of Cd^{2+}, the effect of taurine to stimulate ATP-dependent Ca^{2+} uptake was exclusively affected (Table 1), suggesting that taurine may be acting via ATP-dependent activation of Ca^{2+} channels. At a higher concentration (100 μM), Cd^{2+} was also effective in inhibiting Ca^{2+} uptake in the absence of taurine. The identity of the Ca^{2+} channels involved is unknown.

Table 1. Inhibition of Ca^{2+} uptake by Cd^{2+} in whole retinal preparations in the presence of 1.2 mM ATP

	pmol/µg (mean ± SEM)	
µM Cd^{2+}	control (N = 5)	32 mM taurine (N = 4)
0	0.31 ± 0.03^a	0.59 ± 0.10^b
5	0.25 ± 0.03^a	0.34 ± 0.04^a
100	0.07 ± 0.01^c	0.11 ± 0.01^c

(Different letters denote $p < 0.05$, Student's t-test.)

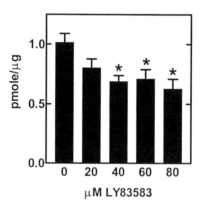

Figure 1. The effects of LY83583 on Ca^{2+} uptake in rat retinal membrane preparations in the presence of 1.2 mM ATP and 32 mM taurine. An asterisk indicates a significant difference ($P < 0.05$) between control values (0 µM LY83583) calculated by one-way ANOVA and the Duncan's multiple range test. Data are expressed as means ± SEM, N = 5-8.

LY83583 inhibits the activation of CNG channels in isolated olfactory receptor neurons[12], acting directly on the channel and also on soluble guanylyl cyclase, the enzyme which catalyzes the formation of the endogenous ligand of the channel (cyclic guanosine-3',5'-cyclic monophosphate, or cGMP). LY83583 inhibited the effects of taurine on ATP-dependent Ca^{2+} uptake in the whole retina preparation at 40-80 µM concentration (Fig. 1) but produced no effects in the absence of taurine (data not shown), suggesting the specific dependence of retinal Ca^{2+} uptake on the activation of CNG channels.

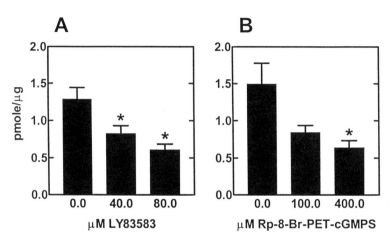

Figure 2. Ca^{2+} uptake in rat ROS in the presence of 1.2 mM ATP and 32 mM taurine. A) The inhibitory effect of LY83583. B) The inhibitory effect of Rp-8-Br-PET-cGMPS. An asterisk indicates a significant difference (P < 0.05) from the respective control values (0 µM LY83583 or 0 µM Rp-8-Br-PET-cGMPS) calculated by one-way ANOVA and the Duncan's multiple range test.

LY83583 exhibited the same antagonism of the effects of taurine when isolated ROS were used (Fig 2A) and again exhibited no effects in the absence of taurine (data not shown). Similar findings were observed with a specific competitive inhibitor of CNG channels was used (Fig. 2B). Rp-8-Br-PET-cGMPS easily passes through the cell membrane easily and competes with cGMP for the same activation site on the CNG channel intracellularly[13]. As with LY83583, no effect was observed in the absence of taurine (data not shown). Data using a CNG channel activator and an inhibitor of cGMP degradation further support the involvement of CNG channel. A membrane permeant analogue of cGMP, dibutyryl cGMP

(DBG), produced a dose-dependent increase in ATP-dependent Ca^{2+} uptake (Fig. 3A). A less potent stimulation was also observed with zaprinast, an inhibitor of cGMP-specific phosphodiesterase (PDE) found in the ROS[14] (Fig. 3B). Significantly, neither cGMP or zaprinast affected the stimulation of ATP-dependent Ca^{2+} uptake by taurine.

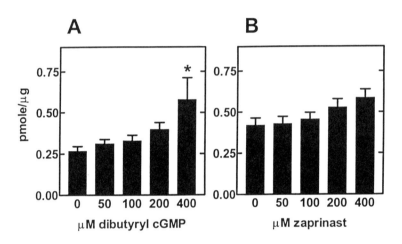

Figure. 3. Ca^{2+} uptake in rat rod outer segments in the presence of 1.2 mM ATP, with 0 mM taurine. A) The stimulatory effects of dibutyryl cGMP. An asterisk indicates a significant difference ($P < 0.05$) from control values (0 μM dibutyryl cGMP) calculated by one-way ANOVA and the Duncan's multiple range test.) B) The effects of zaprinast. Linear regression analyses indicated that the slope for Ca^{2+} uptake was significantly different from zero, sloping upward ($P < 0.01$).

DISCUSSION

The exact nature of the Ca^{2+} uptake measured in retinal tissue is unclear. Ca^{2+} binding to membranes is almost surely a significant component, and Ca^{2+} ion transport into vesicles is assumed to also occur because of the stimulatory effects of ATP. Passive transport is a mechanism that would also be applicable to Ca^{2+} uptake, or at least is a mechanism that has not been disproven. Another possibility is that stimulatory agents like ATP and taurine cause the opening of ion channels which allow Ca^{2+} in the reaction buffer to flow into sealed cellular structures following its gradient. Thus, binding, transport and channel activation all present mechanistic possibilities that need to be considered in understanding and interpreting Ca^{2+} uptake data not only from retinal experiments, but from other tissue types as well.

The involvement of channel activation in Ca^{2+} uptake in the retina has not been addressed before. In these experiments, data from Cd^{2+} experiments suggest that taurine stimulation of Ca^{2+} uptake is dependent on the activation of Ca^{2+} channels. Significantly, Cd^{2+} was effective in antagonizing the taurine stimulation at 5 μM concentration totally but had no effect on ATP-dependent Ca^{2+} uptake (Table 1), suggesting that taurine stimulates a channel system or systems distinct from that which ATP affects. Thus, strictly speaking, taurine is not potentiating the mechanisms that ATP on its own modulates, but rather, taurine activates a separate system or systems in an ATP-dependent manner and total Ca^{2+} uptake becomes significantly increased. ATP alone may be activating Ca^{2+} channels which are less sensitive to Cd^{2+} block as 100 μM Cd^{2+} produced significant inhibition of ATP-dependent Ca^{2+} uptake.

CNG channels are cation channels that allow for the inward flux of Ca^{2+} and Na^+ into the ROS and the specific involvement of CNG channels in the effects of taurine was studied using CNG channel inhibitors. LY83583 potently inhibited CNG channels in patch recordings of olfactory neurons at concentrations as low as 1 μM[12]. In comparison, LY83583 inhibited Ca^{2+} uptake in both whole retinal membranes and isolated ROS at concentrations > or = 40 μM (Fig. 1 and 2A). The difference in potency is in utmost probability due to the inherent advantage patch recording has in measuring Ca^{2+} channel activation as compared to measuring Ca^{2+} uptake in membrane. To verify that the activation of the CNG channel is responsible for a significant portion of Ca^{2+} uptake, a cGMP analogue inhibitor of the CNG channel was used to antagonize the effects of taurine (Fig. 2B). Through all the experiments represented in Figs 1 and 2, ATP-dependent Ca^{2+} uptake was not affected by the different CNG channel inhibitors (data not shown), suggesting that CNG channel modulation is a specific taurine effect. Ca^{2+} uptake was also stimulated with the use of DBG, a cell-permeant analogue of cGMP, and zaprinast, an inhibitor of cGMP degradation, although stimulation was not as pronounced as that caused by taurine.

The various effects may be attributed to other mechanisms than modulation of the CNG channel. For example, the various agents may directly Ca^{2+} binding to membranes, or in the case of LY83583 and Rp-8-Br-PET-cGMPS may inhibit taurine binding or uptake, thus producing specific antagonism of the effects of taurine. But all the data taken together present evidence for the involvement of CNG channel activation in the stimulation of Ca^{2+} uptake in both whole retina and in isolated ROS by taurine.

ACKNOWLEDGMENTS

This study was funded in part by grants from the RGK Foundation of Austin, Texas, and the Taisho Pharmaceutical Co., Ltd. of Tokyo, Japan.

REFERENCES

1. Lombardini, J.B., 1991, Taurine: retinal function. *Brain Res. Rev.* **16**:151-169.
2. Baylor, D., 1996, How photons start vision. *Proc. Natl. Acad. Sci. USA* **93**:560-565.
3. Finn, J.T., Grunwald, M.E., Yau, K.-W., 1996, Cyclic nucleotide-gated ion channels: an extended family with diverse functions. *Annu. Rev. Physiol.* **58**:395-426.
4. Salceda, R., Lopez-Colome, A.M., Pasantes-Morales, H., 1977, Light-stimulated release of [^{35}S]taurine from frog retinal rod outer segments. *Brain Res.* **135**:186-191.
5. Pasantes-Morales, H., Quesada, O., Carabez, A., 1981, Light-stimulated release of taurine from retinas of kainic acid-treated chicks. *J. Neurochem.* **36**:1583-1586.
6. Schmidt, S.Y., 1978, Taurine fluxes in isolated cat and rat retinas: effects of illumination. *Exp. Eye Res.* **26**:529-535.
7. Lopez-Colome, A.M., Pasantes-Morales H., 1980, Effect of taurine on ^{45}CA transport in frog retinal rod outer segments. *Exp. Eye Res* **32**:771-780.
8. Lombardini, J.B.,1985, Effect of taurine on calcium ion uptake and protein phosphorylation in rat retinal membrane preparations, *J. Neurochem.* **45**:268-275.
9. Militante, J.D., Lombardini, J.B., 1998, Pharmacological characterization of the effects of taurine on calcium uptake in the rat retina, *Amino Acids* **15**:99-108.
10. Kuo C.H., Miki N., 1980, Stimulatory effect of taurine on Ca-uptake by disc membranes from photoreceptor cell outer segments. *Biochem Biophys Res Commun* **94**: 646-651.
11. Carbone, E., Swandulla, D., 1989, Neuronal calcium channels: kinetics, blockade and modulation. *Prog. Biophys. Molec. Biol.* **54**:31-58.
12. Leinders-Zufall, T., Zufall, F., 1995, Block of cyclic nucleotide-gated channels in salamander olfactory receptor neurons by the guanylyl cyclase inhibitor LY 83583. *J. Neurophysiol.* **74**:2759-2762.
13. Wei, J-Y, Cohen, E.D., Tan, Y-Y., Genieser, H-G., Barnstable, C.J, 1996, Identification of competitive antagonists of the rod photoreceptor cGMP-gated cation channel: ß-phenyl-1,N^2-etheno-substituted cGMP analogues as probes of the cGMP-binding site. *Biochem.* **35**:16815-16823.
14. Gillespie, P.G., Beavo, J.A., 1989, Inhibition and stimulation of photoreceptor phosphodiesterases by dipyridamole and M&B 22,948. *Mol. Pharmacol.* **36**:773-781.

EFFECTS OF CALMODULIN ANTAGONISTS ON TAURINE-STIMULATED CALCIUM ION UPTAKE IN THE RAT RETINA ARE PARTLY INDEPENDENT OF CALMODULIN ACTIVITY

[1]Julius D. Militante and [1,2]John B. Lombardini

Departments of [1]Pharmacology and [2]Ophthalmology & Visual Sciences,
Texas Tech University Health Sciences Center, Lubbock, Texas 79430, USA

INTRODUCTION

The physiological role of taurine is an intriguing field of study mainly because this free amino acid is present in high millimolar concentrations in many mammalian tissues, including heart, kidney, liver and eye[1]. Taurine has significant effects in several physiological and biochemical model systems, such as protein phosphorylation, ion transport, osmoregulation, neurotransmission and cellular toxicity. Remarkably, the mechanisms of action underlying such effects of taurine have yet to be fully elucidated. One of the more interesting effects of taurine is the regulation of calcium ion (Ca^{2+}) flux in nervous tissue. In general, taurine stimulates high-affinity Ca^{2+} uptake but inhibits low-affinity uptake. In this context uptake is used to describe a process that most probably involves both Ca^{2+} transport and Ca^{2+} binding to membranes. This effect is observed most clearly in retinal tissues (reviewed in reference 2). In both whole retinal preparations and in isolated rod outer segments (ROS), taurine was shown to increase the uptake of Ca^{2+} in an ATP- and bicarbonate-dependent manner at micromolar Ca^{2+}

1

Taurine 4, edited by Della Corte *et al.*
Kluwer Academic / Plenum Publishers, New York, 2000.

concentrations, whereas it inhibited the uptake at millimolar Ca^{2+} concentrations.

The stimulatory effect of taurine on Ca^{2+} uptake in the ROS is antagonized by inhibitors of Ca^{2+}-ion channels, specifically the cyclic-nucleotide-gated (CNG) channels[3]. CNG channels are involved in the generation of visual signals from the retina to the brain[4] and are inhibited by the action of calmodulin (CaM)[5]. CaM is a multifunctional modulatory protein which is activated by the binding of Ca^{2+} ions[6] and is known to modulate the activities of other proteins in the photoreceptor cell[7]. The results of studies in heart sarcolemma suggest that taurine modulates CaM-dependent kinase and adenylyl cyclase activities by antagonizing CaM activity[8,9]. Such findings suggest that taurine may be activating CNG channels, and thus increasing Ca^{2+} uptake, through the inhibition of calmodulin.

The effects of taurine may be dependent on its uptake into the cell. Taurine is known to be transported into retinal cells via a high-affinity uptake system and a separate low-affinity system, although only one, the high-affinity system, is functional in the isolated rat ROS[10]. Significantly, taurine uptake and efflux have been linked to CaM activity in rat cerebral slices[11], rat choroid plexus[12] and a human retinal pigment epithelial (HRPE) cell line[13]. Inactivation of CaM by exposure to trifluoperazine (TFP) was found to inhibit taurine uptake in the choroid plexus[12]. Since taurine appears to be an inhibitor of CaM activity, the CaM-dependent activation of taurine uptake may be a negative feedback mechanism. However, taurine may also bind to cell membranes directly to alter the phospholipid environment[14] and modulate the activity of membrane-bound proteins such as the CNG channels. A mechanism of this type would be independent of CaM functioning as an intermediate-messenger. The interactions between taurine flux, binding and CaM activity require further clarification.

In this study, the effects of known CaM antagonists on the stimulation of Ca^{2+} uptake by taurine were measured in the ROS, in order to provide data on the possible interaction between taurine and CaM in the retina. The CAM antagonists TFP and *N*-(8-aminooctyl)-5-iodonaphthalene-1-sulfonamide (J-8)[15,16] were used to inhibit ATP-dependent Ca^{2+} uptake in the ROS and its stimulation by taurine, as well as the uptake and binding of taurine.

MATERIALS AND METHODS

Materials

[45]Calcium chloride and [^{3}H]taurine were purchased from New England Nuclear (Boston, MA). Taurine and TFP were purchased from Sigma Chemical Co. (St. Louis, MO) and J-8 was obtained from Alexis Corporation (San Diego, CA). Guanidinoethane sulfonate (GES) was synthesized in the laboratory. Ahlstrom glass-fiber filter paper was from Fisher Scientific (Pittsburgh, PA).

Isolation of rod outer segments

ROS were isolated by procedures established previously[3]. Briefly, retinal tissue was dissected from adult rat (Sprague-Dawley) eyes, vortex-mixed for 10-20 seconds and allowed to stand until the retinae settled. The ROS were then collected from the supernatant.

Calcium uptake assay

The isolated ROS were resuspended in Krebs-bicarbonate-Ringer (KBR) buffer for Ca^{2+} uptake experiments. Uptake was measured at 37 °C in the presence of [45]$CaCl_2$, as described previously[3]. For certain experiments, the ROS were added to the incubation tubes in the absence of [45]$CaCl_2$ and exposed to TFP for 5 min (preincubation) before the addition of [45]$CaCl_2$.

Taurine uptake assay

The ROS were resuspended in KBR buffer. The reaction was performed at 37°C in the presence of varying amounts of [^{3}H]taurine and a fixed concentration of ROS, as described previously[10].

Taurine binding assay

For taurine binding experiments, the ROS were resuspended in Krebs-Tris HCl (KT) buffer. The binding assays were performed at 22 °C (room temperature), using procedures described previously[17,18]. The final concentration of [^{3}H]taurine (~2 μCi) ranged from 50 μM - 5 mM.

Statistical analysis

The data were analyzed for statistical significance and regression analyses were performed using GraphPad Prism™ software. Data were analyzed using the Student's t-test, one-way analysis of variance (ANOVA), and linear regression analysis. Duncan's multiple-range test was used for *post-hoc* analysis.

RESULTS

When ROS were preincubated for 5 min with varying concentrations of TFP or J-8 before initiating ATP-dependent Ca^{2+} uptake in the presence of 32 mM taurine, there was a concentration-dependent inhibition of uptake. The effects of J-8 were considerably less than those of TFP (Fig 1). Without preincubation, TFP and J-8 were found to have no effects, and under no conditions did these agents affect the ATP-dependent Ca^{2+} uptake in the absence of taurine (data not shown). The inhibitory effects of TFP could not be overcome by increasing the concentration of taurine, suggesting a non-competitive mechanism for TFP inhibition (Fig 2).

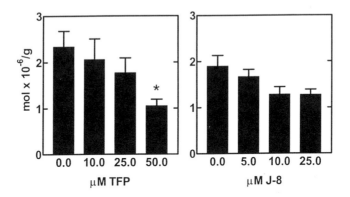

Figure 1. Taurine stimulated ATP-dependent Ca^{2+} uptake in isolated ROS exposed to TFP and J-8. All tubes contained 1.5 mM ATP and 32 mM taurine. ROS were preincubated with TFP or J-8 for 5 min before the initiation of uptake assay. (*Significant difference compared to TFP = 0 with ANOVA and *post-hoc* analysis, $p < 0.05$, N = 3).

At a taurine concentration of 50 μM, both TFP and J-8 treatments produced significant inhibition of high-affinity taurine uptake (Table 1). J-8 was again less effective than TFP. When TFP inhibition was measured over a range of taurine concentrations (10-250 μM), Eadie-Hofstee transformation of the data revealed that the inhibition was non-competitive (data not shown). TFP inhibition of the effects of taurine on ATP-dependent Ca^{2+} uptake showed similar kinetic behavior (Fig. 2).

GES is a taurine analogue that competitively inhibits high-affinity taurine transport in the retina[19,20], and in the ROS[18,19] specifically. If the stimulatory effects of taurine were dependent on taurine uptake, then treatment with GES should antagonize its effects. GES, at a concentration of 32 mM, gave no significant inhibition of the effects of 8, 16 and 32 mM taurine (data not shown). This suggests that the taurine stimulation of ATP-dependent Ca^{2+} uptake is not dependent on taurine uptake.

Table 1. Taurine uptake in ROS (pmol/μg)

TFP (μM)	Activity (Mean ± SEM)		N	J-8 (_M)	Activity (Mean ± SEM)		N
0	0.60	± 0.06	4	0	0.54	± 0.05	3
10	0.46	± 0.05	4	5	0.50	± 0.06	3
25	0.32	± 0.04*	4	10	0.43	± 0.02	3
50	0.18	± 0.03*	4	25	0.31	± 0.03	3

*$P < 0.05$ *vs.* 0 _M TFP, no significant difference between groups for the J-8 experiments (One-way ANOVA, Duncan's multiple-range *post-hoc* test); slope < 1.0 for J-8 treatment indicating a significant concentration-dependent effect (linear regression analysis).

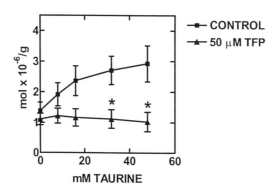

Figure 2. ATP-dependent Ca^{2+} uptake in isolated rod outer segments (ROS) exposed to trifluoperazine (TFP) and varying concentrations of taurine (N = 6). All tubes contained 1.5 mM ATP. ROS were preincubated with TFP for 5 min before the initiation of uptake assay. (*Significant difference compared to respective control with Student's t-test, p < 0.05.)

The effects of TFP on taurine binding to the ROS membrane were studied to delineate the mechanism of action behind the stimulatory effects of taurine on ATP-dependent Ca^{2+} uptake. Fifty μM TFP significantly inhibited the binding of taurine, at concentrations ranging from 50 µM to 5.0 mM (Table 2).

Table 2. Taurine binding (pmol/µg)

Taurine concentration (mM)	Control value	N	Value after 50 µM TFP	N
0.05	0.10 ± 0.18	3	0.27 ± 0.07	3
1.00	7.70 ± 1.32	5	2.84 ± 0.73*	5
2.00	8.95 ± 1.16	5	3.58 ± 0.62*	5
5.00	10.66 ± 1.67	5	5.07 ± 0.78*	5

Values are means $\pm$ SEM. *$P<0.05$ *vs.* respective controls (Student's t-test).

In order to study the relationship between the effects of TFP and the inhibition of CaM activity, Ca^{2+} was excluded from the buffer and intracellular calcium was eliminated by incubating the ROS in a 37 °C water bath for 10 min with 100 µM BAPTA-AM (1,2-bis-[2-aminophen-oxy]ethane-N,N,N',N'-tetraacetic acid acetoxy-methyl ester), a calcium chelator that is membrane permeable[21]. TFP modulation of taurine uptake was preserved in the absence of Ca^{2+}, suggesting that this effect is not mediated through calmodulin inhibition. Taurine uptake was also found to be unaffected by changes in the concentration of $CaCl_2$ in the buffer (0-1000 µM) (data not shown), similar to findings observed in previous studies[18]. Thus, the effect of TFP on taurine transport in the ROS is probably not dependent on its effects on calmodulin.

DISCUSSION

The proposed modulation of retinal CNG channels by taurine[3] might suggest the involvement of CaM in the stimulatory effect of taurine on ATP-dependent Ca^{2+} uptake. This would be consistent with the inhibition of CNG channel activation by CaM binding to the channel[7] and the suggested inhibitory effect of taurine on CaM in sarcolemmal membranes[8,9]. But the idea that taurine inhibits CaM and in turn activates the CNG channels warranted thorough investigation, since the exact mechanism of taurine inhibition of CaM has not been clearly established. The modulation of Ca^{2+}

flux in the retina and other tissues may provide indirect interaction between taurine and CaM, but so far, little data are available about the direct interaction of taurine with the CaM molecule.

TFP and J-8 are widely used pharmacological inhibitors of CaM activity. Initial data which demonstrate the stimulation of Ca^{2+} uptake in the ROS by TFP[18] support the idea that the Ca^{2+} uptake observed in the ROS is dependent on the activation of the CNG channel. However, the stimulation was much less than the stimulation observed with ATP treatment and with combined ATP and taurine treatment. Furthermore, TFP did not increase ATP-dependent Ca^{2+} uptake. Clearly, CaM-independent mechanisms are involved in the modulation of Ca^{2+} uptake in the ROS.

TFP has been used as an inhibitor of both taurine uptake in the choroid plexus[12] and taurine efflux in cortical brain slices[11,22] and, through its use, the CaM-dependent modulation of taurine transport in and out of the cell has been suggested. In this study, TFP treatment was also shown to inhibit taurine uptake. Thus, TFP inhibition of the effect of taurine on ATP-dependent Ca^{2+} uptake may be dependent on TFP behaving as an inhibitor of taurine uptake. However, inhibition of taurine uptake with GES produced no significant effects on the taurine-dependent increase in Ca^{2+} uptake, suggesting that the taurine uptake is not necessary for this action. Thus, taurine may be acting on Ca^{2+} uptake by binding directly to the membrane, and this mechanism may be independent of CaM activity.

J-8 inhibits CaM with almost exactly the same potency as TFP[16]. The relative ineffectiveness of J-8 compared to TFP in inhibiting both ATP-dependent Ca^{2+} uptake and taurine uptake, also suggests that a CaM-independent mechanism may be involved. Finally, but most significantly, the effect of TFP on taurine uptake was observed even in the absence of Ca^{2+}, again indicating that TFP may be acting, at least in part, independently of CaM.

The taurine transport system in the ROS exhibits only high-affinity kinetics[10], and this uptake system is probably saturated at taurine concentrations of less than 1 mM. The high millimolar concentrations of taurine in mammalian tissue suggest physiological function(s) requiring low-affinity interactions. The data presented in this study suggest that the stimulatory effect of taurine is not dependent on the function of the taurine transporter and may instead be dependent on low-affinity binding of taurine to ROS. As TFP was found to decrease low-affinity taurine binding significantly (at taurine concentration up to 5 mM), TFP may be acting in a CaM-independent manner to decrease taurine binding and, in turn, decrease taurine stimulation of ATP-dependent Ca^{2+} uptake.

These data do not invalidate the possibility that taurine may be acting on the CaM molecule directly to produce its effects in the ROS. But they clearly show that taurine may be acting in part, in a CaM-independent manner, most probably by binding to the membrane and modifying the activity of membrane-bound proteins.

ACKNOWLEDGMENTS

These studies were supported in part by grants from the RGK Foundation of Austin, Texas, and the Taisho Pharmaceutical Co., Ltd. of Tokyo, Japan.

REFERENCES

1. Huxtable, R.J. and Sebring, L.A., 1986, Towards a unifying theory for the actions of taurine, *Trends Pharmacol. Sci.* 7: 481-485.
2. Lombardini, J.B., 1991, Taurine: retinal function, *Brain Res. Rev*, **16**:151-169.
3. Militante, J.D. and Lombardini, J.B., 1998, Pharmacological characterization of the effects of taurine on calcium uptake in the rat retina, *Amino Acids* **15**: 99-108.
4. Baylor, D., 1996, How photons start vision. *Proc. Natl. Acad. Sci.* **93**: 560-565.
5. Molday, R.S., 1996, Calmodulin regulation of cyclic-nucleotide-gated channels, *Curr. Opin. Neurobiol.* **6**: 445-452.
6. Niki, I., Yokokura, H., Sudo, T., Kato, M. and Hidaka, H. 1996, Ca^{2+} signaling and intracellular Ca^{2+} binding proteins. *J Biochem* **120**: 685-698.
7. Koch, K.-W., 1995, Control of photoreceptor proteins by Ca^{2+}, *Cell Calcium* **18**: 314-321.
8. Schaffer, S.W., Kramer, J.H., Lampson, W.G., Kulakowski, E. and Sakane, Y., 1983, Effect of urine on myocardial metabolism: role of calmodulin, In *Sulfur Amino Acids: Biochemical and Clinical Aspects* (K. Kuriyama, R.J. Huxtable, and H. Iwata, eds.), Alan R. Liss, New York, pp. 39-50.
9. Schaffer, S.W., Allo, S., Harada, H. and Azuma, J., 1990, Regulation of calcium homeostasis by taurine: role of calmodulin, In *Taurine: Functional Neurochemistry, Physiology, and Cardiology* (H. Pasantes-Morales, D.L. Martin, W. Shain, and R.M. del Rio, eds.), Wiley-Liss, New York, pp. 217-225.
10. Militante, J.D. and Lombardini, J.B. 1999, Taurine uptake activity in the rat retina:protein kinase C-independent inhibition by chelerythrine, *Brain Res.* **818**: 368-3747
11. Law, R.O., 1994, Taurine efflux and the regulation of cell volume in incubated slices of rat cerebral cortex, *Biochim. Biophys, Acta* **1221**: 21-28.
12. Keep, R.F. and Xiang, J., 1996, Choroid plexus taurine transport, *Brain Res.* **715**: 17-24.
13. Ramamoorthy, S., Del Monte, M.A., Leibach, F.H. and Ganapathy, V., 1994, Molecular identity and calmodulin-mediated regulation of the taurine transporter in a human retinal pigment epithelial cell line, *Curr. Eye Res.* **13**: 523-529.

14. Huxtable, R.J., 1989, Taurine in the central nervous system and the mammalian actions of taurine, *Prog. Neurobiol.* **32**: 471-533.
15. MacNeil, S., Griffin, M., Cooke, A.M., Petteit N.J., Dawson, R.A., Owen, R. and Blackburn G.M., 1988, Calmodulin antagonists of improved potency and specificity for use in the study of calmodulin biochemistry, *Biochem. Pharmacol.* **37**: 1717-1723.
16. Craven, C.J., Whitehead, B., Jones, S.K.A., Thulin, E., Blackburn, G.M. and Waltho, P., 1996, Complexes formed between calmodulin and the antagonists J-8 and TFP in solution. *Biochemistry* **35**: 10287-10299.
17. Lombardini, J.B. and Prien, S.D., 1983, Taurine binding by rat retinal membranes, *Exp. Eye Res.* **37**: 239-250.
18. Militante, J.D. and Lombardini, J.B., 1999, The stimulatory effect of taurine on calcium ion uptake in rod outer segments of the rat retina is independent of taurine uptake, *J. Pharmacol. Exp. Ther.* **291**: 383-389.
19. Lake, N. and Cocker, S.E., 1983, In vitro studies of guanidinoethyl sulfonate and taurine transport in the rat retina, *Neurochem. Res.* **8**: 1557-1563.
20. Quesada, O., Huxtable, R.J. and Pasantes-Morales, H., 1984, Effect of guanidinoethane sulfonate on taurine uptake by rat retina, *J. Neurosci. Res.* **11**:179-186.
21. Tsien, R.Y., 1981, A non-disruptive technique for loading calcium buffers and indicators into cells, *Nature* **290**: 527-528.
22. Law, R.O., 1995, Taurine efflux and cell volume regulation in cerebral cortical slices during chronic hypernatremia, *Neurosci. Lett.* **185**:56-59.

SWELLING-INDUCED TAURINE EFFLUX FROM HELA CELLS: CELL VOLUME REGULATION

Ian H. Lambert[1] and Francisco V. Sepúlveda[2]

[1]*The August Krogh Institute, Biochemical Department, Universitetsparken 13, DK-2100, Copenhagen Ø, Denmark and* [2]*ICBM, Facultad de Medicina, Universidad de Chile and Centro de Estudios Científicos de Santiago, Santiago-6760470, Santiago, Chile*

INTRODUCTION

The volume of the cell is an important parameter in regulation of cellular metabolism, secretion, proliferation and programmed cell death[11]. Furthermore, many of the cellular signal transduction mechanisms, which are normally activated by neural, hormonal or autocrine stimulation, also respond to a change in cell volume and thereby elicit changes in membrane transport, metabolism and expression of genes[3].

SWELLING-INDUCED LEAK PATHWAYS

If mammalian cells are exposed to a hypotonic solution, they initially swell as almost perfect osmometers. They reach a maximal degree of cell swelling within the first minutes, and thereafter they regulate their volume back towards the initial value. This back regulation, termed regulatory volume decrease (RVD), involves net loss of KCl and net loss of organic osmolytes. Taurine is an important organic osmolyte in mammalian cells, and loss of taurine is often taken to indicate a reduction in cell volume. In HeLa cells the transport pathway responsible for the net loss of taurine following hypotonic exposure has been proposed to be the volume-sensitive organic osmolyte anion channel (VSOAC) because it accepts ions such as Cl^- as well as a broad range of inorganic osmolytes, such as taurine, sorbitol, and thymidine[1,4]. VSOAC, has been described in different vertebrate cell types

Taurine 4, edited by Della Corte *et al.*
Kluwer Academic / Plenum Publishers, New York, 2000.

and is, with a few exceptions, blocked by dideoxyforskolin, quinine, DIDS and polyunsaturated fatty acids such as arachidonic acid[4].

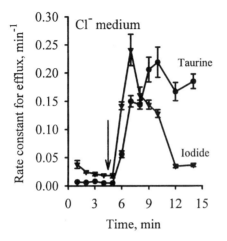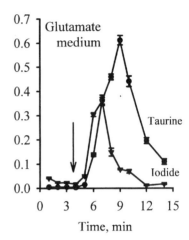

Figure 1. Swelling-induced iodide and taurine efflux. Cells grown to 80% confluence in 35-mm diameter dishes in Dulbecco's modified Eagle's medium containing 5% fetal calf serum were preloaded with ^{125}I and ^{3}H-taurine for 2 h at 37°C. The growth medium was removed and the cells washed 5 times with isosmotic solution. The solution was removed and discarded after the final wash, 0.75 ml of isosmotic solution was added to the dish and at 2 min interval the solution was removed and replaced with another 0.75 ml. The solution removed from the dish was used for estimation of released ^{125}I and ^{3}H-taurine activity. This procedure was repeated throughout the experiment, with the isosmotic solution being replaced with hypoosmotic solution as indicated by the arrows. The radioactivity remaining in the cells at the end of the experiment was determined by disruption of the cells with NaOH. The rate constants for the unidirectional efflux were estimated as the negative slope of a graph where the natural log of the fraction of intracellular radioactivity remaining in the cells was plotted versus time. *Left panel*: The standard isosmotic solution contained in mM: Mannitol 70, NaCl 105, KCl 5, $MgCl_2$ 0.5, $CaCl_2$ 1.3, Tris-HEPES 10, pH 7.4. The hypotonic solution had the same ion concentration as the isotonic solution but contained no mannitol. *Right panel*: The glutamate medium had the same cations as the chloride medium but glutamate was substituted for Cl⁻. The data are reproduced from (18).

However, the hypothesis of a common osmosensitive transport pathway for, taurine and Cl⁻ is questioned. This is the case with the Ehrlich ascites tumor cells, where the swelling-induced taurine efflux pathway differs from the volume-sensitive Cl⁻ channel with respect to its sensitivity towards unsaturated fatty acids and anion channel blockers[8]. From Fig. 1 (left panel) it is seen that in the HeLa cells the time course for activation of the swelling-induced taurine efflux differs from the time course for the swelling-induced iodide efflux (iodide being used as a surrogate for Cl⁻); the iodide efflux is transient with a maximum at about 2 min after reduction in the osmolarity, whereas the taurine efflux is maximal after 4-6 min and remains elevated. That Cl^- efflux activates and inactivates more rapidly than taurine has also been demonstrates in cultured cerebellar astrocytes[13]. Furthermore, by comparing the left and the right panel in Fig. 1 it is also seen that the swelling-induced taurine efflux from HeLa cells is dramatically stimulated by low extracellular Cl^-, whereas the swelling-induced iodide (i.e. Cl^-) efflux is unaffected by substitution of glutamate for Cl^- in the extracellular medium. As DIDS is also a more potent inhibitor of the swelling-induced taurine efflux compared to the swelling-induced Cl^- efflux in HeLa cells[17] it has been proposed that the main swelling-induced transport pathways for taurine and Cl^- in HeLa cells are separate[18].

ATP AND THE SWELLING-INDUCED TAURINE EFFLUX IN HELA CELLS

It has recently been indicated that following hypotonic exposure rat hepatoma cells release ATP via an ATP-binding cassette protein and that ATP, via binding to a purinergic receptor, acts as an autocrine activator of the swelling-induced Cl^- channel[20]. Mechanical stress also induces release of ATP that in the case of Ehrlich cells leads to an increased Ca^{2+} influx and a concomitant activation of an outwardly rectifying whole-cell current[14]. Addition of ATP and UTP in the μM range stimulate taurine release in Ehrlich cells[8]. From Fig. 2 (left panel) it is seen that addition of 5 μM ATP to HeLa cells at the time of hypotonic exposure potentiates the swelling-induced taurine efflux. It is estimated that inclusion of 5 μM and 10 μM ATP in the hypotonic efflux medium increases the rate constant for the swelling-induced taurine efflux significantly from 0.18 ± 0.01 min^{-1} (n = 17) to 0.29 ± 0.02 min^{-1} (n = 13, P<0.001) and 0.28 ± 0.02 min^{-1} (n = 11, P<0.001), respectively. Values are $\pm$ SEM, and the statistical test is a paired t-test. Addition of 5 μM and 10 μM to HeLa cells suspended in isotonic medium increases the rate constant for the taurine efflux from 0.0014 ± 0.0001 min^{-1} (n=12) to 0.0019 ± 0.0004 min^{-1} (n=6) and 0.0030 ± 0.0005 min^{-1} (n=6), respectively, but the increase is not significant at a p < 0.01 level. The effect of ATP on the swelling induced taurine efflux is mimicked by addition of UTP (Fig. 2, right panel).

From Fig. 2 (right panel) it is also seen that exposure of fresh HeLa cells to a hypotonic solution, collected from HeLa cells previously exposed for 10 min, does not stimulate the swelling-induced taurine efflux, indicating that no autocrine (i.e., taurine releasing) factor is released from HeLa cells following hypotonic cell swelling. Furthermore, addition of the ecto-ATPase apyrase to the hypotonic solution does not affect the swelling-induced taurine efflux (Fig. 2, right panel). Thus, although purinergic signalling stimulates the swelling-induced taurine efflux, ATP is not considered as an autocrine factor in the swelling-induced activation of taurine transporting pathways in HeLa cells.

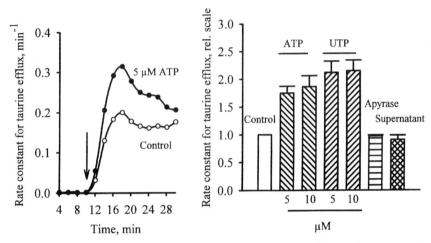

Figure 2. Effect of nucleotides on the swelling-induced taurine efflux. Cells grown to 80% confluence in 35-mm diameter dishes in Dulbecco's modified Eagle's medium containing 10% newborn calf serum were preloaded with ^{14}C-taurine for 2 h at 37°C. The efflux experiments were performed and the rate constant estimated as described in the legend to Fig. 1 with the following exceptions: Samples were 1 ml, the isosmotic solution contained in mM KCl 150, CaCl$_2$ 1.3, MgCl$_2$ 0.5, and HEPES 10, pH 7.4, whereas the KCl concentration in the hyposmotic solution was reduced to 95 mM, with the other components remaining unchanged. Using solutions containing a high concentration of K$^+$ eliminated the normal outward transmembrane K$^+$ gradient, which provides much of the driving force for RVD under physiological conditions[5]. *Left panel*: ATP was prepared in a stock solution containing 0.1 mM EGTA, where the free concentration of ATP was estimated at 10.7 mM using EQCAL software and assuming that ATP binds Mg^{2+} and Ca^{2+} with the same potency. ATP was added at the time of hypotonic exposure as indicated by the arrow. *Right panel*: The maximal rate constant obtained after hypotonic exposure was estimated in HeLa cells in the absence (control) or the presence of (i) ATP or UTP (5 μM, 10 μM), (ii) apyrase (3U/ml) or (iii) a hypotonic "supernatant" collected from HeLa cells pre-exposed for 10 min to the hypotonic solution. Values are given relative to the hypotonic control ± SEM and represent 13 (5 μM ATP), 11 (10 μM ATP), 7 (5 μM UTP), 7 (10 μM UTP), 4 (apyrase) and 4 (supernatant) sets of experiments. The effect of nucleotides was significant at a 0.001 level in a paired Student's *t*-test where experimental values were tested against the hypotonic control value.

ARACHIDONIC ACID METABOLISM AND SWELLING-ACTIVATED TAURINE TRANSPORT IN HELA CELLS

It has previously been demonstrated that the signal cascade, which is activated by the cell swelling and which leads to the activation of the taurine efflux pathway in the Ehrlich cells, involves a phospholipase A_2 (PLA_2) mediated release of arachidonic acid from phospholipids and subsequent oxidation of the fatty acid to leukotrienes via the 5-lipoxygenase (5-LO)[7,8]. From Fig. 3 (left and right panels) it is seen that the swelling-induced taurine efflux from HeLa cells is reduced in the presence of RO 31-4639, which inhibits[2] the pancreatic PLA_2. RO 31-4639 has previously been demonstrated to block the swelling-induced taurine efflux as well as the total RVD response in, e.g., Ehrlich cells[7-9]. From Fig. 3 (right panel) it is also seen that inclusion of the Ca^{2+}/calmodulin antagonist, pimozide, in the flux medium blocks the swelling-induced taurine efflux from HeLa cells. This confirms a previous study in which it was demonstrated that there is a close relationship between the ability of a drug to inhibit the swelling-induced taurine efflux from HeLa cells and its potency as a Ca^{2+}/calmodulin antagonist[5]. Thus, activation of the swelling-induced taurine efflux apparently involves a Ca^{2+}/calmodulin regulated step as well as a yet unidentified PLA_2.

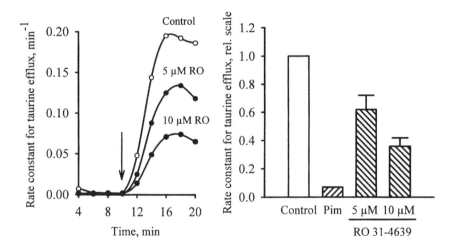

Figure 3. Effect of RO 31-4639 and Pimozide on the swelling-induced taurine efflux. Cells were prepared and the rate constant for the taurine efflux estimated as described in the legend to Fig. 2. *Left panel*: RO 31-4639 (5 μM and 10 μM) was added to the cells 1.5 h before and during the experiment. *Right panel*: Pimozide (10 μM) was present in the isotonic/hypotonic media throughout the experiment. Maximal rate constants after osmotic exposure are given relative to the hypotonic control. The data with pimozide and RO 31-4639 represent 4 and 3 identical sets of experiments, respectively. The effect of pimozide (error bar within the main bar) and 10 μM RO 31-4639 was significant at a 0.007 and 0.02 level, respectively in a paired Student's *t*-test where experimental values were tested against the hypotonic control value.

Activation of PLA_2 generates lysophospholipids as well as free fatty acids. From Fig. 4 (left panel) it is seen that addition of lysophosphatidyl choline (LPC) with palmitic acid in its *sn*-1 position induces a transient increase in the rate constant for the taurine release under isotonic conditions. Substituting stearic acid or oleic acid for palmitic acid in the *sn*-1 position of LPC significantly reduces the taurine releasing effect, whereas lysophosphatidyl ethanolamine, lysophosphatidyl inositol, lysophosphatidyl serine and lyso-phosphatidic acid were almost ineffective as activators of the taurine efflux pathway under isotonic conditions (data not shown). The effect of LPC is not prevented by addition of 10 μM DIDS (Fig. 4, left panel), a concentration which is sufficient to block the swelling-induced taurine efflux (Fig. 4, right panel). Serum depletion has no effect on the LPC induced taurine release, whereas the swelling-induced taurine release is significantly reduced in serum depleted cells. Thus, the maximal rate constant following addition of 10 μM LPC (palmitic acid in *sn*-1) was 0.34 ± 0.03 min^{-1} (n = 5) in normal cells and 0.32 ± 0.04 min^{-1} (n = 5) in serum-depleted cells (cells incubated for 2 h in the absence of serum but in the presence of 5 mM glucose before initiation of the efflux experiment), whereas the maximal swelling-induced efflux was reduced from 0.22 ± 0.01 min^{-1} (n = 24) to 0.05 ± 0.005 min^{-1} (n = 24) by serum depletion. Finally, permeabilization of the membrane by addition of digitonin (5-10 μg/ml) mimics the effects of LPC (data not shown). Thus, the effect of LPC on taurine efflux in HeLa cells is taken to reflect membrane permeabilization.

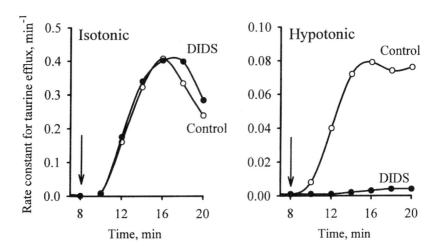

Figure 4. Effect of DIDS on LPC- and swelling-induced taurine efflux. Cells were grown, prepared and the rate constant for the efflux estimated as described in the legend to Fig. 2. with the exception that glucose (5 mM) was substituted for serum in the medium during the 2 h loading with ^{14}C-taurine. DIDS (10 μM) was present in the efflux media throughout the efflux experiment. *Left panel*: LPC (10 μM, palmitic acid in *sn*-1) was added as indicated by the arrow and the efflux followed with time in isotonic solution. *Right panel*: The osmolarity was reduced as described in the legend to Fig. 2 at the time indicated by the arrow. The experiments in the left and right panel each represent 4 sets of identical experiments.

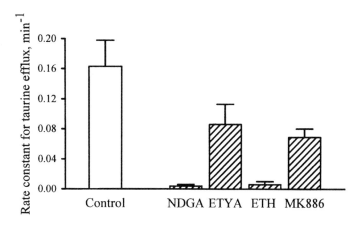

Figure 5. Effect of 5-lipoxygenase inhibition on the swelling-induced taurine efflux. Cells were prepared and the rate constant for the taurine efflux estimated as described in the legend to Fig. 2. NDGA (50 μM), ETYA (50 μM), ETH 615-139 (10 μM) and MK886 (1 μM) were present in the isotonic and hypotonic media throughout the efflux experiment. Values are given as the mean ± SEM of the maximal rate constant obtained after reduction in osmolarity and represent 6 (NDGA), 4 (ETYA), 9 (ETH 615-139) and 4 (MK886) sets of experiments. The effect of the inhibitors is in all cases significant at a 0.01 level in a paired Student's *t*-test, where the experimental value is tested against the hypotonic control value. Similar data have been presented[10].

Figure 5 demonstrates that inhibition of the 5-LO, just like inhibition of PLA$_2$, reduces the swelling-induced taurine efflux in HeLa cells. The 5-LO was in this case blocked by NDGA (a potent antioxidant), ETYA (which looks like arachidonic acid but has triple bonds instead of double bonds and apparently acts as a substrate inhibitor), ETH 615139 (a direct inhibitor[6]) and MK886 (which prevents membrane association of the 5-LO and thereby its activation[15]). 5-LO appears also to be a prerequisite for activation of taurine efflux after osmotic exposure in human fibroblasts[12], in fish erythrocytes[19] and in cerebellar astrocytes[16].

Leukotriene D$_4$ (LTD$_4$) has turned out to be a potent activator of taurine efflux in Ehrlich cells[8] (EC$_{50}$ < 5 nM) and it appears that LTD$_4$ is a second messenger involved in the swelling-induced activation of the taurine efflux pathway in Ehrlich cells[8]. Exposure of HeLa cells to 100 - 400 nM LTD$_4$ does not induce taurine efflux under isotonic conditions and does not enhance the swelling-induced taurine efflux following hypotonic exposure (data not shown). The 5-LO product 5-hydroxoxyeicosatetraenoic acid (5-HETE), on the other hand, enhances the swelling induced taurine efflux in the HeLa cells by a factor of two (EC$_{50}$ ≈ 90 nM; data not shown). However, 5-HETE is an end-product in a cascade and the effect of 5-HETE is therefore taken to reflect the effect of yet another 5-LO product which we still have to identify.

CONCLUSION

It is suggested (i) that following hypotonic exposure, taurine and Cl⁻ leave the HeLa cells via separate main efflux pathways; (ii) the swelling-induced taurine efflux in HeLa cells is increased by addition of extracellular ATP in the µM concentration range, but ATP does not appear to be an autocrine factor in the swelling-induced activation of the taurine efflux pathway; (iii) both a PLA_2 and the 5-LO play a permissive role in the swelling-induced activation of the taurine efflux pathway, and activation involves a Ca^{2+}/calmodulin regulated step; and (iv) LPC is not a second messenger in the swelling-induced activation of the taurine efflux pathway - LPC more likely seems to act as a permeabilizing agent.

ACKNOWLEDGMENTS

The work by IHL was supported by the Novo Nordisk Foundation and the Danish Natural Science Research Council. The work by FVS was supported by grants from Fondecyt 1970244 (Chile), the Volkswagen Stiftung (Germany), International Research Scholars grant from HHMI and a Cátedra Presidencial en Ciencias. Institutional support to the Centro de Estudios Científicos de Santiago from a group of Chilean private companies (AFP Protección, CGE, Codelco, Copec, Empresas CMPC, Gener S.A., Minera Collahuasi, Minera Escondida, Novagas, Business Design Associates, Xerox Chile), Fuerza Aérea de Chile and Municipalidad de Las Condes is also acknowledged.

REFERENCES

1. Hall J.A., Kirk J., Potts J.R., Rae C., Kirk K. 1996. Anion channel blockers inhibit swelling-activated anion, cation, and nonelectrolyte transport in HeLa cells. *Am. J. Physiol.* 271: C579-C588.
2. Henderson L.M., Chappel J.B., Jones O.T. 1989. Superoxide generation is inhibited by phospholipase A2 inhibitors. *Biochem. J.* 264: 249-255.
3. Hoffmann E.K., Pedersen S.F. 1998. Sensors and signal transduction in the activation of cell volume regulatory ion transport systems. In *Cell Volume Regulation*, Ed. F. Lang, Karger, Contrib. Nephrol. 123: 50-78.
4. Kirk K. 1997. Swelling-activated organic osmolyte channels. J. Membrane Biol. 158: 1-16.
5. Kirk J., Kirk K. 1994. Inhibition of volume-activated I- and taurine efflux from HeLa cells by P-glycoprotein lockers correlates with calmodulin inhibition. *J. Biol. Chem.* 269: 29389-29394.

6. Kirstein D., Thomsen M.K., Ahnfelt-Rønne I. 1991. Inhibition of leukotriene biosynthesis and polymorphonuclear leukocyte functions by orally active quinolylmethoxylamines. *Pharmacol. Toxicol* 68: 125-130.

7. Lambert I.H. 1994. Eicosanoids and cell volume regulation. In *Cellular and Molecular Physiology of Cell Volume Regulation*, Ed. K. Strange, CRC Press, Inc: 273-292.

8. Lambert I.H. 1998. Regulation of the taurine content in Ehrlich ascites tumour cells. In *Taurine 3: Cellular and Regulatory Mechanisms*, Eds. S. Schaffer, J.B. Lombardini and R.J. Huxtable, Plenum Publishing Corporation, New York: 269-276.

9. Lambert I.H., Hoffmann E.K. 1991. The role of phospholipase A2 and 5-lipoxygenase in the activation of K and Cl channels and the taurine leak pathway in Ehrlich ascites tumor cells. *Acta Physiol. Scand.* 143 (1): 33A.

10. Lambert I.H., Pedersen S., Hall J.A. 1999. On the role og 5-lipoxygenase in swelling-induced activation of taurine efflux in HeLa cells. Acta Physiol. Scand. 165: CP18..

11. Lang F., Busch G.L., Ritter M., Völkl H., Waldegger S., Gulbins E., Häussinger D. 1998. Functional significance of cell volume regulatory mechanisms. *Physiol. Rev.* 1: 247-306.

12. Mastrocola T., Lambert I.H., Kramhøft B., Rugolo M., Hoffmann E.K. 1993. Volume regulation in human fibroblasts: Role of Ca2+ and 5-lipoxygnease producs in the activation of the Cl- efflux. *J. Membrane Biol.* 136:55-62

13. Pasantes-Morales H., Quesada O., Morán J. 1998. Taurine: An osmolyte in mammalian tissues. In *Taurine 3: Cellular and Regulatory Mechanisms*, Eds. S. Schaffer, J.B. Lombardini and R.J. Huxtable, Plenum Publishing orporation, New York: 209-217.

14. Pedersen S., Pedersen S.F., Nilius B., Lambert I.H., Hoffmann E.K. 1999. Mechanical stress induces release of ATP from Ehrlich ascites tumor cells. *Biochim. Biophys. Acta* 1416: 271-284.

15. Rouzer C.A., Ford-Hutchinson A.W., Morton H.E., Gillard J.W. 1990. MK886, a potent and specific leukotriene biosynthesis inhibitor blocks and reverses the membrane association of 5-lipoxygenase in ionophore challenged leukocytes. *J. Biol. Chem.* 265: 1436-1442.

16. Sánches-Olea R., Morales-Mulia M., Morán J., Pasantes-Morales H. 1995. Inhibition by polyunsaturated fatty acids of cell volume regulation and osmolyte fluxes in astrocytes. *Am. J. Physiol.* 269: C96-C102.

17. Stutzin A., Equiguren A.L., Cid L.P., Sepúlveda F.V. 1997. Modulation by extracellular Cl- of volume-activated osmolyte and halide permeabilities in HeLa cells. *Am. J. Physiol.* 273: C999-C1007.

18. Stutzin A., Torres R., Oporto M., Pacheco P., Eguiguren A.L., Cid L.P., Sepúlveda F.V. 1999. Separate taurine and chloride efflux pathways activated during regulatory volume decrease. *Am J. Physiol.* 277: C392-C402.

19. Thoroed S.M., Fugelli K. 1994. The role of leukotriene D4 in the activation of the osmolality-sensitive taurine channel in erythrocytes from marine fish species. *Acta Physiol. Scand.* 151: 27A.

20. Wang Y., Roman R., Lidofsky S.D., Fitz J.G. 1996. Autocrine signaling through ATP release represents a novel mechanism for cell volume regulation. *Proc. Natl. Acad. Sci.* 93: 12020-12025.

TAURINE FLUXES IN INSULIN DEPENDENT DIABETES MELLITUS AND REHYDRATION IN STREPTOZOTOCIN TREATED RATS

Stephen J. Rose[1], Manju Bushi[2], Inderjeet Nagra[2] and W. Ewart Davies [2]

[1]*Department of Paediatrics, Heartlands Hospital, Birmingham, England*
[2]*Department of Pharmacology, Medical School, University of Birmingham, Birmingham, England*

Abstract: The effect of streptozotocin induced diabetes mellitus and rehydration on brain taurine and brain water content was studied in 4 groups of rats. Two groups of rats with diabetes mellitus were used. In one group, taurine and brain water content were determined following induction of diabetes for one week. In the second group, diabetes was induced for one week but before sacrifice, 15% of body weight of normal saline was introduced into the peritoneum, half at time 0, half 30 minutes later with sacrifice 60 minutes after the first infusion. In two groups of animals (controls), the brain taurine and water content were estimated in normal conditions and after hydration, in exactly the same way as diabetic rats. Brain taurine content was greater in diabetic rats than non-diabetic rats and there was no decrease in brain taurine content within the first hour following rehydration of the diabetic rats. Brain water content was greater in rehydrated diabetic rats than in non-rehydrated diabetic rats but there was no significant change in the brain water content after hydration of non diabetic rats. This suggested that the rapid change in water content of rehydrated diabetic rats was not accompanied by an equally rapid alteration in brain taurine content. This is consistent with the hypothesis that taurine flux could be a major factor in the aetiology of diabetic cerebral oedema. It also allows the development of possible therapeutic options which may increase outward taurine flux from brain cells. Taurine flux is increased by increasing extracellular sodium concentration or

Taurine 4, edited by Della Corte *et al.*
Kluwer Academic / Plenum Publishers, New York, 2000.

decreasing potassium concentration. Phospholemman channels may also influence taurine flux. These may have implications for the optimal method of clinical rehydration undertaken in diabetic ketoacidosis.

INTRODUCTION

Type 1 or insulin dependent diabetes mellitus is a disease of childhood and young adulthood. The peak age of presentation is around 8 to 11 years with usually a 2 to 6 week history of polydipsia and polyuria. On admission, the child is often dehydrated, possibly with 10% loss of body fluid and with total body depletion of sodium and potassium with significant metabolic acidosis from beta lipolysis. The treatment of diabetic ketoacidosis is rehydration with normal saline at the same time as using an intravenous insulin infusion. Unpredictably, a small number of children and young adults suffer the devastating complication of cerebral oedema.[4] This has a high morbidity and high mortality with only a small percentage surviving neurologically intact.[2]

The normal treatment of non diabetic cerebral oedema is via the use of hyperosmolar fluids to remove water from brain tissue. This is relatively unsuccessful in diabetic cerebral oedema suggesting that there are different underlying mechanisms existing in the two conditions.

Taurine acts as a neuronal osmoregulator[5,9] and we suggest that dislocation of taurine and water flux within the brain tissue is a factor in the aetiology of diabetic cerebral oedema.

MATERIALS AND METHODS

Four groups of adult male, albino Wistar rats weighing 189 to 235 g were used. Eight rats were rendered diabetic via the administration of streptozotocin[1] (60 mg/kg) intravenously via the tail vein. Blood glucose was measured on days 5 and 6 and finally on day 8 to determine that a diabetic state had supervened. Blood glucose concentration was determined from tail vein blood with a digital glucometer, Glucometer 2 model 5529.

Four rats were sacrificed on day 8, their brains dissected out and portions placed in pre-weighed sealed containers. The other 4 rats in the group received rehydration therapy of 15% of their body weight given in 2 boluses at time 0, time +30 minutes and then sacrificed at time +60 minutes. Two non-diabetic control groups of 4 rats each were used, non hydrated and hydrated, treated in an analogous way to the study groups.

Taurine content was measured by reversed phase high performance liquid chromatography following modification of a technique previously described by

Hopkins et al[6]. Brain samples for brain water estimation were placed in a watertight container that had previously been weighed and water content determined by freeze drying for at least 48 hours. A total of 80 samples ranging from 15 to 80 milligrams wet weight were obtained from the 4 groups of rats. The dried brain was utilised for brain taurine content measurement.

RESULTS

The brain taurine content in micromols/g dry weight (Table 1) revealed a significantly higher brain taurine content in the brains of diabetic rats and hydrated diabetic rats than in control or hydrated controls ($p < 0.05$).

Table 1. Brain taurine content of different study groups

Study Group	Taurine (µg/g dry weight, n = 4)
Diabetic	24.60 ± 0.10*
Hydrated diabetic	23.97 ± 1.22
Control	20.69 ± 1.21*
Hydrated control	21.73 ± 1.13

*$P < 0.05$

There was no difference in the taurine content of diabetic rats and hydrated diabetic rats. Brain water content (Table 2) demonstrated a significant increase ($p < 0.01$) in hydrated diabetic rats compared with non rehydrated diabetic rats. There was no increase in the water content of hydrated normal controls. The brain water content of hydrated and non-hydrated control animals was similar to the brain water content in re-hydrated diabetic rats.

Table 2. Brain water content of different study groups

Study Group	Water content (%, n = 4)
Diabetic	73.62 ± 2.20*
Hydrated diabetic	82.90 ± 0.92*
Control	84.63 ± 1.29
Hydrated control	77.75 ± 0.52

*$P < 0.01$

DISCUSSION

Diabetic cerebral oedema is an unpredictable and unusual complication of rehydration therapy in clinical diabetic ketoacidosis[4]. Many aetiologies have been proposed which have included the degree of acidosis[3], rapid decrease in hyperglycaemia, changes in plasma osmolality[11] and use of alkali solution[10] but as yet, no one unifying cause has been identified. It is clear, however, that the cerebral oedema in diabetes is different from cerebral oedema in non diabetics in that it responds poorly to conventional therapy with hyperosmolar fluids.

Diabetic ketoacidosis usually develops over several days resulting in a slow increase in serum osmolality which in turn is reflected in a disturbed intracellular osmolality. As serum osmolality increases, there is an inward taurine flux into the brain cells to increase intracellular osmolality in order to maintain cell volume and integrity. This occurs relatively slowly as serum osmolality increases. However, rehydration in diabetic ketoacidosis occurs rapidly with large volumes of hypo-osmolar fluids, 0.9% saline, being administered intravenously within the first hour of treatment with a consequent rapid decrease in serum osmolality. Table 1 demonstrates that there is no change in brain taurine content in the first hour of rehydration. Taurine efflux can therefore not be upregulated in such a short period of time. There is therefore an increase in brain water content secondary to the raised intracellular osmolality with a consequent increase in volume of intracranial contents and cerebral oedema. Water influx is very rapid, but not accompanied by a compensatory loss of taurine from the brain cells. There is thus a dislocation of water flux and taurine flux.

These results are consistent with the hypothesis that the dislocation of water and taurine fluxes in brain tissue may be a factor in the development of diabetic cerebral oedema.

It is interesting to speculate whether these results could suggest an improved therapeutic regimen for diabetic cerebral oedema. The initial therapeutic manoeuvre might be to accelerate the outward taurine flux. Sodium is involved in the egress of taurine. During diabetic ketoacidosis the patient becomes total body sodium depleted and hyperosmolar but the rehydration fluid is 0.9% saline, which although iso-osmolar in normal patients, is hypo-osmolar with respect to the hyper-osmolar state in diabetic ketoacidosis. The first infusion may therefore sensibly be twice normal (1.8%) saline.

Potassium, which is also depleted in diabetic ketoacidosis, reduces taurine egress. However, the omission of potassium from rehydrating fluids has potential consequences secondary to cardiac arrhythmia.

Kowdley et al[7] and Moormon et al[8] indicated that phospholemman may be a taurine channel involved in regulation of cell volume. Therapeutic regimens directed at up regulating the action of phospholemman which may be therapeutically advantageous should be considered, at least, theoretically.

Diabetic cerebral oedema is unpredictable, so there may be a genetic component to its aetiology. As this genetic variability cannot as yet be identified it behoves us to consider altering the therapeutic regimen for diabetic ketoacidosis for all patients in the hope of reducing this catastrophic complication.

REFERENCES

1. Arison, R.N., Ciacco, E.I., Glitzen, M.S., Cassano, J.A., Pruss, M.P., 1967, Light and electron microscopy of lesions in rats rendered diabetic with streptozotocin. *Diabetes* **16**: 51-56.
2. Bello, F.A., Sotos, J.F., 1990, Cerebral oedema in diabetic ketoacidosis in children. *Lancet* **336**: 64.
3. Durr, J.A., Hoffman, W.H., Sklar, A.H., Gammel, T. and Steinhart, C.M., 1992, Correlates of brain oedema in uncontrolled insulin-dependent diabetes mellitus. *Diabetes* **41**: 627-632.
4. Hammond, P., Wallis S., 1992., Cerebral oedema in diabetic ketoacidosis. *Br. Med. J.* **305**: 203-204.
5. Harris, G.D., Lohr, J.W., Fiordalisi, I. and Acara M., 1993, Brain osmoregulation during extreme and moderate dehydration in a rat model of severe diabetic ketoacidosis. *Life Sci.* **53**: 185-191.
6. Hopkins P.C., Kay I.S., and Davies W.E., 1989., A rapid method for the determination of taurine in biological tissue. *Neurochem. Internatl.* **15**: 429-432.
7. Kowdley, G.C., Ackerman S.J., Chen Z., Szabo G., Jones L.R. and Moorman R., 1997., Anion, cation and zwitterion selectivity of phospholemman channel molecules. *Biophys. J.* **72**:141-145.
8. Moorman, J.R., Jones L.R., 1997., Phospholemman: A cardiac taurine channel involved in regulation of cell volume. *Adv.Exp.Med.Biol.* **442**: 219-228.
9. Naegelhus, E.A., Lehmann A. and Ottersen O.P., 1993., Neuronal-glial exchange of taurine during hypo-osmotic stress: a combined immunocytochemical and biochemical analysis in rat cerebellar cortex. *Neuroscience* **54**: 615-631.
10. Rosenbloom A.L., Riley, W.J., Weber, F.T., Malone, J.I. and Donnelly, W.H., 1980., Cerebral oedema complicating diabetic ketoacidosis in childhood. *J.Paediatr.* **96**: 357-61.
11. Silver S.M., Clark, E.C., Schroeder, B.M. and Sterns, R.H. , 1997, Pathogenesis of cerebral oedema after treatment of diabetic ketoacidosis. *Kidney Internatl.* **51**: 1237-1244.

IN VITRO AND *IN VIVO* EFFECTS OF TAURINE AND STRUCTURALLY RELATED SULFUR-CONTAINING COMPOUNDS AGAINST PHENYLHYDRAZINE-INDUCED OXIDATIVE DAMAGE TO ERYTHROCYTES

Prabhat K. Pokhrel and Cesar A. Lau-Cam
College of Pharmacy and Allied Health Professions, St. John's University, 8000 Utopia Parkway, Jamaica, New York 11439, USA.

INTRODUCTION

In vitro and *in vivo* studies conducted on a wide variety of biological systems have verified the ability of taurine (TAU) to counteract the biochemical alterations and damaging effects that follow oxidative stress due to reactive oxygen species generated during the course of normal metabolic events or from exposure to a diverse group of oxidative conditions. For example, TAU was found to decrease the rate of formation of malondial-dehyde (MDA) and the amount of superoxide anion radical formed in a rabbit spermatozoa preparation[1], and to counteract the accumulation of MDA and diene conjugates in rat mesangial cells grown on a high glucose medium[58]. Incorporating TAU into a suspension of frog rod outer segments reduced the levels of MDA and the structural changes caused by exposure to an intense and prolonged illumi-nation[43]. In a model of cataractogenesis by galactosemia, the incubation of intact rabbit lenses in a tissue culture medium containing galactose plus TAU lowered the incidence of lenticular opacification and the MDA content[33]. While cardiomyopathy has been related to TAU deficiency[46], in the presence of this amino acid, isoproterenol-induced lipid peroxidation (LPO) was inhibited and ischemic damage minimized in myocardial cells[5]. Similarly, dietary supplementation with TAU attenuated puromycin aminonucleoside nephropathy in rats while normalizing renal cortical MDA levels and reducing the formation of reactive oxygen species (ROS)[57].

Taurine 4, edited by Della Corte *et al.*
Kluwer Academic / Plenum Publishers, New York, 2000.

In mice, the daily administration of TAU for 15 days reversed the loss of enzymatic and noenzymatic antioxidants and reduced the generation of ROS caused by perchloroethylene in liver and kidney[14]. Culturing isolated rat hepatocytes with lipo-polysaccharide (LPS) and TAU resulted in attenuation of LPS-associated apoptosis and necrosis through the inhibition of nitric oxide and oxygen free radical formation[48]. Moreover, TAU decreased oxygenation-associated LPO and precluded the hypoxia-induced death of rat hepatocytes growing in a calcium-containing medium[39,40]. In rats, oral dosing with TAU after a single dose of carbon tetrachloride, but not before, prevented the accumulation of MDA and the accompanying cellular injury[37,38]. There is also ample evidence to indicate that TAU protects lung cells against oxidant-related injury. Thus, dietary supplementation with TAU plus niacin was reported to ameliorate both amiodarone- and bleomycin-related lung fibrosis and to attenuate the generation of collagen, ROS, MDA equivalents, calcium accumulation, and DNA damage in hamsters[18]. The addition of TAU to the diet of rats treated with cyclophosphamide for 9 days protected lungs against injury, the loss of intracellular antioxidants, and the formation of MDA and hydroperoxides[60]. Moreover, in a study with isolated rat alveolar macrophages, preincubation with TAU decreased LPO and the decline in protein, reduced GSH, and Na^+/K^+-ATPase activity that followed exposure to ozone[6].

Erythrocytes (RBCs) are cells whose membranes contain a variety of proteins interspersed within a lipid bilayer rich in phospholipids containing polyunsaturated fatty acids. Within RBCs there is hemoglobin (Hb), an iron-porphyrin-containing protein that functions as an oxidase and as a peroxidase. Owing to its structural characteristics, chemical composition, and high content in Hb, and to the fact that RBCs circulate through areas of high oxygen tension, these cells are constantly exposed to conditions favoring oxidative stress[2,11,51]. As a result, changes affecting the Hb structure, the content of antioxidant molecules, and the morphology, function and integrity of the membrane are observable in RBCs. One of the most extensively used inducers of oxidative stress in RBCs is phenylhydrazine (PHZ). *In vivo*, this compound causes intravascular hemolysis associated with methemoglobinemia and Heinz body formation as well as a drop in GSH levels, the production of volatile hydrocarbons, and the accumulation of calcium in RBCs[54]. *In vitro*, PHZ affects the oxidative breakdown of Hb to denatured products with major changes in membrane proteins and phospholipids. The mechanism underlying these effects is complex and involves the formation of several reactive intermediates and complexes of PHZ[23,50]. Initially PHZ enters the oxygen binding site of Hb where it reacts via a redox mechanism whereby both Hb and the drug are oxidized[16,17]. While oxyhemoglobin is converted first to species with a higher oxidation state, including, methemoglobin (metHb), and next to a mixture of denatured products referred to as choleglobin and hemichromes[26,29,44], PHZ appears to undergo a reaction with oxyhemoglobin, yielding the phenylhydrazyl radical[12,20] plus H_2O_2, or with molecular oxygen to yield phenyldiazene plus the superoxide anion radical[19]. The precipitation of modified Hbs in the membrane as Heinz bodies is associated with hemolysis, reticulocytosis and a shorter life span of circulating RBCs[27,30]. Although both H_2O_2 and superoxide anion radical have been implicated in the formation of Heinz bodies and the hemolysis that ensues the interaction of PHZ with oxyhemoglobin[8,25,31], evidence supporting a role for other reactive species, including the phenyldiazenyl radical and phenyl radical[21,23] is also available. Likewise, a mechanism based on the peroxidation of unsaturated phospholipids of the RBC membrane by reactive species arising from the oxidation of PHZ has been proposed as an alternative determinant of hemolysis by this arylhydrazine[31,34]. This view is supported by the results of animal studies in which a pretreatment with an antioxidant inhibited both the formation of MDA and hemolysis due to PHZ[25].

The present investigation was undertaken to ascertain (i) whether TAU will exert an antioxidative and, thereby, a protective action in RBCs placed under *in vitro* and *in vivo* conditions leading to oxidative hemolysis; i.e., protection from the chemical and oxidative damage to RBCs resulting from an exposure to PHZ[27]; and (ii) whether the noted actions are dependent on particular structural features of the TAU molecule.

MATERIALS AND METHODS

Chemicals

All the amino acids and sulfur-containing compounds used in the study were obtained from local commercial sources. TAU, hypotaurine (HYTAU), isethionic acid (ISA), N-acetylcysteine (NACYS), L-cysteic acid (CA) and aminomethanesulfonic acid (AMSA) were from Sigma Chemical Co., St. Louis, MO; ethanesulfonic acid sodium salt (ESA) and 1,2-ethanedisulfonic acid (EDSA) were from Aldrich Chemical Co., Milwaukee, WI; β-alanine (BALA) and N-(2-acetamido)-2-aminoethanesulfonic acid (ACES) were from TCI America, Portland, OR; N-methyltaurine (MTAU) was from Pfaltz & Bauer, Waterbury, CT; 2-mercaptoethanesulfonic acid (MESA) was from ICN Biomedicals, Aurora, OH; α-sulfo-β-alanine (ASBA) was from USB, Cleveland, OH; and homotaurine (HMTAU) was from Acros Organics, Pittsburgh, PA. The samples of PHZ, glucose, disodium phosphate, monosodium phosphate, GSH, 2-thiobarbituric acid (TBA), 5,5'-dithiobis-(2-nitrobenzoic acid) (DTNB), metHb, and 1,1,3,3-tetraethoxypropane (TEP) were from Sigma Chemical Co. All other chemicals were of analytical reagent grade and were purchased from Aldrich Chemical Co. or J.T. Baker, Philipsburgh, NJ.

Animals

Male Sprague-Dawley rats, weighing 200-250 g, were purchased from Taconics Farms, Germantown, NY. The animals were housed in plastic cages, 6 per cage, in a temperature-regulated (23 ± 1°C), constant humidity, room. During a 7-day acclimation period, the animals had free access to a pelleted animal diet (Purina Rat Chow, Ralston, Purina, St. Louis, MO) and tap water. The animals used in the *in vivo* studies were fasted for 16 hr prior to an experiment.

Collection of Blood Samples and Preparation of RBC Suspensions

Rats were decapitated just before an experiment, and their blood samples collected into heparinized polyethylene test tubes that were kept immersed in ice. A RBC suspension was prepared according to the method of Sharma and Premachandra[52]. After centrifugation of the blood sample at 700 x g for 10 min, and removal of the plasma and buffy coat by aspiration, the pellet was suspended in ice-cold isotonic phosphate-buffered saline glucose solution (PBSG, 0.01 M phosphate buffer pH 7.4-0.15 M sodium chloride-5 mM of glucose), and the suspension centrifuged at 700 x g for 10 min. After repeating this washing step once, the RBCs were resuspended in PBSG to a hematocrit of 20%, and stored on ice pending an experiment.

Test Solutions for *In Vitro* Studies

Solutions of TAU and the other compounds listed in Tables 1 to 7 were prepared freshly in PBSG to a concentration of 480 mM. A working PHZ solution was prepared on the day of the experiment in PBSG to contain 750 mM, and stored in an amber glass container.

Incubation Procedure for *In Vitro* Studies

To a test tube marked *"Test"*, the RBC suspension (~587 μL) and the test compound solution (~313 μL) were added. After gentle mixing, the mixture was incubated at 37°C for 15 min, mixed with the PHZ solution (100 μL), incubated again at 37°C for 30 min, and centrifuged at 700 x g for 15 min. The supernatant was used for the assay of potassium ions (K^+), and the pellet was used for the assay of MDA, GSH, and catalase. Parallel experiments were carried out with test tubes labeled *"Control Compound"* (containing RBC suspension, a test compound and PBSG), *"Control PHZ"* (containing RBC suspension, PBSG and PHZ) and *"Control Buffer"* (containing RBC suspension and PBSG). In experiments that included BALA, this com-pound was incubated with the RBC suspension at 37°C for 15 min before the addition of a test compound and/or the PHZ solution.

Treatment Solutions for *In Vivo* Studies

Solutions of TAU and the other test compounds used in the study were prepared in 0.9% sodium chloride to contain 480 mM; that of PHZ was prepared in 5 mM phosphate buffer of pH 7.4, and stored in an amber glass container.

Animal Treatments

All administrations were by the intraperitoneal (ip) route. The dose of a sulfur-containing compound was 2.4 mmol/kg, and was given either as a bolus dose 30 min before PHZ, or in two equal portions, one given 30 min before and the other 60 min before PHZ. Animals were sacrificed by decapitation 2 hr after a single dose of PHZ (75 mmol/kg)

Samples from *In Vivo* Studies

Following their collection into heparinized test tubes, the blood samples were centrifuged briefly to separate the plasma from the RBCs. The plasma sample was used for the assay of Hb, metHb and LDH. The RBCs were washed with PBSG as described earlier, and used for the assay of MDA, GSH, and catalase.

Assay of K^+

The amount of K^+ released into the extracellular fluid was measured potentiometrically using a K^+-specific electrode (Cole-Parmer, Vernon Hills, IL). The concentration of electrolyte was derived from a standard curve for K^+ constructed using a commercial solution of KCl.

Assay of Free Hb

The amount of Hb lost into the extracellular medium was measured spectrophotometrically at 540 nm using a commercial assay kit (Procedure No. 525-A, Sigma Chemical Co.) that was based on the oxidation of Hb and its derivatives (excepting sulfHb) to metHb with Drabkin's reagent. The concentration of Hb in the sample was derived from a calibration curve constructed from serial dilutions of a Hb standard preparation (Part No. 525-18, Sigma Chemical Co.).

Assay of MetHb

MetHb was measured spectrophotometrically at 630 nm[49]. In this case the absorbance was read before and after the addition of alkaline KCN to the sample, to convert metHb to cyan-methHb. The amount of metHb present was derived from the difference between the two readings and a calibration curve for metHb prepared on the day of the analysis.

Assay of MDA

MDA was determined by the end-point assay method of Buege and Aust[9], following the reaction of MDA with TBA to yield a colored product that is measured spectrophotometrically at 535 nm. The amount of TBA reactive substances (TBARS) present was calculated by reference to a calibration curve of MDA prepared on the day of the analysis from serial dilutions of a TEP stock solution.

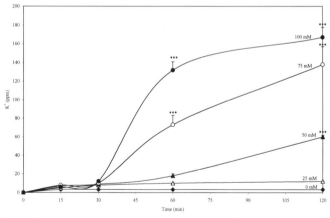

Figure 1. Effects of PHZ (25-100 mM) on the leakage of K^+ from RBCs (20% hematocrit). Values are mean ± SEM (n=6), and they differ significantly from a control treatment, PBSG alone, at ***$p < 0.001$.

Assay of GSH

The intracellular concentration of GSH was measured spectrophotometrically by the method of Ellman[15], which is based on the reaction of a sulfhydryl-containing compound with DTNB to form a color product absorbing at 412 nm. The concentration of the analyte was derived from a calibration curve for GSH.

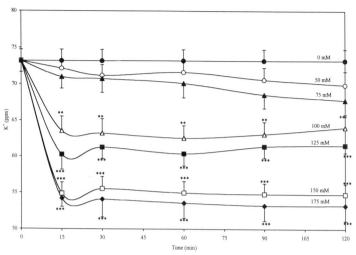

Figure 2. Effects of TAU (20-175 mM) on the leakage of K^+ from RBCs (20% hematocrit). Values are mean ± SEM (n=6), and they differ significantly from PHZ alone at *p<0.05 and **p<0.001.

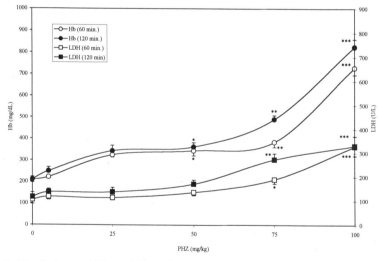

Figure 3. The leakage of Hb and LDH from erythrocytes of rat treated with PHZ (5-100 mg/kg; ip). Values are mean ± SEM (n=6), and they differ significantly from a control treatment, SAL alone, at *p<0.05; **p<0.01; and ***p<0.001.

Statistical Analysis of the Data

The experimental results were analyzed for statistical significance by unpaired Student's t-test, one-way analysis of variance (ANOVA), and Tukey's *post-hoc* test using a commercial software program (StatMost for Windows

2.01™ from Datamost Corporation, Salt Lake City, UT). Differ-ences were considered as statistically significant at p≤0.05. Values were reported as the means ± SEM for n = 6.

Table 1. Effect of sulfur-containing compounds on PHZ-induced K^+ leakage, MDA formation, GSH depletion and catalase activity of RBCs (20% hematocrit) *in vitro*[a,b,c]

Com-pound	K^g (ppm)	MDA (nM/g Hb)	GSH (μM/g Hb)	Catalase (U/g Hb)
PHZ	$72.21 \pm 2.04^{k,j}$	7.87 ± 0.21^j	$6.15 \pm 0.51^{g,j,k}$	$1,922 \pm 185.12^{j,n}$
TAU	$54.85 \pm 1.55^{d,j}$	7.66 ± 0.45^j	7.24 ± 0.54^d	$2,495 \pm 305.81^{f,j}$
HYTAU	$50.45 \pm 2.34^{d,j}$	7.56 ± 0.02^j	7.16 ± 0.3	$2,322 \pm 289.24^{f,j}$
ISA	64.48 ± 1.75^j	7.53 ± 0.02^j	6.85 ± 0.12^d	$2,145 \pm 200.84^{j,m}$
ESA	$41.08 \pm 2.07^{f,j}$	7.49 ± 0.60^j	6.99 ± 0.71	$2,043 \pm 160.34^{j,n}$
AMSA	$40.28 \pm 2.46^{f,j}$	7.63 ± 0.20^j	6.77 ± 0.81^g	$2,108 \pm 204.58^{j,m}$
HMTAU	$50.33 \pm 2.65^{e,j}$	7.25 ± 0.21^j	7.14 ± 0.09	$2,319 \pm 115.87^{f,j}$
NACYS	$5.83 \pm 1.88^{f,n}$	$6.44 \pm 0.18^{d,j}$	$9.25 \pm 0.71^{e,k}$	$2,735 \pm 164.32^{j,n}$
MTAU	$87.50 \pm 4.05^{d,j,n}$	7.36 ± 0.33^j	$4.91 \pm 0.14^{e,h,m}$	$821 \pm 89.58^{f,j,n}$
ASBA	$16.05 \pm 2.43^{f,g,n}$	7.36 ± 0.69^j	8.21 ± 0.09^e	$1,922 \pm 184.38^{j,n}$
MESA	$120.17 \pm 10.98^{f,j,n}$	7.67 ± 0.17^j	$3.57 \pm 0.05^{f,j,n}$	$578 \pm 65.32^{j,n}$
EDSA	$37.35 \pm 2.3^{f,j}$	7.13 ± 0.61^j	6.46 ± 0.71	$2,052 \pm 200.31^{j,n}$
CA	$42.60 \pm 1.56^{f,j}$	7.91 ± 0.24^j	6.85 ± 0.55	$2,201 \pm 209.71^{j,k}$
ACES	$47.62 \pm 4.36^{f,j}$	7.53 ± 0.20^j	6.74 ± 0.12	$1,978 \pm 158.62^{j,m}$
PBSG	$3.28 \pm 0.29^{n,f}$	$3.56 \pm 0.02^{f,n}$	$8.17 \pm 0.84^{d,k}$	$5,615 \pm 500.28^{f,n}$

[a]The concentration of the test compound was 150 mM; that of PHZ was 75 mM; [b]Values represent the mean ± SEM (n = 6). [c]Data were analyzed by unpaired Student's t-test and one-way ANOVA, followed by Tukey's *post-hoc* test. Values differed significantly at [d]p<0.05, [e]p<0.01, [f]p<0.001 vs. PHZ, [g]p<0.05, [h]p<0.01, [j]p<0.001 vs. PBSG and [k]p<0.05, [m]p<0.01, [n]p<0.001 vs. TAU.

RESULTS

In Vitro Experiments

The experimental conditions used to carry out the various *in vitro* studies with PHZ were established on the basis of the release of K^+ from RBCs as a function of the incubation time, concentration of added PHZ and the timing of the pretreatment with the doses of TAU listed in Figs. 1 to 3. From these studies, an incubation time of 60 min, a concentration of TAU (or a related sulfur compound) of 150 mM, and a concentration of PHZ of 75 mM in the RBC suspension were adopted. Because of assay interference by PHZ, other indicators of oxidative membrane damage such as the release of Hb and LDH were not measured in this phase of the study.

As indicated in Table 1, PHZ caused a 22-fold loss in intracellular K^+ relative to a PBSG control ($p<0.001$). This effect was counteracted by TAU and by the majority of TAU analogs and homologs tested alongside (24-78% reduction, $p\leq0.05$), with the potency order being ASBA>EDSA>AMSA≈ ESA>CA>ACES>HYTAU≈HMTAU>TAU). ISA was found not to have a significant protective effect (-11%). Both MTAU and MESA accentuated the action of PHZ (by 21%, $p<0.05$, and 66%, $p<0.001$, respectively). TAU reduced the PHZ-related K^+ release to a greater extent when added to the RBCs separately rather than as a mixture with PHZ (-24% vs. -9%, respectively $p<0.05$; Table 2). In the same study, the protection offered by TAU was not modified by a pretreatment with BALA (Table 2).

Table 2. Effect of TAU and/or BALA on PHZ-induced K^+ leakage, MDA formation, GSH depletion, and catalase activity of RBCs (20% hematocrit) *in vitro*[a,b,c]

Compound	K^g (ppm)	MDA (nM/g Hb)	GSH (µM/g Hb)	Catalase activity (U/g Hb) x 10
PHZ	$72.21 \pm 2.04^{j,k}$	7.87 ± 0.21^j	6.15 ± 0.51	$1,922 \pm 185.12^{j,m}$
TAU/PHZ	$54.85 \pm 1.55^{d,j}$	7.66 ± 0.45^j	$7.24 \pm 0.54^{d,g}$	$2,495 \pm 305.81^{e,j}$
(TAU/PHZ)	65.76 ± 3.9^j	7.72 ± 0.53^j	6.61 ± 0.34	$2,799 \pm 200.98^{d,j,k}$
BALA/PHZ	$70.52 \pm 5.6^{j,k}$	$9.03 \pm 0.84^{d,j}$	6.52 ± 0.55	$2,029 \pm 201.36^{j,m}$
BALA/TAU/PHZ	$57.32 \pm 1.65^{d,j}$	8.64 ± 0.56^j	6.92 ± 0.39	$2,412 \pm 156.63^{j,k}$
PBSG	$3.28 \pm 0.29^{f,n}$	$3.56 \pm 0.28^{f,n}$	8.10 ± 0.85^d	$5,615 \pm 500.21^{f,n}$

[a]The concentration of the test compound was 150 mM; that of PHZ was 75 mM. [b]Values represent the mean $\pm$ SEM ($n = 6$). [c]Data were analyzed by unpaired Student's t-test and one-way ANOVA, followed by Tukey's *post-hoc test*. Values differed significantly at [d]$p<0.05$, [e]$p<0.01$, [f]$p<0.001$ vs. PHZ, [g]$p<0.05$, [j]$p<0.001$ vs. PBSG and [k]$p<0.05$, [m]$p<0.01$, [n]$p<0.001$ vs. TAU.

In the presence of PHZ, the levels of MDA, measured as TBARS, increased about 2.2-fold over that of the PBSG control (Table 1). In this regard, the majority of sulfur compounds tested were found to exert either a weak antagonistic (3-9% reduction: EDSA ≈ HMTAU > ASBA > ESA > HYTAU ≈ ACES ≈ HYTAU ≈ ISA ≈ ACES > TAU ≈ AMSA ≈ MESA; Table 1) or no (CA) action against PHZ-induced MDA formation. The attenuating action of TAU was the same regardless of whether this amino acid was added to the RBC suspension separately (-4%) or together with (-2%) PHZ. It decreased to a nonsignificant extent (~10%) after a pretreatment with BALA (Table 2). By itself, BALA appeared to enhance the formation of MDA due to PHZ (+15%, $p < 0.05$)

The intracellular GSH was significantly reduced by PHZ (-24%, $p < 0.05$, Table 1). In spite of the protective action displayed by most sulfur-containing compounds examined (5-33% increases), the only ones to exert a significant effect were ASBA (+33%, $p < 0.01$) and TAU (+18%, $p < 0.05$). The two notable exceptions were MESA and NMTAU, both of which markedly enhanced the depletion of GSH caused by PHZ ($\geq 20\%$, $p < 0.01$). Although the addition of TAU to the RBC suspension prior to PHZ provided a greater protection than when two compounds were added concurrently (+18% vs. +7.5%, Table 2) the difference was not significant. Furthermore, a pretreatment with BALA resulted in an insignificant decrease in the action of TAU (+18% vs. +12.5%, Table 2). In the absence of TAU, BALA appeared not to oppose the action of PHZ on GSH.

PHZ caused a significant loss (-66%) in RBC antioxidative defense represented by catalase (Table 1). This effect was effectively attenuated by TAU, HYTAU and HMTAU (21-30% increase, $p < 0.001$, Table 1). All other TAU analogs and homologs were either marginally effective (i.e., 3-15% increases: CA, ISA, AMSA, EDSA, ESA, and ACES) or without effect (ASBA). TAU exerted a greater protection when added to the RBC suspension before rather than together with PHZ (+77% vs. +46%, $p < 0.05$, Table 2). This effect was slightly diminished (by ca. 5%, $p < 0.05$, Table 2) when the RBCs were incubated with BALA prior to the addition of TAU. By itself, BALA did not modify the action of PHZ on catalase activity.

Table 3. Effect of various doses of and pretreatment time with TAU on PHZ-induced leakage of Hb, LDL and metHb from RBCs *in vivo*[a,b,c]

Compound	Hb (mg/dL)	LDH (mg/dL)	metHb (mg/dL)
SAL	209.13 ± 19.23^f	115.52 ± 13.44^f	$8.61 \pm 0.86^{f,n}$
PHZ (75 mg/kg)	$491.01 \pm 20.61^{j,n}$	272.19 ± 11.51^h	$113.85 \pm 13.83^{j,m}$
1.2T30/PHZ	$495.20 \pm 27.61^{j,n}$	239.52 ± 22.31^h	$107.61 \pm 15.91^{j,k}$
1.2T60/PHZ	$489.52 \pm 33.21^{j,n}$	210.18 ± 21.51^d	$108.12 \pm 14.08^{j,k}$
2.4T30/PHZ	$389.49 \pm 20.61^{j,n}$	174.19 ± 25.39^e	$96.02 \pm 6.27^{j,k}$
2.4T60/PHZ	$301.51 \pm 30.18^{f,g,k}$	155.21 ± 37.19^f	$87.47 \pm 6.98^{d,j}$
T30/T60/PHZ	215.19 ± 24.56^f	149.59 ± 22.61^f	$77.16 \pm 8.41^{e,j}$
(1.2g1.2 mol /kg)			

[a]TAU was administered as a single 1.2 mol /kg and 2.4 mol /kg dose and in two equal, 1.2 mg/kg, doses, 60 min and 30 min before a single 75 mg/kg dose of PHZ. Blood samples were collected 2 hr after PHZ. [b]Values represent the mean ± SEM (n = 6). [c]Data were analyzed by unpaired Student's t-test and one-way ANOVA, followed by Tukey's *post-hoc test*. Values differed significantly at [d]$p < 0.05$, [e]$p < 0.01$, [f]$p < 0.001$ vs. PHZ, [g]$p < 0.05$, [h]$p < 0.01$, [j]$p < 0.001$ vs. PBSG and [*]$p < 0.05$, [**]$p < 0.01$, [***]$p < 0.001$ vs. TAU.

Table 4. Effect of pretreating rats with a sulfur-containing compound on PHZ-induced leakage of Hb, LDH and metHb from RBCs[a,b,c]

Compound	Hb (mg/dL)	LDH (U/L)	metHb (mg/dL)
SAL	209.13 ± 19.23^f	115.52 ± 13.44^f	$8.61 \pm 0.86^{f,n}$
PHZ	$491.01 \pm 20.61^{j,n}$	$272.19 \pm 24.51^{j,n}$	$113.85 \pm 13.83^{j,m}$
TAU	215.19 ± 24.56^f	$149.59 \pm 22.61^{f,g}$	$77.16 \pm 8.41^{e,j}$
HYTAU	$411.21 \pm 34.11^{d,j,n}$	$199.78 \pm 41.59^{d,h,k}$	$81.57 \pm 10.87^{e,j}$
ISA	$457.33 \pm 51.81^{j,n}$	$220.53 \pm 23.56^{j,k}$	$107.54 \pm 14.45^{j,k}$
ESA	$466.09 \pm 41.25^{j,n}$	$260.18 \pm 21.95^{j,m}$	$110.78 \pm 11.58^{j,k}$
AMSA	$487.87 \pm 38.64^{j,n}$	$244.56 \pm 25.48^{j,k}$	$111.57 \pm 16.28^{j,k}$
HMTAU	$433.59 \pm 39.82^{j,n}$	$234.27 \pm 35.67^{j,k}$	$101.98 \pm 13.54^{j,k}$
NACYS	$335.08 \pm 31.08^{e,j,m}$	$179.34 \pm 13.40^{d,g}$	$79.89 \pm 12.46^{e,j}$
MTAU	$615.35 \pm 50.98^{d,j,n}$	$656.09 \pm 35.47^{f,j,n}$	$146.59 \pm 23.58^{d,j,n}$
ASBA	$581.46 \pm 26.43^{d,j,n}$	$570.18 \pm 10.49^{f,j,n}$	$132.54 \pm 20.14^{j,m}$
MESA	$655.08 \pm 55.45^{e,j,n}$	$701.14 \pm 54.55^{f,j,n}$	$152.07 \pm 18.45^{e,h,n}$
EDSA	$379.24 \pm 29.32^{d,h,n}$	$199.21 \pm 19.25^{d,h,k}$	$93.54 \pm 9.16^{d,j}$
CA	$460.15 \pm 43.02^{j,n}$	$249.54 \pm 21.54^{j,m}$	$112.25 \pm 11.87^{j,m}$
ACES	$478.35 \pm 39.54^{j,n}$	$252.57 \pm 11.21^{j,m}$	$112.56 \pm 13.56^{j,m}$

[a]The sulfur-containing compound was administered in two equal, 1.2 mg/kg, doses, 60 min and 30 min before a single 75 mg/kg dose of PHZ. Blood samples were collected 2 h after PHZ. [b]Values represent the mean ± SEM (n = 6). [c]Data were analyzed by unpaired Student's t-test and one-way ANOVA, followed by Tukey's *post-hoc test*. Values differed significantly at [d]$p<0.05$, [e]$p<0.01$, [f]$p<0.001$ vs. PHZ, [g]$p<0.05$, [h]$p<0.01$, [j]$p<0.001$ vs. PBSG and [k]$p<0.05$, [m]$p<0.01$, [n]$p<0.001$ vs. TAU.

In Vivo Experiments

Prior to the administration of TAU, or of one of its analogs or homologs, to rats, a study was conducted to determine the effects of increasing doses of PHZ (5-100 mg/kg) and of the time of blood sampling following the administration of PHZ (60 min vs. 120 min) on two biochemical parameters of RBC leakiness; LDH and Hb levels. The release of Hb and LDH from RBCs was significantly greater at 120 min after PHZ than at 60 min (Fig. 3). For both cell components, a dose-related change became noticeable only at or above 50 mg/kg of PHZ. Additional preliminary studies were conducted with TAU and PHZ to determine the effect of the dose of and dosing schedule with the amino acid on the release of Hb, LDH and metHb from RBCS. From the results presented in Table 3, it was determined that the degree of protection afforded by TAU was directly related to the dose used (i.e., 2.4 mmol/kg>1.2 mmol/kg), the timing of its administration (i.e., 60 min vs. 30 min before PHZ), and the dosing approach (i.e., divided-dose approach>bolus dose approach). Hence, all subsequent experiments with TAU or a related compound were conducted using those conditions that provided maximum protection against oxidant-related biochemical alterations.

Table 5. Effects of pretreating rats with a sulfur-containing compound on PHZ-induced GSH depletion, MDA formation and change in catalase activity of RBCs[a,b,c]

Compound	GSH (µM/g Hb)	MDA (µM/g Hb)	Catalase activity (U/g Hb) x 10
SAL	$4.32 \pm 0.29^{***,••}$	$3.67 \pm 0.11^{**}$	$2{,}599 \pm 581^{***,••}$
PHZ	$1.63 \pm 0.08^{+++,••}$	$5.24 \pm 0.03^{++,•}$	$992 \pm 149^{+++,•••}$
TAU	$2.91 \pm 0.01^{***,++}$	$3.97 \pm 0.12^{*}$	$1{,}896 \pm 196^{***,++}$
HYTAU	$3.64 \pm 0.04^{***,•}$	$3.91 \pm 0.01^{*}$	$2{,}198 \pm 147^{***}$
ISA	$2.78 \pm 0.08^{**,++}$	$4.22 \pm 0.13^{*}$	$1{,}546 \pm 65^{***,+++,•}$
ESA	$1.74 \pm 0.05^{+++,••}$	$5.12 \pm 0.08^{++,•}$	$897 \pm 80^{+++,•••}$
AMSA	$1.84 \pm 0.04^{+++,•}$	$4.92 \pm 0.09^{+}$	$984 \pm 95^{+++,•••}$
HMTAU	$1.76 \pm 0.08^{+++,••}$	4.62 ± 0.21	$919 \pm 86^{+++,•••}$
NACYS	$3.94 \pm 0.04^{***,••}$	$3.98 \pm 0.02^{*}$	$1{,}798 \pm 99^{***,++}$
MTAU	$1.52 \pm 0.07^{+++,•••}$	$6.24 \pm 0.09^{*,+++,••}$	$358 \pm 30^{***,+++,•••}$
ASBA	$1.74 \pm 0.01^{+++,••}$	$4.58 \pm 0.03^{+}$	$1{,}002 \pm 156^{+++,••}$
MESA	$1.49 \pm 0.05^{+++,••}$	$6.94 \pm 0.01^{**,+++,••}$	$548 \pm 29^{**,+++,•••}$
EDSA	$3.72 \pm 0.01^{***,•}$	$4.21 \pm 0.04^{*}$	$1{,}987 \pm 179^{***,++}$
CA	$1.75 \pm 0.04^{+++,••}$	$5.12 \pm 0.08^{++,•}$	$980 \pm 98^{+++,•••}$
ACES	$1.65 \pm 0.05^{+++,•••}$	$5.19 \pm 0.07^{++,•}$	$891 \pm 108^{+++,•••}$

[a]The sulfur-containing compound was administered in two equal, 1.2 mg/kg, doses, 60 min and 30 min before a single 75 mg/kg dose of PHZ. Blood samples were collected 2 h after PHZ. [b]Values represent the mean ± SEM (n = 6). [c]Data were analyzed by unpaired Student's t-test and one-way ANOVA, followed by Tukey's *post-hoc* test. Values differed significantly at $^{*}p<0.05$, $^{**}p<0.01$, $^{***}p<0.001$ vs. PHZ, $^{+}p<0.05$, $^{++}p<0.01$, $^{+++}p<0.001$ vs. PBSG and $^{•}p<0.05$, $^{••}p<0.01$, $^{•••}p<0.001$ vs. TAU.

Table 4 summarizes the results attained with 2.4 mmol/kg of TAU or a structurally related compound, given in two divided doses, on the release of Hb, LDH and metHb from the RBCs of rats that had received a single 75 mg/kg dose of PHZ 120 min earlier. A significant ameliorating effect on Hb release was observed with TAU (-56%, p<0.001), EDSA (-23%, P<0.01) and HYTAU (-16%, p<0.05), but not with HMTAU, ISA, ESA, CA, ACES and AMSA (≤12% decrease). In contrast, ASBA (+18%), MTAU (+25%) and MESA (+33%), in that order, worsened the action of PHZ (p≤0.05). Similar trends were verified for the actions of TAU and its congeners on the release of both LDH and metHb. In the case of LDH, a significant protection (p≤0.05) was derived from TAU (-45%), HYTAU (-27%) and EDSA (-

27%); with most of the remaining sulfur compounds reducing the loss by 4-19%. In contrast, MESA, MTAU and ASBA markedly accentuated (>100%, p<0.001) the effect of PHZ on LDH. Regarding the release of metHb, significantly lower levels (p≤0.05) were found in the presence of TAU (-32%), HYTAU (-28%), and EDSA (-18%), but not in the presence of most of the other sulfur compounds (≤10% decreases). Again, MTAU (+29%) and MESA (+33.5%), but not ASBA (+16%), intensified the action of PHZ to a significant extent (p<0.01).

Table 6. Effect of pretreating rats with TAU and/or BALA on PHZ-induced leakage of HB, LDH and metHb from the RBCs[a,b,c]

Compound	Hb (mg/dL)	LDH (U/L)	metHb (mg/dL)
SAL	209.13 ± 19.23[***]	115.52 ± 13.44[***,•]	8.61 ± 0.86[***,•••]
PHZ	491.01 ± 20.61[+++,•••]	272.38 ± 24.51[+++,•••]	113.85 ± 13.86[+++,••]
TAU/PHZ	215.19 ± 24.56[***]	149.59 ± 22.61[***,+]	77.16 ± 8.41[**,+++]
BALA/PHZ	495.57 ± 28.96[+++,•••]	269.42 ± 18.79[+++,•••]	110.54 ± 15.24[+++,••]
BALA/TAU/PHZ	220.59 ± 20.57[***]	129.57 ± 18.49[***]	81.57 ± 11.42[**,+++]

[a]BALA was administered as a single dose (150 mM) 30 min before the first dose of TAU. [b]Values represent the mean ± SEM (n = 6). [c]Data were analyzed by unpaired Student's t-test and one-way ANOVA, followed by Tukey's *post-hoc* test. Values differed significantly at [*]p<0.05, [**]p<0.01, [***]p<0.001 vs. PHZ, [+]p<0.05, [++]p<0.01, [+++]p<0.001 vs. PBSG and [•]p<0.05, [••]p<0.01, [•••]p<0.001 vs. TAU.

Table 7. Effect of pretreating rats with TAU and/or BALA on PHZ-induced GSH depletion, MDA formation and catalase activity of RBCs *in vivo*[a,b,c]

Compound	GSH (µM/g Hb)	MDA (nM/g Hb)	Catalase (U/g Hb)
SAL	4.32 ± 0.29[***,••]	3.67 ± 0.11[**]	2599 ± 581[***,••]
PHZ	1.63 ± 0.08[+++,••]	5.24 ± 0.03[++,•]	992 ± 149[+++,•••]
TAU/PHZ	2.91 ± 0.01[***,++]	3.97 ± 0.12[*]	1896 ± 196[***,++]
BALA/PHZ	1.92 ± 0.04[+++,•]	5.17 ± 0.03[++,•]	1156 ± 156[+++,•••]
BALA/TAU/PHZ	3.08 ± 0.03[***,++]	4.01 ± 0.09[*]	1965 ± 201[***,++]

[a]BALA was administered as a single dose (150 mM) 30 min before the first dose of TAU. [b]Values represent the mean ± SEM (n = 6). [c]Data were analyzed by unpaired Student's t-test and one-way ANOVA, followed by Tukey's *post-hoc* test. Values differed significantly at [*]p<0.05, [**]p<0.01, [***]p<0.001 vs. PHZ, [+]p<0.05, [++]p<0.01, [+++]p<0.001 vs. PBSG and [•]p<0.05, [••]p<0.01, [•••]p<0.001 vs. TAU.

Treating rats with TAU or one of its analogs or homologs before an acute exposure to PHZ resulted in a wide spectrum of actions on the intra-cellular levels of GSH and MDA and the activity of catalase (Table 5). Thus, significant protection against the loss of GSH (70-128% increase, p≤0.01) and catalase activity (56-121.5% increase, p<0.001) was derived from HYTAU, EDSA, TAU and ISA, in that order, but not from AMSA, HMTAU, ESA, ASBA, ACES and CA (GSH, 1-13% increases; catalase, 1-10% decreases). In contrast, both MTAU and MESA exerted a worsening effect (GSH, ca. 8% decrease, p≤0.05; catalase, ≥45% decrease, p≤0.01). Although the magnitude of the protective action against PHZ-induced MDA formation attained with the various sulfur compounds varied within a nar-rower range of values (20-25% decreases, p<0.05) than those gathered for GSH and catalase, the general potency orders were quite similar.

As shown in Tables 6 and 7, pretreating rats with BALA did not affect the protective actions of TAU against PHZ-related biochemical alterations in the RBCs relative to rats on TAU-PHZ. Similarly, BALA did not alter the deleterious actions of PHZ on RBCs in the absence of TAU.

For comparative purposes, the actions of TAU and its structurally-related compounds were compared with those of NACYS, a sulfur-containing amino acid known to exhibit good antioxidant properties[4] and to act as a biochemical precursor of GSH[35]. *In vitro*, NACYS was found to protect RBCs against the deleterious effects of PHZ to a greater extent than any of the other test compounds (Table 1). These effects were maintained under *in vivo* conditions (Tables 4 and 5).

DISCUSSION

This study was designed to ascertain the actions of TAU and structurally related sulfur-containing compounds on the membrane permeability and oxidative status of rat RBCs exposed to PHZ under *in vitro* and *in vivo* conditions. PHZ was selected as a model oxidant hemolytic agent because hemolysis and other cellular derangements caused by this compound occur at a relatively slow rate and, therefore, are quite amenable to biochemical analysis. In addition, several sulfur-containing compounds representing analogs and homologs of TAU were tested alongside the parent compound to determine the effects of chain length, N-substitution, deamination, replacement of the amino group by a hydroxyl or sulfhydryl group, and oxidation of the sulfonic acid group on the actions of TAU against the damage of RBCs by PHZ.

PHZ can cause various types of alterations depending on whether they are directed toward the membrane or toward intracellular components such as Hb, GSH and catalase. By acting on the RBC membrane, PHZ can cause drastic changes in membrane composition, fluidity and permeability, partial disturbance of phospholipid organization, and altered cell morphology[41,45]. PHZ will also promote the progressive efflux[42] of Na^+ and K^+. It has been suggested that these effects are the result of (i) LPO of membrane PUFA due to free radicals and H_2O_2[12,13,21,31] resulting from the aerobic interaction of PHZ with RBCs or oxyhemoglobin, even after the addition of superoxide dismutase or catalase[56]; (ii) the attachment of Heinz bodies to membranes[42]; (iii) damage by free radicals generated by the oxidation of dissociated α- and β-globin chains by PHZ[42]; or (iv) a combination of these effects.

In vitro, TAU and most of the chemically-related sulfur compounds were found to attenuate the leakage of K^+ from intact rat RBCS treated with PHZ. From the potency order obtained for the various test compounds, it was verified that except for ISA most of the structural variations of the TAU molecule, including the addition of an α- or β- carboxyl group, replacement of a sulfonic group for the amino group, lengthening or shortening of the carbon chain by one carbon, and reduction of the sulfonic acid group to a sulfinic functionality translated into a greater antagonism of the PHZ effect than that by TAU itself. On the other hand, while N-methylation or the replacement of the amino group by a sulfhydryl group completely abolished the protective action of TAU against K^+ loss, removal of the β-amino group did not. The possibility that TAU and PHZ might directly interact with each other was suggested by the decrease in protection by TAU that occurred when these compounds were added to the RBCs simultaneously rather than individually and 30 min apart.

Two characteristic responses to the oxidative stress caused by PHZ on RBCs are a decrease in the thiol status of the membrane and in intracellular GSH[54]. The negative effect on intracellular GSH was first ascribed to oxidation by an active metHb-H_2O_2 or similar hematin-peroxide complex[7,32]. This concept was later modified to include free H_2O_2 as an additional, if not the only, cause[12]. As it relates to H_2O_2, the antioxidative defense of the RBC represented by GSH peroxidase will be mobilized to remove the peroxide at the expense of the oxidation of GSH to GSSG[12]. However, in the face of an overwhelming excess of H_2O_2, the availability of GSH will soon become exhausted. A similar reduction in red cell GSH has been observed after injecting mice with PHZ, an effect that was related to the production of free radicals[59]. Subsequently, it was verified that PHZ can also promote the release of iron from its complexes in RBCs, iron that can contribute to the oxidation of both Hb and GSH[16] and to the formation of hydroxyl radicals via a Haber-Weiss reaction[22]. Conversely, when GSH is present in sufficiently high concentrations, it may prevent membrane LPO and hemolysis in spite of significant iron release, thus protecting RBCs subjected to oxidative stress[16].

In this study, only a handful of test compounds, including TAU, were able to attenuate the depleting action of PHZ on GSH both *in vitro* and *in vivo*. From the potency order for the *in vitro* effects of these compounds, it was noted that making the chain length one carbon longer, adding a carboxyl group α- to the sulfonic acid end, and reducing the sulfonate group to a sulfinate, did not modify the protective action of TAU, which was in all instances significantly less than that of NACYS. While removal of the amino group or of one methylene group, and the substitution of a sulfonic acid or hydroxyl group for the amino group of TAU lowered the protective effect of the parent compound, N-methylation and substitution of the amino group by a sulfhydryl group heightened the depletion of GSH by PHZ. Under *in vivo* conditions, alterations of the TAU molecule at the amino and sulfonic acid ends, but not to the chain length, contributed to a protective action. Thus, HYTAU, the sulfinic acid analog, ISA, the β-hydroxyl analog, and EDSA, the disulfonated analog, were the only ones capable of antagonizing the depleting action of PHZ. In this regard, it was unexpected to find that HYTAU was as good a protector of intracellular GSH as was NACYS, a biological precursor of GSH.

As an assessment of the extent of LPO by PHZ, the accumulation of MDA was measured as TBARS. Under *in vitro* conditions, none of the sulfur compounds tested except NACYS were able to counteract the peroxidating effect of PHZ. Unlike the results obtained for K^+ leakage and GSH depletion, neither MTAU nor MESA aggravated the action of PHZ. Furthermore, the results for MDA levels observed *in vivo* closely resembled those obtained under the same conditions for GSH, namely that they were inhibited by certain monosulfonic (i.e., TAU, ISA), disulfonic (i.e., EDSA) and sulfinic (i.e., HYTAU) compounds, and enhanced by the β-sulfhydryl (i.e., MESA) and N-methyl (i.e., MTAU) analogs of TAU. The protection offered by TAU and HYTAU was equivalent to that by NACYS. This finding is noteworthy since several earlier studies have shown that in a cell-free system[3,53,55] or in cultured cell preparations[1,10,43]. HYTAU is either a better antioxidant than TAU or the only one endowed with this property.

PHZ is known to decrease the activity of catalase both in the intact RBC and in the crystalline state[12]. In this respect, TAU, HYTAU and HMTAU demonstrated good protective action under *in vitro* conditions; while MESA and MTAU appeared to further increase the loss in enzymatic activity due to PHZ. *In vivo*, HYTAU was more protective than EDSA, NACYS and ISA, with most of the remaining sulfur compounds showing no effect (AMSA, CA) or enhancing the action of PHZ (HMTAU, ACES, ESA, MESA, NMTAU).

By interacting with PHZ in an aerobic environment, Hb undergoes a one electron oxidation to metHb while PHZ is oxidized to phenyldiazene. In turn, metHb can be further converted to oxidation states of altered iron-porphyrin complexes designated as reversible and irreversible hemichromes[28,44]. In this work, TAU was found to prevent the release of Hb into the extracellular medium and the formation of metHb in a dose-related manner, and to the greatest extent among all sulfur-containing compounds tested. In addition, the replacement of the amino group by a sulfonic acid group ((EDSA) and of a sulfinate for a sulfonate functionality (HYTAU) conferred significant protection, while the immediate higher homolog, HMTAU, and the β-hydroxyl analog, ISA, were only marginally protective. ASBA, MTAU and MESA, in that order, aggravated the leakage.

In spite of the wealth of information that is available on the antioxidative actions of TAU, the mechanism by which this compound exerts its beneficial effects remains to be elucidated. Based on their work on canine RBCs, Nakamura *et al.*[36] concluded that TAU can indeed protect cells against oxidant-induced changes in antiperoxidative defenses through a membrane-stabilizing rather than through a direct antioxidant effect. This possibility is tenable since at physiological pH TAU exists as a zwitterion, a state of low lipophilicity that hinders the uptake of the amino acid into RBCs by simple diffusion[24], and as a rinsing of RBCS with PBSG will be sufficient to remove any protection derived from a preincu-bation with the amino acid[47]. The concept of a membrane-based rather than of an intracellular mechanism for TAU has received further support from the present results with BALA, a TAU transport antagonist, which was without effect on the actions of TAU. In this context, finding that the protection of RBCs by TAU disappeared upon N-methylation or the replacement of the amino group by a sulfhydryl one, may imply that these analogs of TAU cannot effectively interact with membrane components, particularly neutral phospholipids[24], to bring about stabilization. In contrast, structural variations of the TAU molecule such as the reduction of the sulfonate to a sulfinate functionality and the replacement of the amino group by a second sulfonate group either preserved or enhanced the beneficial effects of TAU against PHZ-related biochemical alterations. Despite a few discrepancies in the potency and potency order noted for the various sulfur-containing compounds examined here, especially as it relates to the accumulation of MDA, the results gathered under *in vitro* conditions generally agreed with those from *in vivo* experiments. Overall, a greater degree of protection by TAU and some of its congeners was attained *in vivo* than *in vitro*. *In vivo*, the protective actions of TAU were dose-re-lated and they were expressed only above 1.2 mmol/kg, particularly if given 60 min before PHZ. Such differences might reflect differences in the pharmacody-namics of the oxidant between a live test system and a cell suspension; or the possibility that TAU is interacting *in vivo* with a reactive molecule such as glucose to generate an antioxidative glucose-taurine reaction product, or with MDA itself to render it biologically inactive. A future study will be needed to confirm these postulates.

REFERENCES

1. Alvarez, J.G., and Storey, N.T., 1983, Taurine, hypotaurine, epinephrine and albumin inhibit lipid peroxidation in rabbit spermatozoa and protect against loss of motility. *Biol. Reprod.* 29:548-555.

2. Arduini, A., Stern, A., Storto, S., Belfiglio, M., Mancinelli, G., Scurti, R., and Federici, G., 1989, Effect of oxidative stress on membrane phospholipid and protein organization in human erythrocytes. *Arch. Biochem. Biophys.* 273:112-120.

3. Aruoma, O.I., Halliwell, B., Hoey, B.M., and Butler, J., 1988, The antioxi-dant action of taurine, hypotaurine and their metabolic precursors. *Biochem. J.* 256:251-255.

4. Aruoma, O.I., Halliwell, B., Hoey, B.M., and Butler, J., 1989, The antioxidant action of N-acetylcysteine: its reaction with hydrogen peroxide, hydroxyl radical, superoxide and hypochlorous acid. *Free Rad. Biol. Med.* 6:593-597.

5. Azuma, J., Hamaguchi, T., Ohta, H., Takihara, K., Awata, N., Sawamura. A., Harada, H., and Kishimoto, S., 1987, Calcium overload-induced myo-cardial damage caused by isoproterenol and by adriamycin: possible role of taurine in its prevention. In *The Biology of Taurine. Methods and Mechan-isms* (R.J. Huxtable, F. Franconi, and A. Giotti, eds.), Plenum Press, New York, NY, pp. 167-179.

6. Banks, M.A., Porter, D.W., Martin, W.G., and Castranova, V., 1991, Ozone-induced lipid peroxidation and membrane leakage in isolated rat alveolar macrophages: protective effects of taurine. *J. Nutr. Biochem.* 2:308-313.

7. Beutler, E., Robson, M., and Buttenwieser, E., 1957, The mechanism of glutathione destruction and protection in drug-sensitive and non-sensitive erythrocytes. In vitro studies. *J. Clin. Invest.* 36:617-628.

8. Brunori, M., Falcioni, G., Fioretti, E., Giardina, B., and Rotilio, G., 1975, Formation of superoxide in the autoxidation of the isolated α and β chains of human hemoglobin and its involvement in hemichrome precipitation. *Eur. J. Biochem.* 53:99-104.

9. Buege, J.,A., and Aust, S.D., 1978, Microsomal lipid peroxidation, In *Methods of Enzymology*, vol. 52, (S. Fleischer and L. Packer, eds.), Academic Press, New York, NY, pp. 302-310.

10. Cañas, P., Guerra, R., and Valenzuela, A., 1989, Antioxidant properties of hypotaurine: comparison with taurine, glutathione and β-alanine. *Nutr. Rep. Int.* 39:433-438.

11. Clemens, M.R., and Waller, H.D., 1987, Lipid peroxidation in erythrocytes. *Chem. Phys. Lipids* 45:251-268.

12. Cohen, G., and Hochstein, P., 1964, Generation of hydrogen peroxide in erythrocytes by hemolytic agents. *Biochemistry* 3:895-900.

13. Cohen, G., and Hochstein, P., 1965. *In vivo* generation of H_2O_2 in mouse erythrocytes by hemolytic agents. *J. Pharmacol. Exp. Therap.*147:139-143,

14. Ebrahim, A.S., and Sakthisekaran, D., 1997, Effect of vitamin E and taurine treatment on lipid peroxidation and antioxidant defense in perchloroethyl-ene-induced cytotoxicity in mice. *Nutr. Biochem.* 8:270-274.

15. Ellman, G.L., 1959, Tissue sulfhydryl groups. *Arch. Biochem. Biophys.* 82: 70-77.

16. Ferrali, M., Signorini, C., Ciccoli, L., and Comporti, M., 1992, Iron release and membrane damage in erythrocytes exposed to oxidizing agents, phenylhydrazine, divicine and isouramil. *Biochem. J.* 285:295-391.

17. French, J.K., Winterbourn, C.C., and Carrell, R.W., 1978, Mechanism of oxyhaemoglobin breakdown on reaction with acetylphenylhydrazine. *Biochem. J.* 173:19-26.

18. Giri, S.N., and Wang, Q., 1992, Taurine and niacin offer a novel therapeutic modality in prevention of chemically-induced pulmonary fibrosis in hamsters. In *Taurine* (J.B.Lombardini, S.W. Schaffer, and J. Azuma., eds.), Plenum Press, New York, NY, pp.329-340.

19. Goldberg, B., and Stern, A., 1975, The generation of superoxide by the inter-action of the hemolytic agent, phenylhydrazine, with human hemoglobin. *J. Biol. Chem.* 250:2401-2403.

20. Goldberg, B., Stern, A., and Peisach, J., 1976, The mechanism of superoxide anion generation by the interaction of phenylhydrazine with hemoglobin. *J. Biol. Chem.* 251:3045-3051.

21. Goldberg, B., and Stern, A., 1977, The mechanism of oxidative hemolysis produced by phenylhydrazine. *Mol. Pharmacol.* 13:832-839.
22. Gutteridge, J.M.C., 1986, Iron promoters of the Fenton reaction and lipid peroxidation can be released from haemoglobin by peroxides. *FEBS Lett.*201:291-295.
23. Hill, H.A.O., and Thornalley, P.J., 1981, Phenyl radical production during the oxidation of phenylhydrazine and in phenylhydrazine-induced haemo-lysis. *FEBS Lett.* 125:235-238.
24. Huxtable, R.J., and Sebring, L.A., 1986, Towards a unifying theory for the actions of taurine. *Trends Pharmacol. Sci.* 7:481-485.
25. Imanishi, H., Nakai, T., Abe, T., and Takino, T., 1985, Role of free radical scavengers on phenylhydrazine induced hemolysis in rat. *Acta Vitaminol. Enzymol.* 7:71-76.
26. Itano, H.A., Hirota, K., and Hosokawa, K., 1975, Mechanism of induction of haemolytic anemia by phenylhydrazine. *Nature* 256:665-667.
27. Itano, H.A., Hosokawa, K., and Hirota, K., 1976, Induction of haemolytic anemia by substituted phenylhydrazines. *Br. J., Haematol.* 32:99-104.
28. Itano, H.A., and Mannen, S., 1976, Reactions of phenyldiazene and ring-substituted phenyldiazenes with ferrihemoglobin. *Biochim. Biophys, Acta* 421:87-96.
29. Itano, H.A., Hirota, K., and Vedvick, T,S, 1977, Ligands and oxidants in ferrihemo-chrome formation and oxidative hemolysis. *Proc. Natl,. Acad, Sci. USA* 74:2556-2560.
30. Jain, S.K., and Subrahmanyam, D., 1978, Life span of phenylhydrazine-induced reticulo-cytes in albino rats. *Indian J. Exp. Biol.* 16:255-257.
31. Jain, S.K., and Hochstein, P., 1979, Generation of superoxide radicals by hydrazine: its role in phenylhydrazine-induced hemolytic anemia. *Biochim. Biophys. Acta* 586:128-136.
32. Keilin, D., and Hartree, E.F., 1951, Purification of horse-radish peroxidase and comparison of its properties with those of catalase and methemoglobin. *Biochem. J.* 49:88-106.
33. Malone, J.I., Benford, S.A., and Malone Jr., J., 1993, Taurine prevents galactose-induced cataracts. *J. Diab. Comp.* 7:44-48.
34. Misra, H.P., and Fridovich, I., 1876, The oxidation of phenylhydrazine: superoxide and mechanism. *Biochemistry* 15:681-687.
35. Moldeus, P., Cotgreave, I.A., and Berggren, M., 1986, Lung protection by a thiol-containing antioxidant: N-acetylcysteine. *Respiration* 50 (Suppl 1):31-42.
36. Nakamura, T., Ogasawara, M., Koyama, I., Nemoto, M., and Yoshida, T., 1993, The protective effect of taurine on the biomembrane against damage   produced by oxygen radicals. *Biol. Pharm. Bull.* 16:970-972.
37. Nakashima, T., Taniko, T., and Kuriyama, K., 1982, Therapeutic effect of taurine administration on carbon tetrachloride-induced hepatic injury. *Japan. J. Pharmacol.* 32:583-589.
38. Nakashima, T., Takino, T., and Kuriyama, K., 1983, Pretreatment with taurine facilitates hepatic lipid peroxide formation associated with carbon tetrachloride intoxication. *Japan. J. Pharmacol.* 33:515-523.
39. Nakashima, T., Seto, Y., Nakajima, T., Shima, T., Sakamoto, Y., Cho, N., Sano, A., Iwai, M., Kagawa, K., Okanoue, T., and Kashima, K., 1990, Calcium-associated cytoprotective effect of taurine on the calcium and oxygen paradoxes in isolated rat hepatocytes. *Liver* 10: 167-172.
40. Nakashima, T., Shima, T., Sakai, M., Yama, H., Mitsuyoshi, H., Inaba, K., Matsumoto, Y., Sakamoto, K., Kashima, K., and Nishikawa, H., 1996, Evidence of a direct action of taurine and calcium on biological membranes. *Biochem. Pharmacol.* 52: 173-176.
41. Ogiso, T., Ito, Y., Iwaki, M., Nakanishi, K., and Saito, H., 1989, Effect of phenylhydrazine-induced structural alterations of human erythrocytes on basic drug peroxidation. *Chem Pharm, Bull,* 37:430-434.
42. Orringer, E.P., and Parker, J.C., 1977, Selective increase of potassium permeability in red blood cells exposed to acetylphenylhydrazine. *Blood* 50:1013-1021.
43. Pasantes-Morales, H., and Cruz, C., 1985, Taurine and hypotaurine inhibit light-induced lipid peroxidation and protect rod outer segment structure outer segments. *Brain Res.* 330:154-157.

44. Peisach, J., Blumberg, W.E., and Rachmilewitz, E.A., 1975, The demons-tration of ferri-hemichrome intermediates in Heinz body formation following the reduction of oxyhemo-globin A by acetylphenylhydrazine. *Biochim. Biophys. Acta* 393:404-418.
45. Petty, H.R., Zhou, M-J., and Zheng, Z., 1991, Oxidative damage by phenylhydrazine di-minishes erythrocyte anion transport. *Biochim. Biophys. Acta* 1064:308-314.
46. Pion, P.D., Kittlepon, M.D., Rogers, O.R., and Morris, J.G., 1987, Myocardial failure in cats associated with low plasma taurine. *Science* 237:764-768.
47. Pokhrel, P.K., and Lau-Cam, C.A., 1999, Protection by taurine and structurally related sulfur-containing compounds against erythrocyte membrane damage by hydrogen perox-ide. This volume.
48. Redmond, H.P., Wang, J.H., and Bouchier-Hayes, D., 1996, Taurine attenuates nitric oxide and reactive oxygen intermediate-dependent hepatocyte injury. *Arch. Surg.* 131:1280-1287.
49. Rodkey, F.L., and O'Neal, J.D., 1974, Effects of carboxymethemoglobin on the determi-nation of methemoglobin in blood. *Biochem. Med.* 9:261-270.
50. Saito, S., and Itano, H.A., 1981. β-*meso*-Phenylbiliverdin IXα and N-phenylprotoporphyrin IX, products of the reaction of phenylhydrazine with oxyhemoglo-bin. *Proc. Natl. Acad. Sci. USA* 78:5508-5508-5512.
51. Sharabani, M., Plotkin, B., and Aviram, I. (1984) Lipid peroxidation in red blood cell membranes. *Cell. Mol Biol.* 30:329-336.
52. Sharma, R., and Premachandra, B.R., 1991, Membrane bound hemoglobin as a marker of oxidative injury in adult and neonatal red blood cells. *Biochem. Med. Met. Biol.* 46:33-44.
53. Shi, X., Flynn, D.C., Porter, D.W., Leonard, S.S., Vallyathan, V., and Castranova, V., 1997, Efficacy of taurine based compounds as hydroxyl radical scavengers in silica in-duced peroxidation. *Ann. Clin. Lab. Med.* 27:365-373.
54. Stern, A., 1989, Drug-induced oxidative denaturation in red blood cells. *Semin, Hematol*, 26:301-306.
55. Tadolini, B., Pintus, G., Pinna. G.G., Bennardini, F., and Franconi, F., 1995, Effects of taurine and hypotaurine on lipid peroxidation. *Biochem. Biophys. Res. Commun.* 213:820-826.
56. Thornalley, P.J., 1984, The haemolytic reactions of 1-acetyl-2-phenylhydra-zine and hy-drazine: a spin trapping study. *Chem. Biol. Interact.* 50:339-349.
57. Trachtman, H., Del Pizzo, R., Futterweit, S., Levine, D., Rao, P.S., Valder-rama, E.., and Sturman, J.A., 1992, Taurine attenuates renal disease in chronic puromycin aminonucleo-side nephropathy. *Am. J. Physiol.* 262:F117-F123.
58. Trachtman, H., Futterweit, S., and Bienokowski, R.S., 1993, Taurine prevents glucose-induced lipid peroxidation and increased collagen production in cultured rat mesangial cells. *Biochem. Biophys. Res. Commun.* 191:759-765.
59. van Caneghem, P., 1984, Influence of phenylhydrazine on the antioxidant system of the erythrocytes and the liver in mice. *Biochem. Pharmacol.* 33:717-720,
60. Venkatesan, N., and Chandrakasan, G., 1994, In vivo administration of taurine and niacin modulate cyclophosphamide-induced lung injury. *Eur. J. Pharmacol.* 292:75-80.

A TAURINE TRANSPORTER ISOLATED FROM RAT CARDIAC MUSCLE

[1]Jennifer J. Bedford, [2]Jo-Ann Stanton and [1]John P. Leader

Departments of [1]Physiology and [2]Anatomy and Structural Biology, School of Medical Sciences, University of Otago, Dunedin, New Zealand

INTRODUCTION

Taurine is a sulphur-containing amino acid which has been accorded a number of postulated functions[1]. It has long been known to be a compatible solute in invertebrates[2] and in vertebrates[3], where it is the predominant amino-acid in the kidney[4], heart[5] and it also occurs in high concentration in the brain. It plays an important role in bile salt conjugation[6], and it has been implicated as a free radical scavenger[7] and as a neurotransmitter[8].

Taurine is taken up into cells through combination with the cotransporter TAUT which requires the presence of Na^+ and Cl^- ions[9]. The transporter is widely distributed in different tissues and species[10]. The gene coding for TAUT has been cloned from several different tissues[11,12], its amino acid sequence deduced, and a putative structure containing twelve membrane-spanning domains[9] is widely accepted. This places it in a group of transporters mainly associated with the brain, and which appear to function as scavengers of extracellular neurotransmitters.

Consonant with its role as a compatible osmolyte, taurine is accumulated in renal medullary cells exposed to high osmolarities during antidiuresis[4], in heart and brain of rats made experimentally hypernatraemic[13] and also in cultured renal cells (MDCK) when these are exposed to high NaCl[14]. The

Taurine 4, edited by Della Corte *et al.*
Kluwer Academic / Plenum Publishers, New York, 2000.

sequence of molecular events in this latter case is well documented[9]. On exposure to high NaCl, cells shrink and intracellular ionic concentrations rise. Over the next 24 hours, mRNA coding for the transporter rapidly increases within the cells. This is followed by increased uptake of taurine into the cells as new transporter protein is incorporated into the cell membrane.

In cardiac tissue, taurine levels are increased in hypernatraemia[13] and decreased in hypertension and congestive heart failure[15,16]. However, Uchida *et al.*[9] found only weak expression of mRNA coding for the taurine transporter in dog heart. This led them to suggest that taurine accumulation in the heart might involve other processes or mechanisms. For this reason, we undertook a study of taurine transport by cardiac tissue.

METHODS

Experimental treatment of rats

Wistar rats (female, 6 weeks old, 150 gm) (University of Otago Animal Breeding Station) were divided into 2 groups. (i) control: normal drinking and feeding regime; (ii) rats made hyperosmotic with NaCl : no drinking water and daily intraperitoneal injections of 1 mol NaCl[13]. Rats were weighed daily and maintained for 4-6 days.

Tissue analysis

Rats were anaesthetised with ether, and killed by cutting through the neck with secateurs. Blood was collected from the severed neck, immediately centrifuged, and osmotic pressure (Wescor 5500 vapour pressure osmometer), and ion concentration (Na^+ and K^+) (flame photometry) were measured in the serum.

The hearts were removed and quickly transferred to liquid nitrogen. In some cases, a small amount of tissue (about 20 mg fresh weight) was removed from the ventricle, and extracted fresh in 7% perchloric acid for analysis of free amino acids. Picotag analysis and separation of the free amino acids, including taurine, by High Performance Liquid Chromatography (HPLC) was then carried out[17].

Extraction of cardiac poly (A)⁺mRNA

Hearts stored in liquid nitrogen were pulverized in liquid nitrogen and poly (A)⁺mRNA extracted using Tel-Test mRNA-STAT 30 or Trizol (Life Technologies). The yield from one heart was within the range 1-10 μg.

RT-PCR

First strand cDNA was prepared from poly(A)⁺mRNA using Superscript II (Life Technologies) with oligo-(dT) primers. PCR primers identical to those used for amplification of TAUT from rat brain[18] were used, with the following sequences :5'-3' (sense) tgt ctg aaa gac ttc cac aaa gac atc; antisense tca tga ggg ttg ctc tgg agt gaa agg. 2.5 U Amplitaq Gold (Perkin-Elmer) and the following amplification conditions (GeneAmp 2400) were used: heat activated 94°C, 10 minutes; 43 cycles of 94°C for 30 seconds (denaturation), 60°C for 1 minute (annealing), 72°C for 1 minute (extension); 60°C for 10 minutes (final extension) and held at 4°C.

Subcloning the cDNA fragment

The RT-PCR product (1.8 kb) was gel purified and then subcloned using pGEM^R-T vector cloning system (Promega) following the kit's instructions. Confirmation of the result was by PCR using the original primers and by sequencing. These cDNA clones were then used for making cRNA.

Nucleotide sequence analysis

Both the plasmid DNA product and the amplified PCR product were sequenced (Applied Biosystems) (model 373) by the dideoxy nucleotide chain termination method. A SeqEd 1.03 and BLAST search enabled the sequences to be aligned and compared.

cRNA

cRNA was made from the cDNA clones using mMESSAGE mMACHINE (Ambion) capping kit and SP6 RNA polymerase.

The cDNA was linearised with the restriction enzyme Sac II and infilled with T$_4$ DNA polymerase. The enzymes were inactivated by digestion with proteinase K. The purified cDNA was then capped and transcribed using mMESSAGE mMACHINE and SP6 RNA polymerase. The resultant cRNA was used for injection into *Xenopus* oocytes.

cRNA for the gene encoding alkaline phosphatase (American Tissue Culture Collection) was also made. This was injected as a control to confirm that injected cRNA was being expressed by the oocytes[19].

Northern blot analysis

Poly(A)$^+$mRNA was separated by electrophoresis in 1% agarose/formaldehyde denaturing gels and transferred onto a Hybond-N nylon membrane (Amersham). Both DNA and riboprobes were made. The blots were probed with ^{32}P d-CTP pTau (a gift from Dr M. Kwon) by random prime labelling using NE-Blot (New England Biolabs).

Another DNA probe was made from the cDNA clones made from the RT-PCR 1.8 kb purified DNA. The DNA was cloned as above, cut with Sac II restriction enzyme (New England Biolabs) and labelled using NE-Blot.

In each case the blots were hybridised overnight at 42°C, washed at medium stringency (2SSC, 0.5SDS and 0.1SSC, 0.1SDS) and exposed to X-OMAT (Kodak) film.

The relative abundance of the mRNA in each blot was determined by probing the amount of β-actin mRNA on the same membrane.

Microinjection of *Xenopus* oocytes

Adult female *Xenopus* were anaesthetised with MS222 and a portion of the ovary removed from a ventral incision in the abdomen, and transferred to Barth's solution [NaCl, 88 mmol l^{-1}; KCl, 1 mmol l^{-1}; NaHCO$_3$ 4 mmol l^{-1}; HEPES, 12 mmol l^{-1}; Ca(NO$_3$)$_2$, 0.3 mmol l^{-1}; CaCl$_2$, 0.41 mmol l^{-1}; MgSO$_4$, 0.82 mmol l^{-1}; penicillin, 10 μg ml^{-1}., streptomycin sulfate, 10 μg ml^{-1}, gentamycin sulfate, 100μg ml^{-1}; nystatin, 10U ml^{-1}]. Late stage oocytes were separated using collagenase (1mg ml^{-1}) solution. After overnight incubation to separate damaged oocytes, 34 ng nl^{-1} using a Drummond microinjector. Control oocytes were injected with an equal amount of DEPC-treated water. Oocytes were left for 3 days or more to allow expression to develop. Electrophysiological experiments were carried out on eggs left between 3-6 days, using a two-electrode voltage clamp. During initial experiments, oocytes were bathed in a standard NaCl buffer solution [NaCl, 96 mmol l^{-1}mmol l^{-1}; CaCl$_2$mmol l^{-1}; MgCl$_2$, 1 mmol l^1; HEPES, 10 mmol l^{-1}, adjusted to pH 7.4 with NaOH]. In later experiments, HEPES was omitted and the solution was used unbuffered. Oocytes were impaled with microelectrodes bevelled back to give a resistance of 2-5 MegOhm, when filled with 3 mol KCl. Membrane voltage was allowed to stabilize after impalement, usually within 15 minutes, and only those oocytes giving a stable potential of –50 mV or more were used for experiments. Membrane voltage (V$_m$) was then clamped at –50 mV, and steady-state current-voltage relationships were obtained by stepping V_m for 100 ms from the holding potential to various test potentials ranging from -150 mV to $+30$ mV in 20 mV steps. The currents were digitised at 100 μs intervals, and the average current, at 100 ms, of three sweeps was recorded. Control of experiments and analysis of current-voltage relationships was performed using Pclamp6 software.

Uptake of ^{3}H-taurine into oocytes.

3 days after cRNA injection into *Xenopus* oocytes, single oocytes were used for determination of taurine uptake. Oocytes were incubated for 4 hours in standard saline containing ^{3}H-taurine. For experiments with drugs and inhibitors, the bathing solution contained taurine at a concentration of 0.1 mmol, for periods up to 4 hours. At the end of each experiment the oocytes were washed rapidly three times in taurine-free solution, external fluid blotted free, and, after lysing in 10% sodium dodecyl sulphate (SDS), activity was measured in a TopCount scintillation counter (Packard).

Activators of protein kinases used were (i) phorbol 12-myristate 13 acetate (PMA) in DMSO at 1 μmol, (ii) sn- 1,2-dioctanoylglycerol (DOG) in DMSO at 1 μmol, (iii) okadaic acid (OK) in DMSO at 100 nmol, (iv) 8-bromo-adenosinecyclic 3', 5' monophosphate (cAMP) at 100 μmol. In addition, cytocholasin D at 20 μmol, and (viii) colchicine at 20 μmol were used together with PMA to examine the effects of cytoskeletal disruption on taurine uptake.

RESULTS

Serum and tissue analysis in chronic hyperosmolality

Serum osmotic pressure and ion concentrations of cardiac tissue of rats subjected to chronic hyperosmolality are shown in Table 1. Cellular shrinkage by elevation of serum NaCl or glucose over a 4-5 day period is accompanied by an accumulation of cellular osmoles and, in consequence, a partial mitigation of cellular shrinkage. In contrast, elevation of serum osmolality by heightened urea levels following bilateral nephrectomy does not result in either cellular shrinkage nor the accumulation of intracellular ninhydrin positive substance.

Table 1. The effect of chronic hyperosmolality on the composition of rat cardiac muscle. NPS = ninhydin-positive substance. Modified from Bedford and Leader[13].

Treatment	Tissue water Kg kg^{-1} d.wt.		Tissue content mmol kg^{-1} d.wt.			ΣOsm mOsm kg^{-1} d.wt.	ΔOsm mOsm kg^{-1} d.wt.	ΔNPS mmol kg^{-1} d.wt.
		Na	K	Cl	NPS			
Control	3.32±0.04	195±3	352±6	126±3	321±10	971±14		
Hypernatraemia	3.25±0.02	214±4	392±6	159±6	381±11	1069±17	+98	+60
Diabetic	3.25±0.03	163±5	323±6	116±2	388±18	1004±17	+30	+93
Uremic	3.25±0.04	180±6	353±11	132±6	322±15	1141±15	+194	+7

Amino acid analysis of cardiac tissue

Table 2 shows the results of amino acid analysis of rat cardiac ventricle before and after exposure to hyperosmolality. It can readily be seen that in hypernatraemic animals there is a significant increase in several amino acids, but that taurine is not only the most abundant amino acid identified, but also shows the largest increase.

Table 2. The amino acid content [mmol kg^{-1} dry weight] of cardiac muscle of rats made chronically hyperosmolal. Modified from Bedford and Leader [13]

Amino acid	Control	Hypernatraemic	Diabetic	Uremic
Alanine	16.2	21.2	12.6	8.8
Aspartate	25.1	44.7	6.9	7.3
Glutamate	40.2	58.3	58.8	28.9
Glycine	3.6	9.2	3.1	3.4
Histidine	12.9	17.0	6.9	-
Lysine	3.2	12.1	7.1	2.4
Serine	15.8	20.4	17.6	6.5
Threonine	1.2	7.0	2.3	2.7
Taurine	144.8	177.1	144.2	170.0
Total identified	263	367	260	230

Poly(A)$^{+}$mRNA

The amount of poly(A)$^{+}$mRNA extracted from one heart was between 1-10 μg. Initial experiments in which samples of poly(A)$^{+}$mRNA (30-70 ng) were injected into *Xenopus* oocytes showed very weak expression of the taurine transporter as estimated by measurement of current-voltage relations in the presence and absence of taurine.

Sequence of PCR product and cDNA

The purified PC R 1.8 kb product and the cDNA clones derived from this were found to be homologous with that obtained from rat brain[18]. This has 88% homology with that cloned from the canine renal cells (MDCK)[9].

Expression of cRNA in *Xenopus* oocytes

Electrophysiology

Expression of the taurine transporter, assessed electrophysiologically, was weak. Four days after injection of 34 ng of cRNA, oocytes clamped to −150 mV showed an increased current, generally about 20 –30 nA. Only occasionally was this exceeded. This was not affected by increasing either the amount injected or the incubation time. For this reason further experiments were carried out using isotopic flux measurements.

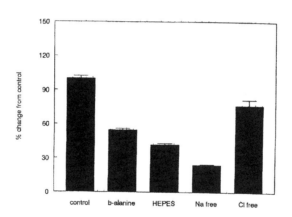

Figure 1. The effects of β-alanine, HEPES buffer, Na-free and Cl-free media on the uptake of ^{3}H-taurine by *Xenopus* oocytes. Results are expressed as % change from control ± S.E.M.

[3]H-taurine uptake into oocytes

(i) In a Na-free bathing medium the uptake of ^{3}H-taurine was found to be between 700 and 1100 fmol oocyte[-1] hr[-1] (n = 36) (Fig 1). This was dependent on the presence of sodium and chloride ions. Replacement of sodium in the bathing medium by choline reduced the taurine flux to 24.1%, while substitution of gluconate for chloride resulted in a reduction of the taurine flux to 76.1% of the control value. Figure 1 also demonstrates that the uptake of taurine was inhibited by the presence of β-alanine; 100 μmol β-alanine resulting in a reduction of the taurine flux to 35% of its original value. In addition, HEPES buffer at 10 mmol strongly inhibited the uptake of taurine, reducing uptake to around 31% of its value in control oocytes. For this reason, HEPES was omitted from salines used in later experiments.

(ii) Uptake of taurine by cRNA-injected oocytes is concentration-dependent. Figure 2 demonstrates that uptake has a K_m of about 20 μmol, a value close to that reported for other taurine transporters.

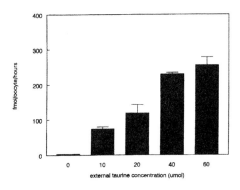

Figure 2. The effect of increasing taurine concentration in the bathing medium on the uptake of ^{3}H-taurine by cRNA-injected *Xenopus* oocytes. Uptake of taurine is expressed as fmol oocyte^{-1} hr^{-1} ± S.E.M.

(iii) Activators of protein kinases.

Figure 3 shows the effect of a number of activators of protein kinases on the uptake of taurine by cRNA-injected oocytes. 8-Br-adenosine-cyclic 3'5' monophosphate, a membrane soluble analogue of cAMP, which activates the cAMP-dependent protein kinase (PKA), at a concentration of 100 μmol in the bathing fluid, caused a reduction in taurine uptake of 51 ± 2%. Similarly, sn-1,2,dioctanoyl glycerol (DOG), an activator of the Ca^{++}/diacylglycerol-dependent protein kinase (PKC) at a concentration of 1 μmol, resulted in a reduction of taurine uptake to 78 ± 3% of control values. A similar result was obtained by the addition of phorbol 12-myristate 13 acetate (PMA)[1 μmol], which is also a potent stimulator of PKC. In this case uptake of taurine was reduced to 56 ± 5% of control values. Finally, the protein phosphatase inhibitor, okadaic acid (OK), at a concentration of 0.1 μmol, caused a reduction in taurine uptake to about 65% of control values.

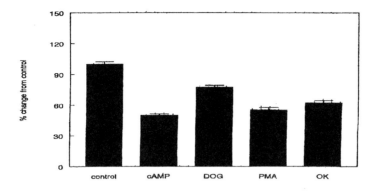

Figure 3. The effects of protein kinase activators on uptake of ³H-taurine in cRNA-injected *Xenopus* oocytes. Results are expressed as % change from control ± S.E.M.

(iv) Disruption of the cytoskeleton.

Both cytochalasin D (20 μmol) and colchicine (20 μmol) caused a large reduction in taurine uptake, to about 20% and 30% of control values respectively (Fig 4). Oocytes exposed to these doses were fragile and easily damaged. When oocytes exposed for 1 hour to cytochalasin D were additionally treated with PMA (1 μmol) nearly all of them disintegrated. The finding that the flux of taurine was not significantly reduced must therefore be treated with caution.

(v) Effect of hypernatraemia upon the expression of TAUT mRNA in cardiac tissue *in vivo*.

Hearts removed from rats made hypernatraemic by intraperitoneal injection of NaCl, or by the addition of taurine to the drinking water, were pulverised in liquid nitrogen and their constituent mRNA extracted and subjected to Northern blot analysis using a DNA probe from a pTau clone (the gift of Dr H.M.Kwon), and a probe derived from the RT-PCR 1.8kb product. Both probes hybridised with a band between 6.8 and 7 kb.

DISCUSSION

It has been known for some years that in hypernatraemia the rat heart accumulates taurine to high levels. It was surprising therefore that when Uchida *et al.*[9] probed dog tissues for mRNA coding for the sodium- and chloride- dependent taurine co-transport which they had isolated from MDCK cells, they found only low levels of mRNA in the heart. This led them to suggest that there might be either another transporter for taurine in heart or another mechanism leading to the substantial accumulation of taurine by this tissue.

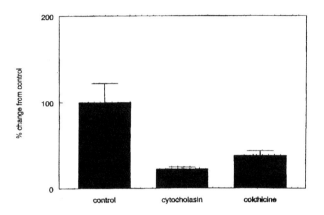

Figure 4. The effects of cytochalasin D and colchicine on uptake of ^{3}H-taurine by cRNA-injected *Xenopus* oocytes following stimulation with phorbol ester. Results are expressed as % unstimulated control ± S.E.M..

Our data confirm that the taurine transporter TAUT is present in rat heart, and that its sequence is homologous to the taurine transporter isolated from rat brain by Smith *et al.*[18]   Preparation of cRNA from rat heart TAUT, followed by injection into *Xenopus* oocytes gave demonstrable uptake of labelled taurine, but the expression was too weak for analysis by electrophysiological methods.

Tracer flux measurements, however, revealed that the taurine transporter had similar properties to those previously described. It showed sodium- and chloride-dependency, and was competitively inhibited by β-alanine.   The K_m of the transporter when expressed in *Xenopus* oocytes was about 20 μmol.

In the presence of activators of PKA and PKC, taurine uptake by oocytes is inhibited by between 30 and 50%. Similar inhibitory effects of PKA and PKC on taurine uptake have been found in a number of cultured cell lines, including Caco-2, HT-29 and LLC-PK$_1$ cells[14,20-22] and in *Xenopus* oocytes expressing mouse TAUT[23].   Involvement of protein phosphorylation is further suggested by the fact that okadaic acid, a potent inhibitor of protein phosphatase,   which   presumably   disrupts   the   pattern   of   protein phosphorylation within cells, also inhibits taurine uptake.

Loo *et al.*[23], on the basis of electrophysiological analysis, conclude that the effect of activation of PKA and PKC is not a direct one on the transporter itself, although several putative phosphorylation sites on the molecule have been noted[9,10].   They suggest that modulation of taurine uptake by oocytes, consequent upon activation by protein kinases, is by an alteration of the number of transporters in the membrane, rather than by alteration in the kinetic properties of inserted transporters. In contrast, Tchoumkeu-Nzouessa and Rebel[22] found that disruption of protein synthesis or interference with microtubular or microfilament function had no effect on the PMA-induced inhibition of taurine transport in cultured glial cells, and that the kinetic properties of the transporter were influenced by phosphorylation. These alternative findings are not mutually exclusive, nor do they rule out the possibility that control mechanisms in different tissues are mediated in different ways. This is underlined by the demonstration that, even though both astrocytes and neurons from primary rat brain cultures demonstrate the presence of taurine uptake from the medium, activation of PKC inhibits taurine uptake by astroglia but has no effect on uptake by neurons.

Hypernatraemia elevates taurine levels in rat heart. However, this accumulation is apparently not the result of increased mRNA production, since this remains unchanged. The increase must therefore be due to either increased activity of existing transporters or recruitment of inactive transporters. Our results to date do not allow distinction between these possibilities.

ACKNOWLEDGEMENTS

This work was supported by the Health Research Council of New Zealand and the Lottery Grants Board (Health Research). We would also like to thank Dr Moo Kwon, John Hopkins University, for the gift of pTAU.

REFERENCES

1. Huxtable, R.J., 1992, Physiological actions of taurine. *Physiol. Rev.* **72**: 101-163.
2. Ballantyne, J.S. and Chamberlain, M.E., 1994, Regulation of cellular amino acid levels. In: *Cellular and Molecular Physiology of Cell Volume Regulation*, (K. Strange, ed.), CRC Press, Boca Raton, Ann Arbor, Tokyo, London, pp. 111-122.
3. Garcia-Perez, A. and Burg, M., 1991, Renal medullary organic osmolytes. *Physiol. Rev.* **71**: 1081-1115.
4. Nakanishi, T. Uyama, O. and Sugita, M., 1991, Osmotically related taurine content in rat renal inner medulla. *Am. J. Physiol.* **261**: F957-F962.
5. Thurston, J.H., Hauhart, R.E. and Naccarato, E.F., 1981, Taurine: Possible role in osmotic regulation of mammalian heart. *Science* **214**: 1373-1374.
6. Gaull, G.E., 1982, Taurine in the nutrition of the human infant. *Acta Paediat. Scand.* (suppl.) **296**: 38-47.
7. Pasantes-Morales, H. and Cruz, C., 1984, Protective effects of taurine and zinc on peroxidation-induced damage in photoreceptor outer segments. *J. Neurosci. Res.* **11**: 303-311.

8. Scheibel, J., Elsasser, T., Brown, B., Dom, R. and Ondo, J., 1984, The stimulation of prolactin secretion by taurine: studies on the site of action. *Brain Res.* **13**: 49-52.
9. Uchida, S., Kwon, M., Yamauchi, A., Preston, A.S., Marumo, F. and Handler, J., 1992, Molecular cloning of the cDNA for an MDCK cell Na^+ and Cl^- dependent taurine transporter that is regulated by hypertonicity. *Proc. Nat.Acad. Sci.* USA **89**: 8230-8234.
10. Liu, Q.R., Lopez-Corcuera, B., Nelson, H., Mandiyan, S. and Nelson, N., 1992, Cloning and expression of a cDNA encoding the transporter of taurine and β-alanine in mouse brain. *Proc. Nat. Acad.Sci.* USA **89**: 12145-12149.
11. Jhiang, M., Fithian, L., Smanik, P., McGill, J., Tong, Q. and Mazzaferri, E.L., 1993, Cloning of the human taurine transporter and characterisation of taurine uptake in thyroid cells. *FEBS Lett.* **318**: 139-144.
12. Ramamoorthy, S., Leibach, F.H., Maheshy, V.B., Han, H., Yang-Feng, T., Blakely, R.D and Ganapthy, V., 1994, Functional characterisation and chromosomal location of a cloned taurine transporter from human placenta. *Biochem. J.* **300**: 893-900.
13. Bedford, J. and Leader, J.P., 1993, Responses of tissues of the rat to anisosmolality *in vivo. Am. J. Physiol.* **264**: R1164-R1179.
14. Jones, D.P., Miller, L.A. and Chesney, R.W., 1990, Adaptive regulation of taurine transport in two continuous renal epithelial cell lines. *Kidney Int.* **38**: 219-226.
15. Fujita, T. and Sato, Y., 1988, Hypotensive effect of taurine. Possible involvement of the sympathetic nervous system and endogenous opiates. *J. clin. Invest.* **82**: 993-997.
16. Kramer, J. Chovan, J. and Schaffer, S., 1981, Effect of taurine on calcium paradox and ischaemic heart failure. *Am. J. Physiol.* **240**: H238-H246.
17. Hubbard, M.J., 1995, Calbindin$_{28kDa}$ and calmodulin are hyperabundant in rat dental enamel cells. Identification of the protein phosphatase calcineurin as a principal calmodulin target and of a secretion-related role for calbindin$_{28kDa}$. *Eur. J. Biochem.* **230**: 68-79.
18. Smith, K.E., Borden, L.A., Wang, C.D., Hartig, P.R., Branchek, TA.. and Weinshank, R.L. 1992, Cloning and expression of a high affinity taurine transporter from rat brain. *Mol. Pharmacol.* **42**: 563-569.
19. Tate, S.S., Urade, R., Micanovic, R., Gerber, L. and Udenfriend, S., 1990, Secreted alkaline phosphatase : an internal standard for the expression of injected mRNAs in the *Xenopus* oocyte. *FASEB J.* **4**: 227-231.
20. Brandsch, M., Miyamoto, K., Ganapathy, V. and Leibach, F.H., 1993, Regulation of taurine uptake in human colon carcinoma cell lines (HT-29 and Caco-2) by protein kinase C. *Am. J. Physiol.* **264**: G939-G946.
21. Ganapathy, V. and Leibach, F.H., 1994, Expression and regulation of the taurine transporter in cultured cell lines of human origin. *Adv. Exp. Med. Biol.* **359**:51-57.
22. Tchoumkeu-Nzouessa, G.C. and Rebel, G., 1996, Regulation of taurine transport in rat astrocytes by protein kinase : role of calcium and calmodulin. *Am. J. Physiol.* **270**: C1022-C1028.
23. Loo, D.F., Hirsch, J.R., Sarkar, H.K. and Wright, E.M., 1996, Regulation of the mouse retinal taurine transporter (TAUT) by protein kinases in *Xenopus* oocytes. *FEBS Lett.* **392**: 250-254.

IDENTIFICATION OF PROMOTER ELEMENTS INVOLVED IN ADAPTIVE REGULATION OF THE TAURINE TRANSPORTER GENE: ROLE OF CYTOSOLIC Ca^{2+} SIGNALING

Xiaobin Han, Andrea M. Budreau, and Russell W. Chesney

Department of Pediatrics, University of Tennessee, and the Crippled Children's Foundation Research Center at Le Bonheur Children's Medical Center, Memphis, TN

INTRODUCTION

The renal tubular epithelium adapts to alterations in the sulfur amino acid composition of the diet[3]. Adaptive regulation of taurine transport following alterations in external taurine concentration has been described in man, cat, mouse, rat, dog, and pig[4,7-9,11-13,15,20]. This adaptive response has been well characterized using an LLC-PK1 cell line derived from the proximal tubule of pig kidney[8]. However, the signal pathway through which the taurine transporter gene is regulated by the availability of dietary taurine is unknown. The calcium ion (Ca^{2+}) is a fundamental agent for cell signaling, and there exist numerous examples of taurine-calcium interactions[14,16]. The present study demonstrates that the Ca^{2+} signal may play an important role in the renal dietary adaptive regulation of taurine transporter gene expression demonstrated in LLC-PK1 cells.

METHODS AND MATERIALS

Cloning of the 5'-Flanking Region of the Rat Taurine Transporter Gene

The rat P1 genomic DNA library was screened (Genome Systems, Inc., St. Louis, MO) by polymerase chain reaction (PCR) using three sets of

Taurine 4, edited by Della Corte *et al.*
Kluwer Academic / Plenum Publishers, New York, 2000.

535

oligonucleotide primers based on the DNA sequence of rat brain taurine transporter cDNA (rB16a)[17]. The primers used for generating the PCR probes were sense primer 5'-GCCAACGCCGCGATCGCCGCCAA-3'/anti-sense primer 5'-CTCCTCGTTTTGCTTGAGAGGC-3'; sense primer 5'-AT GGCCACCAAGGAGAAGCTTCAA-3'/antisense primer 5'GAGAAATGC ACCTCCACCAT-3'; and sense primer 5'-TCAGAGGGAGAAGTGGTCC AGCAAGA-3'/antisense primer 5'-AGCCAGACACAAAACTGGTACCA-3'. A full length (~28 kb) rat taurine transporter gene (TauT) was isolated from the rat P1 library, as determined by Southern blot analysis using 5'-, 3'- and internal probes of rB16a. The 28 kb DNA was digested using Hind III restriction enzyme, subcloned into pBluescript vector (Promega, Madison, WI), and transferred into JM109 bacteria by transformation. The DNA was isolated and analyzed by Southern blot using a 5' probe (5'-AGCCAGGTCCCGGAGTACGA-3') of rB16a generated by 5'-rapid amplification of rat taurine transporter cDNA. A 7.2 kb Hind III fragment containing the 5'-flanking region of the gene was isolated and sequenced using an automatic DNA sequencer.

Construction of 5'-Deletion DNA Constructs

Plasmids containing nested deletions of the proximal 5'-flanking region of TauT were generated by PCR using the 7.2 kb 5'-flanking region DNA as the template. The conditions used were 40 cycles of 1 min of denaturation at 94°C, 1 min of annealing at 58°C, and 1 min of elongation at 72°C. The sense primer designed for PCR contains a unique site for kpnI and the antisense primer contains a unique site for BagI II. The sense primers are 5'-GGGGTACCGAGTTGGGGAGGGA-3'; 5'-GGGGTACCTTACTGAAG GTCACACAG-3'; and 5'-GGGGTACCTTCCCAGGTTTCCGAT-3'. The antisense primer is 5'-AAGATCTTGGCACGGGAGTTCA-3'. PCR products were digested with kpnI and BagI II and re-ligated into the kpnI and BagI II unique sites of pGL3-Basic to generate plasmids containing segments of the taurine transporter gene extending from the transcriptional start site. The constructs were verified by DNA sequencing.

Cell Culture, Taurine Uptake, and Transient Transfection

Cells of the porcine proximal tubule cell line LLC-PK1 (American Type Culture Collection, Rockville, MD) were grown as confluent monolayers in 10 cm diameter tissue culture plates in Dulbecco's minimum essential medium (DMEM):Ham's F_{12} (1:1) with 10% fetal calf serum (FCS) at 37°C in the presence of 5% CO_2 in a humidified incubator. In some experiments, Madin-Darby canine kidney (MDCK) cells, of distal tubule origin, were also used. In experiments using ionomycin, cells were cultured in S-MEM (Ca^{2+}-

free medium) with 10% dialyzed FCS, which does not contain taurine. Taurine uptake was performed as described previously[10]. For transient transfection studies, cells were plated 18 h before transfection and fed with fresh medium 4 h before the procedure. Luciferase reporter plasmids were introduced into the cultured mammalian cells using cationic liposomes (LipofectAMINE, Life Technologies, Grand Island, NY). The transfection was carried out for 16-18 h. Cells were washed twice with phosphate buffered saline (PBS) and incubated in fresh medium for 24-48 h before harvesting. PGL3-control, which contains a luciferase gene driven by the SV40 early region promoter/enhancer, and empty pGL3-Basic vectors were used as positive and negative controls, respectively. To standardize transfection efficiency, 0.1 µg of pRL-CMV vector (pRL Renilla Luciferase control reporter vector, Promega) was co-transfected in all experiments. Transfected cells were harvested 48 h after transfection and lysed in 200 µl of reporter lysis buffer. A luciferase assay was performed using a dual-luciferase assay kit (Promega), and luciferase activity was measured with an Optocomp 1 luminometer (MGM Instruments, Inc., Hamden, CT). Protein concentrations of the cell extracts were determined with the Bradford method (1976), using the Bio-Rad protein assay reagent (Bio-Rad, Hercules, CA).

Manipulation of Taurine and Calcium Availability

Experiments were performed with LLC-PK1 cells and with cells that were transiently transfected with DNA contructs. LLC-PK1 cells were treated with ionomycin (10-80 nM) for 24 h, then harvested for preparation of total cellular RNA. For chelation experiments, EGTA (10 µM) was added 20 min before taurine was added to the medium. Cells were continuously cultured in medium containing 0 or 500 µM taurine for 24 h, then cells were harvested for preparation of total cellular RNA. Northern blot analysis was carried out using a specific RNA probe generated from LLC-PK1 cell taurine transporter cDNA. ß-Actin was used as an internal control for loading. For adaptive regulation experiments, DNA constructs were transiently transfected into LLC-PK1 cells, then cells were cultured in serum-free medium containing 0 µM, 50 µM, or 500 µM taurine for another 48 h. The luciferase assay was performed in cell lysates. To control for transfection efficiency, cells were co-transfected with pRL-CMV vector, and luciferase activity was measured in cell lysates by the dual-luciferase reporter assay system. The basal promoter activity of each construct is represented by relative light output normalized to pRL-CMV control.

Statistical Analysis

The mean data from independent experiments using different cell preparations are reported. Error bars represent standard error of the mean (SEM).
Data were analyzed using Student's *t*-test for unpaired data. Statistical significance was defined as $p < 0.05$.

RESULTS

Regulation of Taurine Uptake by Ionomycin

To test if the concentration of intracellular calcium ion ($[Ca^{2+}]_i$) regulates
taurine uptake by LLC-PK1 cells, cells were treated with the Ca^{2+} ionophore
ionomycin and taurine uptake was measured. Figure 1 shows the dose-
dependent increase in taurine uptake by LLC-PK1 cells in the presence of
ionomycin. The adaptive response of taurine uptake to the concentration of
extracellular taurine was maintained. However, depletion of $[Ca^{2+}]_i$ by
treatment with 10 mM ionomycin and chelation with 500 mM EGTA
blocked the adaptive regulation of the taurine transporter in these cells (Fig.
2).

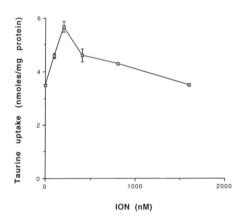

Figure 1. Effect of ionomycin on taurine uptake in LLC-PK1 cells cultured in Ca^{2+}-free medium for 48 h prior to measurement of taurine uptake. Values are mean ± SEM of 3 samples.

Regulation of Taurine Transporter Gene Expression by Ion-
omycin

To further explore the role of the Ca^{2+} signal in the renal adaptive regulation of taurine transporter gene expression by diet, Northern blot analysis

was performed using a specific probe for LLC-PK1 cell taurine transporter mRNA. As shown in Figures 3 and 4, and corresponding to the observed effect on taurine uptake, increasing concentrations of ionomycin caused a dose-dependent increase in mRNA expression, while depletion of intracellular calcium with EGTA blocked the adaptive response.

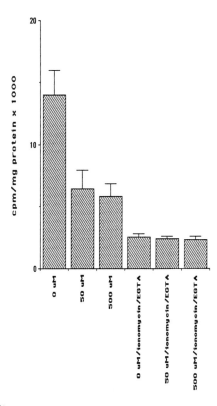

Figure 2. Effect of Ca^{2+} depletion on adaptive regulation of the taurine transporter in LLC-PK1 cells. Cells were cultured in Ca^{2+}-free medium containing 0, 50 mM, or 500 mM taurine with or without ionomycin plus EGTA as indicated for 48 h, then taurine uptake was measured. Values are the mean ± SEM of three samples. *p<0.01 vs control (50 mM taurine).

Determination of Cis-Elements for Renal Adaptive Regulation of TauT

To test whether the 5'-flanking region of TauT contains taurine-responsive element(s), transient transfection studies of luciferase reporter constructs (derived from pGL3-Basic) p-124, p-189, p-269, p-574, p-963 and p-1532 were performed in LLC-PK1 cells. To examine the effect of medium taurine on TauT promoter activity, transfected cells were cultured in serum-free medium containing 0 μM, 50 μM, or 500 μM taurine for 48 h, then a luciferase assay was performed. As shown in Figure 5, TauT promoter activity was found to be adaptively regulated by medium taurine in cells transfected with constructs p-963 and p-1532. In these cells, TauT promoter activity was more than doubled when cultured in 0 μM taurine, but decreased by half when cultured in medium containing 500 μM taurine, as compared with control (50 μM taurine). These results indicate that at least one <u>t</u>aurine <u>re</u>sponsive <u>e</u>lement **(TREE)** exists between -574 and -1532 of the 5'-flanking region of the rat taurine transporter gene (TauT).

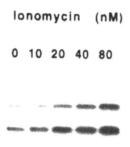

Figure 3. Up-regulation of taurine transporter mRNA by ionomycin. Northern blot analysis of LLC-PK1 total cellular RNA after treatment with ionomycin (10-80 nM) for 24 h. Lane 1: 0 nM ionomycin (ION); lane 2: 10 nM ION; lane 3: 20 nM ION; lane 4: 40 nM ION; lane 5: 80 nM ION. ß-Actin was used as an internal control for loading.

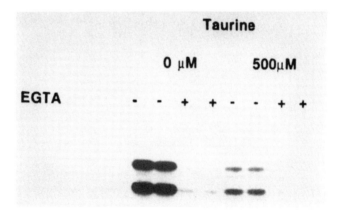

Figure 4. Effect of chelation of calcium with EGTA on adaptive regulation of taurine transporter in LLC-PK1 cells. Northern blot analysis of LLC-PK1 total cellular RNA. Lane 1-2: 0 μM taurine; lane 3-4: 0 μM taurine/EGTA; lane 5-6: 500 μM taurine; lane 7-8: 500 μM taurine/EGTA. ß-Actin was used as an internal control for loading.

Differential Regulation of TauT Promoter by Ionomycin and cAMP in MDCK and LLC-PK1 Cells

To determine whether Ca^{2+} signaling is involved in the renal adaptive regulation of TauT, LLC-PK1 and MDCK cells transfected with p-1532 were tested. Conditions for cell culture and transfection were identical to those described in the legend to Figure 3. In this study, transfected cells were treated with ionomycin (50 nM) and cAMP (500 mM) for 24 h in culture. As shown in Figure 6, ionomycin (50 nM) increased the promoter activity of TauT in LLC-PK1 cells as compared with control, but had no effect on TauT promoter activity in MDCK cells. In contrast, TauT promoter activity was increased by cAMP in MDCK cells, but not in LLC-PK1 cells.

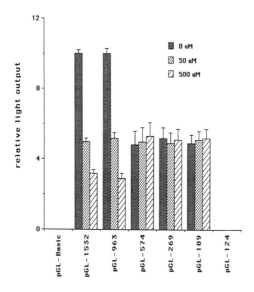

Figure 5. Adaptive regulation of TauT promoter activity by medium taurine. DNA constructs were transiently transfected into LLC-PK1 cells, then cells were cultured in serum-free medium containing 0 μM, 50 μM, or 500 μM taurine for 48 h. The luciferase assay was performed with the cell lysates. To control for transfection efficiency, cells were co-transfected with pRL-CMV vector, and luciferase activity was measured by the dual-luciferase reporter assay system. The basal promoter activity of each construct is represented by relative light output normalized to pRL-CMV control.

DISCUSSION

In most biologic systems, Ca^{2+} is a fundamental agent for cell signaling, and there exist numerous examples of taurine-calcium interactions. For instance, studies have shown that taurine regulates myocardial Ca^{2+} homeostasis through direct or indirect modulation of several key calcium transport systems[16]. Long-term exposure of heart myocytes to taurine decreased both nuclear and cytosolic Ca^{2+} without significantly changing either nuclear or cytosolic Na^+ levels[14]. Changes in cytoplasmic Ca^{2+} levels are involved in the regulation of a number of mammalian genes. The recovery process following injury to renal proximal tubules involves the up-regulation of several genes that promote cell viability and enhance cell survival[1]. The increase in intracellular Ca^{2+}, while cytotoxic in some cases, appears to be the activating signal for the expression of genes involved in the stress response to hypoxia[18]. In epithelial cells, among others, increases in calcium activate expression of the protooncogenes c-myc, c-fos, and c-jun[19]. Dominguez *et al*[5] found that elevation of Ca^{2+} caused by the calcium ionophore A23187 activated expression of the GLUT1 gene in LLC-PK1 cells.

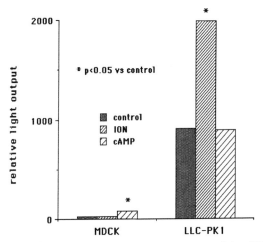

Figure. 6. Effect of ionomycin and cAMP on TauT promoter activity. DNA constructs were transiently transfected into LLC-PK1 cells 2 h before ionomycin and/or cAMP were added, then cells were cultured in medium containing 50 μM taurine for 48 h. The luciferase assay was performed using cell lysates. To control for transfection efficiency, cells were co-transfected with pRL-CMV vector, and luciferase activity was measured by the dual-luciferase reporter assay system. The basal promoter activity of each construct is represented by relative light output normalized to pRL-CMV control.

Previous studies have shown that adaptive regulation of the taurine transporter occurs at the transcriptional level in both MDCK and LLC-PK1 cells[7]. However regulation of taurine uptake by medium osmolarity is only observed in MDCK cells[10], suggesting that the mechanisms for adaptive regulation of gene expression may be different. In these studies, the transient transfection model was used to investigate the roles of ionomycin and cAMP on activation of TauT gene transcription in the two cell lines. Elevated cytosolic Ca^{2+} induced by ionomycin up-regulated transcription of the TauT promoter in LLC-PK1 cells but not in MDCK cells, while cAMP increased transcription of the TauT promoter in MDCK cells but not in LLC-PK1 cells, illustrating that different mechanisms are involved in the signaling pathways.

Our findings suggest that the 5'-flanking region of the rat taurine transporter gene (TauT) contains at least one *cis* element that is involved in the adaptive regulation of TauT expression by dietary taurine through cytosolic Ca^{2+} signaling. Addition of a calcium ionophore increased expression of the taurine transporter in a dose-dependent manner in LLC-PK1 cells, while chelation of calcium with EGTA blocked the adaptive regulation of taurine transporter gene expression in the cells. This calcium signaling appears to act at a site between nucleotides -574 and -964 in the 5'-flanking region upstream of the gene. Since taurine is known to decrease both nuclear and cytosolic Ca^{2+}, the absence of taurine may result in increased Ca^{2+} entry leading to the up-regulation of taurine transporter expression.

Hypoxic injury to the kidney is associated with aminoaciduria and taurinuria, and with enhanced Ca^{2+} entry[6]. Conversely, in the presence of high taurine, calcium entry may be reduced, and taurine transporter expression may be down-regulated. The present study demonstrates that the Ca^{2+} signal plays an important role in the renal dietary adaptive regulation of taurine transporter gene expression demonstrated in LLC-PK1 cells. The response of the rat renal taurine transporter gene to Ca^{2+} stress may represent a role for adaptive regulation of the taurine transporter in the recovery of renal cells from injury.

REFERENCES

1. Bashan, N., Burdett, E., Guma, A., Sargeant, R., Tumiati, L., Liu, Z., and Klip, A., 1993, Mechanisms of adaptation of glucose transporters to changes in the oxidative chain of muscle and fat cells, *Am. J. Physiol.,* 264(2 Pt 1):C430-C440.
2. Bradford, M., 1976, A rapid and sensitive method for the quantitation of microgram quantities of protein utilizing the principle of protein-dye binding, *Anal. Biochem.,* 72:248-254.
3. Chesney, R.W., Gusowski, N., and Friedman, A.L., 1983, Renal adaptation to altered amino acid intake occurs at the luminal brush border membrane, *Kidney Int.,* 24:588-594.
4. Chesney, R.W., Scriver, C.R., and Mohyuddin, F., 1976, Localization of the membrane defect in transepithelial transport of taurine by parallel studies in vivo and in vitro in hypertaurinuric mice, *J. Clin. Invest.,* 57:183-193.
5. Dominguez, J.H., Song, B., Liu-Chen, S., Qulali, M., Howard, R., Lee, C.H., and McAteer, J., 1996, Studies of renal injury: II. Activation of the glucose transporter (GLUT1) gene and glycolysis in LLC-PK1 cells under Ca^{2+} stress, *J. Clin. Invest.,* 98:395-404.
6. Greene, E.L. and Paller, M.S., 1994, Calcium and free radicals in hypoxia/reoxygenation injury of renal epithelial cells, *Am. J. Physiol.,* 266:F13-F20.
7. Han, X., Budreau, A.M., and Chesney, R.W., 1997, Adaptive regulation of MDCK cell taurine transporter (pNCT) mRNA: Transcription of pNCT gene is regulated by external taurine concentration, *Biochim. Biophys. Acta,* 1351:296-304.

8. Han, X., Budreau, A.M., and Chesney, R.W., 1998, Molecular cloning and functional expression of an LLC-PK1 cell taurine transporter that is adaptively regulated by taurine, *Adv. Exp. Med. Biol.,* 442:261-268.

9. Jessen, H. and Jacobsen, C., 1997, Adaptive regulation of taurine and beta-alanine in a human kidney cell line from the proximal tubule, *Biochim. Biophys. Acta,* 1325:309-317.

10. Jones, D.P., Jiang, B., and Chesney, R.W., 1994, Regulation of taurine transport by external taurine concentration and medium osmolality in renal tubular cells in culture, *Adv. Exp. Med. Biol.,* 359:131-138.

11. Jones, D.P., Miller, L.A., and Chesney, R.W., 1995, The relative roles of external taurine concentration and medium osmolality in the regulation of taurine transport in LLC-PK1 and MDCK cells, *Pediatr. Res.,* 37:227-32.

12. Park, T., Rogers, Q.R., Morris, J.G., and Chesney, R.W., 1989, Effect of dietary taurine on renal taurine transport by proximal tubule brush border membrane vesicles in the kitten, *J. Nutr.,* 119:1452-60.

13. Pickett, J.P., Chesney, R.W., Beehler, B., Moore, C.P., Lippincott, S., Sturman, J., and Ketring, K.L., 1990, Comparison of serum and plasma taurine values in Bengal tigers with values in taurine-sufficient and -deficient domestic cats, *J. Am. Vet. Med. Assoc.,* 196:342-346.

14. Rao, M.R. and Tao, L., 1998, Effects of taurine on signal transduction steps induced during hypertrophy of rat heart myocytes, *Adv. Exp. Med. Biol.,* 442:137-144.

15. Satsu, H., Watanabe, H., Arai, S., and Shimizu, M., 1997, Characterization and regulation of taurine transport in Caco-2, human intestinal cells, *J. Biochem.,* 121:1082-1087.

16. Schaffer, S.W., Ballard, C., and Azuma, J. 1994. Mechanisms underlying physiological and pharmacological actions of taurine on myocardial calcium transport. *In* Taurine in Health and Disease. R. J. Huxtable and D. V. Michalk, editors. Plenum Press, New York. 171-180.

17. Smith, K.E., Borden, L.A., Wang, C.D., Hartig, P.R., Branchek, T.A., and Weinshank, R.L., 1992, Cloning and expression of a high affinity taurine transporter from rat brain, *Mol. Pharmacol.,* 42:563-569.

18. Trump, B.F. and Berezesky, I.K., 1992, The role of cytosolic Ca^{2+} in cell injury, necrosis, and apoptosis, *Curr.Opin. Cell Biol.,* 4:227-232.

19. Yamamoto, N., Maki, A., Swann, J.D., Berezesky, I.K., and Trump, B.F., 1993, Induction of immediate early and stress genes in rat proximal tubule epithelium after injury: The significance of cytosolic ionized calcium., *Renal Fail.,* 15:163-171.

20. Zelikovic, I., Chesney, R.W., Friedman, A.L., and Ahlfors, C.E., 1990, Taurine depletion in very low birth weight infants receiving prolonged parenteral nutrition: Role of renal immaturity, *J. Pediatr.,* 116:301-306.

SUPPRESSION OF BLEOMYCIN-INDUCED INCREASED PRODUCTION OF NITRIC OXIDE AND NF-kB ACTIVATION BY TREATMENT WITH TAURINE AND NIACIN

S. N. Giri, G. Gurujeyalakshmi and Y. Wang

Department of Molecular Biosciences, School of Veterinary Medicine, University of California Davis, CA 95616, USA

INTRODUCTION

Bleomycin (BL)- induced pulmonary fibrosis is initially characterized by alveolar inflammation, influx of inflammatory cells followed by progressive proliferation of septal cells, increased production of septal matrix and loss of lung architecture[1,2]. The process of cellular injury in lung fibrosis is thought to be mediated by oxygen radicals produced by infiltrating inflammatory cells. Peroxynitrite is a potent oxidant produced by the rapid reaction of nitric oxide (NO) and superoxide. In addition, BL itself is capable of producing superoxide and hydroxy radicals by binding to DNA:Fe^{2+} and forming a DNA: Fe^{2+} :BL complex which undergoes redox cycling and generates these reactive oxygen species (ROS)[3,4]. Recent evidence suggests that macrophage and neutrophil-derived ROS may have important roles in lung inflammation and fibrosis by stimulating the production of proinflammatory and fibrogenic cytokines that mediate enhanced fibroproliferative response[5-7] . The roles of these cytokines in the BL-induced lung fibrosis have been well characterized and it is commonly

Taurine 4, edited by Della Corte *et al.*
Kluwer Academic / Plenum Publishers, New York, 2000.

understood that not a single cytokine but a network of cytokines controls the inflammatory and fibrotic responses[8].

A role for NO in the physiologic processes of the lung is suggested by the findings of nitric oxide synthase (NOS) activity in lung tissues and by demonstration of vasodilatory effects of inhaled NO in pulmonary vessels and airways and by observations that lung epithelial cells produce factor(s) capable of causing smooth muscle relaxation. In lung tissues, NOS has been localized and identified in two different forms: constitutive form present in endothelial cells[9] and brain[10] and inducible form found in macrophages[11]. Several studies have shown that inducible nitric oxide synthase (iNOS) mRNA expression, protein and nitric oxide production can be induced in macrophages and neutrophils by specific stimuli. It has been demonstrated by Huot and Hacker that macrophages activated in vivo by BL secrete nitrite spontaneously and this secretion can be blocked by a substrate-specific analogue of the L-arginine-dependent effector mechanism, N^6-monomethylarginine[12].

The production of proinflammatory and profibrogenic cytokines in macrophages is controlled in part at the level of gene trascription by a number of DNA binding proteins which interact with specific sequence motifs in the promoter region of the gene. A widely distributed DNA binding nuclear factor kB (NF-kB) which is normally sequestered in the cytoplasm as an inactive multiunit complex bound to inhibitory protein, IkB-α, is known to regulate the production of many cytokines[13]. A number of stimuli including ROS can activate this complex by causing phosphorylation and degradation of IkB-α and thus allowing the translocation of the active dimer into the nucleus, where it binds to the promoter region of genes such as IL-1α, IL-6 and TNF-α containing the NF-kB motif and stimulates the expression of their genes[14,15].

Since a large part of bleomycin (BL)-induced lung damage is due to an excess production of ROS either by macrophages and neutrophils or by the DNA:Fe^{2+}: BL complex, we hypothesized that intratracheal (IT) instillation of BL increases the production of nitric oxide and cytokines secondary to upregulation of inducible nitric oxide synthase (iNOS) and activation of NF-kB in the lungs, respectively; and the antifibrotic effects of the combined treatment with taurine and niacin as demonstrated in our laboratories[16,17] may reside in their ability to suppress the BL- induced upregulation of iNOS and activation of NF-kB. In order to test this hypothesis, we investigated the effects of saline or BL instillation on the expression of iNOS mRNA, protein, NO production, NF-kB activation, IkB-α levels and changes in the levels of some cytokines during the course of development of lung fibrosis with and without taurine and niacin treatment.

METHODS

Animal model

A single dose BL-mouse model of lung injury and fibrosis has been employed in the present study. Briefly, all experiments were carried out in male C57-B6 mice weighing 25-28 g (Simonsen, Gilroy, CA). The mice had access to water and either pulverized Rodent Laboratory Chow 5001(Purina Mills, St. Louis, MO) or the same pulverized chow containing niacin (N) 2.5% (w/w) and taurine (T) 1% in water. Animals were randomly divided into four experimental groups: saline-instilled (SA) with a control diet (CD) and drinking water (SA+CD); saline-instilled with taurine in drinking water and niacin in diet (SA+TN); BL-instilled with the control diet and drinking water (BL+CD); and BL-instilled with taurine in drinking water and niacin in diet (BL+TN). The animals were fed these diets starting 3 days before the intratracheal (IT) instillation and continuing through out the course of the experiment. After mice were anesthetized with ketamine and xylazine, either 50 µl sterile isotonic saline or 0.1 unit of bleomycin sulphate in 50 µl saline per mouse was IT instilled.

BALF collection and lung processing

The mice were sacrificed by an overdose of sodium pentobarbital at 1, 3, 5, 7, 14 and 21 days after IT instillation and broncho-alveolar lavage was carried out as previously described[18]. Briefly, the lung was lavaged with 1 ml of isotonic sterile saline four times. The recovery of the lavaged fluid ranged from 3.0 to 3.6 ml. After the lavage, the lung was dissected out, freeze-clamped and then stored at –80 °C. The BALF was centrifuged at 4 °C for 10 min at 1500 rpm. The supernatant was gently aspirated and stored at -80 °C until used for nitrite and cytokine determination. In another set, the animals were killed by decapitation and the lungs were quickly removed, freeze-clamped and then dropped in liquid N_2 and stored at -80 °C until used for mRNA analysis.

Lung hydroxyproline content

Collagen deposition was estimated by determining the hydroxyproline content of the whole lung. The lung was excised, homogenized, and hydrolyzed in 6N HCl for 16-18 hr at 110 °C. Hydroxyproline content was measured by the colorimetric method of Woessner[19]. Data are expressed as μg of hydroxyproline / lung.

Nitrite assay

Nitrite in the BALF was determined by a spectrophotometric method based on the Griess reaction[20]. Prior to analysis, nitrate in the samples was reduced to nitrite using the enzyme nitrate reductase (Boehringer Mannheim Corp., Indianopolis, IN) 0.1units/ml, in the presence of 50 μM NAPDH, 5 μM flavin-adenine dinucleotide (FAD), 6 μg/ml lactate dehydrogenase and 0.2 mM sodium pyruvate. A total of 150 μl sample was mixed with 150 μl of Griess' reagent (1% sulfanilamide in 5% phosphoric acid and 0.1% napthylenediamide-dihydrochloride in water), incubated at room temperature for 5 min and optical density of the reaction product was read on a plate reader at 550 nm. Nitrite concentrations in the samples were determined from a standard curve generated by different concentrations of sodium nitrate after reduction as the samples.

Cytokine assay

IL-α, TNF-α and TGF-β in the BALF supernatant were assayed by specific enzyme-linked immunosorbent assay kits, obtained from Genzyme. The sensitivities for different cytokines were as follows: IL-α, 15 pg/ml; TNF-α, 15 pg/ml; and TGF-β, 50 pg/ml.

RNA extraction and RT-PCR

Total RNA from the lung was isolated using the RNeasy total RNA extraction protocol (Qiagen, Chatsworth, CA) according to the manufacture's description. The PCR primers for GAPDH and NOS in message amplification were obtained from Clontech Laboratories (Palo Alto, CA). The following 5'-primer and 3'-primer sequences were employed:

GAPDH: (5') primer 5' TGAAGGTCGGTGTGAACGGATTTGGC 3'
           (3') primer 5' CATGTAGGCCATGAGGTCCACCAC 3'
NOS: (5') primer 5' CCCTTCCGAAGTTTCTGGCAGCAGC 3'
           (3') primer 5' GGCTGTCAGAGCCTCGTGGCTTTGG 3'

First strand cDNA synthesis was performed using an Advantage RT-for-PCR kit (Clontech, Palo Alto, CA). After the completion of the first-strand cDNA synthesis, the samples were diluted 1:5 and the PCR reactions were carried out as described previously[21]. Briefly, PCR was performed at 94 °C with initial denaturation for 5 min followed by 30-35 cycles of amplification at 94 °C (45 sec), 60 °C (45 sec) and 72 °C (2min). Finally, the samples were extended at 72 °C for 7 min. The specificity of amplification was checked by assessing whether a fragment of the expected size had been obtained with the positive control. Equal amounts of the PCR-amplified products (10 µl) were run on a 2% agarose ethidium bromide-stained gel. The relative intensity of RT-PCR products for each gene was determined by Bio-Rad Image Analyzer (Bio-Rad, Hercules, CA).

Tissue protein extraction

Frozen tissue samples were minced and homogenized in protein extraction buffer (20 mM HEPES, pH 7.5, 1.5 mM PMSF). Homogenized samples were transferred to a microfuge tube, adjusted to a final concentration of 0.4 M sodium chloride, and centrifuged at 9,000 xg at 4 °C for 30 min. Supernatants were collected and added to an equal volume of protein extraction buffer containing 20% glycerol and 0.4 M sodium chloride[22]. Protein concentrations of the tissue extracts were determined by the Bio-Rad reagent (BioRad, Richmond, CA).

Determination of NF-kB activation

NF-kB activity in lung nuclei of BL-instilled mice in BL+CD and BL+TN groups was determined by electrophoretic mobility shift assays using the Promega gel shift assay system (Promega, Madison, WI). DNA binding activity was determined following incubation of 20 µg of tissue proteins at room temperature for 20 min with ^{32}P-labeled double stranded oligonucleotide containing the NF-kB binding motif (5'-AGTTGAGGGGACTTTCCCAGGC-3'). The incubation mixture included 50 µg/ml of poly(dl-dC) in a binding buffer (4% glycerol, 0.5 mM EDTA, 0.5 mM DTT, 1 mM MgCl$_2$, 50 mM NaCl, 10 mM Tris-HCl, pH 8.0). The DNA/protein complexes were analysed on 6 % polyacrylamide gels in 0.5x Tris/Borate/EDTA buffer (0.0445 Tris, 0.0445 M Borate, 0.001 M EDTA). The specificity of binding was determined by the addition of an excess amount of the same unlabeled oligonucleotide (100-fold). Nonspecific competitions were similarly performed using an unlabeled oligonucleotide probe encompassing an AP-2 transcription factor site. The gels were autoradiographed on X-ray film.

Western blot analysis of NOS and IkB-α proteins

Whole lung tissue was homogenized in lysis buffer (10 mM HEPES, pH 7.9, 150 mM NaCl, 1 mM EDTA, 0.5 mM PMSF, 1 µg/ml leupeptin, 1 µg/ml aprotenin and 1 µg/ml pepstatin) on ice. Homogenates were centrifuged at 9000 xg at 4 °C for 30 min to remove cellular debris. Protein concentrations of the lung tissue were determined by the Coomassie blue-dye binding assay (Bio-Rad Laboratories, Hercules, CA). Total cellular proteins (40 µg) were separated on SDS-PAGE gels (4-20 % Tris-glycine minigels) and transferred to PVDF membrane and immunoblotted as described previously[23]. Non-specific binding sites were blocked with Tris-buffered saline (Tris 100 mM, NaCl 0.9 %, pH 7.5., Tween-20 0.1%, (TBS-T) and 5 % non-fat dry milk at room temperature for 18 h. Membranes were then incubated in a primary antibody (0.2 µg/ml) of rabbit polyclonal anti-NOS (Santa Cruz Biotechnology) in TBS-T. After 4 washes in TBS-T, membranes were incubated in a 1:5000 dilution of horseradish peroxidase conjugated anti-rabbit IgG (Santa Cruz Biotechnology). Immunoreactive NOS proteins were detected by enhanced chemiluminescence (ECL).

IkB-α proteins were immunoprecipitated from lung homogenates with agarose conjugates specific for IkB-α (Santa Cruz Biotechnology, Santa Cruz, CA). Total cellular protein, (200 µg) was immunoprecipitated with 10 µg of IkB-α antibody agarose conjugate. Immunoprecipitates were processed according to the manufacturer's instructions. Aliquots of immunoprecipitates were separated on SDS-PAGE gels (4-20% Tris-glycine minigels) and transferred to PVDF membrane and immunoblotted as described previously[23]. Membranes were incubated in 1:1000 dilution of a primary antibody rabbit polyclonal anti-lkB-α (Santa Cruz Biotechnology) in TBS-T. After 4 washes in TBS-T, membranes were incubated in a 1:5000 dilution of horseradish peroxidase conjugated anti-rabbit IgG (Santa Cruz Biotechnology). Immunoreactive IkB-α proteins were detected by enhanced chemiluminescence (ECL). IkB-α protein was quantitated on a scanning densitometer (Shimadzu model CS-9301 PC, Columbia, MD).

Statistics

Treatment - related differences were evaluated using a two way ANOVA, followed by pair wise comparisons using the Newman-Keuls test. Statistical significance was considered at $P \leq 0.05$.

RESULTS

We had previously demonstrated that the combined treatment with taurine and niacin reduced the BL-induced lung injury and fibrosis in hamsters[16,17]. In the present study , we confirmed our above findings in mice by demonstrating significant increases in the lung hydroxyproline content in BL + CD groups at 14 and 21 days after BL instillation as compared to mice in SA + CD and SA + TN control groups; and the ability of the combined treatment with taurine and niacin in significantly attenuating these increases in BL + TN groups as compared to their respective BL+ CD groups at both times (Fig.1).

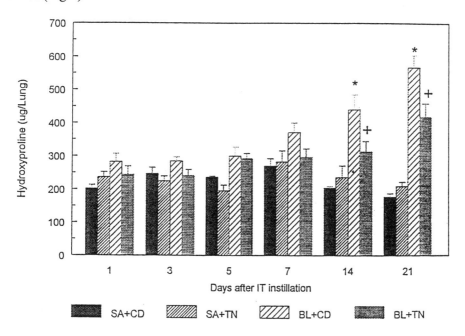

Figure 1. Effects of TN on BL-induced lung fibrosis. Hydroxyproline content in the lung homogenates was measured at the indicated time points after IT instillation of BL or saline. Each value represents the mean ± SEM of five animals. (*) Significantly higher (P<0.05) than all other groups at the corresponding times, and (+) significantly lower than the BL+CD groups at the corresponding times.

In order to test our hypothesis that an excess production of NO and some cytokines such as IL-1α, TNF-α and TGF-β is involved in the BL- induced lung fibrosis and the antifibrotic effect of the combined treatment with taurine and niacin resides in their ability to suppress their increased production, we measured their levels in the BALF. As shown in Fig. 2, the NO levels in the BALF were significantly increased in BL + CD groups as compared to controls by 200%, 230%, 210%, 230%, 250%, and 350% at 1, 3, 5, 7, 14, and 21 days, respectively. Treatment with taurine and niacin significantly decreased the production of NO in the BALF from mice in BL + TN groups as compared to the corresponding BL + CD groups. In order to correlate the NO levels in the BALF with the lung levels of iNOS mRNA and protein, they were analysed using RT-PCR and Western blot techniques, respectively.

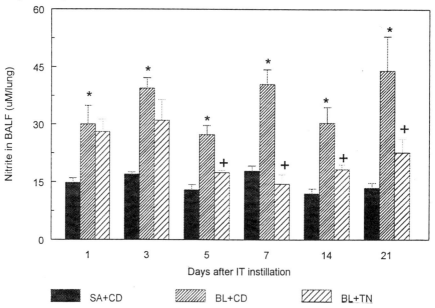

Figure 2. Effects of TN on NO production in BALF recovered from BL or saline-instilled mice. NO level was determined (as nitrite) in BALF at the indicated time points. Each value represents the mean ± SEM of five animals. (*) Significantly higher (P<0.05) than all other groups at the corresponding times, and (+) significantly lower than the BL+CD groups at the corresponding times.

It is interesting that the instillation of BL caused an increased expression of iNOS mRNA (Fig.3) and protein (Fig. 4) in BL+ CD groups as compared to mice in SA+ CD control groups and treatment with taurine and niacin significantly suppressed the over- expression of mRNA and protein in BL + TN groups.

The IL-α levels in the BALF from mice in BL + CD groups were increased by 2 and 3 fold at 3 and 5 days as compared to SA + CD control groups at the corresponding times, respectively, and treatment with taurine and niacin prevented these increases in BL + TN groups at these times (Fig. 5).

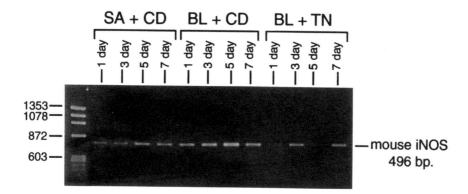

Figure 3. Effects of TN on BL-induced iNOS mRNA transcripts from lung tissues. Total RNA was isolated at the indicated time points after IT instillation of BL or saline and RT-PCR was performed with primers specific for iNOS.

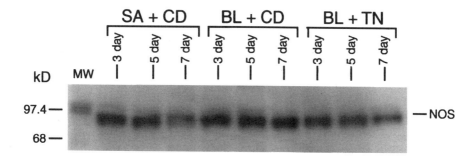

Figure 4. Effects of TN on lung iNOS protein expression during BL-induced lung fibrosis. Western blot analysis showing a time-dependent effects of TN on BL-induced increased iNOS protein expression at the indicated time points in mouse lung. Data are representative gel of three independent experiments.

The TNF - α levels in the BALF from mice in BL + CD groups remained elevated from day 1 through day 21, as compared to SA + CD control groups at the corresponding times and treatment with taurine and niacin decreased the TNF-α levels in BL+ TN groups (Fig. 6).

The TGF-β levels in the BALF from mice in BL + CD groups were also significantly increased by several fold at varying times after IT instillation of BL and treatment with taurine and niacin decreased these levels in BL + TN groups (Fig. 7).

The kinetics of NF-kB activation during BL- induced lung fibrosis was studied to explain the increased and decreased levels of cytokine in the BALF from mice in BL+ CD and BL+ TN groups, respectively.

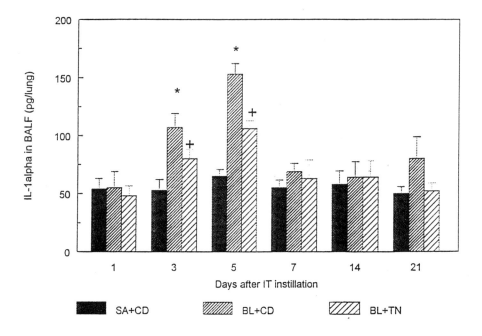

Figure 5. Effects of TN on IL-α production in BALF recovered from BL or saline-instilled mice. IL-α level was determined in BALF at the indicated time points. Each value represents the mean ± SEM of five animals. (*) Significantly higher (P<0.05) than all other groups at the corresponding times, and (+) significantly lower than the BL+CD groups at the corresponding times.

The lung nuclear extract from mice in SA + CD control group exhibited a basal level of NF-kB DNA binding activity. However, the level of NF-kB activation or its nuclear localization was much higher in BL+ CD groups than the control groups.

The nuclear localization of NF-kB in BL + CD groups was increased within one day and thereafter it increased progressively and peaked by day 14 and remained elevated until the end of the study (Fig. 8). However, the combined treatment with taurine and niacin inhibited the nuclear localization of NF- kB in BL + TN groups through the entire period of the study. Since this combined treatment with suppressed the BL- induced increased localization of NF- kB in the lung from mice in BL + TN groups, we measured the levels of IkB- α, a regulatory protein for NF- kB in whole lung homogenate.

It was interesting that the levels of IkB -α protein were elevated in BL + TN groups as opposed to depletion in BL + CD groups (Fig. 9).

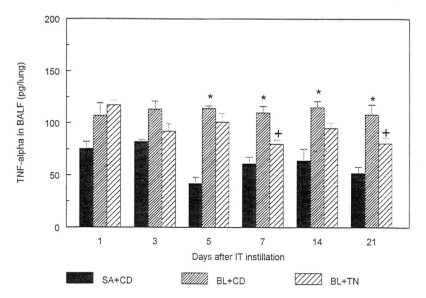

Figure 6. Effects of TN on TNF-α production in BALF recovered from BL or saline-instilled mice. TNF-α level was determined in BALF at the indicated time points. Each value represents the mean ± SEM of five animals. (*) Significantly higher (P<0.05) than all other groups at the corresponding times, and (+) significantly lower than the BL+CD groups at the corresponding times.

DISCUSSION

One of the widely accepted mechanisms for BL- induced lung fibrosis is that it binds to DNA: Fe^{2+} and forms a complex. This DNA: Fe^{2+} : BL complex undergoes recycling and generates ROS including superoxide and hydroxy radicals[3,4] . The superoxide radicals are known to react rapidly with the NO produced by activated macrophages in response to BL- induced lung injury and fibrosis and produce a potent oxidant, peroxynitrite.

An excess production of NO as found in the BALF from mice during the course of development BL- induced lung fibrosis in BL +CD groups is consistent with the findings of other investigators[12] and also consistent with the presence of a higher level of NO in the exhaled breath of patients suffering from active fibrosing alveolitis [24] .

It is interesting that the increased levels of NO in the BALF from mice in BL + CD groups having a higher amount of hydroxyproline, an index of lung fibrosis, than the controls correlated well with an increased expression of lung iNOS message and protein in these mice.

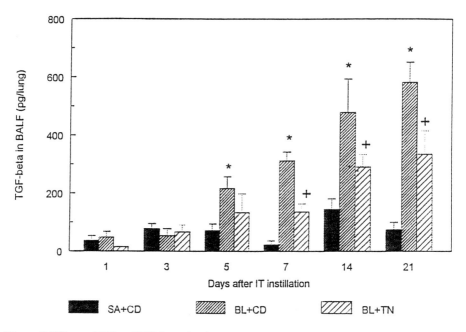

Figure 7. Effects of TN on TGF-β production in BALF recovered from BL or saline-instilled mice. TGF-β level was determined in BALF at the indicated time points. Each value represents the mean ± SEM of five animals. (*) Significantly higher (P<0.05) than all other groups at the corresponding times, and (+) significantly lower than the BL+CD groups at the corresponding times.

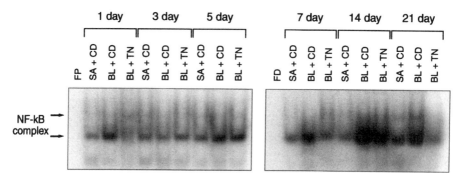

Figure 8. Effects of TN on the kinetics of lung NF-kB activation in BL or saline-instilled mice. The lung protein extracts were prepared at the indicated time points and analyzed for NF-kB activity using a 22-mer double stranded oligonucleotide probe containing a NF-kB site. Each lane represents protein extract from SA+CD, BL+CD, or BL+TN group. FP = free probe; Arrow head = NF-kB complex. Data are representative gel of three independent experiments.

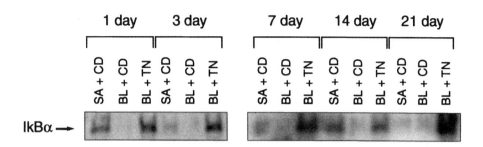

Figure 9. Effects of TN on lung IkB-α protein expression during BL-induced lung fibrosis. Western blot analysis showing a time-dependent effects of TN on BL-induced depletion of IkB-α protein expression at the indicated time points in mouse lung. Data are representative gel of three independent experiments.

These findings suggest that the synthesis of iNOS is increased at the gene expression level during the course of development of lung fibrosis in response to IT instillation of BL in mice. The ability of the combined treatment with taurine and niacin to block the BL- induced increased production of NO in the BALF from mice in BL +TN groups secondary to down- regulation of iNOS reveals a novel mechanism for the antifibrotic of this combination. Our findings support the data reported by Schuller- Levis et al that taurine offers protection against oxidant –induced lung injury by inhibiting the production of TNF-α and NO[25,26] . However, it is not clear how IT instillation of BL upregulates the synthesis of iNOS in the lungs and treatment with taurine and niacin down- regulates the same at the gene expression level.

The BL- induced production of ROS explains our finding of NF-kB activation in the lungs of mice in BL + CD groups. This is based on the findings that the ROS generated in response to a variety of stimuli activate the NF- kB by phosphorylating and degrading the inhibitory protein IkB-α which normally exists as a complex with NF–kB in the cytoplasm[14,15]. The degradation of IkB-α allows the translocation of active dimer of NF-kB into the nucleus where it binds to the promoter region of genes of IL- α and TNF-α and stimulates their production secondary to up- regulation of their genes.

The results of the present study are consistent with the above hypothesis since mice treated with BL in BL+ CD groups had higher levels of IL-α, TNF -α and TGF-β in the BALF than the mice in SA + CD control groups and the mice treated with taurine and niacin which suppressed the BL–induced activation of NF-kB in BL + TN groups had reduced levels of these cytokines in the BALF. These results support the findings reported by other investigators that the release of proinflammatory and profibrogenic cytokines plays a central role in the BL –induced lung fibrosis[5-7] ; and the ability of the combined treatment with taurine and niacin to block the BL-induced release of these cytokines secondary to down-regulation of their genes in response to suppression of NF-kB activation, as presented here, constitutes one of the novel mechanisms for their antifibrotic effect. Although NF-kB appears to play a critical role in cytokine- mediated inflammation by upregulating the transcription of a specific set of cytokine genes in response to stimuli, it is not known what role NF– kB plays in the differential production   or in coordination of production of various cytokines.

Our data suggest that nuclear translocation of the transcription factor NF –kB is important for BL-induced lung injury and fibrosis and provide evidence for its activation being linked with the lung injury produced by this drug in mice. The activation of NF–kB is thought to occur secondary to proteolytic degradation of IkB-α. The results presented at this symposium support the role of IkB-α in vivo since activation of NF–kB during the course of BL- induced lung fibrosis was accompanied by depletion of lung levels of IkB-α in mice in BL + CD groups and combined treatment with taurine and niacin which prevented the activation, preserved the IkB-α levels and manifested antifibrotic effect in BL + TN groups. It is not known whether taurine and niacin administered alone will produce a similar effect. However, it should be noted that these two compounds were found to produce antifibrotic effect independently in a single dose BL- hamster model of lung fibrosis[27,28] . In addition, taurine[26] is shown to have antioxidant effect against O_3 and niacin[29] is shown to offer protection against paraquat which inflicts lung injury by way of generating superoxide radicals. Our findings are consistent with the findings of other in vitro and in vivo studies in which treatments with antioxidants such as N-acetylcysteine[30], pyrrlidine dithiocarbamate[31] and epigallocatechin- 3- gallate[32] were found to block the activation of NF-kB by inhibiting the signal trasduction- induced phosphorylation of IkB- α.

CONCLUSIONS

The activation of transcription factor NF- kB appears to play a central role in the BL- induced lung injury and fibrosis and the antifibrotic effect of the combined treatment with taurine and niacin resides in their ability to suppress the NF - kB activation by preserving the IkB- α protein. In absence of NF- kB activation, some proinflammatory and profibrogenic cytokines are not released and this will retard the progression of the inflammatory and fibrotic events. In addition, the combined treatment with taurine and niacin inhibits the production of NO by down-regulating the synthesis of BL-induced over-expression of lung iNOS. The results of the present study have revealed novel strategies to develop drugs which can specifically inhibit the activation of NF–kB and over-expression of iNOS in the lungs and the availability of such drugs may prove to be therapeutically efficacious for the treatment of lung inflammation and fibrosis.

ACKNOWLEDGEMENTS

These studies were supported by the research grants awarded to Shri N. Giri by the National Institute of Health, NHLBI, Grant # RO1-HL- 56262. the expert typing skill of Mrs. Evett Kilmartin is gratefully acknowledged.

REFERENCES

1   Crouch, E., 1990 Pathobiology of pulmonary fibrosis. *Am J Physiol* **259**: 159-184.
2   Clark, J.G., Overton, J.E., Marino, B.A., Uitto, J. and Starcher, B.C., 1980   Collagen synthesis in bleomycin-induced pulmonary fibrosis in hamsters.   *J Lab Clin Med* **96**: 943-953.
3   Caspary W.J., Lanzo, D.A., and Niziak, C., 1982, Effect of deoxyribonucleic acid on the production of reduced oxygen by bleomycin and iron. *Biochemistry* **21**: 334-338.
4   Sugiura, Y. and Kikuchi, T., 1978, Formation of superoxide and hydroxy radicals in iron (II)-bleomycin-oxygen system: Electron spin resonance detection by spin tapping. *J Antibiot* **31**: 1310-1312.
5   Phan, S.H. and Kunkel, S.L., 1992, Lung cytokine production in bleomycin-induced pulmonary fibrosis. *Exp Lung Res* **18**: 29-43.
6   Piguet, P.F., Collart, M.A., Grau, G.E., Kapanchi, Y. and Vassali, P., 1989, Tumor necrosis factor/cachectin plays a key role in bleomycin-induced pneumopathy and fibrosis. *J Exp Med* **170**: 655-663.
7   Scheule, R.K., Perkins, R.C., Hamilton, R. and Holian, A., 1992, Bleomycin stimulation of cytokine secretion by the human alveolar macrophages. *Am J Physiol* **262**: 386-L391.
8   Smith, R.E., Strieter, R.M., Phan, S.H. and Kunkel, S.L., 1996, C-C Chemokines: Novel mediators of the profibrotic inflammatory response to bleomycin challenge. *Am J Respir Cell Mol Biol* **15**: 693-702.
9   Kobsik, L., Bredt, D.S., Lowenstein, C.J., Drazen, J., Gaston, B., Sugarbaker, D. and Stamler, J., 1993, Nitric oxide synthase in human and rat lung: Immunocytochemical and histochemical localization. *Am J Respir Cell Mol Biol* **9**: 371-377.
10 Sessa, W.C., Harrison, Luthin, D.R., Pollock, J.S., Lynch, K.R., 1993, Genomic analysis and expression patterns reveal distinct genes for endothelial and brain nitric oxide synthase. *Hypertension* **21**: 934-938.
11 Sherman, M., Aberhard, E.E., Wong, V.Z., Griscavge, J.M., and Ignarro, L.J., 1993, Pyrrolidine dithiocarbonate inhibits induction of rat nitric oxide synthase in rat alveolar macrophages. *Biochem Biophys Res Commun* **191**: 1301-1308.

12 Huot, A.E. and Hacker, M.P., 1990, Role of reactive nitrogen intermediate production in alveolar macrophage-mediated cytostatic activity induced by bleomycin lung damage in rats. *Cancer Res* **50**: 7863-7866.

13 Blackwell, T.S. and Christman, W.W., 1997, The role of nuclear factor-kB in cytokine gene regulation. *Am J Respir Cell Mol Biol* **17**: 3-9.

14 Bauerle, P.A. and Henkel, T., 1994, Function and activation of NF-kappa B in the immune system. Ann rev Immunol **12**: 141-179.

15 Schreck, R., Reiber, P. and Baeuerle, P.A., 1992, Reactive oxygen intermediates as apparently widely used messengers in the activation of the NF-kB transcription factor and HIV-1. *EMBO J* **10**: 2247-2258.

16 Wang, Q., Giri, S.N., Hyde, D.M. and Li, C., 1991, Amelioration of bleomycin-induced pulmonary fibrosis in hamsters by combined treatment with taurine and niacin. *Biochem Pharmacol* **42**: 1115-1122.

17 Gurujeyalakshmi, G., Iyer, S.N., Hollinger, M.A. and Giri, S.N., 1996, Procollagen gene expression is down-regulated by taurine and niacin at the transcriptional level in the bleomycin hamster model of lung fibrosis. *J Pharmacol Exp Therap* **227**: 1152-1157.

18 Giri, S.N., Hollinger, M.A. and Schiedt, M.J., 1981, The effects of paraquat and superoxide dismutase on pulmonary vascular permeatiblity and edema in mice. *Arch Evin Health* **36**: 149-154.

19 Woessner, Jr., J.F., 1961, The determination of hydroxyproline in tissue and protein samples containing small proportions of this imino acid. *Anal Biochem* **93**: 440-444.

20 Ding, A.H., Nathan, C.F. and Stuehr, D.J., 1988, Release of reactive nitrogen intermediates from mouse peritoneal macrophages. *J Immunol* **141**: 2407-2412.

21 Gurujeyalakshmi, G., Hollinger, M.A. and Giri, S.N., 1998, Regulation of transforming growth factor-β_1 mRNA expression by taurine and niacin in the bleomycin hamster model of lung fibrosis. *Am J Respir Cell Mol* **18**: 334-342.

22 Choi, A.M.K., Sylvester, S., Otterbein, L and Holbrook, N.J., 1995 Molecular responses to hyperoxia in vivo: Relationship to increased tolerance in aged rats. *Am J Respir Cell Mol Biol* **13**: 74-82.

23 Gurujeyalakshmi, G., Hollinger, M.A. and Giri, S.N., 1999, Pirfenidone inhibits PDGF isoforms in the bleomycin hamster model of lung fibrosis at the translational level. *Am J Physiol* **276** (Lung Cell Mol Physiol 20): L311-L318.

24 Paredi, P., Kharitonov, S.A., Loukides, S., Pantelidis, P., DuBois, R.M. and Barnes, P.J., 1999, Exhaled nitric oxide is increased in active fibrosing alveolitis. *Chest* **115**: 1352-1356.

25 Schuller-Levis, G., Quinn, M.R., Wright, C. and Park, E., 1994, Taurine protects against oxidant-induced lung injury: Possible mechanism(s) of action. *Adv Exp Med Biol* **359**: 31-39.

26 Schuller-Levis, G., Gordon, R.E., Park, E., Pendino, K.J. and Laskin, D.L., 1994, Taurine protects rat bronchioles from acute ozone-induced lung inflammation and hyperplasia. *Exp Lung Res* **21**: 877-888.

27 Wang, Q., Giri, S.N., Hyde, D.M. and Nakashima J.M., 1989, Effect of taurine on bleomycin-induced lung fibrosis in hamsters. *Proc Soc Exp Biol Med* **190**: 330-338.

28 Wang, Q., Giri, S.N., Hyde, D.M., Nakamshima, J.M. and Javadi, I., 1990, Niacin attenuates bleomycin-induced lung fibrosis in the hamster. *J Biochem Toxicol* **5**: 13-2239.

29 Brown, O.R., Heitkamp, M. and Song, C.S., 1981, Niacin reduces paraquat toxicity in rats. *Science* **212**: 1510-1512.

30 Leff, J.A., Wilke, C.P., Hypertson, B.M., Shanley, P.F., Beehler, C.J. and Repine, J.E., 1993, Postinsult treatment with N-acetylcysteine decreases IL-1 induced neutrophil influx and lung leak in rats. *Am J Physiol* **265**: L501-L506.

31 Nathens, A.B., Bitar, R., Davreux, C., Mujard, M., Marshall, J.C., Dackiw, A.P.B., Watson, R.W.G. and Rotstein, O.D., 1997, Pyrrolidine dithiocarbamate attenuates endotoxin-induced acute lung injury. *Am J Respir Cell Mol Biol* **17**: 608-616.

32 Lin, Y.L. and Lin J.K., 1997, (-)-Epigallocatechin-3-gallate blocks the induction of nitric oxide synthase by down-regulating lipopolysaccharide-induced activity of transcription factor nuclear factor-kB. *Mol Pharmacol* **52**: 465-472.

CYTOPROTECTIVE EFFECT OF TAURINE AGAINST HYPOCHLOROUS ACID TOXICITY TO PC12 CELLS

Sean Kearns and Ralph Dawson, Jr.
Department of Pharmacodynamics, College of Pharmacy, University of Florida, Gainesville, FL

Abstract: Taurine has been shown to be an effective scavenger of hypochlorous acid (HOCl). The role of HOCl is well established in tissue damage associated with reperfusion injury mediated by neutrophils. The role of HOCl in CNS injury and inflammatory reactions has not been well established. Myeloperoxidase activity is present in the CNS and it has been associated with ischemic injury. The aim of the present study was to determine the cytotoxicity of HOCl in a neuronal cell line (PC12) and the ability of taurine to prevent or reverse neurotoxicity. PC12 cells were grown in 96 well plates at a plating density of ~100,000 cells per well. HOCl was made up fresh from NaOCl for each experiment and the concentration verified spectrophotometrically. PC12 cells were exposed to HOCl for 1 hour in phosphate-buffered saline. Taurine was added at the time of HOCl treatment and in some experiments a post-treatment with taurine was performed by adding 1 or 10 mM taurine to the culture media (RPMI 1640). The cells were allowed 24 hours to recover and viability was determined using a tetrazolium-based (MTT) assay. The first series of experiments evaluated the toxicity of HOCl and the efficacy of taurine to protect PC12 cells. HOCl at 50 µM reduced PC12 cell viability by 50% and 150 µM reduced viability to <25% of control levels. Taurine (0.5-20 mM) was tested for cytoprotection against 150 µM HOCl and PC12 cells treated with 0.5 mM taurine exhibited only a 20% reduction in viability compared to untreated controls. Taurine concentrations of 1 mM or higher provided nearly 100% protection against HOCl. A second study was performed comparing taurine to β-alanine, glutathione and isethionic acid. HOCl (100 µM) reduced viability to $25 \pm 1\%$ of controls and taurine, β-alanine and glutathione at 1 mM provided nearly complete protection. In contrast, isethionic acid, which lacks an amino group, failed to provide protection. Taurine (1 or 10 mM) added after 50 µM HOCl treatment did not provide any protection and PC12 cell viability was

Taurine 4, edited by Della Corte *et al.*
Kluwer Academic / Plenum Publishers, New York, 2000.

reduced to <39% of controls. In contrast, if taurine (50 μM) was present during the HOCl treatment and 1 mM taurine was added after the treatment, PC12 cell viability was 80 ± 5% of controls. A combination of 250 μM taurine during the HOCl treatment and 1 mM taurine post-treatment produced 100% protection. These results clearly show that taurine is an efficient scavenger of HOCl and can prevent neuronal damage caused by HOCl. Since myeloperoxidase expression in the CNS is increased by ischemia, one function of taurine released during an ischemic event may be to scavenge HOCl and provide neuroprotection.

INTRODUCTION

Hypochlorous acid (HOCl) is the toxic product of the enzyme myeloperoxidase (MPO). MPO catalyses the formation of HOCl from chloride and H_2O_2 and can also catalyse the formation of other oxidants[1,2]. Activated neutrophils and microglia express MPO and can generate HOCl[3,4]. Neutrophils have been implicated in tissue injury that accompanies post-ischemic events[5-7]. Toxicity arising from MPO-derived oxidants has also been implicated in Alzheimer's disease[4,8]. HOCl is a strong oxidant that can damage lipids, proteins and DNA[9,10]. The neurotoxic potential of HOCl generated by activated neutrophils or microglia in the CNS has received limited attention relative to other types of oxidative injury.

HOCl can react with superoxide to generate hydroxyl radicals and can also react with ferrous iron to generate more reactive species[11]. Reduced glutathione (GSH), ascorbic acid and the amino group of endogenous compounds can serve to scavenge HOCl[10,11]. Reaction of HOCl with the amino group of amino acids yields chloramines which themselves can be cytotoxic and more long-lived than HOCl[3,12,13]. Considerable evidence now exists that taurine acts as a major scavenger of endogenously generated HOCl[1,2,14]. In fact, the taurine chloramine (TauCl) formed from the reaction of HOCl with taurine has been shown to act as a cellular signalling molecule that can downregulate the expression of a number of gene products (iNOS, TNF-α, PGE$_2$, & COX-2) involved in inflammatory reactions[15-17]. Neutrophils contain a high (mM) concentration of taurine[18] and ischemia and a number of neurotoxic treatments can elevate extracellular taurine concentrations in the CNS[19-21]. Thus, taurine may serve an important role in the CNS in reducing host damage mediated by HOCl during an inflammatory response.

The present experiments were undertaken to test the hypothesis that taurine could prevent the direct toxicity of HOCl to neurons. We performed dose-response studies for both HOCl and taurine. Taurine was also administered pre and post HOCl insult and evaluated for its neuroprotective actions. Preliminary studies have also been undertaken to examine the cytoprotective actions of taurine analogs.

METHODS

PC12 cells were cultured in RPMI media supplemented with 10% horse serum, 5% fetal bovine serum, and penicillin/streptomycin. PC12 cells have been used extensively as a model system to test for neuroprotection[22] . PC12 cells were obtained from ATCC and cultures were used between passage 12 through 27. The cells were cultured until confluent and then plated 24 hours before use in experiments into 96 well plates treated with rat tail collagen. Dilutions based on cell counts using a hemocytometer were made to obtain an overall plating density of approximately 150,000 cells per well in a 96 well plate. HOCl was obtained from a stock 4% sodium hypochlorite solution and the concentration of available HOCl was determined for each experiment spectrophotometrically based on the extinction coefficient. Taurine was dissolved in phosphate buffered saline (PBS, pH=7.4) and diluted to obtain the needed concentrations. All amino acid analogues were tested at a concentration of 1 mM and dissolved in PBS. All solutions were made fresh the day of the experiment. The order of treatment addition was always taurine or analogue followed by HOCl treatment. The cells were exposed to treatments for one hour followed by aspiration of the experimental treatments. Controls received PBS only and experimental plates were treated with HOCl ± taurine or taurine analogs. Complete media was then added and the cells were placed back into the incubator for a 24-hour recovery period. For pre-treatment experiments taurine (1 or 10 mM) supplemented RPMI media was added the day before the insult. For post-treatments taurine supplemented media was added after HOCl was removed and incubated for the 24-hour recovery period. Cell viability was determined using the MTT assay[23]. Briefly, 24 hours after HOCl treatment cells were treated with 100 uL per well MTT (30 mg/60 ml PBS) and 100 uL complete media and allowed to incubate for 3-4 hours. The MTT was then aspirated and the cells treated with 100 uL isopropyl alcohol/HCl for 5 min. to solubilize the MTT. The plates were then read at 550 nm and optical densities for the wells were obtained. Individual treatment conditions were applied to 12 wells of the 96 well plate per experiment and experiments were replicated 2-6 times. Dose-response studies were replicated 4-6 times and pre- and post-treatment studies reported are the average of two independent experiments. The data from studies with taurine analogs are the average of 3 independent replications.

RESULTS

HOCl effectively killed PC12 cells at concentrations as low as 50 μM and taurine (20 mM) provided nearly complete protection even from 150 μM HOCl (Figure 1).

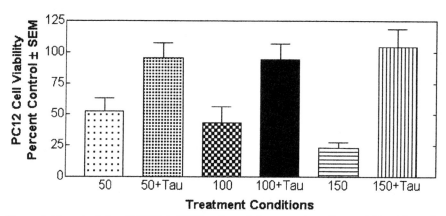

Figure 1. Neurotoxicity of HOCl to PC12 cells and the protective effects of 20 mM taurine (Tau).

The efficacy of taurine to prevent the toxicity of 150 μM HOCl to PC12 cells was tested using a range of taurine concentrations. Taurine could significantly reduce HOCl toxicity at concentrations as low as 0.5 mM and concentrations at or above 1 mM provided complete protection (Figure 2).

A series of experiments were undertaken to ascertain if incubation with taurine after an HOCl (50 μM) insult could enhance PC12 survival post-treatment. Post-treatment for 24 hours with either 1 or 10 mM taurine added to the culture media after the HOCl treatment failed to reduce cell death (Table 1). In contrast, concurrent treatment with low concentrations of taurine (50 or 250 μM) in the presence of HOCl (50 μM) followed by post-treatments with 1 or 10 mM taurine did result in significant protection (Table 1). Pre-treatment of PC12 cells for 24 hours with either 1 or 10 mM taurine added to the culture media did not prevent HOCl (50 μM) toxicity (data not shown). An initial experiment where the PC12 cultures were not rinsed with PBS after the pre-treatment resulted in protection, but this was likely due to residual taurine present from the pre-incubation media.

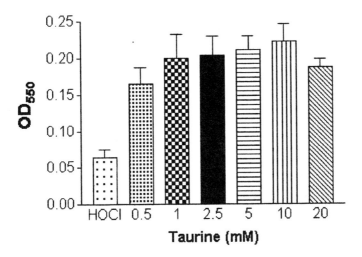

Figure 2. Neuroprotection provided by increasing concentrations of taurine against 150 μM HOCl. PC12 cell viability is indicated by increasing absorbance (OD) of MTT at 550 nm

Table 1. Effects of taurine post-treatment on PC12 cell viability

Treatment Conditions	Percent of Control
Control	100 ± 0.5
HOCl (50 μM)	39 ± 0.5
Taurine Post-treatment (1 mM)	32 ± 0.1
Taurine Post-treatment (10 mM)	37 ± 4
Taurine (50 μM) + 1 mM Post-treatment	80 ± 5
Taurine (50 μM) + 10 mM Post-treatment	86 ± 0.1
Taurine (250 μM) + 1 mM Post-treatment	92 ± 11
Taurine (250 μM) + 10 mM Post-treatment	108 ± 3

Data are expressed as percent of control ± SEM and are the averaged results from 2 independent experiments.

The final series of experiments evaluated the effects of taurine analogs on PC12 survival after treatment with 100 μM HOCl. Analogs were all tested at a concentration of 1 mM since this concentration of taurine had previously been shown to elicit maximal protection. Taurine, β-alanine or glutathione when co-incubated with 100 μM HOCl produced nearly complete protection of PC12 cells (Table 2). In contrast, isethionic acid which lacks an amino group failed to provide any protective effects.

Table 2. Effect of taurine analogs on PC12 cell viability

Treatment Conditions	Percent of Control
Control	100 ± 5
HOCl (100 µM)	25 ± 1
Taurine (1 mM)	84 ± 6
β-Alanine (1 mM)	110 ± 2
Isethionic Acid (1 mM)	26 ± 5
Glutathione (1 mM)	91 ± 2

Data are expressed as mean $\pm$ SEM and are the average of 2-5 independent replications.

CONCLUSION

Taurine was an effective scavenger of HOCl as evidenced by its cytoprotective actions in PC12 cells. As previously shown[10-12], taurine can directly scavenge HOCl and prevent subsequent damage by other reaction mechanisms[24]. HOCl was found to be quite cytotoxic to PC12 cells with only a one hour exposure to a 50 µM concentration producing greater than a 50% reduction in cell viability. This is comparable to previous reports of HOCl-induced cell lysis in murine macrophage-like tumor cells that occurred at threshold dose of 25-35 µM[9]. Taurine at concentrations as low as 50 µM combined with a 1 mM post-treatment could produce significant neuroprotection against 50 µM HOCl. The direct scavenging properties of taurine appeared essential since post-treatment alone with even 10 mM taurine could not rescue PC12 cells. It is unclear why the pre-treatment with taurine that should have elevated intracellular taurine content failed to result in any protective effects. HOCl-induced damage may have occurred prior to a significant injury-mediated increase in cellular taurine release.

Studies with taurine analogs produced the expected results with both glutathione and β-alanine attenuating HOCl toxicity via the scavenging actions of the amino functional group. In contrast, isethionic acid (2-hydroxyethane sulfonic acid) had no protective effects since it lacks an amino group. The stoichiometry of the scavenging reaction for taurine and HOCl is 1:1[10], thus dietary treatments with taurine which easily produce serum taurine concentrations in excess of 1 mM[25] could potentially reduce tissue injury caused by MPO-mediated HOCl release. Several previous reports have shown that taurine can protect various cell types from neutrophil-mediated injury[6,26]. Our results suggest that one mechanism for taurine's cytoprotective effects is to scavenge HOCl that is one of the products of the respiratory burst in activated immune cells.

ACKNOWLEDGMENTS

These studies were supported in part by a grant from Taisho Pharmaceutical Company. The authors would like to acknowledge the technical assistance of Steve Messina and Baerbel Eppler.

REFERENCES

1. Weiss, S.J., Klein, R., Slivka, A. and Wei, M., 1982, Chlorination of taurine by human neutrophils: Evidence for hypochlorous acid generation. *J. Clin. Invest.* **70**: 598-607.
2. Marquez, L.A. and Dunford, H.B., 1994, Chlorination of taurine by myeloperoxidase; kinetic evidence for an enzyme-bound intermediate. *J. Biol. Chem.* **269**: 7950-7956.
3. Thomas, E.L., Grisham, M.B. and Jefferson, M.M., 1983, Myeloperoxidase-dependent effect of amines on functions of isolated neutrophils. *J. Clin. Invest.* **72**: 441-454.
4. Reynolds, W.F., Rhees, J., Macccciejewski, D., Paladino, T., Sieburg, H., Maki, R.A. and Masliah,E., 1999, Myeloperoxidase polymorphism is associated with gender specific risk for Alzheimer's disease. *Exper. Neuro.* **155**: 31-41.
5. Smith, J.K., Grisham, M.B., Granger, D.N. and Korthuis, R.J., 1989, Free radical defense mechanisms and neutrophil infiltration in postischemic skeletal muscle. *Am. J. Physiol.* **256**: H7889-H793.
6. Raschke, P., Massoudy, P. and Becker, B.F., 1995, Taurine protects the heart from neutrophil-induced reperfusion injury. *Free Rad. Biol. Med.* **19**: 461-471.
7. Hudome, S., Palmer, C., Roberts, R.L., Mauger, D., Housman, C. and Towfighi, J., 1997, The role of neutrophils in the production of hypoxic-ischemic brain injury in the neonatal rat. *Pediatr Res* **41**: 607-616.
8. Jolivalt, C., Leninger-Muller, B., Drozdz, R., Naskalaski, J.W., and Siest, G., 1996, Apolipoprotein E is highly susceptible to oxidation by myeloperoxidase, an enzyme present in the brain. *Neurosci. Ltrs.* **210**: 61-64.
9. Schraufstatter, I.U., Browne, K., Harris, A., Hyslop, P.A., Jackson, J.H., Quehenberger, O. and Cochrane, C.G., 1990, Mechanisms of hypochlorite injury of target cells. *J. Clin. Invest.* **85**: 554-562.
10. Prutz, W.A., 1996, Hypochlorous acid interactions with thiols, nucleotides, DNA, and other biological substrates. *Arch. Biochem. Biophys.* **332**: 110-120.
11. Folkes, L.K., Candeias, L.P., and Wardman, P., 1995, Kinetics and mechanisms of hypochlorous acid reactions. *Arch. Biochem. Biophys.* **323**: 120-126.
12. Zgliczynski, J.M., Stelmaszynska, T., Domanski, J. and Ostrowski, W., 1971, Chloramines as intermediates of oxidation reaction of amino acids by myeloperoxidase. *Biochim. Biophys. Acta,* **233**: 419-424.
13. Cantin, A., 1994, Taurine modulation of hypochlorous acid-induced lung epithelial cell injury in vitro. *J. Clin. Invest.* **93**: 606-614.
14. Cunningham, C., Tipton, K.F., and Dixon, H.B.F., 1998, Conversion of taurine into *N*-chlorotaurine (taurine chloramine) and sulphoacetaldehyde in response to oxidative stress. *J. Biochem.* **330**: 939-945.

15. Marcinkiewicz, J., Grabowska, A., Bereta, J. and Stelmaszynska, T., 1995, Taurine chloramine, a product of activated neutrophils, inhibits the generation of nitric oxide and other macrophage inflammatory mediators. *J. Leukoc. Biol.* **58**: 667-674.

16. Liu, Y., Tonna-DeMasie, M., Park, E., Schuller-Levis, G. and Quinn, M.R., 1998, Taurine chloramine inhibits production of nitric oxide and prostaglandin E2 in activated C6 glioma cells by suppressing inducible nitric oxide synthase and cyclooxygenase-2 expression. *Mol. Brain Res.* **59**: 189-195.

17. Redmond, H.P., Stapleton, P.P., Neary, P. and Boucher-Hayes, D., 1998, Immunonutrition: The role of taurine. *Nutrittion* **14**: 599-604.

18. Green, T.R., Fellman, J.H., Eicher, A.L., and Pratt, K.L., 1991, Antioxidant role and subcellular location of hypotaurine and taurine in human neutrophils. *Biochem. Biophys. Acta* **1073**: 91-97.

19. Huxtable, R.J., 1989, Taurine in the central nervous system and the mammalian actions of taurine. *Prog. Neurobio.* **33**: 471-533.

20. Seki, Y., Feustel, P.J., Keller, R. W., Trammer, B.I. and Kimelberg, H.K., 1999, Inhibition of ischemia-induced glutamate release in rat striatum by dihydrokinate and an anion channel blocker. *Stroke.* **30**: 43-440.

21. Saransaari, P. and Oja, S.S., 1997, Enhanced taurine release in cell damaging conditions in the developing and ageing mouse hippocampus. *Neuroscience* **79**: 847-854.

22. Borowitz, J.L., Kanthasamy, A.G., Mitchell, P.J., and Isom, G.E., 1993, Use of PC12 cells as a neurotoxicological screen: characterization of anticyanide compounds. *Fund. Appl. Toxicol.* **20**: 133-140.

23. Mosman, T., 1983, Rapid colormetric assay for cellular growth and survival: application to proliferation and cytotoxicity assays. *J. Immunol. Meth.* **65**: 55-63.

24. Kozumbo, W.J., Agarwal, S. and Koren, H.S., 1992, Breakage and binding of DNA by reaction products of hypochlorous acid with aniline, 1-naphthylamine, or 1-naphthol. *Toxicol. Appl. Pharmacol.* **115**: 107-115.

25. Dawson, R., Liu, S., Eppler, B., and Patterson, T., 1999, Effects of dietary taurine supplementation or deprivation in aged male Fischer 344 rats. *Mech. Aging Dev.* **107**: 73-91.

26. Finnegan, N.M., Redmond, H.P., and Bouchier-Hayes, D.J., 1998, Taurine attenuates recombinant interleukin-2-activated, lymphocyte-mediated endothelial cell injury. *Cancer* **82**: 186-199.

EFFECT OF TAURINE AND β-ALANINE ON MORPHOLOGICAL CHANGES OF PANCREAS IN STREPTOZOTOCIN-INDUCED RATS

Kyung Ja Chang

Department of Food and Nutrition, Inha University, Inchon, 402-751, Korea

Abstract: In order to determine the effects of taurine supplementation or depletion on the morphological changes of pancreatic β-cells in streptozotocin-induced diabetic rats, Sprague-Dawley male rats were fed the purified diets supplemented with 1, 2 or 3% taurine or 5% β-alanine in their drinking water for 7 weeks. After 3 weeks, diabetes was induced by streptozotocin injection (50mg/kg body-weight).   Pancreatic morphology was observed by transmission electron microscopy. The pancreatic β-cell of the non-diabetic (CO) group had the many secretory granules, rough endoplasmic reticulum and rod shaped mitochondria. However, the β-cells of non taurine-supplemented diabetic (EO) group were severely damaged, showing depleted secretory granules. In the 1% taurine-supplemented diabetic group, the β-cells were less damaged compared to the EO group and had some apparently normal secretory granules, but most of rough endoplasmic reticulum and mitochondria was destroyed. The β-cell of 2% taurine-supplemented diabetic group had swollen rough endoplasmic reticulum, round-shaped mitochondria and some apparently normal secretory granules. The β-cell of 3% taurine-supplemented diabetic group was little different from that of non-diabetic group. The pancreatic β-cell of taurine-depleted diabetic group was not destroyed but had many small secretory granules which appeared immature. This was reflected in the blood glucose concentrations of this group.   Therefore, taurine may prevent insulin-dependent diabetes by protection of the pancreatic β-cell and may also preserve normal secretory granules. From these results, taurine supplementation may be recommended for prevention and treatment of diabetes.

Taurine 4, edited by Della Corte *et al.*
Kluwer Academic / Plenum Publishers, New York, 2000.

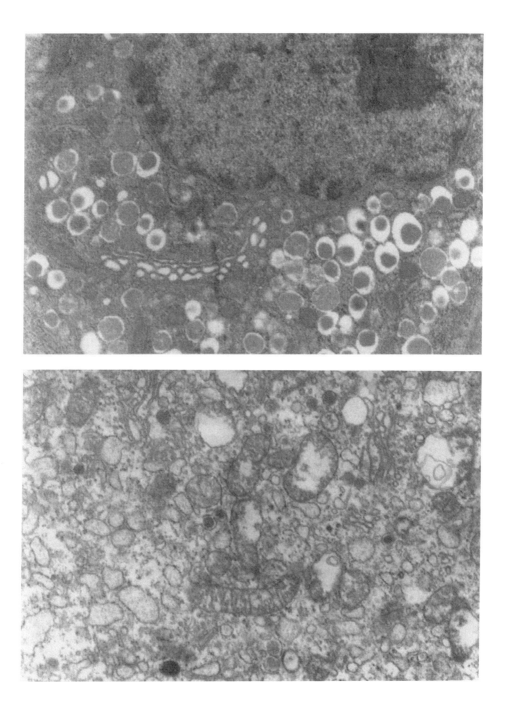

Fig. 1 A (Upper), B (Lower). Pancreatic β-cells of rats X 10,000. A: Control group (CO), B: Diabetic group (E0).

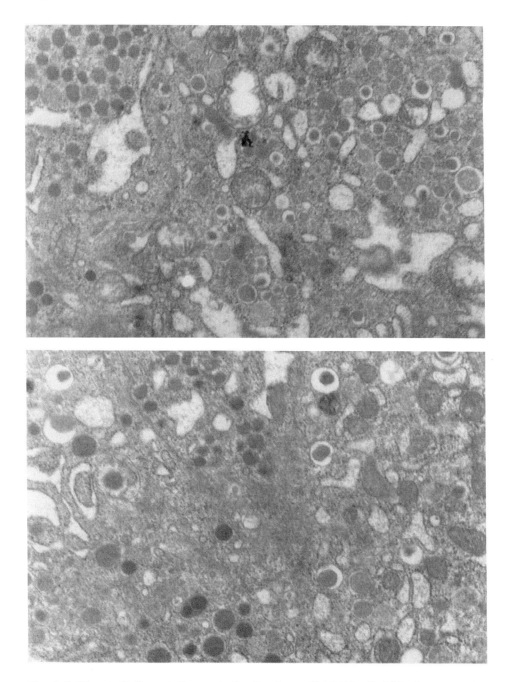

Fig. 1 C (Upper), D (Lower). Pancreatic ß-cells of rats X 10,000. C: Diabetic group + 1% Taurine (E1), D: Diabetic group + 2% Taurine (E2).

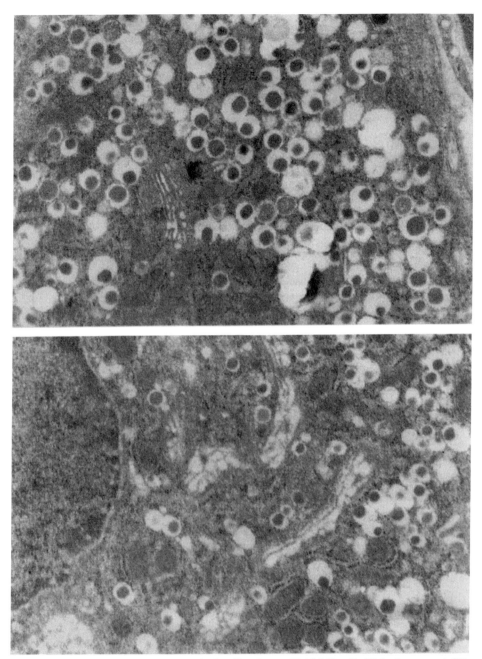

Fig. 1 E (Upper), F (Lower). Pancreatic ß-cells of rats X 10,000. E: Diabetic group + 3% Taurine (E3), F: Diabetic group + 5% ß-Alanine (EA).

INTRODUCTION

Streptozotocin has been found to cause hyperglycemia by inducing selective damage to pancreatic β-cells which have insulin-secreting granules[1,2]. This provides a reliable animal model of insulin-dependent diabetes mellitus[3]. Taurine (2-aminoethanesulfonic acid) is present in high concentrations in mammalian tissues including pancreas[4,5]. It has been demonstrated that hyperglycemia in rats that results from cold exposure is significantly suppressed by pretreatment with taurine. It has also been suggested that pancreatic taurine may play important physiological roles in protecting the function and/or structure of β-cells against pancreato-toxic substances such as streptozotocin[7,8]. The purpose of this study was to determine the effects of taurine supplementation and taurine depletion on the morphological changes of pancreatic β-cells in streptozotocin-induced diabetic rats.

MATERIALS AND METHODS

Six groups, each of 6 rats were used. Details of the treatments and diets are given Fig. 1 and Table 1 of the accompanying paper[8]. After 7 weeks the pancreas of each rat was removed and tissue samples were fixed in Bouin solution for paraffin sections, and prefixed in 2.5 % glutaraldehyde and postfixed in 1 % osmium tetroxide. Following dehydration, the samples were treated with Epon 812. Thin sections were stained with uranyl acetate and lead citrate and examined with a Jeol-100CX electron microscope.

RESULTS AND DISCUSSION

Morphological changes of rat pancreatic β-cells were observed by transmission electron microscopy (Fig 1). The ultrastructure of the islet cells of the normal non-diabetic group of rats (CO) was similar to that reported by others[9-12]. The β-cell of the non-diabetic group had many secretory granules, rough endoplasmic reticulum and rod shaped mitochondria. However, in the diabetic group (EO) it was severely damaged and the secretory granules were depleted. In the diabetic group that received 1% taurine-supplementation (E1), pancreatic β-cell was less damaged, compared to the E0 group and had some of normal secretory granules, but most of rough endoplasmic reticulum and mitochondria had been destroyed. The β-cell of the 2% taurine-supplemented diabetic group (E2) had swollen rough endoplasmic reticulum, round shaped mitochondria and some normal secretory granules. The β-cell of the 3% taurine-supplemented diabetic group (E3) was little different from that of the non-diabetic group. However, the β-cell of taurine-depleted diabetic group (EA) was not destroyed but had many small secretory granules which appeared to be immature. This was reflected in the blood glucose concentrations of this group[8]. Taken together with the data in the accompanying paper[8], these results indicate that taurine may prevent insulin-dependent diabetes by protection of the pancreatic β-cell and preserving normal secretory granule function. Thus, taurine supplementation may be recommended for prevention and treatment of diabetes.

ACKNOWLEDGEMENTS

We thank Dong-A Pharmaceutical Co. for the donation of taurine and Dr. I.S. Park for technical help in using the electron microscope.

REFERENCES

1. Rerup, C.C., 1970, Drugs producing diabetes through damage of the insulin secreting cells. *Pharmacol. Rev.* **22**:485-518.
2. Gaulton, G.N., Schwartz, J.L., and Eardley, D.D., 1985, Assessment of the diabetogenic drugs alloxan and streptozotocin as models for the study of immune defects in diabetic mice. *Diabetologia* **28**:769-775.
3. Zafirova, M., Jablenska, R., Popov, A., Goranova, I., Vassileva, E., Duhault, J., Marquie, G., and Petkov, P., 1991, Morphological characteristics of the endocrine pancreas in alloxandiabetes after cyclosporin A administration. *Cellular and Molecular Biology* **37**:585-596.
4. Wright, C. E., Tallen, H.H., Lin, Y. Y., Gaull, G.E., 1986, Taurine Biological update. *Ann Rev. Biochem* **55**:427-453.
5. Briel, G., Gylfe, E., Hellman and Neuhoff, V., 1972, Microdetermination of free amino acids in pancreatic islets isolated from obese-hyperglycemic mice. *Acta. Physiol. Scand.* **84**:247-253.
6. Nakagawa, N. and Kuriyama, K., 1975, Effect of taurine on alteration adrenal functions induced by stress. *Jap. J. Pharmac* **25**:737-746.
7. Tokunaga, H., Yoneda, Y. and Kuriyama, K., 1979, Protective actions of taurine against streptozotocin-induced hyperglycemia. *Biochemical Pharmacology* **28**:2807-2811.
8. Chang, K.J. and Kwon, W., 2000, Immunohistochemical localization of insulin in pancreatic β-cells of taurine-supplemented or taurine-depleted diabetic rats. *This volume.*
9. Orchi, L., 1974, A portrait of the pancreatic β-cell. *Diabetologia* **10**:163-187.
10. Fukuma, M., 1974, Electron microscope studies in the granule release from the rat pancreatic β-cells in organ culture. *J. Electron Microscopy* **23**:167-183.
11. Rickert, D.E., Fischer, L.J., Burke, J. P., Redick., J.A., Erlandsen, S.L., Parsons, J. A., and Van Orden, L.S., 1976, Cyproheptadine-induced insulin depletion in rat pancreatic beta

cells: Demonstration by light and electron microscopic immunocytochemistry. *Horm. Metab. Res.* **8**:430-434.

12. Hanai, N., 1984, Morphological and immunocytochemical study of rat pancreatic beta cell changes induced by cyclizine. *J. Appl. Toxicol.* **4**:308-314.

IMMUNOHISTOCHEMICAL LOCALIZATION OF INSULIN IN PANCREATIC β-CELLS OF TAURINE-SUPPLEMENTED OR TAURINE-DEPLETED DIABETIC RATS

Kyung Ja Chang and Woojung Kwon
Department of Food and Nutrition, Inha University, Inchon, 402-751, Korea

Abstract: The purpose of this study was to observe the effects of taurine supplementation or depletion on the immunohistochemical localization of insulin in pancreas of streptozotocin-induced diabetic rats. Male Sprague-Dawley rats were fed for 7 weeks with a purified diet that was supplemented with 0, 1, 2 or 3% taurine in their drinking water. To induce taurine depletion, rats were treated with 5% β-alanine in their drinking water. After 3 weeks, diabetes was induced by streptozotocin injection (50mg/kg body-weight). The pancreatic tissue was stained immunocytochemically, using an antibody to insulin, and examined by light microscopy. The insulin levels in pancreatic β-cells of the diabetic group that received no taurine-supplement were significantly decreased, compared to the non-diabetic group. The levels of insulin in β-cell of 1% and 2% taurine-supplemented diabetic groups were significantly higher than those of the diabetic group, whereas the levels in the group receiving 3% taurine were not significantly different from that of non-diabetic rats. Therefore, it may be suggested that taurine protect pancreatic β-cells against destruction by ptozotocin injection in a dose-dependent way.

INTRODUCTION

Diabetes mellitus is a complex metabolic disorder which is characterized by hyperglycemia. It is caused by either a deficiency or a defective action of insulin[1]. Diabetogenic drugs, such as alloxan and streptozotocin have provided reliable animal models of insulin-dependent diabetes mellitus[2]. These drugs cause destruction of pancreatic β-cells, which contain insulin-

Taurine 4, edited by Della Corte *et al.*
Kluwer Academic / Plenum Publishers, New York, 2000.

secreting granules, and induce hyperglycemia[3,4]. Taurine (2-aminoethane-sulfonic acid) is a β-amino acid that is present in high concentrations in mammalian tissues. It has been suggested that pancreatic taurine may protect the function and structure of pancreatic β-cell drugs such as streptozotocin[6]. The purpose of this study was to determine the dose-effects of taurine supplementation and the effects of taurine depletion on the immunohistochemical localization of insulin in the pancreas of streptozotocin-induced diabetic rats.

MATERIALS AND METHODS

Seventy-two male Sprague-Dawley rats, weighing about 90-100g each, were acclimatized for 4 days prior to their use in experiments. The room was maintained at 23-27 °C with a light cycle between 09:00 and 21:00 h. The animals were housed individually in stainless-steel wire-floored cages with free access to food and water. As shown in Figure 1, rats were divided into 6 groups by randomization. For 7 weeks the animals were fed the purified experimental diets and supplemented with 1% taurine in drinking water for 7 weeks. Three weeks after feeding the experimental diet (Table 1), diabetes was induced by a single thigh subcutaneous injection of streptozotocin (50 mg /kg body weight) after 16 h fasting of the diabetic groups. Daily fasting blood-glucose levels were determined with a reagent strip and a blood glucose sensor (ExacTech, Medicense, Inc. U.S.A.) The pancreas of rat was removed and tissue samples were fixed with Bouin solution for paraffin embedding. Serial thick paraffin sections were stained immunohistochemically for insulin using a polyclonal antibody and PAP staining kits by PAP technique[7], and examined with a light microscope. All values represent the mean $\pm$ S.E. All statistical analyses were carried out by Duncan's multiple-range test using the SAS program.

RESULTS AND DISCUSSION

During the experimental period diabetic rats had significantly lower body weight gain compared to that of the non-diabetic group and there were no significant differences in body weight gain between taurine-depleted and taurine-supplemented groups (Table 2). However, the taurine-depleted group (EA) had significantly lower body weight compared to those of the other groups from week 1 to week 3 of the experiment[8]. Also it had been reported that taurine-depletion using 5% β-alanine or 1% guanidinoethyl sulfate causes growth retardation[9,10], which was confirmed in this study.

Fasting blood glucose concentrations of the diabetic (E0) and 1% taurine-supplemented diabetic (E1) groups were significantly different compared to those of non-diabetic (CO) and EA groups (Table 2). However, fasting blood glucose concentrations of 2% taurine-supplemented (E2) and 3% taurine-supplemented (E3) groups were not significantly different compared to those of the CO and EA groups. These results were similar to those from oral glucose tolerance tests at 3 weeks after streptozotocin injection[8].

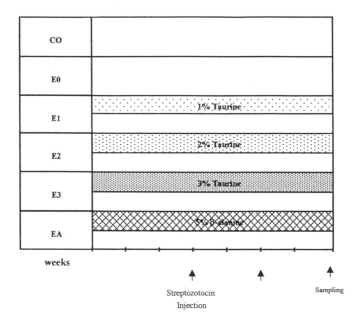

CO : Control group
E0 : Diabetic Group
E1 : Diabetic Group + 1% Taurine
E2 : Diabetic Group + 2% Taurine
E3 : Diabetic Group + 3% Taurine
EA : Diabetic Group + 5% β -alanine

Figure 1. Experimental design

Although the EA group showed slow growth and some death from stress due to streptozotocin injection, fasting blood glucose concentrations of this group were almost normal. This seems to have effects on food and water intake as well as urine volume[8]. However, further studies and discussion, particularly in relation to the immunohistochemical results are needed.

Table 1. Composition of the experimental diet (g/100g diet)

Ingredients	Before streptozotocin injection	After streptozotocin infection
Corn starch	45.0	60.0
Sucrose	15.0	-
Casein	20.0	20.0
α-Cellulose	5.7	7.0
Vitamin mixture*	1.0	1.0
Salt mixture[+]	4.0	4.0
DL-methionine	0.3	0.3
Corn oil	9.0	4.7
Olive oil	-	3.0

*Composition of Vitamin mixture, g/kg mixture: Vit. A Acetate (500,000 IU per g) 1.8g, Vit. D Concentrate (500,000 IU per g) 0.2125g, α-tocopherol (1000 IU per g, 50%) 11g, Ascorbic acid 45g, Choline Chloride (50%) 150g, Menadione (50%) 4.5g, Inositol 5g, P-aminobenzoic acid (PABA) 5g, niacin 4.25g, Riboflavin 1g, Thiamin hydrochloride 1g, Pyridoxine hydrochloride 1g, Calcium pantothenic acid 3g, Biotin (2%) 1g, Folic acid 0.09g, Vit. B_{12} (1%) 0.135g and Dextrose to 1kg.
[+]Composition of mineral mixture, g/kg mixture: $CaHPO_4$ 500g, NaCl 74g, K_2SO_4 52g, Potassium citrate monohydrate 220g, MgO 24g, Manganous carbonate (43-48% Mn) 3.5g, Ferric citrate (16-17% Fe) 6g, Zinc carbonate 1.6g, Cupric carbonate (53-55% Cu) 0.3g, KIO_3 0.01g, Chromium potassium sulfate 0.55g, Na_2SeO_3 0.01g, Sucrose to 1kg.

Table 2. Effects of taurine supplementation and taurine depletion on body weight gain and fasting blood glucose concentrations

Group	Body weight gain (g)	Fasting blood glucose concentration (mg/dl)
CO	225.92 ± 10.35^a	81.30 ± 11.19^a
E0	81.38 ± 20.13^b	213.38 ± 49.78^a
E1	115.52 ± 18.37^b	206.43 ± 47.95^a
E2	109.36 ± 14.41^b	110.33 ± 29.36^{ab}
E3	128.83 ± 12.75^b	167.80 ± 62.48^{ab}
EA	110.67 ± 29.94^b	71.25 ± 15.43^b

Values are Mean $\pm$ S.E. Means with different letters are significantly different at $p<0.05$ by Duncan's multiple range tests. CO: Control group, E0: Diabetic group, E1: Diabetic group +1% taurine, E2: Diabetic group +2% taurine, E3: Diabetic group + 3% taurine, EA: Diabetic group +5% β-alanine.

The distribution of insulin-immunoreactive cells in pancreatic islets from the non taurine-supplemented diabetic group (E0) was significantly decreased, compared to that of non-diabetic group (CO) (Table 3, Figure 2). Furthermore, the distribution of insulin-immunoreactive cells in pancreatic islets from the taurine-depleted diabetic group (EA) was the lowest among all the groups. However, the distributions of insulin-immunoreactive cells in pancreatic islets from the 1% and 2% taurine-supplemented diabetic groups (E1, E2) were significantly less decreased compared to that of the non taurine-supplemented diabetic group (E0). In the 3% taurine-supplemented group (E3), the distribution of insulin-immunoreactive cells in pancreatic islets was not significantly different from that of non-diabetic group (CO).

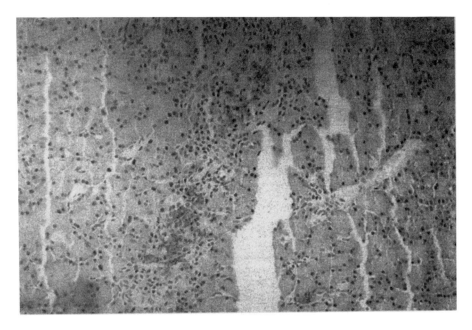

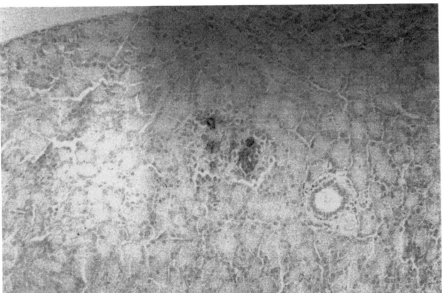

Fig. 2 A (Upper), B (Lower). Insulin-immunoreactive cells of pancreatic islets of rats X 10,000. A. Control group (CO), B. Diabetic group (E0).

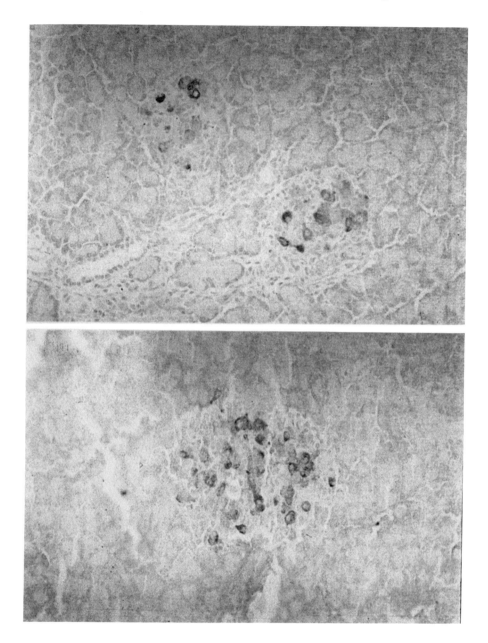

Fig. 2C (Upper), D (Lower). Insulin-immunoreactive cells of pancreatic islets of rats. X 10,000, C. Diabetic group + 1% taurine (E1), D. Diabetic group + 2% taurine (E2).

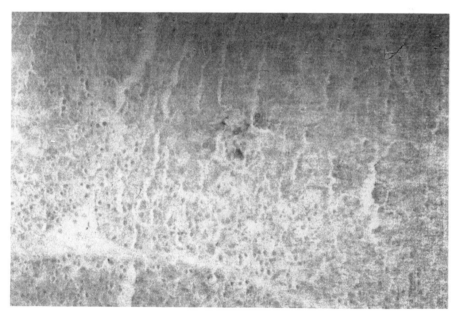

Figure 2 E (Upper) F (Lower). Insulin-immunoreactive cells of pancreatic islets of rats X 10,000. E. Diabetic group + 3% taurine (E3), F. Diabetic group + 5% ß-alanine (EA).

Table 3. Effects of taurine suppplementation and taurine depletion on distribution of insulin-immunoreactive cells in pancreatic islets

Group	Distribution of insulin-immunoreactive cells
CO	85.30 ± 2.43^a
E0	43.92 ± 1.73^c
E1	58.90 ± 5.59^b
E2	60.00 ± 4.49^b
E3	82.04 ± 2.68^a
EA	32.36 ± 3.42^d

Values are Mean $\pm$ S.E. Means with different letters are significantly different at p<0.05 by Duncan's multiple range test. CO: Control group, E0: Diabetic group, E1: Diabetic group +1% taurine, E2: Diabetic group +2% taurine, E3: Diabetic group +3% taurine, EA: Diabetic group +5% β-alanine.

Using antibodies to insulin, clear brown color reaction products were observed in pancreatic islet cells from all the groups, as has been already demonstrated by other investigators[2,11,12]. In case of the diabetic groups, the staining intensity became gradually fainter in a dose-dependent way (E3>E2>E1>EA). Therefore, it may be suggested that taurine protects pancreatic β-cells against destruction by streptozotocin injection in a dose-dependent way.

ACKNOWLEDGEMENTS

This study was supported by the '97 research grant of Inha University. We thank Dong-A Pharmaceutical Co. for the donation of taurine and Dr.T.S. Hwang for technical help with the immunohistochemical reactions.

REFERENCES

1. Mann, J., 1998, Diabetes mellitus. In *Essentials of Human Nutrition* (J. Mann and A. S. Truswell, eds), Academic Press, New York, pp.327-338.
2. Zafirova, M., Jablenska, R., Popov, A., Goranova, I., Vassileva, E., Duhault, J., Marquie, G. and Petkov, P., 1991, Morphological characteristics of the endocrine pancreas in alloxan diabetes after cyclosporin A administration. *Cellular and Molecular Biology* **37**:585-596
3. Rerup, C.C., 1970, Drugs producing diabetes through damage of the insulin secreting cells. *Pharmacol. Rev* **22**:485-518.
4. Gaulton, G.N., Schwartz, J.L., and Eardley, D.D., 1985, Assessment of the diabetogenic drugs alloxan and streptozotocin as models for the study of immune defects in diabetic mice. *Diabetologia* **28**:769-775.
5. Wright, C. E., Tallen, H.H., Lin, Y. Y., Gaull, G.E., 1986, Taurine: Biological update. *Ann Rev. Biochem.* **55**:427-453.

6. Tokunaga, H., Yoneda, Y. and Kuriyama, K., 1979, Protective actions of taurine against streptozotocin-induced hyperglycemia. *Biochemical Pharmacology* **28**:2807-2811.

7. Sternberger, L. A., 1979, The unlabelled antibody peroxidase-antiperoxidase (PAP) method. In *Immunocytochemistry, 2nd ed.* (L. A. Sternberger eds.), John Wiley and Sons Ins., New York, pp.104-167.

8. Chang, K. J., 1999, Effects of taurine and β-alanine on blood glucose and blood lipid concentrations in streptozotocin-induced diabetic rats. *Korean J. Nutrition* **32**:213-220.

9. Lake, N., 1982, Depletion of taurine in the adult rat retina. *Neurochem. Res.* **7**:1385-1390.

10. Lake, N., and Marte, L., 1988, Effects of β-alanine treatment on the taurine and DNA content of the rat heart and retina. *Neurochem . Res.* **13**:1003-1006.

11. Rickert, D.E., Fischer, L.J., Burke, J. P., Redick., J.A., Erlandsen, S.L., Parsons, J. A., and Van Orden, L.S., 1976, Cyproheptadine-induced insulin depletion in rat pancreatic beta cells: Demonstration by light and electron microscopic immunocytochemistry. *Horm. Metab. Res.* **8**:430-434.

12. Hanai, N., 1984, Morphological and immunocytochemical study of rat pancreatic beta cell changes induced by cyclizine. *J. Appl. Toxicol.* **4**:308-314.

TAURINE ATTENUATES THE INDUCTION OF IMMEDIATE-EARLY GENE EXPRESSION BY PDGF-BB

[1]Keisuke Imada, [1]Takaaki Takenaga, [1]Susumu Otomo, [2]Yu Hosokawa and [2]Masayuki Totani

[1]*Pharmacological Evaluation Laboratory, Taisho Pharmaceutical Co., Ltd., Ohmiya, Japan:*
[2]*Division of Maternal and Child Health Science, The National Institute of Health and Nutrition, Tokyo, Japan*

INTRODUCTION

Taurine is a most abundant amino acid distributed widely in mammalian tissues, and contributes to cell homeostasis by the osmoregulatory activity, the regulation of protein phosphorylation, calcium modulation and several other actions[1]. It has been reported that taurine prevents the development of atherosclerosis in cholesterol-fed rabbits without any lipid-lowering effects, however the mechanism is unknown[2].

Platelet-derived growth factor (PDGF) is a major mitogen, and PDGF-BB, one of the subtypes of PDGF, plays a critical role for the pathogenesis of atherosclerosis. PDGF-BB promotes both of the migration and the proliferation of vascular smooth muscle cells, resulting in the thickening of aortic vessel-walls[3]. Thus we examined whether taurine prevents the promotion of atherosclerosis by PDGF-BB to reveal the mechanism of anti-atherosclerosis effect of taurine.

Taurine 4, edited by Della Corte *et al.*
Kluwer Academic / Plenum Publishers, New York, 2000.

589

MATERIALS AND METHODS

Cell cultures

Balb/3T12-3, NIH/3T3 and A7r5 (rat aortic smooth muscle cell) from ATCC were cultured in Dulbecco's modified eagle medium (DMEM) containing with 10% (v/v) Fetal Bovine Serum (FBS) (or 10% (v/v) Calf Serum (CS) for NIH/3T3) under 95% air-5% CO_2 condition. For the experiments, cells were seeded in the culture dishes or plates at the concentration of 5×10^4 cells/ml. After 24 hr their media were changed to starvation media (DMEM containing 0.5% (v/v) FBS or 0.5% (v/v) CS) and further incubated for 24 hr to be in a resting phase, and then treated with taurine and other reagents.

RNA extraction and Northern blot analysis

After 45 min of treating the cells with taurine and other reagents, cells were washed with phosphate buffered saline (PBS) and the total RNA was extracted by acid guanidinium-phenol-chloroform (AGPC) method[4]. Ten μg of total RNA was subjected to Northern blot analyses for c-fos mRNA and GAPDH mRNA. Relative expression of c-fos were quantified by the densitometric scanning, and revised by that of GAPDH (NIH Image version 1.60).

Reverse transcriptation-polymerase chain reaction (RT-PCR)

Ets-1 mRNA expression were determined by RT-PCR. Four μg of total RNA was subjected to reverse transcriptase reaction, and then the Ets-1 cDNA was amplified by PCR (30 cycles) using the primers; 5'-CAGGCACTGAAAGCTACCTT-3' for forward and 5'-TGAAAGATGACTGGCTGCTC-3' for reverse. PCR products (684 bp) were subjected to agarose gel electrophoresis.

RESULTS AND DISCUSSION

One of the early immediate genes, c-fos is well known as a gene induced by several cytokines, growth factors, and other mitogenic stimuli[5]. Thus we examined the effect of taurine on the c-fos expression stimulated by several treatments including PDGF-BB. As shown in Fig. 1, taurine attenuated the induction of c-fos expression by PDGF-BB or the calcium ionophore A23187, but not by 20% (v/v) serum or the typical protein kinase C activator, phorbol 12-mylistate 13-acetate (PMA), in Balb/3T12-3 cells, which are a mouse fetal fibroblast cell line.

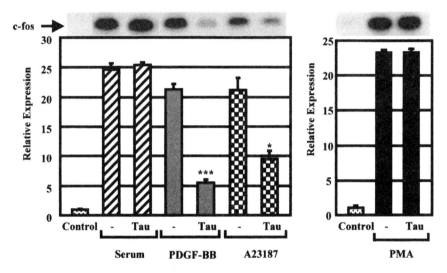

Figure 1. Effect of Taurine on the c-fos expression induced by serum, PDGF-BB, A23187 or PMA in Balb/3T12-3 cells. Cells in a resting phase were treated with taurine (10 mM) followed by the treatment with serum (20%), PDGF-BB(10 ng/ml), A23187 (5 µM) or PMA (100 nM), and c-fos mRNA was determined by Northern blot analysis. The relative expression was quantified by densitometric scanning, taking the control cells as 1. *:significantly different from the cells treated with A23187 (student's t-test, p<0.05) ***: significantly different from the cells treated with PDGF-BB (student's t-test, p<0.001)

Subsequently, we focused on the action of PDGF-BB using NIH/3T3 cells, which are well known to be responsive to PDGF-BB. Taurine attenuated the c-fos expression by PDGF-BB in a dose-dependent manner (data not shown), and other taurine-related compounds such as β-alanine or GABA did not change the c-fos expression induced by PDGF-BB (Fig. 2A). On the other hand, taurine did not attenuate the c-fos induction by PDGF-AA, which is another subtype of PDGF (Fig. 2B). We also confirmed the effect of taurine in A7r5 cells, which are rat aortic vascular smooth muscle cell line.

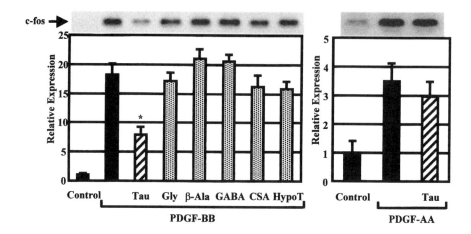

Figure 2. Effect of Taurine and the related compounds on the expression of c-fos induced by PDGF-BB or PDGF-AA in NIH/3T3 cells. Cells in a resting phase were treated with taurine (Tau) or the other related compounds; glycine (Gly), β-alanine (β-Ala), γ-aminobutyric acid (GABA), cysteinesulufinic acid (CSA) or hypotaurine (HypoT) (10 mM each) followed by the treatment with PDGF-BB (10 ng/ml) or PDGF-AA (10 ng/ml). After 45 min incubation, c-fos mRNA was determined by Northern blot analysis. The relative expression was quantified by densitometric scanning, taking the control cells as 1. *: significantly different from the cells treated with PDGF-BB (Dunnett's test, p<0.05)

Taurine attenuated the c-fos expression by PDGF-BB. As with NIH/3T3 cells, long-term incubation with taurine before PDGF-BB treatment attenuated the taurine effect. Thus it is suggested that taurine ubiquitously attenuates the c-fos expression by PDGF-BB in the same fashion. We also examined the Ets-1 expression by PDGF-BB. Ets-1 is also one of the immediate early gene and contributes to the expression of, among others, the matrix-degrading enzyme/stromelysin[6,7]. Ets-1 expression was confirmed as a 684 bp PCR product induced by PDGF-BB or PMA. Taurine suppressed the expression induced by PDGF-BB, not by PMA (Fig. 3).

Finally, the effect of taurine on the cell proliferation by PDGF-BB was examined. Cells in a resting phase were treated with taurine and PDGF-BB, and 72 h after treatment, cell numbers were counted under microscopy. As shown in Fig. 4, taurine suppressed the cell proliferation promoted by PDGF-BB.

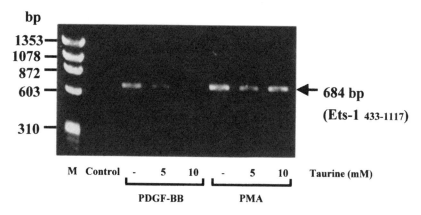

Figure 3. Effect of Taurine on the Ets-1 expression induced by PDGF-BB or PMA in A7r5 cells. Cells in a resting phase were treated with various concentration of taurine followed by the treatment with PDGF-BB (10 ng/ml) or PMA (100 nM), and Ets-1 mRNA was determined by RT-PCR.

In this study, we demonstrated that taurine selectively attenuates the expressions of c-fos and Ets-1 induced by PDGF-BB, but not by PDGF-AA. This suppressive effect appears to be independent of PKC pathway. Calcium uptake might be involved in the effect of taurine, because taurine also attenuated the c-fos expression induced by the calcium ionophore A23187. It has been reported that taurine suppresses the expressions of c-fos and c-jun induced by angiotensin II in cardiac cells[8]. Thus the point of the action of taurine might be on the MAP-kinase pathway and calcium-dependent pathway such as calmodulin kinase. Taurine is naturally taken-up into the cells through the taurine transporter[1], and thus regulation of taurine transporter is considered to be important. Further studies are needed to determine how taurine acts on gene expression.

It is noteworthy that taurine did not attenuate the c-fos expression induced by PDGF-AA, which is non-responsive in the development of atherosclerosis[3]. Specific suppression by taurine on the c-fos expression induced by PDGF-BB is very likely to underlie its action against atherosclerosis. Indeed, taurine suppressed the cell proliferation promoted by PDGF-BB in vascular smooth muscle cells. These results suggest that taurine is effective in the suppression of vessel-wall thickening. Taken together, these data suggest that taurine might be a drug for the prevention of coronary heart disease.

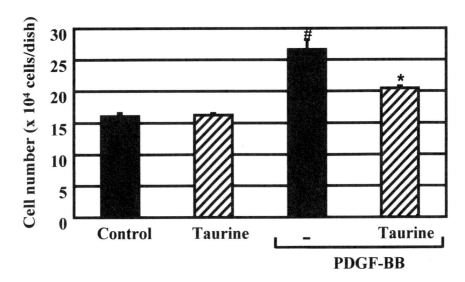

Figure 4. Effect of Taurine on the cell proliferation promoted by PDGF-BB in A7r5 cells. Cells in a resting phase were treated with various concentration of taurine followed by the treatment with PDGF-BB (10 ng/ml). After 72 hr, cell numbers were counted under microscopy. #: significantly different from the control cells (Student's t-test, $p<0.05$) *: significantly different from the cells treated with PDGF-BB (Student's t-test, $p<0.05$)

REFERENCES

1. Huxtable, R.J., 1992, Physiological actions of taurine, *Physiol. Rev.* **72**:101-163
2. Petty, M.A., Kintz, J. and DiFrancesco, G.F., 1990, The effect of taurine on atherosclerosis development in cholesterol-fed rabbits, *Eur. J. Pharmacol.* **180**:119-127
3. Hughes, A.D., Clunn, G.F., Refson, J. and Demoliou-Mason, C, 1996, Platelet-derived growth factor (PDGF): actions and mechanisms in vascular smooth muscle, *Gen. Pharmacol.* **27**:1079-1089
4. Chomczynski, P. and Sacchi, N., 1987, Single-step method of RNA isolation by acid guanidinium thiocyanate-phenol-chloroform extraction, Anal. Biochem. 162:156-159
5. Distel, R.J. and Spiegelman, B.M., 1990, Protooncogene c-fos as a transcription factor, *Adv. Cancer Res.* **50**:37-55
6. Wasylyk, B., Wasylyk, C., Flores, B., Begue, A., Leprince, D. and Stehelin, D., 1990, The c-ets proto-oncogenes encode transcription factors that cooperate with c-Fos and c-Jun for transcriptional activation, *Nature* **346**:191-193
7. Wasylyk, C., Gutman, A., Nicholson, R. and Wasylyk, B., 1991, The c-Ets oncoprotein activates the stromelysin promoter through the same elements as several non-nuclear oncoproteins, *EMBO J.* **10**:1127-1134
8. Takahashi, K., Azuma, M., Taira, K., Baba, A., Yamamoto, I., Schaffer, S.W. and Azuma, J., 1997, Effect of taurine on angiotensin II-induced hypertrophy of neonatal rat cardiac cells, J. *Cardiovascul. Pharmacol.* **30**:725-730.

POST-OPERATIVE MONITORING OF CORTICAL TAURINE IN PATIENTS WITH SUBARACHNOID HEMORRHAGE: A MICRODIALYSIS STUDY

Enrico De Micheli, Giampietro Pinna, Alex Alfieri, Giovanna Caramia, Loria Bianchi*, Maria A. Colivicchi*, Laura Della Corte* and Albino Bricolo
Dipartimento di Neurochirurgia, Università ed Ospedale di Verona, Italia
*Dipartimento di Farmacologia Preclinica e Clinica «M. Aiazzi Mancini», Università di Firenze, Italia

Abstract: Intracerebral MD enables the retrieval of endogenous substances from the extracellular fluid (ECF) of the brain and has been demonstrated to be a sensitive technique for early detection of subtle vasospasm-induced neurometabolic abnormalities in patients with subarachnoid hemorrhage (SAH). The aim of this study was to monitor cortical extracellular concentrations of energy metabolism markers, such as glucose and lactate, neurotransmitter amino acids, such as glutamate, aspartate, GABA and taurine to identify any neurochemical patterns of cerebral ischemia. A prospective clinical study was conducted on a group of 16 patients with non-severe SAH operated on within 72 hours after initial bleeding. Following aneurysm clipping, an MD catheter was inserted in the cortical region where vasospasm could be expected to develop, and perfused with artificial CSF at 0.3 µl/min flow rate. Dialysate was collected every 6 hours and then analyzed on High Performance Liquid Cromatography (HPLC) for glucose, lactate, pyruvate, glutamate, aspartate, GABA and taurine. Mean ECF taurine concentrations ranged from 1.4 ± 0.7 to 12.3 ± 7.8 µmol/l in single patients: global mean value was 5.8 ± 3.8 µmol/l. In this series, the highest absolute taurine value was 25.7 µmol/l, observed in a patient who developed clinical and radiological signs of cerebral ischemia. Nine patients presented clinical disturbances related to cerebral vasospasm. In this setting, representing a mild-to-moderate hypoxic condition, MD data demonstrated that lactate is the most sensitive marker of cellular energy imbalance. Increased lactate levels positively correlated with glutamate ($P<0.0001$),

Taurine 4, edited by Della Corte *et al.*
Kluwer Academic / Plenum Publishers, New York, 2000.

aspartate (P<0.0001), GABA (P<0.0001) and taurine (P<0.0001) concentrations. These results suggest that also in humans increased taurine levels reflect a condition of cellular stress. This study confirms that MD is a sensitive technique to reveal subtle metabolic abnormalities possibly resulting in cell damage.

INTRODUCTION

Microdialysis is a new technique which allows the continuous acquisition of samples from a limited area of the brain with good temporal resolution. This method enables the retrieval of critical biochemical substances from the extracellular fluid (ECF) of the brain, thus representing a valid tool for studying the mechanisms of cerebral secondary ischemia in humans[9,10,12].

One of the most common causes of unfavorable outcome following subarachnoid aneurysmal hemorrhage (SAH) is cerebral vasospasm-induced delayed ischemia, often leading to reversible or permanent neurological deficits as a consequence of neuronal damage or death[5,11]. Cerebral vasospasm is the narrowing of the largest cerebral arteries, which occurs in response to subarachnoid bleeding. Vasospasm incidence generally peaks between 5 and 7 days after hemorrhage and often causes a reduction in regional blood flow, which may lead to focal ischemia. Often a focal rather than a diffuse phenomenon, vasospasm development is not always detected by Transcranial Doppler evaluation nor routine neuroradiological exams, therefore, a technique such as MD may be important for direct neurochemical monitoring.

The objectives of this study were: 1) to test the sensitivity of MD to detect the neurochemical changes due to vasospasm in patients with SAH; and 2) to identify patterns of neurochemical markers which may assist in predicting the development of cerebral ischemia. Patients with non-severe SAH were studied to determine if a correlation exists between MD data, clinical course and outcome, blood flow velocity and Computed Tomography (CT) imaging.

SUBJECTS AND METHODS

Subjects

Inclusion criteria for this study was based on the following: 1) patient age (>18 years); 2) admission in good neurological condition[3] (Grades I-III and World Federation of Neurological Surgeons Scale[15] Grades 1-3); 3) SAH diagnosis by clinical history and confirmed by CT; 4) angiography demonstrating a saccular aneurysm; 5) early surgical intervention (<72 hours after SAH).

Methods

During hospitalization, patients underwent repeated CT and TCD exams; flow velocity was calculated based on the middle cerebral artery (MCA) mean value, ipsilateral to the MD catheter[8,14]. MCA flow velocity values above 80 cm/sec were considered pathologic[13]. Patient outcome was evaluated based on neurological examination according to the Glasgow Outcome Scale[4] (GOS) and CT scans performed 6 and 12 months after SAH.

During surgery, a 10 mm flexible microdialysis catheter with an external diameter of 0.5 mm (CMA/70 custom catheter, CMA/Microdialysis, Solna, Sweden) was inserted in the subfrontal (patients with AcoA and ACA aneurysm) or temporal (ICA and MCA) cortex, and perfused with artificial CSF (140mM Na^+, 2.7 mM K^+, 1.2mM Ca^{2+}, 0.9mM Mg^{2+}, 147mmol Cl^-) at a rate of 0.3 μl/min using a microinjection pump (CMA 106, CMA/Microdialysis, Solna, Sweden). An equilibration period of 60 minutes without sampling was allowed after catheter implantation. The dialysate was collected every 6 hours.. Microdialysis samples were divided into aliquots and analyzed enzymatically on two different High Performance Liquid Cromatography (HPLC) systems. Glucose, lactate and pyruvate were analyzed using an enzymatic technique (CMA 600 Analyzer, CMA/Microdialysis, Solna, Sweden). Amino acids were fluorimetrically detected following precolumn derivatization with o-phthaldialdehyde as described by Bianchi et al.[1] The study was approved by the Hospital Ethics Committee and patients gave informed consent.

Statistics

The results were analyzed by simple regression analysis using commercially available software (Stat View 5.0, Abacus Concepts Inc., Berkeley, CA.; SPSS 8.0; SPSS Inc., Chicago, IL.). Differences with a probability value of less than 0.05 were considered statistically significant.

RESULTS

Sixteen patients were included in this study. Clinical characteristics of the patients are illustrated in Table 1a and b. Nine patients presented vasospasm-related clinical disturbances during the microdialysis period (Cases 1-3, 6, 9,

10, 14-16); transient in 8 patients, and persistent in 1 (Case 2). In these cases, MD data showed a progressive increase of lactate and decrease of glucose. This trend was also seen in patients with increased blood flow velocity but with an uneventful clinical course.

Table 1 a. Clinical characteristics of 16 patients with non-severe SAH who underwent post-operative intracerebral microdialysis[§]

Case	Admission Grade (H & H)	Aneurysm Location	Surgery Day	Catheter Location	° MD Days
1	III	ACoA	1	L F	9
2	II	R ICA	3	R T	11
3	III	L ICA	2	L T	9
4	II	L MCA	2	L T	11
5	II	L ACA	2	L F	2
6	II	ACoA	3	L T	6
7	II	ACoA	2	R F	9
8	II	ACoA	3	L F	9
9	II	R MCA	2	R T	11
10	II	ACoA	3	R F	8
11	II	R ICA	2	R T	9
12	II	R MCA	3	R T	6
13	II	R ICA	2	R T	6
14	II	ACoA	1	R F	5
15	III	ACoA	3	R F	12
16	II	ACoA	2	R F	5

[§]Abbreviations: ACA = anterior cerebral artery; ACoA = anterior communicating artery; F = Frontal; H & H = Hunt & Hess; ICA = internal carotid artery; L = left; MCA = middle cerebral artery; R = right; T = Temporal; TCD = transcranial doppler.
° MD Days refers to duration of microdialysis measurement
* Blood flow velocity values > 120 cm/sec.

In 9 patients (Cases 8-16), ECF pyruvate levels were also measured; Lactate/Pyruvate (L/P) ratios were found to be above the reported «normal» threshold of 25^{10} in 8 cases, while no correlation between L/P ratio and clinical status was found.

Glutamate levels (mean 4.4 µmol/l) in this study are comparable to those previously found in SAH patients with favorable outcome[10]. In addition, glutamate was also found to positively correlate with lactate (P< 0.0001), but not with G/L ratio or L/P ratio, as in severe SAH[10]. The highest glutamate values found were in those patients who later developed an ischemic area adjacent to the catheter site (Case 10) (Fig. 1).

Mean taurine levels in individual patients (Table 2) range from 1.4 to 12.27 µmol/l. Taurine was also measured in a patient who underwent surgery for an aneurysm which had not bled: in this case which we consider as a control, the mean value resulted as the lowest mean value of taurine in the study (0.84 µmol/l).

Table 1 b. Clinical characteristics of 16 patients with non-severe SAH who underwent post-operative intracerebral microdialysis[§]

Case	TCD Vasospasm	Clinical Vasospasm	Ischemia on CT	Outcome (GOS[12])
1	* Yes	Yes		GR
2	* Yes	Yes	Yes	Md
3	Yes	Yes		GR
4	* Yes			GR
5				GR
6		Yes		GR
7	Yes			GR
8	Yes			GR
9	* Yes	Yes	Yes	GR
10	Yes	Yes	Yes	GR
11				GR
12	Yes			GR
13	* Yes			GR
14	* Yes	Yes		GR
15		Yes	Yes	GR
16	Yes	Yes		GR

[§]Abbreviations: CT = computerized tomography; GOS[12] = Glasgow Outcome Scale; GR = good recovery; Md = moderate disability; TCD = transcranial doppler.
* Blood flow velocity values > 120 cm/sec.

Table 2. Taurine levels of the cortical ECF in SAH patients

Patient	Mean	SD	Min	Max
1	6.3	1.42	4.06	10.07
2	3.75	1.52	1.82	9.31
3	6.77	0.98	4.3	9.56
4	1.75	0.92	0.55	3.47
5	4.35	0.96	6.25	3.61
6	6.19	0.9	4.65	7.16
7	5.18	1.01	3.43	8.31
8	4.13	0.79	2.91	5.96
9	7.01	4.69	2.65	25.71
10	2.48	1.21	1.19	6.8
11	6.48	1.42	3.99	9.39
12	10.43	2.68	6.76	13.73
13	12.27	7.79	2.44	21.38
14	4.42	2.79	2.04	11.32
15	9.8	1.5	7.75	16.26
16	1.4	0.7	0.6	3.33
CTRL	*0.84*	*0.86*	*0.2*	*3.12*

Abbreviations: SD = standard deviation; CTRL = control patient operated for cerebral aneurysm without subarachnoid hemorrhage.

The highest absolute values for taurine were found in a patient who later developed an ischemic area adjacent to the catheter site (Case 9) (Fig. 2).

In all patients, CT scans were obtained before, during and after the microdialysis study, and at the 3-month follow-up. In 4 cases (Cases 2, 9, 10,

15), sequential CT showed a delayed development of a persistent hypodense lesion (cerebral infarction). In 2 patients, in which the cerebral infarct was adjacent to the catheter site, MD revealed increasing glutamate (Case 10) and taurine (Case 9) levels, eventually reaching the highest values found in this study.

Clinical outcome according to the Glasgow Outcome Scale (GOS)[4] was assessed at 6 and 12 months after SAH. Fifteen patients made a good recovery, and one was evaluated as moderately disabled (Case 2).

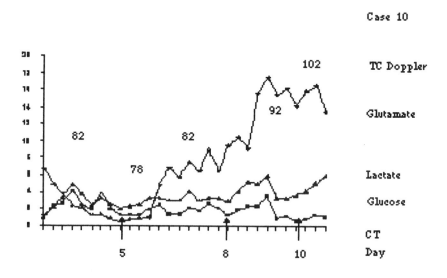

Figure 1. Case 10: MD Profile. Glucose (mmol/l), lactate (mmol/l) and glutamate (μmol/l) levels in the ECF. Glutamate was found to rise much earlier than blood flow velocity (Day 9) and CT (Day 10) became pathologic.

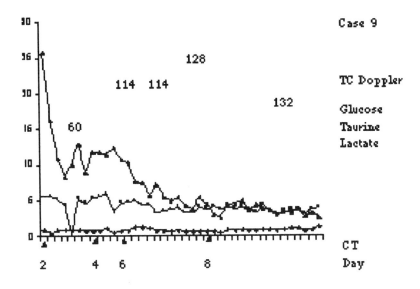

Figure 2. Case 9: MD Profile. Glucose (mmol/l), lactate (mmol/l) and taurine (µmol/l) levels in the ECF. High taurine levels were present when blood flow velocity was still normal and much earlier than ischemia was visible on CT (Day 8).

DISCUSSION

Aneurysmal subarachnoid hemorrhage is a cerebrovascular event with high morbidity and mortality[6,7]. The major causes related to poor outcome are the effects of initial bleeding, surgical complications and delayed ischemic deterioration due to cerebral vasospasm, with the latter representing the leading cause of disability and death in these patients and accounting for nearly 30% of unfavorable outcomes[6,7,11]. Since arterial vasospasm is more often a focal rather than diffuse phenomenon, cerebral blood flow velocities, which are measured with TCD from the large cerebral arteries, may not always reflect microcirculatory alterations. Intracerebral microdialysis is a more sophisticated technique to monitor metabolic changes in cerebral areas which are considered at risk for secondary ischemia development[10,12]. Since it is generally accepted that vasospasm rarely occurs immediately after SAH, monitoring the fluctuations in the ECF levels of neurochemical markers may be useful in predicting impending cerebral hypoxic and/or ischemic phenomena.

The physiopathology of vasospasm-induced ischemic damage is still poorly understood; yet two mechanisms, progressive cellular energy failure and massive extracellular release of excitotoxic amino acids, have been suggested to play a crucial role[2,12]. ECF glucose levels indicate the availability of the major substrate for brain energy metabolism, the supply being directly dependent on blood flow. Interstitial lactate level is a marker of cellular oxygen deficiency which indicates cellular anaerobic energy production. In response to mild to moderate hypoxia, when the oxidative phosphorylation is preserved, glycolysis increases with a higher production of both lactate and pyruvate and a less marked increase of the L/P ratio. In this condition, lactate has proven to be a more sensitive marker of cellular energy imbalance than the L/P ratio[10], as confirmed also in this study, where a significant positive correlation between glutamate and lactate and between taurine and lactate was found. Excitatory amino acids are of particular importance for ischemic neuronal damage, and their massive release into the extracellular space may exert excitotoxic effects eventually leading to cell death[2,12]. Our study suggests that sustained high levels of glutamate and taurine, when associated with increasing lactate production, may predict the development of irreversible ischemic processes.

The results of this study suggest the following: cerebral metabolic perturbation occurs to a certain degree also in non-severe SAH patients with uneventful clinical course and good outcome; in mild-to-moderate hypoxia, lactate is a more sensitive marker of the cellular energy metabolism state than the L/P ratio; in hypoxic conditions, positive correlation between glutamate and lactate and between taurine and lactate exist; sustained elevated levels of glutamate and taurine, when associated with increased lactate production, may represent a pattern predicting an impending condition of cellular ischemia.

This study demonstrates the sensitivity of intracerebral microdialysis in revealing subtle metabolic abnormalities in patients with non-severe SAH. Consequently, MD may prove to be a useful neurochemical monitoring tool for early detection of impending spasm-induced cerebral ischemia in routine clinical practice.

ACKNOWLEDGEMENTS

The authors wish to thank Victoria M. Praino for her assistance to the manuscript. This work was partly supported by UE, BMH1-CT-1402 and COST D8 & D13, and by MURST (Italy).

REFERENCES

1. Bianchi L, Della Corte L, Tipton KF (1999) Simultaneous determination of basal and evoked output levels of aspartate, glutamate, taurine and 4-aminobutyric acid during microdialysis and from superfused brain slices. *J. Chromatogr. B* **723**:47-59.
2. Choi, D.W., 1988, Glutamate neurotoxicity and diseases of the nervous system. *Neuron* 1:623-634.
3. Hunt, W.E., Hess, R.M., 1968, Surgical risk as related to time of intervention in the repair of intracranial aneurysms. *J Neurosurg.* **28**:14-19.
4. Jennet, B., Bond, M., 1975, Assessment of outcome after severe brain damage. A practical scale. *Lancet* **1**:480-484.
5. Kassell, N.F., Sasaki, T., Colohan, A.R.T., et al., 1985, Cerebral vasospasm following aneurysmal subarachnoid hemorrhage. *Stroke* **16**:562-572.
6. Kassel, N.F., Torner, J.C., Haley, E.C., Jr., et al., 1990, The international cooperative study on the timing of aneurysm surgery Part 1: Overall management results. *J Neurosurg.* **73**:18-36.
7. Kassel, N.F., Torner, J.C., Jane, J.A., et al., 1990, The international cooperative study on the timing of aneurysm surgery. Part 2: Surgical results. *J Neurosurg.* **73**:37-47.
8. Lindegaard, K.F., Nornes, H., Bakke, S.J., et al., Cerebral vasospasm diagnosis by means of angiography and blood velocity measurements. *Acta Neurochir.* **100**:12-24.
9. Persson, L., Hillered, L., 1992, Chemical monitoring of neurosurgical intensive care patients using intracerebral microdialysis. *J Neurosurg.* **76**:72-80.
10. Persson, L., Valtysson, J., Enblad, P., et al., 1996, Neurochemical monitoring using intracerebral microdialysis in patients with subarachnoid hemorrhage. *J Neurosurg.* **84**:606-616.
11. Saveland, H., Hillman, J., Brandt, L., et al., 1992, Overall outcome in aneurysmal subarachnoid hemorrhage: a prospective study from neurosurgical units in Sweden during a 1-year period. *J Neurosurg.* **76**:729-734.
12. Saveland, H., Nilsson, O., Boris-Moller, F., et al., 1996, Intracerebral microdialysis of glutamate and aspartate in two vascular territories after aneurysmal subarachnoid hemorrhage. *Neurosurgery* **38**:12-20.
13. Seiler, R.W., Grolimund, P., Aaslid, R., et al., 1986, Cerebral vasospasm evaluated by transcranial ultrasound correlated with clinical grade and CT-visualized subarachnoid hemorrhage. *J Neurosurg.* **64**:594-600.
14. Sloan, M.A., Haley, E.C. Jr, Kassel, N.F., et al., 1989, Sensitivity and specificity of transcranial Doppler ultrasonography in the diagnosis of vasospasm following subarachnoid hemorrhage. *Neurology* **39**:1514-1518.
15. Teasdale, G., Knill-Jones, R.P., Lindsay, K.W., 1983, Clinical assessment of SAH. *J. Neurosurg.* **59**:550-555.

TAURINE INTAKE AND EXCRETION IN PATIENTS UNDERGOING LONG TERM ENTERAL NUTRITION

K. H. Cho[1], E. S. Kim[1] and J. D. Chen[2]

[1]Dept. Food Science and Nutrition, Dankook University, Seoul 140-714, Korea, [2]Institute of Sports Medicine, The Third Teaching School of Clinical Medicine, Beijing Medical University, Beijing 100083, China

Abstract: The purpose of this study was to investigate whether serum concentration and urinary excretion of taurine are influenced by marginal taurine intake. Twenty one male patients (75 to 95 years old), suffering from coronary heart disease, multiple cerebral infarction, cancer, subdural hematoma or respiratory failure were grouped according to duration of tube feeding (group one, 5.9 ± 2.9; group two, 14.8 ± 2.3; group three 48.0 ± 22.7, mean $\pm$ SD, months). The mean intake of taurine was 347.0 ± 25.6, 339.8 ± 25.6 and 337.1 ± 26.9 µmol/day (mean $\pm$ SEM) in group one, two and three, respectively. The fasting serum taurine levels were 106.5 ± 9.6, 95.0 ± 9.9 and 56.8 ± 11.0 µmol/L (mean $\pm$ SEM) in group one, two and three, respectively. Taurine level in group three patients was significantly lower than that of group one and two ($p<0.05$). The twenty-four hour urinary taurine excretion was 776.1 ± 176.7, 782.4 ± 245.3 and 388.3 ± 169.3 µmol/day (mean $\pm$ SEM) in group one, two and three, respectively. These results suggest that marginal taurine intake in patients receiving long term tube feeding could result in taurine deficiency.

INTRODUCTION

Plasma and urine taurine concentrations in infants fed on taurine-free formulas as well as in children and adults maintained on long term parenteral nutrition are reduced[1,2]. Infants undergoing taurine-free parenteral nutrition and children fed on taurine-free formulas are prone to develop retinal abnormalities[2] and immature brainstem auditory evoked-responses[3]. Thus, taurine has been considered a conditionally essential nutrient in humans[4]. So far, however, studies on taurine status in patients receiving long term enteral nutrition (EN) have been not performed. This study was aimed at investigating whether serum concentration and urinary excretion of taurine are influenced by marginal taurine intakes.

SUBJECTS AND METHODS

Subjects

Twenty one male patients, aged 75 to 95 years, suffering from coronary heart disease, multiple cerebral infarction, cancer, subdural hematoma or respiratory failure, were grouped according to the duration of tube feeding, as shown in Table 1.

Table 1. Characteristics of old aged male patients undergoing long term enteral nutrition (EN)

	Group 1 (n=10)	Group 2 (n=5)	Group 3 (n=6)
Males	10	5	6
Age (y)*	82.1 ± 5.2 (75-95)	85.2 ± 6.0 (77-88)	85.2 ± 6.2 (77-93)
EN duration (mo)*	5.9 ± 2.9 (1-9)	14.8 ± 2.3 (12-17)	48.0 ± 22.7 (27-89)
Multiple cerebral infarction	4	-	4
Coronary heart disease	2	4	-
Subdural hematoma	-	-	1
Cancer	3	-	1
Respiratory failure	1	1	-

* Figures represent mean value ± SD; ranges in brackets.

Their daily taurine intake ranged between 27.7 and 63.4 mg from tube feeding solutions. None of the patients received any oral feeding during the study period.

This study was performed after the approval of the military hospital authorities in Beijing, China, and the informed consent of all the subjects.

Methods

Enteral nutrition solutions

The blended diet for EN was designed to provide the RDA of vitamins and trace elements with an intake of 1500 to 2000 kcal corresponding to 25kcal/kg ideal body weight. The protein was 43.1g/1000kcal.

Taurine analysis in blood and urine

At the end of different EN periods blood samples were withdrawn from the patients after at least 8 hours fasting. Serum was obtained by allowing a 5ml blood sample to clot in a serum separator tube for 30 min, and centrifuging it at 11,000 x g for 10 min[5]. Serum and 24-hour urine samples (0.5 ml) were deproteinised by adding 0.5ml of 5% sulphosalicylic acid (SSA). The mixture was centrifuged at 11,000 x g for 10 min and the supernatant quickly frozen and stored at $-70°C$ until analysis. Taurine analysis was performed with an amino acid analyser (HITACHI 835) equipped with a single-column, 3-buffer lithium citrate elution system. All samples were analysed within 2 weeks after collection.

Statistical analysis

The significance of differences among serum and urine taurine concentrations in the groups of patients was checked by ANOVA and Tukey test. The correlation between age or EN duration and blood and urine contents was assessed by the linear regression analysis. Significance level was set at $P<0.05$.

RESULTS

Daily intake, urinary excretion, balance and serum concentration of taurine in patients at the end of different EN periods are reported in Table 2.

The mean intake of taurine was similar for the three groups of patients, while the serum taurine level in group three was significantly lower by about 50%, as compared to the values foured in group one and two. Moreover, in the same group of patients the twenty-four hour urinary taurine excretion showed a trend towards a decrease, although not significant. The balance between taurine intake and urinary excretion was negative in groups one and two while it was close to zero in group three. This finding suggests that the endogenous taurine synthesis proceeds at a lower rate in group three patients.

Table 2. Intake, excretion, balance and serum concentration of taurine in patients undergoing enteral nutrition

	Group 1 (n=10)	Group 2 (n=5)	Group 3 (n=6)
Intake (μmol/day)	347 ± 25.6[#]	339.8 ± 25.6	337.1 ± 26.9
Excretion (μmol/day)	776.1 ± 176.7	782.4 ± 245.3	388.3 ± 169.3
μmol/g creatinine	1295.6 ± 409.7	1245.5 ± 305.8	805.1 ± 307.0
Balance (μmol/day)	-429.4 ± 170.4	-442.6 ± 257.5	-51.2 ± 152.6
Serum concentration (μmol/L)	106.5 ± 9.6	95.0 ± 9.9	56.8 ± 11.0*

[#] mean ± SEM; * concentration of serum in group 3 was significantly lower than that in group 1 and 2 (p<0.05).

When analysing the data from all patients as a whole, trying to correlate serum concentration or urinary excretion of taurine to the age of patients or the duration of EN, not significant relationships were found (see, Figs.1, 2 and 3).

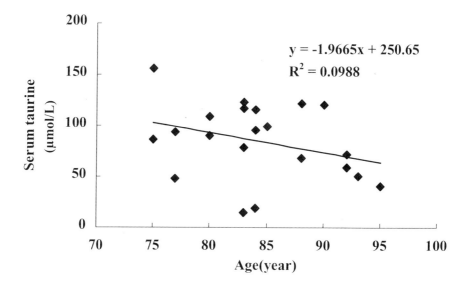

$$y = -1.9665x + 250.65$$
$$R^2 = 0.0988$$

Figure 1. Correlation between age of patients and serum taurine concentration

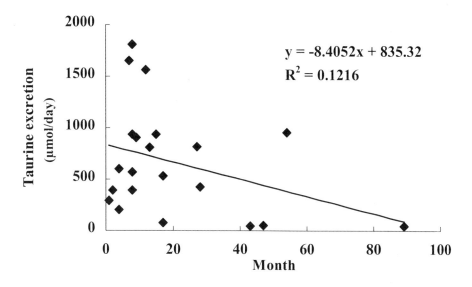

Figure 2. Correlation between EN duration and daily urinary taurine excretion

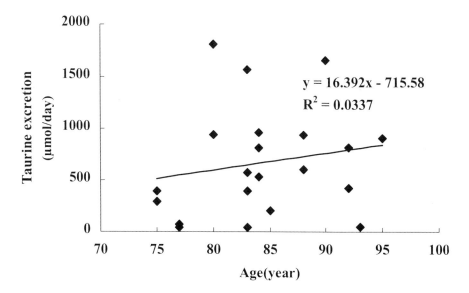

Figure 3. Correlation between age of patients and daily urinary taurine excretion

DISCUSSION

This study was aimed to investigate whether serum concentration and urinary excretion of taurine are influenced by marginal taurine intakes. The present results show that daily taurine intake, serum taurine level and urinary taurine excretion were considerably lower as compared to what is reported in the literature[6-10]. It is possible that both old age and severe diseases from which the patients were suffering had contributed to this discrepancy. However, the long lasting EN might have played a great role, in consideration of the fact that, although taurine is synthesised in humans, the rate of its synthesis is inadequate to maintain normal serum taurine concentrations in case of chronically low taurine intakes.

The actual daily intake of taurine is estimated to range from 319.6 to 3,196.4 μmol (40 to 400mg in the population at large)[8]. While taurine intake by British male omnivores averages 607 μmol/day (76mg/day)[7] that by Chinese typical males (18-45 years old, 60kg body weight, light physical activity) ranges between 268 and 637 μmol/day (33.5 to 79.7mg/day)[6]. Park et al[9] reported that taurine intake by Korean females, 20 years old is between 329.2 ± 68.7 (lactoovovegetarian) and 927.0 ± 180.6 μmol/day (omnivores).

Normal whole blood mean concentration of taurine ranges between 160 and 320 μmol/L[10] while normal serum concentration is considered to be 126.3 ± 32.1 μmol /L[5].

Laidlaw et. al.[11] reported that urinary taurine excretion usually amounts to 903 ± 580 μmol/day (mean ± SD). The reported normal range of daily taurine excretion spans from 220 to 1,850 μmol[12]. A normal food diet supplying 335.6 μmol of taurine per day (42 mg/day) resulted in a mean urinary daily excretion of 575 μmol (72 mg) in a group of young women[13]. The urinary taurine excretion in Korean was significantly lower in lactoovovegetarians (589 ± 69 μmol/day) as compared to omnivore females (1,569 ± 290 μmol/day, mean ± SD)[9]. Urinary taurine excretion and plasma taurine concentration were found to be much lower in vegetarians fed on a diet from which taurine was missing as compared to nonvegetarian control subjects[11]. In normal nutrition situations, excess dietary taurine is excreted in the urine[5]. In situations where supply of taurine is limited, however, the kidney acts to conserve taurine by increasing reabsorption and decreasing its excretion[14,15]. In case of reduced dietary taurine intake, as it happens in infants receiving taurine-free formulas, patients maintained on TPN, and vegetarians lacking taurine in their diet, normally the compensation is supported by enhanced renal reabsorption and reduced urinary taurine excretion[1,11].

It is interesting that the balance of taurine in group three patients is closer to zero as compared to either group one or two.

A marginal taurine intake in patients undergoing long term enteral nutrition could result in taurine deficiency. Old aged patients in such condition have a nutritional requirement for taurine. Special consideration to this fact, therefore, should be devoted in preparing tube feeding solutions for enteral nutrition. An enteral feeding product fortified with taurine in would be capable of restoring or improving blood and urinary levels of taurine patients undergoing long term enteral nutrition.

Our findings strongly suggest that taurine supplementation of EN solutions for old aged patients undergoing long-term enteral nutrition is required.

ACKNOWLEDGEMENTS

This work was supported by grants from KOSEF and Dong-A Pharmaceutical Co. Ltd., Korea.

REFERENCES

1. Gaull G.E., Rassin D.K., Raiha N.C.R., and Heinonen K., 1977, Milk protein quantity and quality in low-birth-weight infants. III. Effects on sulfur amino acids in plasma and urine. *J. Pediatr.*, **90**:348-355
2. Geggel H.S., Ament M.E., Heckenlively J.R., Martin D.A., and Kopple J.D., 1985, Nutritional requirement for taurine in patients receiving long-term parenteral nutrition. *N. Engl. J. Med.*, **312**:142-146
3. Dhillon S.K., Davies W.E., Hopkins P.C., and Rose S.J., 1996, Effects of dietary taurine on auditory function in full term infants. *Adv. Exp. Med. Biol.*, **402**:507-514
4. Laidlaw S.A., and Kopple J.D., 1987, New concepts of the dispensible amino acids. *Am. J. Clin. Nutr.*, **46**:593-605
5. Paauw J.D., and Davis A.T., 1990, Taurine concentrations in serum of critically injured patients and age-and sex-matched healthy control subjects. *Am. J. Clin. Nutr.*, **52**:657-60
6. Zhao X., Jia J., and Lin Y., 1998, Taurine content in Chinese food and daily taurine intake of Chinese men. *Adv. Exp. Med. Biol.*, **402**:501-505
7. Rana S.K., and Sanders T.A.B., 1986, Taurine concentrations in the diet, plasma, urine and breast milk of vegetarian compared with omnivores. *Br. J. Nutr.*, **56**:17-27
8. Hayes K.C., and Trautwein E.A., 1994, Taurine, In: Modern nutrition in health and disease., Shils M.E., Olson J.A., and Shike M. eds. 8th ed., Lea and Febiger, Philadelphia, pp.477-485
9. Park T., Kang H., and Sung M., 1999, Taurine intake, plasma taurine levels and urinary excretions in lactoovovegetarians and omnivores in Korea. International Taurine Symposium 1999, Abstract, p26
10. Trautwein E.A., and Hayes K.C., 1990, Taurine concentrations in plasma and whole blood in humans: estimation of error from intra- and inter-individual variation and sampling technique. *Am. J. Clin. Nutr.*, **52**:758-764
11. Laidlaw S.A., Shultz T.D., Cecchino J.T., and Kopple J.D., 1988, Plasma and urine taurine levels in vegetarians. *Am. J. Clin. Nutr.*, **47**:660-663

12. Jacobsen J.G., and Smith L.H. Jr., 1968, Biochemistry and physiology of taurine and taurine derivatives. *Physiol. Rev.,* **48**:424-511
13. Thompson D.E., and Vivian W.M., 1977, Dietary-induced variations in urinary taurine levels of college women. *J. Nutr.,* **107**:673-679
14. Rozen R, and Scriver C.R., 1982, Renal transport of taurine adapts to perturbed taurine homeostasis. *Proc. Natl. Acad. Sci. USA.* **79**:2101-2105
15. Chesney R.W., Gusowski N, and Dabbagh S., 1985, Renal cortex taurine content regulates renal adaptive response to altered dietary intake of sulphur amino acids. *J. Clin. Invest.* **76**:2213-2221

DOES THE TAURINE TRANSPORTER GENE PLAY A ROLE IN 3P-SYNDROME?

Xiaobin Han, Andrea M. Budreau, Russell W. Chesney and John A. Sturman

Department of Pediatrics, University of Tennessee, and the Crippled Children's Foundation Research Center at Le Bonheur Children's Medical Center, Memphis, TN

INTRODUCTION

Taurine has been shown to be essential for the development and survival of mammalian cells, especially cells of the cerebellum and retina[18]. Notably, taurine concentration reaches the mM range in the CNS[20]. The highest intracellular concentration of taurine is found in neonatal and early postnatal brain, suggesting a developmental role. Sturman's group found that the surviving F1 offspring of taurine-deficient female cats have a large number of neurologic defects, including degeneration of retinal pigmented epithelium and ocular tapetum, delayed cerebellar granule cell division and migration, and abnormal cerebral cortical development. Taurine also appears to optimize the proliferation and differentiation of human fetal cerebral brain cells in culture. Addition of taurine to the medium improves neuron growth, neurite expansion, and neuronal survival[2].

Taurine is considered a conditionally essential amino acid in man, as the enzymatic activity of cysteine sulfonic acid decarboxylase (CSAD), which catalyzes the formation of taurine from cysteine, is low in humans, and even lower or absent in term and preterm infant liver[28]. This ontogeny of catalytic enzyme activity is also true in other primates[21]. In children with short gut syndrome, taurine is an essential component of total parenteral nutrition (TPN) solutions to prevent a defined retinopathy, which is reversed only by taurine administration. Human milk also contains high concentrations of taurine[11]. We have shown previously that taurine is a semi-conditionally essential amino acid in very low birth weight premature infants, especially those who cannot ingest food and are obliged to be fed by TPN[28]. Plasma taurine values in preterm infants fall to almost non-detectable values,

Taurine 4, edited by Della Corte *et al.*
Kluwer Academic / Plenum Publishers, New York, 2000.

comparable to those in the taurine-deprived cat. Restoration and maintenance of normal taurine values in preterm infants and infants with short gut syndrome receiving TPN requires the specific addition of taurine to TPN solutions, especially since cysteine is relatively insoluble and, hence, is not a reliable source of taurine[6,7].

The phenotypic expression of taurine deficiency in cats is similar to the deletion 3p-chromosome (3p25-pter) syndrome in humans, which is expressed with complex neurological and renal malformations. Profound growth failure, microcephaly, characteristic craniofacial features, mental and developmental retardation, and renal abnormalities have been reported in patients with 3p-syndrome[12,13,16,23,27]. Cardiovascular abnormalities and both renal and pulmonary cell carcinomas have also been associated[8,22,25] with deletion of the short arm of chromosome 3.

The striking similarities between the characteristics of patients with 3p-syndrome and taurine-deficient kittens led us to postulate that the taurine transporter gene (TauT) may play an important role in mammalian neurological and renal development.

METHODS

Animals

Female common cats were fed a completely defined taurine-free synthetic diet (BioServe, Frenchtown, NJ) for at least 6 months prior to mating. The cats were severely taurine-depleted, with plasma taurine concentrations of less than 1 µmole/100 ml (0.1 µM) Other females (controls), fed the same diet supplemented with 0.05% taurine, maintained plasma taurine levels of approximately 25-50 µM. Male cats were fed the taurine-supplemented diet except during mating with females receiving the taurine-free diet. Surviving F1 kittens were sacrificed at the time of weaning (8 weeks after birth) and used in this study.

Histology

For sectioning, kidneys were fixed in 10% phosphate-buffered formalin and B-5 solution, paraffin embedded, sectioned, and stained with hematoxylin and eosin.

RESULTS

Evidence of renal damage in taurine-deficient kittens

A recent observation in the F1 generation of inbred, taurine-deficient cats provides further evidence of the role of taurine deficiency in renal damage. Taurine-deficient female cats were bred with taurine-deficient males. Surviving F1 generation kits showed blindness, ataxia, cerebellar abnormalities and kyphosis, as well as evidence of renal scarring with small contracted kidneys. Renal size is influenced by taurine deficiency (Table 1, Fig. 1). The small kidneys showed irregular scars, sclerosed glomeruli, epithelial atypia, drop-out of tubules, and advanced cortical atrophy. In addition, cats fed the taurine-deficient diet had a flattened tubular epithelium, reduced mitochondria at the apex of the tubule, simplification of the tubule and glomerular hypertrophy, as compared with age-matched controls. Distal tubule effects were marked in cats fed the taurine-deficient diet for 2 years. Histologic examination of these kidneys revealed extensive cortical and medullary scarring, and glomerular hypertrophy with interstitial fibrosis (Figure 2). These cat kidneys resemble the autopsy findings in patients with chromosome 3p-deletion syndrome[1].

Table 1. Renal size in cats on taurine-deficient and -sufficient diets

Age	Taurine Status	Renal Size (g)
8 weeks	0.0%	5.35 ± 0.57*
	0.5%	7.55 ± 0.68
Adult (18 weeks)	0.0%	10.86 ± 1.70**
	0.5%	23.1 ± 1.2
2 years (life-long)	0.0%	3.6

p value versus control: *$p < 0.05$; **$p < 0.01$

Phenotypic Characteristics

A remarkable similarity exists between human infants with deletion of bands p25-pter of chromosome 3 and the F1-generation of taurine-deficient kits (Table 2). These chromosome 3p-deletion children have a specific syndrome of craniofacial manifestations, prenatal growth delay, hypotonia, developmental retardation, cataracts and cleft palate. In addition, they have renal anomalies including renal hypoplasia, renal malposition (pelvic kidneys), and cortical cysts with increased interstitial connective tissue[1]. Genitourinary anomalies also include small cortical cysts and bladder hypoplasia. The facial features of the taurine-deficient kit and the band p25-

pter human infant show ptosis, telecanthus, a slight mongoloid slant, down-turned and low set ears, epicanthal folds and, in some cases, a cleft lip[12,13,16,23,26]. Postaxial polydactyly has also been found in both the cat and human conditions[1].

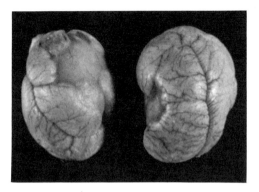

Figure 1. Small, contracted kidney from taurine-deficient kitten showing renal scarring (left) as compared with normal (right)

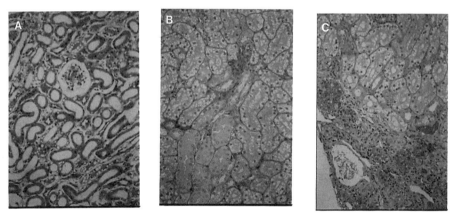

Figure 2. Kidney histology of taurine-deficient and normal kittens. a) Kidney from an 18-week-old normal kitten (400x), b) Kidney from an 18-week-old taurine-deficient kitten showing glomerular hypertrophy occupied by eosinophilic proteinacious material (400x). c) Kidney from an 18-week-old taurine-deficient kitten displaying enlarged glomeruli and bands of interstitial fibrosis (400x).

DISCUSSION

The taurine transporter has been cloned recently from several species and tissues, including canine kidney cells[24], rat brain[17], mouse brain[10], human thyroid cells[9], human placenta[15], and porcine kidney cells[5]. The genes encoding the taurine transporter of different species and tissues share a high degree of identity. The gene is located on the central region of mouse chromosome 6, and on human chromosome 3p21-25, where a conserved linkage group of genes was found between mouse and man[14].

Table 2. Similarity between 3p-syndrome patients and taurine-deficient kittens

	3p-deletion	taurine-deficient
growth retardation	+	+
mental/developmental retardation	+	+
triangular face	+	+
abnormal nose	+	+
low set or malformed ears	+	+
renal abnormalities	+	+
cardiovascular abnormalities	+	+

It has been demonstrated that in patients with 3p-syndrome, deletion of 3p25-pter is associated with profound growth failure, characteristic facial features, retinal changes, and mental retardation, suggesting that deletion of the taurine transporter gene might contribute to some phenotypic features of 3p-syndrome[14]. As described, the F1 generation of inbred, taurine-deficient kits showed a characteristic facial appearance, blindness, ataxia, cerebellar abnormalities and kyphosis, as well as severe neonatal renal damage which progresses to renal scarring. These observations led us to postulate that the taurine transporter gene (TauT) may play an important role in mammalian brain and renal development and differentiation. As the renal tubular epithelium controls the total body pool of taurine, altered taurine transport may result in reduced intracellular taurine, which in turn may contribute to renal cell death and nephron drop-out comparable to that seen in animal models of developmental renal disease, such as the p53 overexpressing mouse[4].

Certain pediatric renal diseases, such as chronic renal failure, may initially occur as a result of events *in utero* or during the neonatal period, important periods for kidney development and differentiation. Factors contributing to taurine homeostasis during development are the interplay between endogenous synthesis and diet, whereas demand is reflected in the specific requirements of the various tissues coupled with conservation or loss of taurine via excretion. The human infant in the first months of life (0-6 months) has a reduced capacity to synthesize taurine from methionine due to an ontologically-related reduction in cysteine sulfinic acid. We have shown that the total body pool size of taurine is controlled by the renal taurine transporter, which is adaptively regulated by dietary taurine level. A limited amount of taurine may also be transported from the extracellular into the intracellular space of kidney cells by other channels[3], such as band-3. In the cat deprivation model, tissue taurine could not be totally depleted but was reduced to about 30% of normal[19], suggesting that in the presence of subnormal levels of taurine renal development does not proceed normally.

REFERENCES

1. Beneck, D., Suhrland, M.J., Dicker, R., Greco, M.A., and Wolman, S.R., 1983, Deletion of the short arm of chromosome 3: a case report with necropsy findings, *J. Med. Genet.,* 21:307-310.
2. Chen, X., Pan, Z.L., Liu, D.S., and Han, X., 1998, Effect of taurine on human fetal neuron cells: proliferation and differentiation, *Adv. Exp. Med. Biol.,* 442:397-403.
3. Fievet, B., Borgese, F., and Motais, R., 1995, Expression of band 3 anion exchanger induces chloride current and taurine transport: structure-function analysis, *EMBO J,* 1:5158-5169.
4. Godley, L.A., Eckhaus, M., Paglino, J.J., Owens, J., and Varmus, H.E., 1996, Wild-type p53 transgenic mice exhibit altered differentiation of the ureteric bud and possess small kidneys, *Genes Devel.,* 110:836-850.
5. Han, X., Budreau, A.M., and Chesney, R.W., 1998, Molecular cloning and functional expression of an LLC-PK1 cell taurine transporter that is adaptively regulated by taurine, *Adv. Exp. Med. Biol.,* 442:261-268.
6. Helms, R.A., Chesney, R.W., and Storm, M.C., 1995, Sulfur amino acid metabolism in infants on parenteral nutrition, *Clin. Nutr.,* 14:381-387.
7. Helms, R.A., Christensen, M.L., Storm, M.C., and Chesney, R.W., 1995, Adequacy of sulfur amino acid intake in infants receiving parenteral nutrition, *J. Nutr. Biochem.,* 6:462-466.
8. Hung, J., Kishimoto, Y., Sugio, K., Virmani, A., McIntire, D.D., Minna, J.D., and Gazdar, A.F., 1995, Allele-specific chromosome 3p deletions occur at an early stage in the pathogenesis of lung carcinoma, *JAMA,* 273:558-563.
9. Jhiang, S.M., Fithian, L., Smanik, P., McGill, J., Tong, Q., and Mazzaferri, E.L., 1993, Cloning of the human taurine transporter and characterization of taurine uptake in thyroid cells, *FEBS,* 318:139-144.
10. Liu, Q.R., Lopez-Corcuera, B., Nelson, H., Mandiyan, S., and Nelson, N., 1992, Cloning and expression of a cDNA encoding the transporter of taurine and b-alanine in mouse brain, *Proc. Natl. Acad. Sci. USA,* 89:12145-12149.
11. Lucas, A., Morley, R., Cole, T.J., and Gore, S.M., 1994, A randomised multicentre study of human milk versus formula and later development in preterm infants, *Arch. Dis. Child.,* 70:F141-F145.
12. Mowrey, P.N., Chorney, M.J., Venditti, C.P., Latif, F., Modi, W.S., Lerman, M.I., Zbar, B., Robins, D.B., Rogan, P.K., and Ladda, R.L., 1993, Clinical and molecular analyses of deletion 3p25-pter syndrome, *Am J. Med. Genet.,* 46:623-629.
13. Narahara, K., Kikkawa, K., Murakami, M., Hiramoto, K., Namba, H., Tsuji, K., Yokoyama, Y., and Kimoto, H., 1990, Loss of the 3p25.3 band is critical in the manifestation of del(3p) syndrome: Karyotype-phenotype correlation in cases with deficiency of the distal portion of the short arm of chromosome 3, *Am. J. Med. Genet.,* 35:269-273.

14. Patel, A., Rochelle, J.M., Jones, J.M., Sumegi, G., Uhl, G.R., Seldin, M.F., Meisler, M.H., and Gregor, P., 1995, Mapping of the taurine transporter gene to mouse chromosome 6 and to the short arm of human chromosome 3, *Genomics,* 1:314-317.
15. Ramamoorthy, S., Leibach, F.H., Mahesh, V.B., Han, H., Yang, F.T., Blakely, R.D., and Ganapathy, V., 1994, Functional characterization and chromosomal localization of a cloned taurine transporter from human placenta, *Biochem. J.,* 300:893-900.
16. Ramer, J.C., Ladda, R.L., and Frankel, C., 1989, Two infants with del(3)(p25pter) and a review of previously reported cases, *Am. J. Med. Genet.,* 33:108-112.
17. Smith, K.E., Borden, L.A., Wang, C.D., Hartig, P.R., Branchek, T.A., and Weinshank, R.L., 1992, Cloning and expression of a high affinity taurine transporter from rat brain, *Mol. Pharmacol.,* 42:563-569.
18. Sturman, J.A., 1988, Taurine in development, *J. Nutr.,* 118:1169.
19. Sturman, J.A., French, J.H., and Wisniewski, H.M., 1985, Taurine deficiency in the developing cat: persistence of the cerebellar external granule cell layer, *Adv. Exp. Med. Biol.,* 179:43-52.
20. Sturman, J.A., Hepner, G.W., Hofmann, A.F., and Thomas, P.J. 1976. Taurine pool sizes in man: Studies with ^{35}S-taurine. *In* Taurine. R. J. Huxtable, editor. Raven Press, New York. 21-23.
21. Sturman, J.A., Wen, G.Y., Wisniewski, H.M., and Neuringer, M.D., 1984, Retinal degeneration in primates raised on a synthetic human infant formula., *Int. J. Dev. Neurosci.,* 2:121-130.
22. Szücs, S., Muller-Brechlin, R., DeRiese, W., and Kovacs, G., 1987, Deletion 3p: the only chromosome loss in a primary renal cell carcinoma, *Cancer Genet. Cytogenet.,* 26:369-373.
23. Tazelaar, J., Roberson, J., Van Dyke, D.L., Babu, V.R., and Weiss, L., 1991, Mother and son with deletion of 3p25-pter, *Am. J. Med. Genet.,* 39:130-132.
24. Uchida, S., Kwon, H.M., Yamauchi, A., Preston, A.S., Marumo, F., and Handler, J.S., 1992, Molecular cloning of the cDNA for an MDCK cell Na(+)- and Cl(-)-dependent taurine transporter that is regulated by hypertonicity [published erratum appears in Proc Natl Acad Sci U S A 1993 Aug 1; 90(15):7424], *Proc. Natl. Acad. Sci. U S A,* 89:8230-8234.
25. Whang-Peng, J., Kao-Shan, C.S., and Lee, E.C., 1982, Specific chromosome defect associated with human small-cell lung cancer: deletion 3p(14-23), *Science,* 215(8):181-182.
26. Wieczorek, D., Bolt, J., Schwechheimer, K., and Gillessen-Kaebach, G., 1997, A patient with interstitial deletion of the short arm of chromosome 3 (pter-p21.2::p12-qter) and a CHARGE-like phenotype, *Am. J. Med. Genet.,* 69:413-417.
27. Witt, D.R., Biedermann, B., and Hall, J.G., 1985, Partial deletion of the short arm of chromosome 3 (3p25 - 3pter). Further delineation of the clinical phenotype, *Clin. Genet.,* 27:402-407.
28. Zelikovic, I., Chesney, R.W., Friedman, A.L., and Ahlfors, C.E., 1990, Taurine depletion in very low birth weight infants receiving prolonged parenteral nutrition: Role of renal immaturity, *J. Pediatr.,* 116:301-306.

EXTRACELLULAR LEVELS OF TAURINE IN TUMORAL, PERITUMORAL AND NORMAL BRAIN TISSUE IN PATIENTS WITH MALIGNANT GLIOMA: AN INTRAOPERATIVE MICRODIALYSIS STUDY

Enrico De Micheli[1], Alex Alfieri[1], Giampietro Pinna[1], Loria Bianchi, Maria A. Colivicchi, Alessia Melani, Felicita Pedata, Laura Della Corte and Albino Bricolo[1]

Dipartimento di Farmacologia Preclinica e Clinica "M. Aiazzi Mancini", Università di Firenze & [1]Dipartimento di Neurochirurgia, Università di Verona, Italy

INTRODUCTION

Available information on the metabolism of brain malignant gliomas is scarce and mostly derived from studies on animal models. Human data are limited to studies employing positron emission tomography (PET) and proton magnetic resonance spectroscopy (MRS), using amino acids synthesised with positron-emitting labels to image tumors[1,2], or analysing tissue specimens obtained at surgery or autopsy. Such studies have reported an increased accumulation of free amino acids in brain tumors, mainly due to increased carrier-mediated active transport in their supporting vasculature, expression of a more active metabolism related to increased cell proliferation, rather than to disruption of the blood brain barrier[3-5].

Taurine 4, edited by Della Corte *et al.*
Kluwer Academic / Plenum Publishers, New York, 2000.

Higher concentrations of taurine are present in metabolically active tissues, such as retina, brain, heart, neutrophils and eosinophils[6]. In the brain, one of the many functions ascribed to taurine is osmoregulation, caracterised by taurine release with accompanying water in response to brain tissue edema. High resolution MRS studies have shown that malignant glioma tissues contain higher levels of taurine than benign astrocytomas or normal brain[7]. Increased taurine levels have been found in other malignant tumors and taurine has been considered as a MRS malignancy marker for colon cancer[8].

The aim of the present work was to investigate whether the concentrations of endogenous taurine and other amino acids in the extracellular fluid of human brain tumor tissue (TT), a condition where cell proliferation causes brain swelling and edema, were different from those in the adjacent parenchima (AP) or in the normal brain tissue (NBT), using intraoperative microdialysis in patients undergoing surgery for brain glioblastoma resection.

MATERIALS AND METHODS

The work involving human subjects, performed at the Stereotaxic Neurosurgery Unit, Department of Neurosurgery, University of Verona, Italy, was approved by the Hospital Ethical Committee, and complies with the European community guiding policies and principles for experimental procedures.

Fifteen patients (7 males, 8 females; mean age 50 ± 7 and 47 ± 6, respectively) with cortical glioblastoma, who had not previously undergone surgery, radiotherapy, chemotherapy, nor cerebral biopsy, were included in this study. Diagnosis was confirmed by histopathology, according to the W.H.O. classification. Ki-67 immunoreactivity (%) was histologically assessed as an index of the degree of tumor cell proliferation[9]. At surgery for tumor resection, a flexible microdialysis catheter (CMA 70, Solna, Sweden) was inserted in the tumor tissue (TT), the parenchima adjacent to the tumor (AP), and the normal brain tissue (NBT). Following 20 min equilibration period, extracellular fluid from each of the 3 regions was collected every 20 min, using a microinfusion pump (CMA 106, Solna, Sweden) operating at a rate of 2 μl/min.

The concentrations of taurine, and other endogenous amino acids in the microdialysate were measured by HPLC separation followed by fluorimetric detection of their p-orthophtalaldehyde (OPA) derivatives[10]. Statistical analysis of differences between experimental groups was performed by ANOVA for repeated measures followed by Fisher's LSD.

RESULTS

All analysed amino acids, with the exception of glutamate and aspartate showed significantly (P<0.01) higher extracellular concentrations within the tumor tissue as compared to normal brain tissue (Fig. 1). Valine showed significantly higher extracellular concentrations also within the peritumoral tissue.

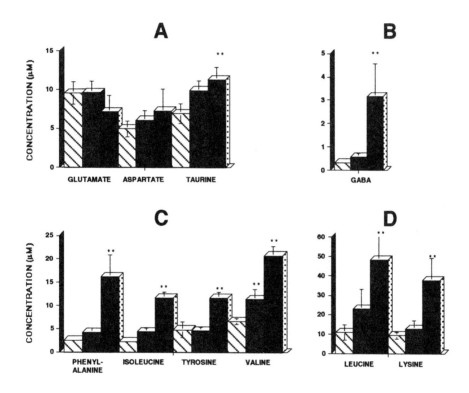

Figure 1. Extracellular concentrations (μM, mean ± s.e.m., n=15) of endogenous glutamate, aspartate and taurine (panel A), GABA (panel B), phenylalanine, isoleucine, tyrosine and valine (panel C), leucine and lysine (panel D) in NBT (striped columns), AP (grey columns) and TT (black columns). **Statistically significant difference (P<0.01) from NBT.

Among the amino acids that were higher in the extracellular fluid derived from TT, taurine was the only one showing a statistically significant correlation with the degree of cell proliferation (r = 0.747, P<0.01).

DISCUSSION

To our knowledge this is the first *in vivo* study applying microdialysis to the measurement of endogenous amino acids in the extracellular fluid of human cerebral gliomas.

The present data provide evidence that glioblastoma growth is associated with increased extracellular concentrations of endogenous amino acids, confirming data obtained by PET/radiolabelling or *ex vivo* by analysing tissue specimens obtained at surgery or autopsy by MRS or HPLC analysis[1,2,11,12]. This may be partly due to blood brain barrier breakdown, however, the up regulation of amino acid transporter expression in their supporting vasculature has been suggested to play a major role[3-5].

The observations that TT and AP taurine extracellular levels were significantly elevated and that TT taurine levels were correlated to the degree of cell proliferation suggest a possible association of increased taurine with edema. Indeed cell proliferation leads to peritumoral edema as a result of deficient perfusion due to elevation of regional tissue pressure, with bioptic analysis of peritumoral edema showing decreased energy charge potential and elevated levels of lactate[13]. Release of taurine associated with edema has been observed in several pathological conditions, i.e. hyperammonemia and hepatic encephalopathy[14]. Once released taurine may favour inhibitory neurotransmission contributing to the pathological or protective mechanisms active in such conditions.

Nakamura et al.[15] have reported that human glioma cells express a high-affinity taurine transporter and that the protein kinase C (PKC) signal transduction system plays an important role in the regulation of taurine transport and intracellular taurine concentrations in these cells. Since PKC activity is significantly increased in human malignant glioma cells[15], and their proliferation rate may be modulated by the PKC signal transduction system[16,17], it is possible that the proliferation rate-related taurine release is a result of a PKC-mediated inhibition of taurine uptake in malignant glioma.

ACKNOWLEDGEMENTS

This work was supported by MURST (IT) and COST D8 (UE).

REFERENCES

1. Kawai, K., Fujibayashi, Y., Saji, H., Yonekura, Y., Konishi, J., Kubodera, A. and Yokoyama, A., 1991, A strategy for the study of cerebral amino acid transport

using iodine-123-labeled amino acid radiopharmaceutical: 3-iodo-alpha-methyl-L-tyrosine, *J. Nucl. Med.* **32**:819-824.

2. Schober, O., Meyer, G.J., Stolke, D. and Hundenshagen, H., 1985, Brain tumor imaging using C-11-labeled L-methionine and D-methionine, *J. Nucl. Med.* **26**:98-99.

3. Kuwert, T., Morgenroth, C., Woesler, B., Matheja, P., Palkovic, S., Vollet, B., Samnick, S., Maasjosthusmann, U., Gildehaus, F.-J., Wassmann, H. and Schober, O., 1996, *Eur. J. Nucl. Med.* **23**:1345-1353.

4. Ogawa, T., Miura, S., Murakami, M., Iida, H., Hatazawa, J., Inugami, A., Kanno, I., Yasui, N., Sasajima, T. and Uemura, K., 1996, *J. Nucl. Med.* **23**:889-895.

5. Miyagawa, T., Oku, T., Uehara, H., Desai, R., Beattie, B., Tjuvajev, J. and Blasberg. R.J., 1998,"Facilitated" amino acid transport is upregulated in brain tumors, *J. Cereb. Blood Flow* **18**:500-509.

6. Wright, C.E., Tallan, H.H., Lin, Y.Y. and Gaull, G.E., 1986, Taurine: biological update, *Ann. Rev. Biochem.* **55**:427-453.

7. Peeling, J. And Sutherland, G., 1992, High-resolution ^{1}H NMR spectroscopy studies of extracts of human cerebral neoplasm, *Magn. Res. Med.* **24**:123-136.

8. Moreno, A., Rey, M., Montane, J.M., Alonso, J. And Arus, C., 1993, ^{1}H NMR spectroscopy of colon tumors and normal mucosa biopsies: elevated taurine levels and reduced polyethylenglycol absorption in tumors may have diagnostic significance. *NMR Biomed* **6**:111-118.

9. Schlueter, C., Ducrow, M., Wholenberg, C., Becker, M.H.G., Key, G., Flad, H.D. and Gerdes, J., 1993, The cell proliferation-associated antigen-antibody Ki-67: a very large, ubiquitous nuclear protein with numerous repeated elements, representing a new kind of cell cycle-maintaining protein. *J. Cell Biol.* **123**:513-522.

10. Bianchi L, Della Corte L, Tipton KF (1999) Simultaneous determination of basal and evoked output levels of aspartate, glutamate, taurine and 4-aminobutyric acid during microdialysis and from superfused brain slices. *J. Chromatogr. B* **723**:47-59.

11. Kinoshita, Y. and Yokota, A., 1997, Absolute concentrations of metabolites in human brain tumors using in vitro proton magnetic resonance spectroscopy, *NMR Biomed.* **10**:2-12.

12. Shibasaki, T., Uki, J., Kanoh, T., Kawafuchi, J., 1979, Composition of free amino acids in brain tumors, *Acta. Neurol. Scand.* **60**:301-311.

13. Go KG, Krikke AP, Kamman RL, Heesters MA (1997) The origin of lactate in peritumoral tissue as measured by proton-magnetic resonance spectroscopic imaging. *Acta Neurochir. (Suppl)* **70**:173-175.

14. Hilgier, W. and Olson, J.E., 1994, Brain ion and amino acid contents during edema development in hepatic encephalopathy. *J. Neurochem.* **62**:197-204.

15. Nakamura H, Huang SH, Takakura K (1996) High-affinity taurine uptake and its regulation by protein kinase C in human glioma cells. *Adv. Exp. Med. Biol.* **442**:377-384.

16. Couldwell WT, Uhm JH, Antel JP, Yong VW (1992) Enhanced protein kinase C activity correlates with the growth rate of malignant gliomas in vitro. *Neurosurgery* **29**:880-886.

17. Couldwell WT, Antel JP, Yong VW (1992) Protein kinase C activity correlates with the growth rate of malignant gliomas: Part II. Effects of glioma mitogens and modulators of protein kinase C? *Neurosurgery* **31**:717-724.

THE TREATMENT OF AMMONIA POISONING BY TAURINE IN COMBINATION WITH A BRONCHOLYTIC DRUG

A. Zemlyanoy, Yu. Lupachyov, A. Tyaptin, P. Torkounov, M. Varlashova and N. Novosyolova

Department of Comparative Neurochemistry, Institute of Evolutionary Physiology, Biochemistry RAS. St-Petersburg, M.Torez av.44

INTRODUCTION

Our interest in ammonia toxic injury is based on the ability of ammonia to cause pulmonary edema. Lung edema is a problem in emergency medicine and clinical toxicology. At the present, there is no real progress in treatment of toxic lung edema. For experimental treatment, we use substances with antioxidant, antihypoxic and antitoxic effects[1,4] because of the significant role of oxidative processes in pulmonary edema formation[2,3]. This has led us to use the sulfur-containing substances, unithiol and taurine.

Unithiol is a modified dimercaprol (British Antilewisite, BAL) in which a hydroxyl group has been changed to a $-SO_3$ group. We examined the ability of unithiol and taurine to modulate the effects of other medicinal compounds. In a separate series of experiments, taurine application has been examined in combination with broncholytic preparations (sympathomimetics, such as naphtysine). We have chosen the inhalation route to expose drugs directly to injuried lung tissues.

The purpose of this study was to evaluate the effects of taurine and unithiol aerosol inhalation on the intoxication development caused by inhaled ammonia.

MATERIALS AND METHODS

White outbred male rats of 200 ± 20 g body wieght were exposed to 1.70 $g/min/m^3$ ammonia inhalation. The ammonia dose range was equal to LCt_{99}.

Taurine 4, edited by Della Corte *et al.*
Kluwer Academic / Plenum Publishers, New York, 2000.

Taurine and unithiol aerosol particle diameter was in the range of 2.9-3.7 μm. Ten minutes after ammonia poisoning, aerosol treatment was began and continued for 10 min. The dose of taurine and unithiol which rats received during the treatment was 24 mg/kg. There were the following experimental groups: control, placebo-control, and two animals groups which received taurine and unithiol. The animals of the control group were left untreated after ammonia inhalation. The animals of the placebo-control group were treated with distilled water aerosol. The treatment was carried out once. After the treatment, the animals were observed for 4 days. To evaluate the effects of drugs, we used such signs as viability, mortality and the calculated coefficient of the intoxication severity (measured in relative units) and lung weight coefficient.

For mortality analysis, the cases of poisoned animals death were summarized. For analysis of dynamics and distributions of death of experimental animals, the whole period of observation during 4 days was divided into 7 periods: 1: animal death during poisoning exposure or during the first 60 min after exposure; 2: death of animals within 2-8 h of exposure; 3: death within 9-24 h; 4: death within 1-2 days; 5: death within 2-3 days; 6: death within 3-4 days; 7: death occured after the observation period or the animal survived. Each of 7 periods was attached the definite number in the order of decreasing. The summation of mentioned numbers in frame of one experimental animal group permited to evaluate the average level of injury expressed in numbers, for each particular group of animal there was coefficient of the intoxication severity. The degree of lung edema was easily evaluated by calculation of lung weight coefficient factor on following formula:

$$LWC = (\text{lung weight} / \text{body weight}) \cdot 1000$$

Data obtained from experimental and control groups were compared using Student's t-criterion.

RESULTS AND DISCUSSION

Ammonia inhalation by rats results in all animals dying within 4 days in the absence of treatment. The coefficient of intoxication severity was 5.4, and lung weight coefficient factor was 17.9. The treatment of poisoned animal by the antioxidant, unithiol aerosol, halved the mortality rate, to an average level of severity of 3.2 units with a lung weight coefficient of 12.5 units. The treatment of poisoned animals by taurine aerosol did not result in a significant decrease in the investigated parameters (Table 1).

Table 1. Influence of taurine, naphthisin and unitiol aerosols on viability, mortality and the coefficient of the intoxication severity in rats after ammonia poisoning

Group	Dynamics of mortality, hours							n	Mortality %	Coefficient of intoxication severity (Average of ranks)
	1	8	24	48	72	96	>96			
				Ranks						
	7	6	5	4	3	2	1			
1. Control	2	1	2	1	1	-	-	7	100 ±14	5. 3 ± 0. 57
2. Placebo-control	3	-	1	-	1	-	1	6	100 ± 17	5. 4 ± 1. 03
3. Unitiol 24 mg/kg	-	1	2	-	-	-	3	3	50 ± 22	3. 2 ± 1.0*
4. Taurine 24 mg/kg	1	-	1	-	2	-	2	5	83 ± 17	3. 8 ± 0. 87
5. Naphthysin 12 mg/kg	-	-	1	3	1	-	1	6	100 ± 17	3. 5 ± 0. 56
6. Taurine 24 mg/kg & naphthysin 12 mg/kg (combination)	-	-	-	-	1	1	-	2	33 ± 21*	1. 5 ± 0. 34*

Agents were given by inhalation. Number of rats: 7 in control group, 6 in others; n: number of dead rats in group; *significant difference (p<0.05) in comparison with control group

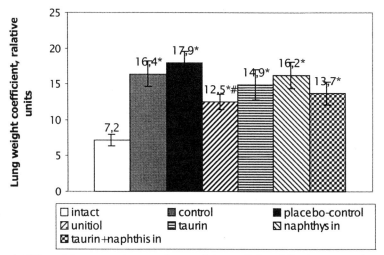

Figure 1. Effect of drug aerosols on rat lung weight coefficient following ammonia exposure. *Significantly different (p<0.05) in comparison to intact animals; #Significantly different (p<0.05) in comparison to control animals.

We did not observe any positive effect with application of naphtisine inhalation. The treatment of ammonia poisoning by inhalation of a combination of taurine and naphtisine resulted in decreasing of mortality to one third in comparison with control, a reduction of coefficient of intoxication severity down to 1.5 units, and lung weight coefficient factor down to 13.7 units.

The results of experiments have not confirmed our expectations in regard to the possible efficacy of taurine. Unithiol, an antioxidant, had an effective positive effect on ammonia poisoning. However, our results demonstrate the opportunity for taurine application after ammonia inhalation poisoning in combination with drugs of other groups. We assume that the further study the route of taurine application in combination with other medicines will be reasonable and promising.

CONCLUSIONS

1. Taurine aerosol application showed no significant positive effect against ammonia poisoning and toxic lung edema. Taurine and naphthysin combination aerosol application gave a significant positive effect against ammonia lung injury.

2. Unithiol inhalation is a prospective method for treatment against the ammonia poisoning.

3. Unitiol effect against ammonia poisoning shows a significant role of oxidation processes in ammonia intoxication development. Probably, unithiol as an antioxidant decreases oxidation processes, thus reducing the signs of ammonia poisoning.

4. The drug inhalation route may offer considerable opportunity aganist toxic lung edema. This new approach could be of major interest for clinical practice.

REFERENCES

1. Huxtable R.J. 1989, Taurine in the central nervous system and the mammalian actions of taurine. *Progr. Neurobiol.* 32:471-533.
2. Park E., Schuller-Levis G., and Quinn M.R. 1995, Taurine chloramine inhibits prodactions of nitric oxide and TNFα in activated RAW 264.7 cells by mechanisms that involve transcriptional and translational events. *J. Immunol.* 154:4778-4784.
3. Schuller-Levis G., Quinn M.R., Wright C., and Park. E. 1994, Taurine protect against oxidant-induced lung injury: possible mechanism(s) of action. *Adv. Exp. Med. Biol.* 359:159-169.
4. Torkounov P., and Sapronov N. 1997, Cardioprotective effect of taurine. *Exp. Clin. Pharmacol.* 60:72-77.

CONCLUDING COMMENTS

Giampietro Sgaragli, [1]Keith Tipton and [2]Laura Della Corte
Istituto di Scienze Farmacologiche, Università degli Studi di Siena, Via Piccolomini 170, Siena,
[1]Department of Biochemistry, Trinity College, Dublin and Dipartimento di Farmacologia
[2]Preclinica e Clinica "M. Aiazzi Mancini", Università degli Studi di Firenze.

It would be an impossible task to try to summarize a meeting such as this. Even to select "highlights", which would perforce be an entirely personal interpretation, would take many pages. However, it is appropriate to thank our Plenary Lecturers, Diana Conte Camerino, Stephan Schaffer and Jang-Yen Wu, for so ably conveying their own excitement for different aspects of taurine research and setting the scene for many of the other presentations. Ryan Huxtable, who has been a constant stimulator and organizer of so many taurine meetings, deserves our thanks for his "Keynote Lecture"; a somewhat misleading description of a wide-ranging presentation that covered so many different keynotes.

Several aspects of the diverse actions of taurine were clarified but others remain elusive. For example, the complexity of the interaction of taurine with its binding sites and how these cooperate to give the physiological and biochemical effects of taurine, remains obscure. It is to be hoped that, by the time of the next meeting (Taurine 5), specific ligands for these sites will have been developed.

Several presentations stressed the nutritional value of taurine to the human, under conditions that included pregnancy, early development, exercise and diabetes as well as for protection against ionizing radiation and the ageing process. Because of this it is important that recommended dietary intake levels are established for taurine.

As participants in the meeting we should, perhaps, thank oneanother for the excellence of the science presented and the stimulating discussions that arose from them.

There has been substantial progress since Taurine 3 and the pace of development is such that Taurine 5, in two years time, should prove to be an exciting occasion.

APPENDIX: SOME TAURINE ANALOGUES REFERRED TO IN THIS BOOK

(a) Some metabolites and natural analogues

TAURINE
2-aminoethane-sulfonic acid $NH_2\text{-}CH_2\text{-}CH_2\text{-}SO_3H$

N-chlorotaurine $NH(Cl)\text{-}CH_2\text{-}CH_2\text{-}SO_3H$
(TAURINE-
MONOCHLOROAMINE)

2-hydroxyethane-sulfonic $HO\text{-}CH_2\text{-}CH_2\text{-}SO_3H$
acid
(ISETHIONIC ACID)

2-aminoethane-sulfinic acid $NH_2\text{-}CH_2\text{-}CH_2\text{-}SO_2H$
(HYPOTAURINE)

2-amino-3-sulfo-propionic $HOOC\text{-}CH(NH_2)\text{-}CH_2\text{-}SO_3H$
acid
(CYSTEIC ACID)

cysteinesulfinic acid $HOOC\text{-}CH(NH_2)\text{-}CH_2\text{-}SO_2H$

γ-L-glutamyl-taurine $COOH\text{-}CH(NH_2)\text{-}CH_2\text{-}CH_2\text{-}CO\text{-}NH\text{-}CH_2\text{-}CH_2\text{-}SO_3H$
(LITORALON)

2-amino-3-hydroxy-1- $NH_2\text{-}(OH\text{-}CH_2)CH\text{-}CH_2\text{-}SO_3H$
propanesulfonic acid

N-(2,3-dihydroxy-n- $NH\text{-}(CH_2\text{-}CHOH\text{-}CH_2OH)\text{-}CH_2\text{-}CH_2\text{-}SO_3H$
propyl)taurine

Decanoylsarcosyl-taurine	$CH_3-(CH_2)_8-CO-N(CH_3)-CH_2-CO-NH-CH_2-CH_2-SO_3H$
N-(1-carboxyethyl)-taurine	$COOH-(CH_3)CH-NH-CH_2-CH_2-SO_3H$
cerilipin	$R_2-CH(OH)-CO_2-CH(R_1)-CH_2-CO-NH-CH((CH_2)_3NH_2)$ $CO-NH-CH_2-CH_2-SO_3H$
γ-aminobutyric acid (GABA)	$NH_2-CH_2-CH_2-CH_2-COOH$

(b) Substitutions and simple modifications

-NH$_2$ GROUP MODIFIED

N-methyltaurine	$NH(CH_3)-CH_2-CH_2-SO_3H$
N,N-dimethyltaurine	$N(CH_3)_2-CH_2-CH_2-SO_3H$
N,N,N-trimethyltaurine	$^+N(CH_3)_3-CH_2-CH_2-SO_3^-$
guanidinoethanesulfonic acid	$NH=C(NH_2)-NH-CH_2-CH_2-SO_3H$
guanidinoethanesulfinic acid	$NH=C(NH_2)-NH-CH_2-CH_2-SO_2H$
N-(2-acetamido)-2-aminoethanesulfonic acid	$CH_3COCH_2NH-CH_2-CH_2-SO_3H$

-SO$_3$ GROUP MODIFIED

2-aminoethylphosphonic acid	$NH_2-CH_2-CH_2-PO_3H_2$
2-aminoethylarsonic acid	$NH_2-CH_2-CH_2-AsO_3H_2$
3-aminopropionic acid (β-ALANINE)	$NH_2-CH_2-CH_2-COOH$
ethanolamine-O-sulphate	$NH_2-CH_2-CH_2-O-SO_3H$

CARBON CHAIN-LENGTH ALTERED

aminomethanesulfonic acid $\quad NH_2\text{-}CH_2\text{-}SO_3H$

3-aminopropanesulfonic acid $\quad NH_2\text{-}CH_2\text{-}CH_2\text{-}CH_2\text{-}SO_3H$
(HOMOTAURINE)

(c) Cyclic analogues

pyridine-3-sulfonic acid

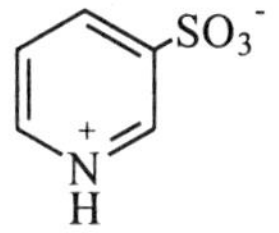

piperidine-3-sulfonic acid

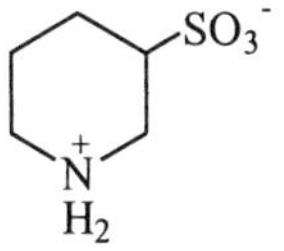

aniline-2-sulfonic acid

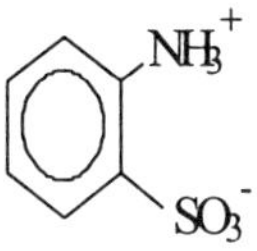

($\pm$)-2-aminocyclohexanesulfonic acid
(*cis* and *trans*)

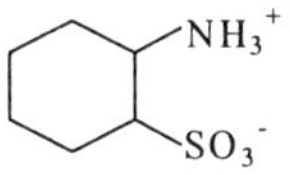

N-[1'-aza-cyclohepten-2'-yl]-2-
aminoethane-sulfonic acid
and
N-[1'-aza-cyclopenten-2'-yl]-2-
aminoethane-sulfonic acid

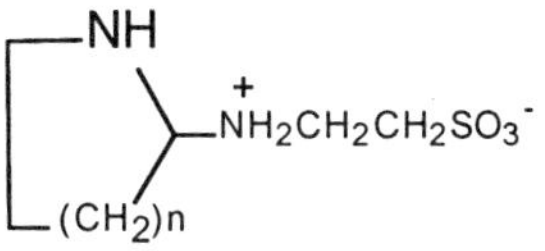

N-[1'-aza-cyclohepten-2'-yl]-3-
aminopropane-sulfonic acid
and
N-[1'-aza-cyclopenten-2'-yl]-3-
aminopropane-sulfonic acid

—NH

—NH2CH2CH2CH2SO3⁻

—(CH2)n

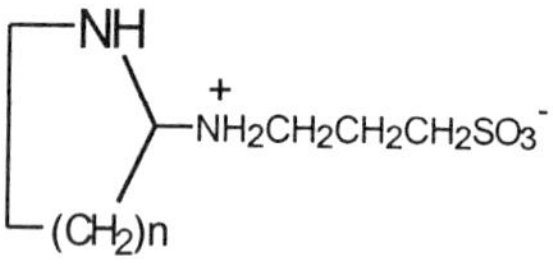

2-aminocyclopentanesulfonic acid

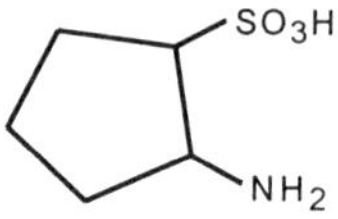

SO3H

NH2

piperazine-*N,N'*-bis(2-ethanesulfonic acid)
R= -CH₂CH₂SO₃⁻

NR

RN

quinoline-8-sulfonic acid

N

SO3H

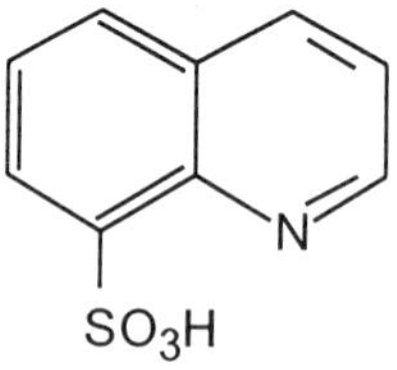

1,2,3,4-tetrahydroquinoline-8-sulfonic
acid

N
H

SO3H

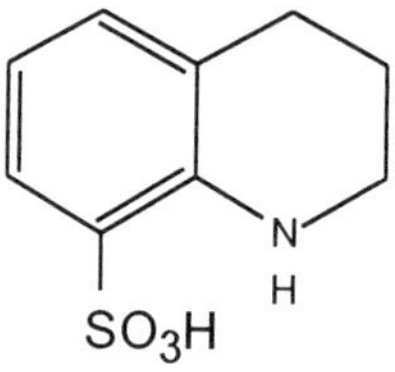

norbornene derivatives

3-amino-bicyclo[2.2.1]heptane-
2-sulfonic acid

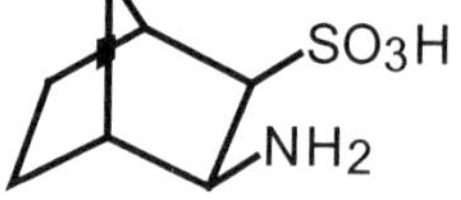

SO3H

NH2

cis

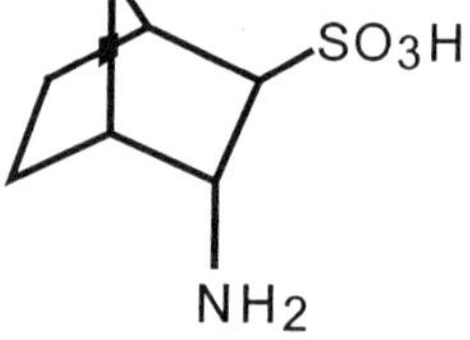

SO3H

NH2

trans

(d) Others

6-aminomethyl-3-methyl-4H-1,2,4-
benzotiadiazine-1,1-dioxyde
(TAG)

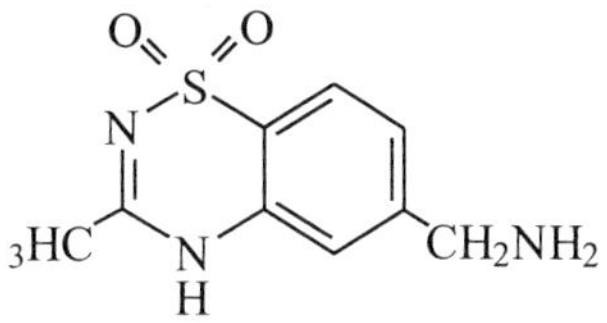

aminoacetic acid
(GLYCINE)

$NH_2\text{-}CH_2\text{-}COOH$

retinylidentaurine
(TAURET)

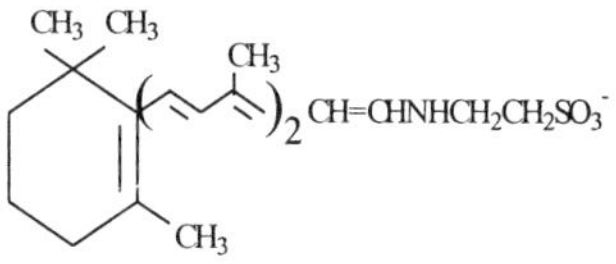

Taurocholic acid

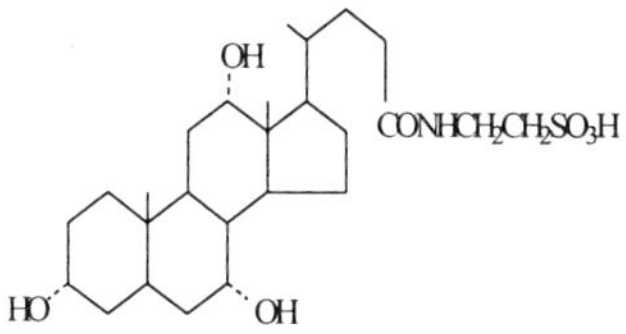

3-acetamido-1-propanesulfonic acid
(calcium salt)
(ACAMPROSATE)

$[CH_3\text{-}CO\text{-}NH\text{-}(CH_2)_3\text{-}SO_3^-]_2\ (Ca^{2+})$

INDEX